A GUIDE TO ORGANOPHOSPHORUS CHEMISTRY

A GUIDE TO ORGANOPHOSPHORUS CHEMISTRY

LOUIS D. QUIN
Distinguished Visiting Professor
Department of Chemistry
University of North Carolina at Wilmington

and

Professor Emeritus, Duke University and University of Massachusetts

STRUCTURAL ILLUSTRATIONS BY
GYÖNGYI SZAKÁL QUIN
Research Scientist
University of North Carolina at Wilmington

A John Wiley & Sons, Inc., Publication
New York / Chichester / Weinheim / Brisbane / Singapore / Toronto

This book is printed on acid-free paper. ∞

Copyright © 2000 by John Wiley & Sons. All rights reserved.

Published simultaneously in Canada.

No part of this publication may be reproduced, stored in a retrieval system or transmitted in any form or by any means, electronic, mechanical, photocopying, recording, scanning or otherwise, except as permitted under Section 107 or 108 of the 1976 United States Copyright Act, without either the prior written permission of the Publisher, or authorization through payment of the appropriate per-copy fee to the Copyright Clearance Center, 222 Rosewood Drive, Danvers, MA 01923, (978) 750-8400, fax (978) 750-4744. Requests to the Publisher for permission should be addressed to the Permissions Department, John Wiley Sons, Inc., 605 Third Avenue, New York, NY 10158-0012, (212) 850-6011, fax (212) 850–6008, E-Mail: PERMREQ@WILEY.COM.

For ordering and customer service, call 1-800-CALL-WILEY

Library of Congress Cataloging-in-Publication Data:
Quin, Louis D., 1928–
  A guide to organophosphorus chemistry / by Louis D. Quin.
    p. cm.
  Includes index.
  ISBN 0-471-31824-8 (alk. Paper)
  1. Organophosphorous compounds.
  I. Title.
QD305.P46Q56 2000
547′.07–dc21   99-43429

Printed in the United States of America.

10 9 8 7 6 5 4 3 2 1

CONTENTS

Preface / ix

1 THE NATURE OF ORGANOPHOSPHORUS CHEMISTRY / 1

2 SOME GENERAL CONSIDERATIONS OF ORGANOPHOSPHORUS COMPOUNDS / 8

2.A. Organization of Compounds by the Bonding to Phosphorus / 8
2.B. Oxidation State Relationships / 10
2.C. Stability of the Carbon–Phosphorus Bond / 11
2.D. Importance and Characteristics of the Phosphoryl (P=O) Group / 12
2.E. 1,2-Dipoles (Ylides) Containing Phosphorus / 15
2.F. The 5-Coordinate State of Phosphorus / 17
2.G. The POH to P(O)H Rearrangement with 3-Coordinate Phosphorus / 20
2.H. Phosphorus Bonds to Heteroatoms (X, S, OR, NR_2) / 21
2.I. Analogies between the Chemistry of Phosphorus Compounds and that of Related Elements (N, S, C) / 23
2.J. Incorporation of Phosphorus in Rings / 25
2.K. Reactive Intermediates and Reaction Mechanisms / 27
2.L. Safety in the Handling of Organophosphorus Compounds / 34

2.M. Important Literature Resources in Phosphorus Chemistry / 35
References / 37
Appendix 2.1. Some Notes on IUPAC Nomenclature / 41

3 THE COMMON 3-COORDINATE FUNCTIONS (σ^3, λ^3) / 44

3.A. Halophosphines / 44
3.B. Alkoxy and Amino Derivatives of Phosphines / 57
3.C. Phosphines / 68
3.D. Biphosphines / 86
References / 88
Appendix 3.1. Syntheses of Pertinent Compounds in *Organic Syntheses* and *Inorganic Syntheses* / 93

4 THE 4-COORDINATE PHOSPHINE OXIDES, OTHER CHALCOGENIDES, AND PHOSPHONIUM SALTS / 95

4.A. Phosphine Oxides / 95
4 B. Phosphine Sulfides, Selenides, and Tellurides / 113
4.C. Quaternary Phosphonium Salts / 117
4.D. Heterophosphonium (Quasiphosphonium) Salts / 123
References / 129
Appendix 4.1. Syntheses of Pertinent Compounds in *Organic Syntheses* and *Inorganic Syntheses* / 132

5 THE ACIDS OF ORGANOPHOSPHORUS CHEMISTRY AND THEIR DERIVATIVES / 133

5.A. Structures and General Characteristics / 133
5.B. Derivatives of Phosphorus Acids / 138
5.C. The Formation of Phosphorus Acids with C—P Bonds / 141
5.D. Synthesis of Phosphoric Acid Derivatives / 153
5.E. Sulfur Derivatives of Phosphorus Acids / 153
5.F. Carbanion Formation and Reactivity at the α-Carbon of Phosphonates / 158
References / 162
Appendix 5.1. Syntheses of Pertinent Compounds in *Organic Syntheses* and *Inorganic Syntheses* / 166

6 PHOSPHORUS-31 NMR SPECTROSCOPY / 169

6.A. Some General Considerations / 169

6.B. The Literature of ^{31}P NMR Spectroscopy / 172
6.C. General Trends and Structural Influences on ^{31}P NMR Shifts / 173
6.D. ^{31}P NMR Shifts of Phosphines / 180
6.E. ^{31}P NMR Shifts of Halophosphines / 188
6.F. ^{31}P NMR Shifts of 3-Coordinate Oxy, Thioxy, and Amino Compounds / 189
6.G. ^{31}P NMR Shifts of Tertiary Phosphine Oxides and Sulfides / 191
6.H. ^{31}P NMR Shifts of Quaternary Phosphonium Salts / 192
6.I. ^{31}P NMR Shifts of Heterophosphonium Ions and Reaction Intermediates / 193
6.J. ^{31}P NMR Spectra of Phosphorus Acids with P—C Bonds and Some Derivatives / 196
6.K. ^{31}P NMR Shifts of Phosphoric Acid Derivatives / 198
References / 201

7 OTHER SPECTROSCOPIC TECHNIQUES IN ORGANOPHOSPHORUS CHEMISTRY / 204

7.A. ^{1}H NMR Spectroscopy / 204
7.B. ^{13}C NMR Spectroscopy in Organophosphorus Chemistry / 212
7.C. Infrared Spectroscopy / 217
7.D. Mass Spectrometry of Organophosphorus Compounds / 219
7.E. Specialized Spectroscopic Methods Useful in Organophosphorus Chemistry / 223
References / 229

8 HETEROCYCLIC PHOSPHORUS COMPOUNDS / 233

8.A. General / 233
8.B. The Fundamental Ring Systems / 234
8.C. Multicyclic Phosphorus Ring Systems / 265
References / 268

9 OPTICALLY ACTIVE ORGANOPHOSPHORUS COMPOUNDS / 272

9.A. General / 272
9.B. Racemic Mixture Separation Techniques / 274
9.C. Optically Active Phosphorus Compounds from Transformations of Other P-Chiral Precursors and from Asymmetric Synthesis Methods / 282
9.D. Kinetic Resolution of Phosphorus Compounds by Enzymes / 298

9.E. Summary of Some Stereospecific Reactions Occurring at Phosphorus Functions / 299

References / 303

10 THE LOW- AND HIGH-COORDINATION STATES OF PHOSPHORUS / 307

10.A. General / 307

10.B. The 1-Coordinate States (σ^1, λ^3 and σ^1, λ^1) / 308

10.C. The 2-Coordinate States with Double Bonds to Phosphorus (σ^2, λ^3) / 317

10.D. The 3-Coordinate State (σ^3, λ^5) / 333

10.E. Compounds with 5- and 6-Coordinate Phosphorus / 339

References / 345

11 ORGANOPHOSPHORUS CHEMISTRY IN BIOLOGY, AGRICULTURE, AND TECHNOLOGY / 351

11.A. Biological Phosphates / 351

11.B. The Natural Products Chemistry of the C—P Bond / 357

11.C. Organophosphorus Compounds in Medicine / 362

11.D. Agricultural Applications of Organophosphorus Compounds / 367

11.E. The Importance of Organophosphorus Compounds in Technology / 374

11.F. Some Major Uses of Organophosphorus Compounds as Reagents in Organic Synthesis / 379

References / 384

Index / 387

PREFACE

Organophosphorus chemistry is a very broad and exciting field, with many opportunities for research or applications development. This book is my personal reflection, from my 50 years of experience, on the fundamentals of the field and to some extent on where future development might take place. It is written as a survey to aid the student or practicing chemist with little prior exposure to the field, for I know how difficult it is for "beginners" to get a grasp on its fundamentals. What I have written is what I would like my own students to know, and I hope this proves to be true of other research directors as well. Inevitably, there will be aspects that may not receive the amount of attention other researchers might prefer and that may be treated too superficially. Writing a concise summary of a very large field is indeed a daunting task and requires discipline in not going too deeply into any particular area, to maintain the size of the book at a reasonable (and affordable) level. There are many lines of important research being followed in major laboratories throughout the world, and it is with regret that it has not been possible to mention all of this activity and to recognize the many contributions of my colleagues in research. It needs to be recognized that the topics in this book are not presented in the depth of a review article, and much of the literature cited has been selected primarily for illustrative purposes. At all times, however, readers are referred to appropriate monographs or review articles that will provide the greater coverage that might be needed. My goal then can be viewed as preparing the reader for the next level of study of organophosphorus chemistry.

    This book has been written during a period of appointment at the University of North Carolina at Wilmington, and I am truly grateful for the support given me by the Department of Chemistry and for the warmth of my reception by the faculty. The entire manuscript has been read by John Verkade, Iowa State University, and I am grateful to him for making valuable suggestions and corrections. Eli Breuer, Hebrew University of Jerusalem, kindly provided important information on the use of phosphorus compounds as drugs. I also thank my colleague at Duke University, Donald B. Chesnut, for many illuminating discussions on computational phosphorus chemistry and especially the interpretation of ^{31}P NMR chemical shifts. But my major source of support has come from my wife, Gyöngyi, who not only drew all of the many equations and structures but also assisted in some library work and was always enthusiastic about this project. Finally, I acknowledge my debt of gratitude for the excellent work of all of my former students and postdoctoral research associates both at Duke University and at the University of Massachusetts at Amherst, for without them and the background and stimulation provided by their many contributions, this book would never have been undertaken.

LOUIS D. QUIN

White Oak Bluffs
Stella, North Carolina

A GUIDE TO ORGANOPHOSPHORUS CHEMISTRY

CHAPTER 1

THE NATURE OF ORGANOPHOSPHORUS CHEMISTRY

The organic chemistry of phosphorus is based on the existence of numerous stable functional groups that contain the carbon-phosphorus bond or that are organic derivatives of inorganic phosphorus acids. It is a very large and currently highly active field, of interest to academic and industrial chemists alike. There are many books on various aspects of organophosphorus chemistry already in existence, and the vast research literature is well covered by these works. There is, however, no recent book that constitutes an introduction to the subject for those wishing to familiarize themselves with it or one that is designed as an aid for those planning to conduct research in the area and who need assistance in mastering the fundamentals of structure and synthesis, reactivity and mechanism, applicable spectral techniques, etc., before going into the research literature. This book is presented to chemists with these needs and is written at a level that should make it useful for those with no prior exposure to organophosphorus chemistry, a cohort that might include advanced undergraduates, graduate students, and research workers in industrial and government laboratories. Indeed, it is rare for undergraduate students to be exposed to this huge and important area; the typical organic chemistry textbook mentions the Wittig olefin synthesis that employs phosphorus ylides as reactants, and perhaps one or two other reactions, and gives displays of the natural phosphoric acid esters vital to life processes. There are, however, many other functional groups that lead to fascinating structural possibilities and reactions frequently unlike those of close neighbors in the Periodic System, notably nitrogen and sulfur. There are now numerous commercial applications of organophosphorus compounds and cases in which these compounds are used as valuable reagents in the synthesis of nonphosphorus organic compounds, altogether making organophosphorus chemistry a lively and intellectually stimulating field in which to conduct research.

2 THE NATURE OF ORGANOPHOSPHORUS CHEMISTRY

As with much of organic chemistry, organophosphorus chemistry had its beginnings in the nineteenth century. Some early syntheses were performed by P. E. Thénard and especially A. W. von Hofmann, but it was the pioneering work in the laboratory of Karl Arnold August Michaelis at the University of Rostock in Germany, from 1874 to 1916, that led to the discovery and characterization, to the extent possible at that time, of some of the major functional groups and of the methods for their synthesis that we still use today. Also of critical importance to the early development of phosphorus chemistry was the extensive work of the school of Aleksandr Erminingel'dovich Arbusov in Russia starting around the turn of the century. The names of these two great pioneers are forever to be associated with organophosphorus chemistry through reactions that bear their names. Phosphorus chemistry remains especially strong in these countries, although the current activity is truly international. Indeed, at the most recent of the triennial series of International Conference on Phosphorus Chemistry, chemists from more than 40 countries gave presentations. The field took on a more practical aspect in the post–World War II period, motivated significantly by the discovery of powerful insecticidal activity in certain phosphoric acid and phosphonic acid esters. By 1970, some 100,000 organophosphorus compounds were known, and many thousands have been made since that time. The increase in research activity has not lessened over the intervening years, and many new uses for organophosphorus compounds have been discovered. We can summarize these uses as follows, but must defer a fuller discussion to the final chapter of this book, after the groundwork has been presented that will lead to appreciation of their structures and synthetic approaches.

- Agricultural chemicals, including insecticides, herbicides, and plant growth regulators
- Medicinal compounds, including anticancer, antiviral, and antibacterial agents and agents for the treatment of bone diseases
- Catalyst systems based on metal-coordinated tertiary phosphines used for many industrial processes (Oxo hydroformylation, olefin hydrogenation, Reppe olefin polymerization, etc.); a new application is the accomplishment of asymmetric syntheses from the use of complexes of optically active phosphines
- Flame retardants for fabrics and plastics
- Plasticizing and stabilizing agents in the plastics industry
- Selective extractants of metal salts from ores, especially those of uranium
- Additives in the petroleum products field
- Corrosion inhibitors

Perhaps new uses are yet to be discovered, for the field of organophosphorus chemistry is by no means mature.

Many important fundamental discoveries about structure, properties, and synthesis have been made in the "modern" era of phosphorus chemistry that started after World War II. A particularly good example is provided by the area of heterocyclic phosphorus chemistry, which had received little attention during the period when rapid developments were being made in heterocyclic nitrogen, oxygen, and sulfur chemistry. The intensive effort of recent years has made phosphorus a recognized partner in heterocyclic chemistry, easily evidenced by the extensive treatment of phosphorus heterocycles in several chapters devoted to them in the second edition of *Comprehensive Heterocyclic Chemistry*.[1] The same fundamental monocyclic systems of N, O, S chemistry, such as are illustrated by structures **1–5**, are now well recognized.

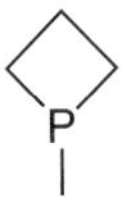

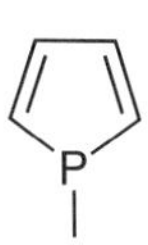

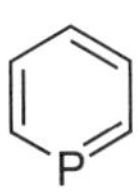

Although hundreds of P-containing ring systems are now known, this field is far from developed, and new ring systems await discovery.

Striking also are the discoveries in the late twentieth century of compounds in which phosphorus is bonded in other than the familiar triply connected (e.g., tertiary phosphines, **6**, and phosphites, **7**) and quadruply connected forms (e.g., trialkyl phosphates, **8**; phosphonic acids, **9**; and quaternary phosphonium salts, **10**). These are shown as obeying the octet rule of valence, although as will be seen in chapter 2,D the P—O group is not well represented by the structure shown, and is more commonly written as P=O.

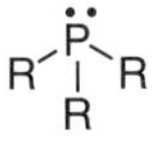

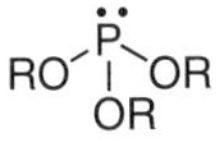

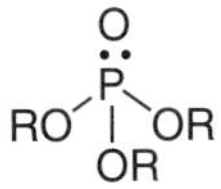

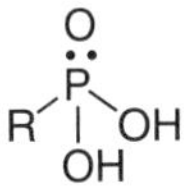

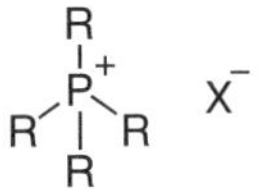

The first breakage of the octet rule in singly bonded structures occurred convincingly in the 1960s when it was recognized that phosphorus could form stable structures (as well as reaction intermediates or transition states) with five (**11** and **12**) and even six bonds (**13**).

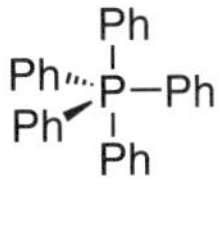

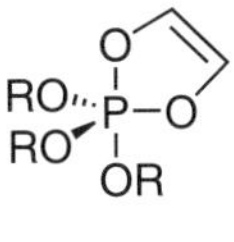

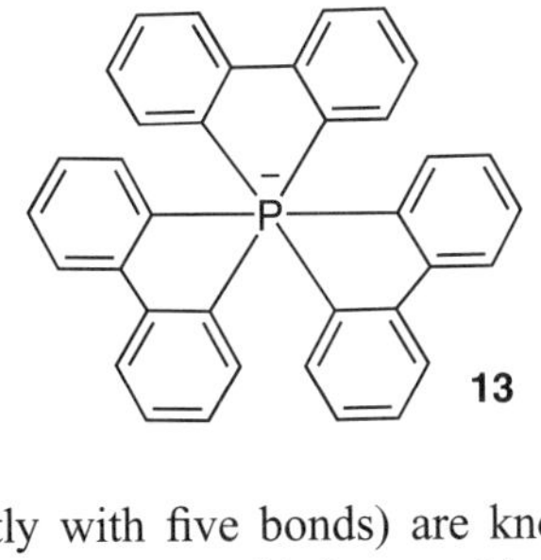

Now hundreds of such compounds (mostly with five bonds) are known. Even more unexpected was the discovery that phosphorus could form stable molecules with multiple bonds of the familiar $p_\pi-p_\pi$ type, a structural feature that for years dogma had stated could not exist in stable form to elements of the second row of eight elements and below. Indeed, when constructed properly, some surprisingly simple compounds with these features have become known, and this has led to to the addition of many new functional groups to phosphorus chemistry, with much remaining to be learned about them. Some examples of these new structures are shown as structures **14–17**.

These structures, and some other types to be seen later, are described as having *low coordination*, a concept that will be developed further. In compounds **14–17**, this is taken to mean the presence of fewer than the normal number of σ-bonds for a molecule having a lone electron pair.

Phosphorus chemistry has shared in the enormous benefits that spectroscopic techniques offer in the determination of structure and the characterization of compounds. Nuclear magnetic resonance has played its usual dramatic role, not only by the measurement of the ^1H and ^{13}C NMR spectra but also because phosphorus itself is NMR active (^{31}P has a spin of $\frac{1}{2}$) and gives characteristic absorptions. In addition, ^{31}P couples with ^1H and ^{13}C, and much structural information is provided by the coupling constants. Anyone working with organophosphorus compounds would be well advised to include characterization by ^{31}P NMR among the techniques employed.

Finally, we should recognize the dramatic discoveries of the involvement of natural organic derivatives of phosphoric acid in biological systems. Detailed treatment of the role and metabolism of these compounds lies in the realm of biological chemistry, but laboratory techniques for their synthesis and characterization rest solidly on the groundwork provided by organophosphorus chemistry. The most important of the natural phosphates are simply esters of the parent phosphoric acid or of diphosphoric (pyrophosphoric) or triphosphoric acid, but this grossly understates the complexity of the molecular structures involved. Thus compound **18** is an example of a monophosphate; it is a member of the phospholipid family. Similarly, compound **19** illustrates the triphosphate structure; it is the well-known adenosine triphosphate (ATP).

Phosphorus natural products chemistry has been expanded in the last few decades to include some compounds with direct carbon-phosphorus bonds. 2-Aminoethylphosphonic acid (**20**) is the most commonly encountered compound in Nature. These products are particularly prominent among marine invertebrates but are found in other life forms as well (including humans, in minute amounts). Even bacterial fermentation broths can yield compounds with the C—P bond; some are surprisingly good antibiotics, as is true for structure **21**, which has been commercialized as Fosfomycin. Compound **22**, similarly isolated, is an effective antifungal agent.

$$H_2N-CH_2-CH_2-\underset{\underset{OH}{|}}{\overset{\overset{O}{\|}}{P}}-OH$$

20

$$\underset{H}{\overset{CH_3}{\diagdown}}C-\underset{H}{\overset{}{C}}\diagup\underset{OH}{\overset{\overset{O}{\|}}{P}}\diagdown OH$$
(with O bridging the two carbons)

21

$$H_2N-\underset{\underset{NH}{\|}}{\overset{}{C}}NH(CH_2)_3CH-\overset{\overset{O}{\|}}{C}NH-\underset{\underset{H_3C}{|}}{N}-\underset{\underset{OH}{|}}{\overset{\overset{O}{\|}}{P}}-CHCOOCH_3$$
with NHR on the CH

22

To cover such a huge and intricate field in a book of limited size requires considerable condensation of the material and avoidance of excessive detail, especially in the treatment of synthetic methods and functional group properties. To make up for these deficiencies, the reader will be referred to the pertinent reference books or review articles, where more complete treatment of the topic can be found. To accomplish its goals, this book is organized in the following way. Chapter 2 presents some ideas on structure, bonding, and general characteristics of phosphorus compounds that overarch the field and will be an aid in appreciating the more specific material that follows. The great majority of phosphorus compounds have functional groups that are based on the phosphorus atom with three covalent bonds with an unshared electron pair and on derivatives formed by utilization of the electron pair. The synthesis and chemical properties of these numerous functional groups are described in the next three chapters, which indeed present much of the core of organophosphorus chemistry. The application of spectroscopic techniques in the study of phosphorus compounds is covered in chapters 6 and 7; the first of these is devoted entirely to ^{31}P NMR spectroscopy, in accord with the great importance of this technique in practical work with organophosphorus compounds. With mastery of functional group chemistry and of spectroscopic techniques, we are then ready to move into the more specialized subjects of heterocyclic phosphorus chemistry (chapter 8) and of stereochemical properties of phosphorus compounds (chapter 9). Chapter 10 surveys the newly discovered phosphorus functional groups in low-coordination states and also hypervalent structures based on the presence of five or six covalent bonds to phosphorus. Finally, chapter 11 outlines the practical side of organophosphorus chemistry, including biological phosphates and natural C−P compounds, as well as some uses of organophosphorus compounds.

There are two features of the book that require preliminary comment. In constructing the book in the way noted, numerous cases arise in which repetition of a subject will occur. Thus a reaction that is described as a property of a functional group may later appear as a method of synthesis for another group or groups. Rather than simply supplying a cross-reference, a brief introduction to the reaction is again provided, to make the account more immediately readable. A second point is that the chapters defining the major functional groups (chapters 3 to 5), include a list of the syntheses that are described in *Organic Syntheses* and *Inorganic Syntheses*. Many important processes are included in these reports, and they provide much practical information on how particular reactions are run in the laboratory.

In a book of this limited size and scope, it is not possible to cover every aspect of organophosphorus chemistry or to give recognition to all of the many advances that have been made in recent years. In particular, there is no discussion of some very new work in areas that have become fashionable in general organic chemistry. For example, there is no mention of the preparation of phosphorus macrocycles and phosphorus derivatives of calixerenes, cyclodextrins, and other frameworks of interest in supramolecular chemistry, which are proving to be of value in metal binding. At the July 1998 International Conference on Phosphorus Chemistry at least five papers were presented in this area. (The abstracts are published in *Phosphorus, Sulfur and Silicon*.) The preparation of derivatives of fullerenes is another fashionable and intriguing venture in organic chemistry, and phosphorus reagents are finding use here also. An example offered at the same conference is that of Cheng and co-workers[2] who bonded ylide and phosphonate groups to fullerene. Also should be mentioned the appearance of the combinatorial method of synthesis in organophosphorus chemistry. An excellent example was published by Fathi and co-workers[3], who generated libraries of phosphoramidates of biological significance.

Finally, an application of the extremely important contemporary concept of "green" chemistry in regard to industrial phosphorus syntheses needs to be mentioned. The great majority of phosphorus compounds are synthesized from phosphorus chlorides as initial reactants; the by-product problem is severe in both the formation and reactions of these halides, and the safety of the environment can be threatened in these operations. As an alternative, the use of elemental phosphorus, both the white and red allotropes, as a starting material is receiving attention; for many years chemists have explored the potential of the element in synthesis, and a large number of successful reactions have been carried out with it. The most important reactions are based on alkylation or arylation with halides, on the attack by a great variety of nucleophiles, on oxidative processes in the presence of organic reactants, and on attack by free radicals. A few of these are mentioned at appropriate places in this book, but much more is known than will be found here. Review articles[4] reveal the wide range of reaction possibilities that exist. Some of the earliest syntheses (from the 1800s) of organophosphorus compounds were performed with elemental phosphorus, but evidence that this synthetic aproach has again entered the thinking of chemists, especially because of the reduction of the industrial by-product problem, can be found in a spate of publications from the late 1990s. (At the 1998 conference four papers were concerned with this topic.) The review literature on elemental phosphorus forms an excellent basis for those considering "green" organophosphorus syntheses.

REFERENCES

1. A. R. Katritzky, C. W. Rees, and E. F. V. Scriven, eds., *Comprehensive Heterocyclic Chemistry* II, Pergamon Press, Oxford, UK, 1996.
2. C.-H. Cheng, S.-C. Chuang, K. C. Santosh, and D.-D. Lee, paper presented at the 1998 International Conference on Phosphorus Chemistry, Cincinnati, Ohio, July 12–17, 1998.
3. R. Fathi, M. J. Rudolph, R. G. Gentles, R. Patel, E. W. MacMillan, M. S. Reitman, D. Pelham, and A. F. Cook, *J. Org. Chem.* **61**, 5600 (1996).
4. M. M. Rauhut, *Top. Phosphorus Chem.* **1**, 1 (1964); M. Grayson, *Pure Appl. Chem.* **9**, 193 (1964); L. Maier, *Fortschr. Chem. Forsch.* **19**, 1 (1971); C. Brown, R. F. Hudson, and G. A. Wartew, *Phosphorus Sulfur* **5**, 67 (1978).

CHAPTER 2

SOME GENERAL CONSIDERATIONS OF ORGANOPHOSPHORUS COMPOUNDS

2.A. ORGANIZATION OF COMPOUNDS BY THE BONDING TO PHOSPHORUS

In beginning a study of organophosphorus chemistry, it is important to recognize the many structural possibilities that are centered on this element and to become familiar with the names of the more common functional groups. Presented here is a convenient classification scheme that has evolved over the last few years. It is based on the description of the functional groups by the number of attached atoms and by the extent of any multiple bonding that may be present. We will use this classification scheme in later chapters.

Until about 1950, all reported organophosphorus compounds had either three atoms or four atoms attached directly to phosphorus and were known, respectively, as tricovalent (or trivalent) or tetracovalent (or tetravalent or sometimes pentavalent if the phosphoryl group P=O was present). In the last three decades, however, compounds with one, two, five and six atoms attached to phosphorus have been discovered; some of these have multiple bonds that formerly had been thought to be incapable of existence. These new types of compounds have rapidly attained a prominent place in phosphorus chemistry. In the organizational scheme necessitated by these new developments, phosphorus is first assigned a *coordination number*, which is a statement of the number of directly attached atoms (or the number of σ bonds). Thus phosphines of the formula R_3P, formerly called tricovalent, are now described as 3-coordinate compounds. The coordination number is designated by the use of the Greek letter sigma with a superscript; a phosphine is, therefore, described as σ^3. The Greek letter lambda is used to describe the total number of bonds, including π-bonds, and thus represents the valence of phosphorus. The common phosphines are, therefore, more fully described as σ^3, λ^3. However, phosphines with a double bond to phosphorus (R—P=CH$_2$) are now known; these are 2-coordinate (two σ-bonds) with a total of three bonds and are designated σ^2, λ^3. Similarly, an alkyl ester of phosphoric acid, (RO)(OH)$_2$P=O, has four attached atoms to P and is

TABLE 2.1. An Organization of Some Common Organophosphorus Species

Designation	Structure	Name of class
	A. Coordination number 1	
σ^1, λ^1	R—P	phosphinidenes
σ^1, λ^3	R—C≡P	phosphaalkynes
	B. Coordination number 2	
σ^2, λ^3	R$_2$C=PR	phosphaalkenes
	RO—P=O	oxophosphines
	R—P=S	thioxophosphines
	R—N=P	iminophosphines
	RP=PR	diphosphenes
σ^2, λ^2	R$_2$P$^+$	phosphenium cations
	C. Coordination Number 3^a	
σ^3, λ^3	R$_3$P	phosphines
	RPX$_2$	alkylphosphonous dihalides
	R$_2$PX	dialkylphosphinous halides
	RP(OR)$_2$	dialkyl phosphonites
	(RO)$_3$P	trialkyl phosphites
σ^3, λ^5	R—PO$_2$	dioxophosphoranes
	RO—PO$_2$	alkyl metaphosphates
	RP(O)(=CH$_2$)	methyleneoxophosphoranes
	RP(=CR$_2$)$_2$	bis(methylene)phosphoranes
	D. Coordination Number 4^a	
σ^4, λ^4	R$_4$P$^+$	phosphonium ions
σ^4, λ^5	RP(O)(OH)$_2$	phosphonic acids
	R$_2$P(O)(OH)	phosphinic acids
	R$_3$P(O)	phosphine oxides
	RO—P(O)(OH)$_2$	alkyl phosphates
	(RO)$_2$P(O)(OH)	dialkyl phosphates
	(RO)$_3$P(O)	trialkyl phosphates
	E. Coordination Number 5^b	
σ^5, λ^5	R$_5$P	phosphoranes
	F. Coordination Number 6^b	
σ^6, λ^6	R$_6$P$^-$	no common name

^{a} To simplify the presentation, the many structural possibilities for which S, Se, or RN= replace (O) and RS or R$_2$N replaces RO are not shown.
^{b} RO, RS R$_2$N, and halogen may replace R.

4-coordinate; the multiple bond to oxygen leads to the description σ^4, λ^5. The anhydride RO—P(=O)$_2$ (a metaphosphate) has a lower coordination number (3) and two multiply bonded oxygens, leading to the designation σ^3, λ^5. The σ, λ convention is of great help in presenting an organizational table of organophosphorus compounds. From Table 2.1, we can gain further understanding of the term *low-coordination*, mentioned in chapter 1. Any compound with a sigma number <3 is said to be of low coordination, and any compound with a lambda number of 5 but a sigma value <4 is likewise classed as a low-coordination species.

2.B. OXIDATION STATE RELATIONSHIPS

Little use is made in organophosphorus chemistry of the calculation of oxidation numbers. Some obvious relationships between oxidation states do exist, however. Thus 3-coordinate phosphorus species with a lone electron pair invariably are capable of oxidation to the phosphoryl structure. The literature refers to all such 3-coordinate species as having oxidation state P(III), regardless of whether P bears an electron-attracting group such as halogen, or an electron-releasing group, such as H or methyl, and makes little use of the conventional calculation of oxidation number where O is -2 and H is $+1$. Thus the usual calculations would assign oxidation number $+3$ to phosphorus in a trialkyl phosphite but -3 to PH$_3$, yet both are considered to have the P(III) state. Conversions of phosphines to the higher oxidation state in phosphine oxides (R$_3$P=O), of alkylphosphonous dichlorides (RPCl$_2$) to the phosphonic dichlorides (RP(O)Cl$_2$), etc., are commonplace. The oxidized products in the past have been assigned oxidation state P(V), and this has led to a description of these products as being pentavalent. They are, however, tetracovalent and are better described in contemporary work as 4-coordinate. In practice, the P(III) to P(V) conversion is generally quite easy, but the reduction of the oxide back to the 3-coordinate form is much more difficult. Procedures for these changes are discussed in appropriate chapters.

Another obvious relationship is that between 4-coordinate functional groups with a P—H bond and the corresponding oxidized form with P—OH. Again, the oxidation (Eq. 2.1) is quite easy, whereas the reduction is quite difficult. This transformation is typified by the oxidation of dialkyl phosphites, more properly known as H-phosphonates, to the corresponding dialkyl phosphates.

$$(RO)_2\overset{\overset{O}{\|}}{P}-H \xrightarrow{KMnO_4} (RO)_2\overset{\overset{O}{\|}}{P}-OH \qquad (2.1)$$

A second example is provided by the oxidation of alkyl alkylhydrogenphosphinates to alkyl alkylphosphonates.

$$(RO)\underset{\underset{R}{|}}{\overset{\overset{O}{\|}}{P}}-H \xrightarrow{(O)} (RO)\underset{\underset{R}{|}}{\overset{\overset{O}{\|}}{P}}-OH \qquad (2.2)$$

Superficially, these conversions seem to involve transformations among two 4-coordinate forms rather than of P(III) to P(V). Mechanistic evidence suggests, however, that the true species being oxidized is the 3-coordinate form with a lone pair on P, formed by a proton shift from P to O (Eq. 2.3). This point is discussed further in section 2.G.

$$\left[(RO)_2\overset{O}{\underset{\|}{P}}-H \rightleftharpoons (RO)_2\overset{..}{P}-OH \right] \xrightarrow{(O)} (RO)_2\overset{O}{\underset{\|}{P}}-OH \qquad (2.3)$$

2.C. STABILITY OF THE CARBON–PHOSPHORUS BOND

The thermal stability of the C–P bond is quite high, regardless of the phosphorus functionality. The heat of dissociation of the 4-coordinate C–P bond is generally accepted to be about 65 kcal/mol, and there is never any difficulty in handling most alkyl and aryl phosphorus compounds even at moderately high temperatures. Phosphine oxides and phosphonates are especially stable and can be safely heated to temperatures of 150 to 200°, or higher in some cases. Three-coordinate compounds, while possessing stable C–P bonds, may have other forms of thermal instability, such as dehydrohalogenation of halo derivatives or rearrangement of alkoxy derivatives. There are, however, some special bonding situations that can lead to low thermal stability even in phosphine oxides and phosphonates. For example, certain unsaturated bridged ring systems may undergo retrocycloaddition reactions in which a C–P bond is broken[1] (Eq. 2.4). Small rings are also susceptible to thermal cleavage[2] (Eq. 2.5).

(2.4)

(2.5)

Cleavage of a C–P bond can also occur in certain reactions; a few examples of special cases where this can take place will be seen in later chapters on functional group chemistry (e.g., the action of NaOH on certain quaternary salts and trihalomethylphosphoryl compounds, and in the Wittig, Horner, and Wadsworth-Emmons reactions for coupling ylides, phosphine oxides, and phosphonates, respectively, with aldehydes and ketones). In general, however, one can consider the C–P bond to have good stability in most chemical reactions and feel confident that it will survive the experimental conditions to which it is subjected, including acid and base hydrolysis, as well as oxidizing conditions. Even strong oxidizing agents in solution are generally without effect on the C—P bond in 4-coordinate structures. Exceptions are rare and require special structural situations, as in the case of heterocycles with severely strained bond angles; here peroxyacids can cleave the C—P bond with insertion of oxygen[3] (Eq. 2.6), a process resembling the familiar Baeyer–Villiger reaction of ketones.

$$\text{Ph-P(O)-C(Me)_2-C(Me)_2 (ring)} \xrightarrow{ArCO_3H} \text{Ph-P(O)-O-C(Me)_2-C(Me)_2 (ring)} \quad (2.6)$$

2.D. IMPORTANCE AND CHARACTERISTICS OF THE PHOSPHORYL (P=O) GROUP

The phosphoryl group, which for the moment we will represent as P=O, is of profound importance in organophosphorus chemistry. It occurs in several common functional groups (Table 2.1), and many thousands of compounds containing the P=O group are known, more than any other functional group type. It is easily formed and resists chemical modification; where it can be formed or retained in a chemical reaction, it will exert a directing effect on the outcome of the reaction by this strong preference. It is the great thermal stability of this bond, with a dissociation energy of 128 to 139 kcal/mol,[4a] that accounts for these properties and leads to major differences between structurally similar nitrogen and phosphorus compounds. Thus amine oxides are formed from amines only by special oxidizing reagents, are easily reduced by a great variety of reagents, and are of relatively low thermal stability compared to phosphine oxides. Phosphines, on the other hand, can be oxidized by almost any oxidizing agent, even oxygen of the atmosphere, and phosphine oxides require special or very strong reducing agents for reversion to the phosphine (e.g., lithium aluminum hydride and silicon hydrides). Another striking characteristic of the phosphoryl group is its occurrence in a range of very stable acid structures (see Table 2.1 and chapter 5). Again, a marked difference from nitrogen chemistry is present; acids based on nitrogen functional groups with bonds to carbon are rare and of little importance. Thus phosphonic acids, $RP(O)(OH)_2$, are of great stability and among the most important of organophosphorus compounds, yet the corresponding nitrogen structure does not appear to have any significance.

The great stability of the phosphoryl group is derived from a form of multiple bonding; this is the key distinguishing feature that accounts for the differences from nitrogen chemistry, for this possibility does not exist for the first row of eight elements. The nature of this multiple bonding, clearly manifested in many measurable properties, has been a subject of study and conjecture for many years; views continue to evolve on the proper description of this bonding. Many studies have been performed to determine the effects of the multiple bonding; it is clear that it can be quite variable, depending on the other substituents on phosphorus. Strongly electronegative groups on phosphorus increase the multiple bonding; this is readily seen from infrared spectral studies, by which a comparison of the energy of the stretching of the P=O bond with different substitution can be made.[4a] Thus in $F_3P=O$ the stretching frequency is 1418 cm^{-1}, whereas in $Me_3P=O$ it is 1170 cm^{-1}; it is at intermediate frequencies for substituents with electronegativities between the extremes of F and C. Another strong indication of multiple bonding is provided by measurements of the P to O internuclear distance; this is typically 1.475 to 1.490 Å, much shorter than is the P to O single bond (1.60 Å for a P−O−P unit[4a]). Other physical data indicating multiplicity in the phosphoryl bond are summarized in a recent review.[4a]

In early literature, and occasionally even today, one sees the rather inadequate, simple structure **1** for a phosphoryl compound, implying the presence of an octet of electrons and the coordinate covalent (or semipolar) bond to phosphorus (as is *required* of nitrogen).

$$R_3P \longrightarrow O$$
1

An old and still quite common description of the phosphoryl bond accounts for the multiplicity by invoking the concept of "backbonding" of a pair of electrons from oxygen into one of the vacant 3d-orbitals of phosphorus, and this has led to a simple resonance expression (Eq. 2.7) for the structure.

$$R_3\overset{+}{P}-\overset{-}{O} \quad \longleftrightarrow \quad R_3P=O \qquad (2.7)$$
2

The double-bond structure **2** is by no means to be likened to that of a conventional double bond of carbon to carbon, or carbon to oxygen, which, of course, is drawn the same way yet is based on interaction of a p-orbital on each atom. Their chemistry is totally different. Double bonds to carbon, for example, are well known to undergo numerous simple addition reactions to give stable adducts; there is no counterpart for this behavior by the phosphoryl group, which is singularly (but not totally) unreactive. Structure **2**, however, with the implied presence of a multiple bond has served us well over the years; it is deeply entrenched throughout phosphorus chemistry, and its use is not likely to decline rapidly even as radically different views of the true bonding emerge.

Representing the phosphoryl group by structure **2** creates another problem. In the last few years, evidence has been obtained for the existence of a true p_π-p_π double bond between phosphorus and oxygen. This occurs when phosphorus is 2-coordinate, as in the σ^2, λ^3 structure RO−P=O. This true phosphorus-oxygen double bond will be discussed in chapter 10.

The concept of the phosphoryl multiple bonding being derived from backbonding of oxygen electrons into a phosphorus 3d-orbital has been in wide use for many years. Much has been written about it, and descriptions of the theory may be found in various monographs, (e.g., those listed in reference 5), or in most older texts on inorganic chemistry. In more recent theoretical studies (e.g., reference 6), where the power of modern computational methods has been employed, it has been shown that there is no role for d-orbitals in forming valence bonds, and the bonding picture assumed for many years for the phosphoryl group has to be abandoned. Excellent reviews that deal with this problem, and which advocate a new view of the bonding, have been published by Kutzelnigg[7] and by Gilheany.[4a,8] According to these views, the well-accepted multiple bonding in the common phosphoryl group is derived from the interaction of a 2p orbital on oxygen with an antibonding orbital of the R_3P moiety. This would be a form of negative hyperconjugation and is developed especially by the work of Reed and Schleyer.[9] With this new interpretation, the involvement of d-orbitals on phosphorus in the phosphoryl group, so widely employed in the past for valence bonding, is diverted into the role of assisting in the bonding by polarizing other existing valence orbitals. In his influential reviews, Gilheany concludes that there is probably no better simple way to represent the bond in a molecule than by the presently-used P=O, but with a new understanding of its meaning.

The story does not end here, however; two calculations, both published in 1998, by Chesnut[10] and by Dobado, et al.,[11] that use the Bader "Atoms in Molecules" theory support a modified view of the bonding. There is the usual σ bond between P and O, which leads to a strong ionic charge on each atom; it is then the interaction between these charged atoms that strengthens and shortens the P−O bond, giving the effect one would expect from formal multiple bonding. This thinking would lead to a preference for representing the phosphoryl bond as P^+-O^-, almost full circle to the early form drawn in the resonance of Eq. 2.7. It is of interest that this form was favored to account for recent experimental electron paramagnetic resonance (EPR) data;[12] the authors point out that no double bond can be present and express the structure of the radical observed as $R_3P^+-O\cdot$.

Another theoretical approach,[13] using the electron localization function (ELF) method, has led to a picture of the bonding in $H_3P=O$ as a model in which the three lone pairs on oxygen are on the side opposite of phosphorus, staggered with respect to the three σ bonds, again speaking against any sort of formal multiple bonding. Chesnut and Savin[13] present an excellent summary of recent theoretical work on the phosphoryl problem. At this writing, it is too early to tell if the new views will influence the way the phosphoryl group is treated by experimentalists, because the d-orbital bonding concept has been so useful and pervasive. In this book, we will continue to draw it as P=O, but with the understanding that the multiple bonding lacks a second bond component in the valence bond sense from a back-donation effect and is indeed ionic.

Regardless of the precise description of the bonding in the P=O group, there is no doubt that the bond is rendered short, strong, and highly polar by whatever effects may be present. This view is entirely consistent with much experimental data on the general characteristics of phosphoryl compounds. Thus they have high dipole moments (4.51 D in $Ph_3P=O^{4a}$) and a very strong tendency to form hydrogen bonds with its oxygen; oxygen also exhibits nucleophilicity to strong electrophiles. For example, phosphine oxides react with anhydrous hydrogen halides[4b] to form $(HO)R_3P^+$, and a few are known to react with chlorotrimethylsilane[14] to form $(Me_3SiO)R_3P^+$. The H-bonding property can lead to remarkable water solubility effects; phosphine oxides especially have very high solubility in water, and the presence of a $Me_2P=O$ substituent can solubilize a chain or ring of a considerable number of carbons. The water solubility is, however, diminished in phosphoryl compounds with alkoxy or amino substituents. Phosphoryl compounds can function as highly polar aprotic solvents; the case of hexamethyl phosphoramide, $(Me_2N)_3P=O$, is especially well known. Similarly, the easily prepared cyclic phosphine oxide **3** has been found to have value as a polar aprotic solvent; its dielectric constant is 53.[15]

Me—P=O (cyclic)

3

Other spectroscopic and chemical properties of phosphoryl groups will be encountered in later chapters. We understand a great deal about the group's chemistry; with fresh starts made in the theoretical studies, perhaps a final view, accepted by all, on the nature of the bonding will be forthcoming. It should be added that the same new considerations could apply to the thiophosphoryl and selenophosphoryl groups. Just as with the phosphoryl group itself, we will continue to express these groups as P=S and P=Se, with the same reservations about their validity.

2.E. 1,2-DIPOLES (YLIDES) CONTAINING PHOSPHORUS

There are some very important structures in organophosphorus chemistry that can be formally described as 1,2-dipoles, in which a phosphorus atom with a unit positive charge is attached directly to another atom bearing a unit negative charge. When the negative atom is carbon, the structure is more commonly called an ylide; this is the key reactant in the familiar Wittig olefin synthesis. The structure and chemistry of ylides are the subject of extensive reviews.[4c,4d,16] We face here the same problem in the electronic representation as we did in the case of the phosphoryl group; the formula of the dipolar ylide is **4**, but we must consider if a neutral form with multiple C–P bonding (**5**, called an ylene) contributes importantly to its true character through resonance (Eq. 2.8).

$$R_3\overset{+}{P}-\overset{-}{C}H_2 \longleftrightarrow R_3P=CH_2 \qquad (2.8)$$
$$\quad\;\, \mathbf{4} \qquad\qquad\qquad \mathbf{5}$$

The bonding situation is discussed in depth by Kutzelnigg,[7] Johnson,[16] Gilheany,[8] and Bachrach and Nitsche.[4d] Many structural data are available, and numerous spectral observations have been made. It is of interest that free rotation occurs about the P–C bond, and that spectral evidence supports strong carbanionic character on C. The similarity to the problem of the phosphoryl group should be apparent. In earlier days, the ylene might have been considered to involve backbonding of the carbanionic lone pair to a phosphorus d-orbital, but this view should be discarded in light of demonstrations of the non-importance of phosphorus d-orbitals in forming multiple bonds.[6,8] To the extent that any multiple bonding is present in ylides, it may better be described by the negative hyperconjugation model involving interaction of oxygen electrons with antibonding orbitals as proposed for the P=O group or by the concept of ionic interaction in the same way as was proposed for the phosphoryl group.[10,11,13] To the organic chemist, however, the chemistry of ylides is quite definitely that of a species with a strong carbanionic character, easily seen in the Wittig reaction with a carbonyl compound (Eq. 2.9). This reaction is discussed in greater detail in chapter 11.

$$R_3\overset{+}{P}-\overset{..}{\overset{-}{C}}H_2 + O=CR'_2 \longrightarrow R_3\overset{+}{P}-CH_2-\overset{\overset{O^-}{|}}{C}R'_2 \longrightarrow R_3P=O + CH_2=CR'_2 \quad (2.9)$$

In most literature, the ylide structure alone is shown, and this will be done here. It must be remembered, however, that what makes the ylide stable is the secondary interaction between the carbanionic center and positive phosphorus, making the P−C bond short and strong.

Ylides of different types will be found in other studies of organophosphorus chemistry. An important example is that of the heterocyclic phosphinine system. When phosphorus is 2-coordinate, cyclic electron delocalization from interaction only of p-orbitals is present, and the system is aromatic (**6**). The phosphinine ring can also be found with σ^4, λ^4 bonding as in **7**, but aromaticity in its usual form making use of cyclic p-orbital interactions cannot be present. The system does have extensive delocalization, as seen in the resonance form **8**, and is quite stable.[17] Compounds with this structure were, in fact, known before the true aromatic σ^2, λ^3 form as in **6** was discovered.

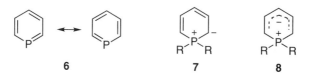

      6   **7**   **8**

The same problem in bonding appears in other derivatives of phosphorus. Thus phosphines (R_3P) can be converted to iminophosphoranes (also known as phosphine imines), which are widely expressed as $R_3P=NR$ and are seen to be nitrogen counterparts of phosphine oxides.[16] Again, the true meaning of the double bond is unclear, but current thinking speaks against any formal multiple valence bond being present. The double-bonded structure should be viewed in the same way that the double-bonded representation of the phosphoryl is viewed; there is no backbonding into d-orbitals, and the shortening and strengthening of the P−N bond may arise from interaction of the charged atoms seen in structure **9** of the resonance hybrid (Eq. 2.10). We will continue to represent the iminophosphorane structure with the familiar P=N but recognize its impreciseness.

$$R_3P=NR \longleftrightarrow R_3\overset{+}{P}-\overset{-}{NR} \qquad (2.10)$$

9

There are some extremely important compounds that are organic derivatives of the inorganic phosphonitrilic chloride (also known as chlorocyclotriphosphazene) ring system. The ring is composed of three iminophosphorane units and may require bonding treatment of the same nature. Thus, although the ring system is generally drawn as in structure **10**, a proper alternative may prove to be structure **11** to avoid the suggestion that formal multiple bonding is present, if it can be agreed that a reinterpretation of the bonding is in order here. There does seem to be a form of cyclic delocalization in this 6-membered ring, interpreted for many years as involving d-orbitals, that makes it quite stable.

10   **11**

2.F. THE 5-COORDINATE STATE OF PHOSPHORUS

One of the special properties of phosphorus that is of great importance is its ability to accept more than the usual complement of 8 bonding electrons, thus acquiring 5-coordinate character with 10 electrons, or 6-coordinate with 12 electrons. This property of phosphorus was not firmly established until 1948, when the compound Ph_5P was synthesized and characterized by Wittig and Rieber.[18] Now there are many compounds known with this structural feature, which is also known to appear frequently in reaction mechanisms as a transient intermediate or transition state (see section 2.K). An understanding of the characteristics of the 5-coordinate state is necessary in the beginning of any study of organophosphorus chemistry. The subject will be further developed in chapter 10; some useful reading at this time would be the reviews of Smith[19a] and Burgada and Setton.[4e]

The geometry of the 5-coordinate state in phosphoranes (R_5P) is well established, from many X-ray crystallographic investigations, to be that of the trigonal bipyramid (**12**), with longer bonds to the two apical positions than to the three equatorial positions. The bond angle to the apical atom from an equatorial atom is 90°, whereas the equatorial bond angle is 120°. The square pyramid **13** is an alternative for the disposition of the five bonds, but has been encountered so far mostly among some

12   **13**

spirocyclic phosphoranes[20a,21] and will not be further discussed. 5-Coordinate structures can, however, show deviation from the pure trigonal bipyramid (TBP) form in the direction of the square pyramid.[21]

The bonding of the TBP was early recognized to be consistent with the inclusion of the phosphorus d_{z^2} orbital to give sp^{3}d hybridization, but the TBP geometry does not *prove* this hybridization, and here again dispute exists on the relevance of d-orbitals in forming valence bonds. Although the sp^{3}d bonding picture is a convenient description that has been in use for many years, an objection to it is that the d_{z^2} orbital is diffuse and of rather high energy, and alternative bonding theories have been sought for several years. A discussion of this problem may be found in reference 7. The Rundle view,[22] favored by Kutzelnigg,[7] employs a bond between three atoms (in this case, the central P atom and the two apical atoms) that contains only four electrons. In valence bond terminology, the three-center four-electron bond would be expressed by the resonance form $X^- \text{---} P^+ - X \leftrightarrow X - P^+ \text{---} X^-$. Thus X has one-half of a negative charge, and P is covalent with a positive charge. This accounts for the apical bonds being longer than the equatorial bonds. Molecular orbital (MO) theory indicates the bonding to consist of one bonding MO, one nonbonding MO and one antibonding MO. Because the first two MO's are occupied by the four electrons, each P–X linkage can be considered as a half bond.

One argument that had been used to support the d-orbital picture is that the most stable structures are formed with the most electronegative elements, consistent with the idea that the diffuseness of the d-orbital is lessened by electron attraction and better orbital overlap becomes possible. It is also well known that the electronegative elements prefer to form bonds at the apical position, because the apical sites are more polarizable and the bonds are longer and weaker than the equatorial sites. Indeed, this is an important consideration in phosphorane chemistry; the most stable structures have one or more strongly electronegative elements present, especially fluorine, and compounds with only alkyl groups on phosphorus are unstable. Experience has led to the following order for the apical preference (apicophilicity) of some common substituents:

$$F > H > CF_3 > PhO > Cl > MeS > MeO > Me_2N > Me > Ph$$

Electronegativity is not the only factor involved; the steric size of the substituent is important, and smaller groups are more able to occupy the more crowded (smaller bond angle, 90°) apical positions than are larger groups. π-Conjugative effects are also thought to be important; π-donors such as R_2N and RS favor the equatorial position, where interaction with the acceptor orbital perpendicular to the equatorial plane is allowed. π-Accepting substituents prefer the apical position.

Much experimental evidence supports the concept that phosphoranes can be "fluxional," meaning that groups are not held firmly in a given apical or equatorial position but are constantly interchanging positions in isomeric structures. It is generally not possible to isolate isomeric phosphoranes that differ in the apical or equatorial disposition of substituents, because this isomerization is usually very rapid at room temperature. It can, however, be retarded by reducing the temperature, allowing spectra for individual isomers to be recorded. An early example[23] is the molecule Et_2NPF_4, which gives but one ^{19}F NMR signal at 25° (δ ^{19}F −10, $^1J_{PF}$ = 851 Hz) for the F atoms rapidly equilibrating between apical and equatorial positions, but at −85°, it gives two signals, in the ratio 1:1, with δ ^{19}F −17.3 ($^1J_{PF}$ = 793) and −5.2 ($^1J_{PF}$ = 916, J_{FaxFeq} = 70 Hz) being assigned to the apical and equatorial F, respectively. The mechanism by which this intramolecular isomerization occurs has been much studied, and for the most part that known as Berry pseudorotation (BPR) has found general acceptance. The BPR mechanism involves the bending of bonds (a vibrational process) into new positions in this fashion: holding one equatorial atom in place, the two apical bonds are bent into the equatorial positions of a new trigonal bipyramid, and at the same time two equatorial bonds are bent into apical positions. The equatorial atom held in place is referred to as the pivot for the interchanges. The process is easily understood by examining the movement of atoms in Eq. 2.11.

$$(2.11)$$

The application of BPR to a real molecule is illustrated by Eq. 2.12.

$$(2.12)$$

The alternative mechanism is called Turnstile Rotation (TR) and also involves bond bending to accomplish the same interchange of groups. There are some cases with cyclic structures in which the BPR process seems physically impossible, but TR allows the interchange needed. TR is more difficult to visualize (Eq. 2.13); it involves rotating a pair consisting of one apical and one equatorial atom (called the duo) relative to the other three atoms (the trio), thus the name turnstile. Bond bending is needed to get the atoms into the form of a new TBP. As illustrated in Eq. 2.13, atoms 2 and 4 are the rotating duo; bending of atoms 1 and 3 to form a 90° angle takes place simultaneously so that 3 will take an apical position, and 2 bends to take the other apical position. Note that one group (1) remains in its original equatorial position; the others have exchanged positions, just as in the BPR mechanism. Both BPR and TR processes require little energy to achieve the bendings needed and thus occur rapidly.

$$(2.13)$$

2.G. THE P—OH TO P(O)H REARRANGEMENT WITH 3-COORDINATE PHOSPHORUS

There is a form of tautomeric behavior in phosphorus chemistry that can be attributed to the marked tendency to form the phosphoryl group when possible. It became known as early as the beginning of the century that the replacement of the three chlorine atoms of phosphorus trichloride by hydroxy groups to form phosphorous acid (H_3PO_3) did not in fact give the expected structure $(HO)_3P$ but instead the dihydroxy compound $(HO)_2P(O)H$, in which the hydrogen of one OH group had migrated to phosphorus, thus creating the phosphoryl group. Similar behavior was observed for the dialkyl esters of phosphorous acid that can result from the reaction of alcohols with PCl_3 in the absence of a base (Eq. 2.14).

$$PCl_3 + ROH \longrightarrow RO-P(OR)-OR \xrightarrow[-RCl]{HCl} RO-P(OR)-OH \rightleftharpoons RO-P(OR)(=O)-H \quad (2.14)$$

Many studies over the years have revealed that there is an equilibrium between the two forms, but with only a tiny amount of the "enol" (P—OH) form present. The enol content is in fact likened to that of the enol form of acetone in water, about $10^{-4}\%$. The same phenomenon prevails in any case in phosphorus chemistry when an OH group is placed on 3-coordinate phosphorus. There are only two accepted exceptions (vide infra) to the statement that the P—OH form never dominates or even has a significant concentration in the equilibrium, and it is safe to say that a structure thought to have hydroxy on 3-coordinate phosphorus must be written in the isomeric phosphoryl form. This property has some important consequences in organophosphorus chemistry. Thus the phosphonous acids, formally written as structure **14** where they appear to be dibasic, are in fact monobasic and are more properly expressed as the tautomer **15**. True derivatives of phosphonous acids are known, however, such as $RPCl_2$ and $RP(OR)_2$, in which no tautomeric rearrangement is possible. Hydrolysis of these derivatives probably gives structure **14** as an intermediate, which rearranges to the observed "keto" tautomer (Eq. 2.15).

$$R-P(Cl)_2 \xrightarrow{H_2O} R-P(OH)_2 \rightleftharpoons R-P(=O)(H)(OH) \quad (2.15)$$
$$\qquad\qquad\qquad\qquad \mathbf{14} \qquad\qquad \mathbf{15}$$

Similar behavior is exhibited by the phosphinous acids. The parent acid may be expressed as structure **17**, but it has no real existence; it is known in derivative form (e.g., ester **16**) but hydrolysis of this ester (Eq. 2.16), or of the corresponding P—Cl compound, gives the tautomeric form that is seen to be a neutral compound called a secondary phosphine oxide.

$$R-P(R)-OR' \xrightarrow{H_2O} R-P(R)-OH \rightleftharpoons R-P(=O)(H)(R) \quad (2.16)$$
$$\quad\mathbf{16} \qquad\qquad \mathbf{17}$$

One final example is that of hypophosphorous acid, H_3PO_2; it is written not as $(HO)_2PH$ but as $(HO)PH_2(O)$. Some organophosphorus derivatives such as the monoesters are based on this structure (e.g., $(RO)P(O)H_2$).

One of the exceptions to the rule stated above is found when two (but not one) trifluoromethyl groups are present on phosphorus as in compound **18**;[24] apparently electron attraction from phosphorus is so pronounced that the lone pair is not available for acceptance of hydrogen from OH. The other exceptional type of compound has the imido group bonded to phosphorus as in **19**.[25] In both **18** and **19**, the characteristic infrared stretching band of P=O (see chapter 7.C) is missing, and ^{31}P shifts (chapter 6) are in the 3-coordinate, not the 4-coordinate, region.

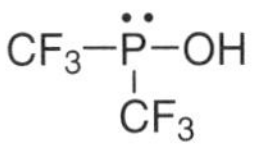

Further information on the tautomeric behavior of 3-coordinate phosphorus acids may be found in the review literature.[26]

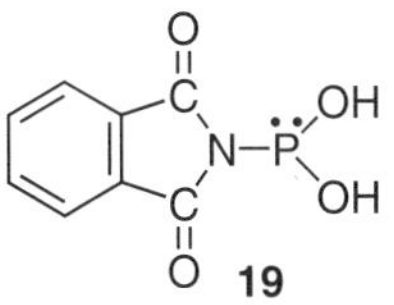

2.H. PHOSPHORUS BONDS TO HETEROATOMS (X, S, OR, NR₂)

Some of the great diversity of compounds in phosphorus chemistry comes about from the high thermal stability of phosphoryl functionalities in which a single bond to one (or more) of the electronegative atoms O, N, or halogen (X), as well as to S, is present. These structures are treated as derivatives of the corresponding acids. To illustrate with the case of an alkylphosphonic acid ($R-P(O)(OH)_2$) as the parent, the following structural types, among others, are known:

$RPOX_2$
$RPO(OR)_2$
$RPO(NR_2)_2$
$RPO(NHR)_2$
$RPO(X)(NR_2)$
$RPO(OR)(NR_2)$
$RPO(SR)_2$.

Somewhat less stable are functions in which both halogen and alkoxy are present; they are known compounds but can undergo elimination of alkyl halide on heating with the formation of polymeric material (Eq. 2.17).

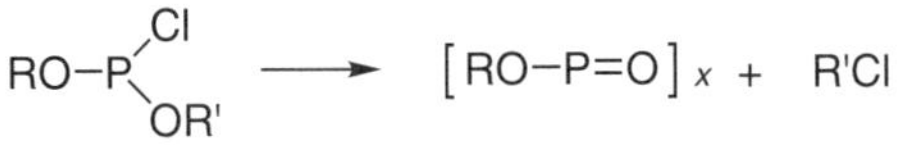

$$RO-P{<}^{Cl}_{OR'} \longrightarrow [RO-P=O]_x + R'Cl \qquad (2.17)$$

The dissociation energies (kcal/mol)[20b] of some of these atomic combinations are P−Cl, 79; P−Br, 63; P−F, 126; P−O, 86; and P−N, 55. These values are considerably higher than those for comparable nitrogen bonds (e.g., N−Cl, 46; N−F, 65; N−O, 50). An additional difference from the nitrogen compounds is that the phosphorus atom, because of its low electronegativity (2.2, with N 3.0), bears a greater positive charge; it is prone to attack by nucleophiles, with displacement of anionic species. Thus "positive halogen" is never created from phosphorus compounds in the way that it is from some nitrogen halides.

The thermal stability of the bond from 3-coordinate phosphorus to the heteroatoms is also adequate to ensure that the numerous structural organizations of phosphoryl compounds will exist here also, making possible a series of compounds, such as the phosphonous acid derivatives RPX_2, $RP(OR)_2$, $RP(NR_2)_2$, $RP(SR)_2$. Compounds with mixtures of these substituents are also well known. Alkoxy substituents have a special property, however; at high temperatures, or with catalysis by alkyl halides, they can undergo a rearrangement to phosphoryl compounds, a process referred to as the Michaelis-Arbusov rearrangement (Eq. 2.18). These rearrangements are of great practical synthetic importance and will be described more fully in chapter 5.

$$RO-P(OR)(OR) \longrightarrow RO-P(=O)(R)(OR) \qquad (2.18)$$

Exposing 3-coordinate halides with P-alkyl groups to high temperatures can lead to decomposition by loss of hydrogen halide; aryl-substituted halides, however, are quite stable thermally.

In many of the phosphoryl derivatives, sulfur or selenium may take the place of the multiply bonded oxygen atom. Thus several families of thiono (P=S) derivatives, with the same single bonds as in the comparable phosphoryl compounds, are known, such as the following for a thiophosphonic acid, $RP(S)(OH)_2$:

$RP(S)X_2$
$RP(S)(OR)_2$
$RP(S)(NR_2)_2$
$RP(S)(NR_2)X$

The strength of the P=S bond is considerable (90 kcal/mol[20b]) but, of course, less than that of the extremely stable P=O group (130 kcal/mol).

Over the years, assigning names to the numerous functionalities that can result from these many substitutions on phosphorus has been troublesome, and more than one system has been in use. Standardization, however, has been provided by the International Union of Pure and Applied Chemistry (IUPAC);[27] and with the exception of some common names for simple structures, its rules (which occasionally offer alternative names) are those that are now generally employed. *Chemical Abstracts* uses a very similar, but not identical, system. Some features of the IUPAC-

2.I. ANALOGIES BETWEEN THE CHEMISTRY OF PHOSPHORUS COMPOUNDS AND THAT OF RELATED ELEMENTS (N, S, C)

approved nomenclature are presented for reference purposes in Appendix 2.1. For the most part, however, we can avoid many of the intricacies of nomenclature in the discussions in this book.

Because nitrogen and phosphorus are in the same group in the Periodic System, it is natural to inquire to what extent are their functionalities similar, and of what aid would knowledge of nitrogen chemistry be in understanding phosphorus chemistry. The same may be asked about sulfur, the neighbor in the same row to the right of phosphorus.

In comparing phosphorus to nitrogen, some basic differences must be remembered. Phosphorus is much less electronegative than nitrogen (2.2 vs. 3.0), and its greater size causes bond lengths to be considerably larger (P—C, 1.84 Å in phosphines; N—C, 1.70 Å in amines). This creates a difference in the steric environment around these atoms. A more profound difference is in the ability of phosphorus to participate in a form of multiple bonding to oxygen that is absent in nitrogen chemistry; this leads to a great variety of structural types (e.g., phosphoric, phosphonic, and phosphinic acids and derivatives) that are quite unknown for nitrogen.

Many examples are known of differences between comparable P and N compounds, as in the much easier oxidation of phosphines than amines and in the much greater stability of phosphine oxides than amine oxides. Frequently, the paths followed in a reaction with a given reagent can be quite different; thus, with OH$^-$, quaternary phosphonium salts undergo cleavage of a C—P bond to form tertiary phosphine oxides (chapter 4.C), whereas ammonium salts undergo Hofmann degradation to give tertiary amines. Other differences will be noted here, but there are indeed some useful comparisons that can be made. These are found in the 3-coordinate condition. Phosphines may be primary, secondary, or tertiary and resemble the corresponding amines in stability and basicity. The phosphines are also good nucleophiles, as are amines. In phosphines, the barrier to pyramidal inversion is much higher, typically 35 kcal/mol, and many examples of pyramidal, optically active phosphines are known (e.g., **20**; see also chapter 9). This is not possible for amines, in which the pyramidal inversion is extremely rapid, as in **21**.

We have already noted that bonds to atoms such as X, O, N, and S are common in phosphorus chemistry but, with the exception of O, are much less so in nitrogen chemistry.

SOME GENERAL CONSIDERATIONS OF ORGANOPHOSPHORUS COMPOUNDS

The resemblance to sulfur chemistry is found in the higher oxidation states, because sulfur also can form highly stable multiple bonds to oxygen. Phosphonic ($RP(O)(OH)_2$) and sulfonic ($RS(O)_2OH$) acids thus have rather similar characteristics, in being stable crystalline solids of considerable acidity. Phosphonic acids do differ, of course, in being dibasic. Alkyl sulfates and phosphates likewise are similar.

The analogy to sulfur is continued in the structures with no OH groups: tertiary phosphine oxides ($R_3P=O$) are very stable and highly polar solids (generally), just as are the sulfones (R_2SO_2).

There are a number of structures in lower coordination states with unsaturation in which a superficial resemblance can be found between phosphorus and nitrogen. The properties of the species differ drastically, however. In general, the nitrogen structures are stable, whereas the phosphorus species require special structural features to achieve stability, or in some cases can be detected only as transients. Thus the well-known nitrate and nitro groups have phosphorus counterparts that cannot be isolated, although their existence as transients has been established (chapter 10.D). These species (σ^3, λ^5) are the alkyl metaphosphates, $RO-PO_2$, and alkyl dioxophosphoranes, $R-PO_2$. Another σ^3, λ^5 phosphorus species is typified by structure **22**, which resembles a nitrone **23**; in structure **22**, stability is imparted by the presence of the sterically demanding substituents on C, and in their absence the substances do not exist.

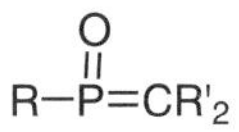

$$\underset{\textbf{22}}{\overset{\overset{\displaystyle O}{\|}}{R-P}=CR'_2} \qquad \underset{\textbf{23}}{\overset{\overset{\displaystyle O}{\uparrow}}{R-N}=CR'_2}$$

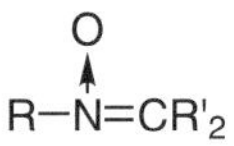

There are several other low-coordination phosphorus species that can be stabilized by steric or electronic effects and that have stable well-known nitrogen counterparts. Some are shown below and will be seen again in chapter 10.

$R-P=CR_2$ (phosphaalkene) and $R-N=CR_2$ (imine)
$RO-P=O$ (alkyl phosphenite) and $RO-N=O$ (nitrite)
$R-P=NR$ (iminophosphines) and $R-N=NR$ (azo)
$R-C\equiv P$ (phosphaalkyne) and $RC\equiv N$ (nitrile)

Phosphorus plays a major role in life processes as, of course, do both nitrogen and sulfur, but the structures in which these elements are found are totally different. Phosphorus is only known in 4-coordinate form, as phosphoric acid derivatives (esters, amides, and anhydrides), and to a much lesser extent as phosphonic and phosphinic acid derivatives (chapter 11). There are no counterparts to the vast number of nitrogen heterocycles or amines of Nature, or to the thiol, sulfide, and heterocyclic forms of sulfur. A sulfonic acid, however, has a phosphorus counterpart; taurine (**24**), which is a component of some lipids, resembles 2-aminoethylphosphonic acid (**25**), is widely distributed in lower life forms, and is bound in lipids as well as in glycoproteins.

$$\underset{\mathbf{24}}{H_2N-CH_2-CH_2-\overset{O}{\underset{O}{\overset{\|}{\underset{\|}{S}}}}-OH} \qquad \underset{\mathbf{25}}{H_2N-CH_2-CH_2-\overset{O}{\underset{OH}{\overset{\|}{P}}}-OH}$$

It has become apparent in recent years that there can be a remarkable similarity in some reactions of the carbon–carbon double bond and of the phosphorus–carbon double bond. This point was well made by Appel in a 1990 review of the chemistry of the newly discovered C=P bond[28a] (see chapter 10). In 1998, a book was published that extended the analogy to include the C−P triple bond, heterocyclic systems, and the electron deficient phosphinidene species (R−P); its title, *Phosphorus: The Carbon Copy*,[29] clearly suggests the theme of the book. In considerations to follow, we will see some of the properties of the phosphorus species that lead to the analogy with unsaturated or electron deficient carbon compounds. For the present, one example will be given that makes the case rather compellingly. There is a phosphorus counterpart to the well-known Cope [3,3]sigmatropic bond shift of dienes, wherein the C=P bonds of the diene act in the same way as C=C bonds do. The "phospha-Cope" equilibration of 1,3,4,6-tetraphosphahexa-1,5-diene (**26**) with the isomer formed by opening the 3,4-bond (**27**) is shown in Eq. 2.19.[28a] Several related examples are known.

(2.19)

2.J. INCORPORATION OF PHOSPHORUS IN RINGS

An important part of contemporary phosphorus chemistry is concerned with the synthesis and properties of cyclic systems containing this element. Heterocyclic phosphorus chemistry was slow to develop; much was known about N and S heterocycles before the first P heterocycle with C−P bonds, PhP(CH$_2$)$_5$, was synthesized in 1915,[30] and progress in the field was very slow until the last 3–4 decades. Now a great number of ring systems are known, and research in this area continues at a rapid pace. Some representative ring systems will be discussed in chapter 8, but a general introduction is useful at this point because heterocyclic systems will be encountered in chapters to follow on the synthesis and properties of various phosphorus functionalities.

Phosphorus occurs in saturated rings with carbon in all sizes from 3 to 10 or more members and with the element in all of the many possible 3- and 4-coordinate functional groups. Unsaturated derivatives are also known; of special interest is the 5-membered phosphole (**28**) and 6-membered phosphinine (**29**) systems, because of their similarity to pyrrole and thiophene and to pyridine, respectively. As will be seen, the phosphole system lacks the pronounced aromatic character of pyrrole and thiophene; phosphinines are, however, known that have appreciable aromaticity.

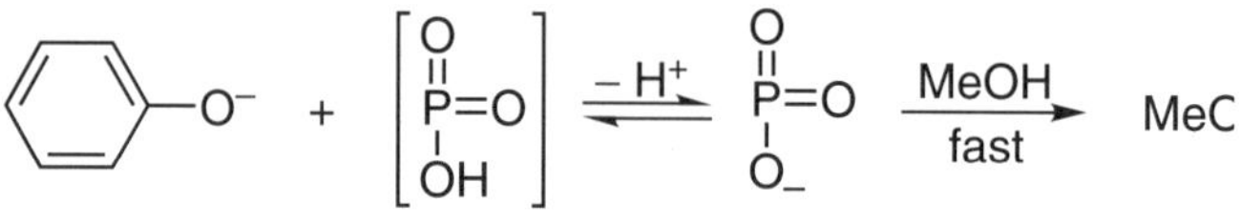

$$\text{Ph-O}^- + \begin{bmatrix} \text{O} \\ \| \\ \text{P=O} \\ | \\ \text{OH} \end{bmatrix} \underset{}{\overset{-H^+}{\rightleftharpoons}} \begin{matrix} \text{O} \\ \| \\ \text{P=O} \\ | \\ \text{O}_- \end{matrix} \xrightarrow[\text{fast}]{\text{MeOH}} \text{MeO}$$

36

Because phosphorus bonds to O, N, and S are also quite stable, these atoms can be found along with P in heterocyclic systems. Many of the numerous possible structures are known. To illustrate, a collection of the known saturated 6-membered rings containing one phosphorus atom (others may also be present) is given in Scheme 2.1. Adding to the number of possibilities is the capability of P to exist in its various functional groups; these are not shown in Scheme 2.1, where only the ring skeleton is given. When functional groups are added, the total number of structural types far exceeds that for rings based on a single nitrogen atom.

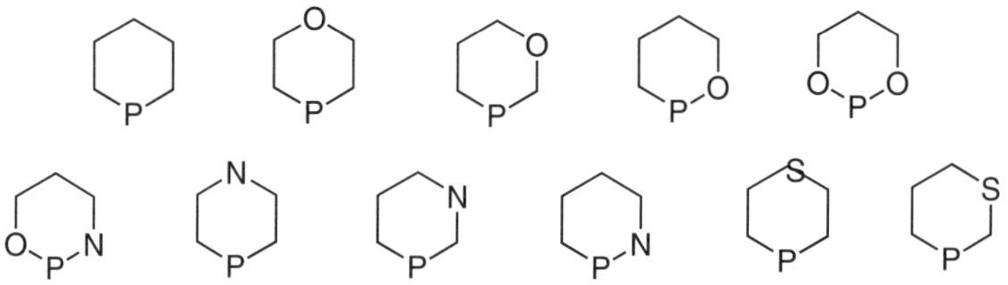

Scheme 2.1

In general, ring structure does not greatly modify the properties expected for a particular functional group. Ring shapes resemble those of carbon compounds; thus a four-membered ring is nonplanar with the butterfly shape, and the six-membered ring is in the chair shape. The size of the phosphorus atom is such that 1,3-diaxial interactions are diminished. In the 3-coordinate condition, phosphorus has a stable pyramidal structure, which leads to the existence of stable cis, trans isomers (e.g., **30** and **31**), a possibility unknown for nitrogen except in certain three-membered rings.

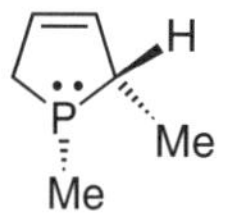

30 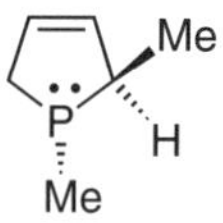 **31**

Numerous phosphorus heterocycles are known with a second ring fused to the P-containing ring and with bridges across the ring. Altogether, the number of heterocyclic phosphorus compounds now known has become quite large and continues to grow.

2.K. REACTIVE INTERMEDIATES AND REACTION MECHANISMS

In this section, the nature of reactions occurring on phosphorus is introduced in brief form; many reaction mechanisms will be presented in later chapters that will expand on the ideas presented here. A monograph by Kirby and Warren[31] is an excellent source of general mechanistic information. There is a similarity to some of the reactive intermediates of carbon chemistry; phosphorus species resembling carbanions, carbocations, and free radicals do exist but, of course, have their own unique properties and significance.

The hydrogen on phosphines is acidic (e.g., Ph_2PH has pK_a 22) and phosphide anions (RHP^- or R_2P^-) are readily formed by reaction with a strong base (the latter ion also by other methods, seen in chapter 3). Metallic phosphides are much more stable than carbanions and can be isolated and employed as nucleophiles in various reactions.

Cationic species (phosphenium ions, reviewed in reference 28b) are much more difficult to generate and are known in only special cases. The first examples were reported in 1972.[32] In both structures, and in many of the ions prepared later, stability was provided by the presence of an amino group on P, which aided in dispersing the positive charge by resonance as in Eq. 2.20. Stabilization can also be provided by RS- substituents and by certain benzophosphole structures.[33] The phosphinous chloride t-Bu_2PCl failed to give a phosphenium ion with $AlCl_3$, forming instead the chlorophosphonium salt t-$Bu_2PCl_2{}^+AlCl_4{}^-$ as a major product. This structure, whose origin is not clear, was proved by X-ray analysis;[33b] an earlier report had indicated the observation of the 1:1 complex of the reactants.[33c]

$$\text{(2.20)}$$

It is possible that phosphenium ions may appear as transient reactive intermediates in certain reactions, but this has been difficult to prove. There are several reactions known in which a phosphorus halide, in the presence of aluminum chloride, attacks an unsaturated carbon, as in Equations 2.21[34] and 2.22.[35]

$$RPCl_2 \xrightarrow[CH_2Cl_2]{AlCl_3} [RPCl_2 \cdot AlCl_3] \xrightarrow{PhC\equiv CPh} \quad \text{(2.21)}$$

$$CH_2=CH-CMe_3 \xrightarrow[PCl_3]{AlCl_3} \quad \text{(2.22)}$$

Although it is convenient to consider these reactions as involving a phosphenium ion (Cl_2P^+ and $RPCl^+$, respectively) formed by aluminum chloride removing chlorine from phosphorus, which then attacks the unsaturated carbon, there is in fact no direct proof of these species being involved. The ^{31}P NMR spectra of solutions of phosphorus chlorides (except PCl_3) and aluminum chloride, while revealing the presence of Lewis intermolecular complexes, contain no signals attributable to a phosphenium ion, although it may be present in undetectable amounts. It is possible also that nonbonding attraction of $AlCl_3$ for Cl may act to polarize the P—Cl bond, rendering phosphorus partially positive and thus more electrophilic.

Phosphorus cations of structure $RR'P(O)^+$, a "phosphacylium ion," could also conceivably be produced by removal of a negative group from 4-coordinate phosphoryl derivatives. This possibility has been only recently demonstrated[36] in the case of a thiophosphoryl derivative; as seen in Eq. 2.23, dissociation of the mixed anhydride **32** to a phosphaacylium cation **33** (probably strongly solvated) in ionizing solvents could account for reactivity effects observed, such as the electrophilic substitution of anisole (Eq. 2.23) or the retention of stereochemistry in a reaction with a nucleophile, rather than the usual inversion that accompanies substitution at phosphoryl centers. There is as yet, however, no evidence that such cationic intermediates play any role in common reactions of phosphoryl or thiophosphoryl species, although the possibility remains open.

(2.23)

There are two types of free radicals based on phosphorus,[37] and both are involved in useful synthetic processes. Abstraction of a hydrogen atom from a primary or secondary phosphine by attack of a radical such as ROO· or by ultraviolet light can give the species RHP· or $R_2P·$ as intermediates. The subject has been reviewed.[4f] A typical synthetic process using such a radical is shown in Eq. 2.24, where tertiary phosphines with a long alkyl chain are the products.

$$R_2P-H \xrightarrow[\text{initiator}]{\text{UV or}} R_2P·$$ (2.24)

$$R_2P· + CH_2=CHR \longrightarrow R_2PCH_2\dot{C}HR \xrightarrow{R_2PH} R_2PCH_2CH_2R + R_2P·$$

Radicals can also be formed as intermediates by addition of other radicals to tertiary phosphines. Such products have the structure $R_4P·$ with nine electrons and are referred to as phosphoranyl radicals. An example of their involvement in a useful process (describable as a free radical Michaelis–Arbusov reaction) is seen in Eq. 2.25.

$$\text{Ph} \cdot + (\text{EtO})_3\text{P:} \longrightarrow \text{Ph}\dot{\text{P}}(\text{OEt})_3 \longrightarrow \text{PhP(O)(OEt)}_2 + \text{Et} \cdot \quad (2.25)$$

The lone pair on 3-coordinate phosphorus, particularly in phosphines, is involved as a nucleophilic center in many reactions. These are highly important processes in organophosphorus chemistry and are discussed in chapter 3.C.

Nucleophilic displacements of P-substituents are common among phosphoryl and thiophosphoryl compounds, and the mechanism of this process has been extensively studied. The important feature is the ability of phosphorus to accept an electron pair of the nucleophile, creating a 5-coordinate intermediate (or transition state). The intermediate collapses to the original tetrahedral geometry by expulsion of one of the ligands. Stable 5-coordinate molecules are in fact known, as was discussed in section 2.F, and their geometry is well established to be that of the trigonal bipyramid. When the trigonal bipyramid is created as a reaction intermediate, the incoming group takes up an apical position. When collapse of the adduct occurs, it is the group in the other apical position that departs. Unless there are any special constraints from a ring structure, the trigonal bipyramid is in a state of rapid equilibrium with the several isomeric structures that could result from interchange of the substituents through the Berry pseudorotation process. This process is illustrated in Eq. 2.26 with the nucleophilic substitution of chlorine from a phosphinic chloride. This mechanism will be seen to be of the utmost importance in organophosphorus chemistry. When the process is applied to an optically active phosphorus compound as in Eq. 2.26, it is found that the product has the inverted geometry in its new tetrahedral form.

$$\underset{\text{O}}{\overset{\text{Cl}}{\text{Me}-\text{P}-\text{Et}}} \xrightarrow{\text{EtONa}} \underset{\text{OEt}}{\overset{\text{Cl}}{\text{Me}\cdots\text{P}-\text{Et}}} \longrightarrow \underset{\text{OEt}}{\overset{\text{O}}{\text{Me}-\text{P}\cdots\text{Et}}} \quad (2.26)$$

Special situations arise from geometrical constraints when phosphorus is part of a small ring system, and cases of retention are known. An excellent discussion of these and many other facets of the 5-coordinate state and the nucleophile substitution process can be found in the monograph of Holmes.[38]

Three-coordinate phosphorus also may contain groups that can be displaced by nucleophiles. These groups may be halogen, alkoxy, amino, and thioxy but the most commonly encountered substituent is chlorine. The mechanism of the displacement from 3-coordinate phosphorus is less well studied but may also involve addition of the incoming nucleophile to phosphorus, to form a transition state (thus an S_N2-like process) or intermediate. This intermediate may be viewed as the familiar trigonal bipyramid with the original electron pair on phosphorus occupying one of the positions of this structure. Presumably here also the group to depart must be in the apical position. The process is illustrated in Eq. 2.27 with the reaction of a phosphinous chloride and sodium ethoxide. Mechanisms for the displacement reactions from 3-coordinate phosphorus have been reviewed.[4g]

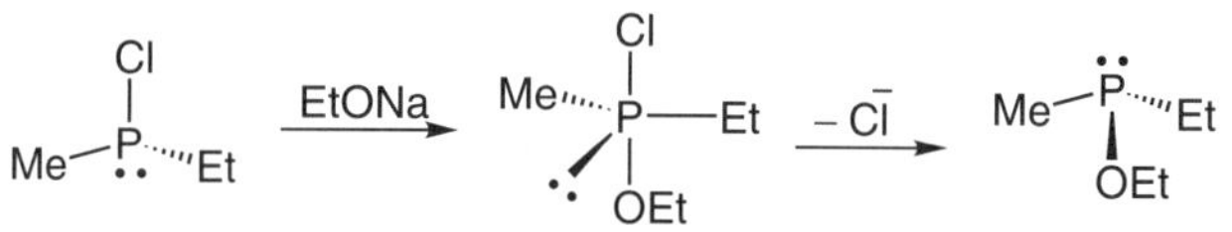

$$\text{Me}-\overset{\text{Cl}}{\underset{..}{\text{P}}}\text{..Et} \xrightarrow{\text{EtONa}} \text{Me}\overset{\text{Cl}}{\underset{\underset{..}{\text{OEt}}}{\overset{|}{\text{P}}}}\text{—Et} \xrightarrow{-\text{Cl}^-} \text{Me}-\overset{..}{\underset{\text{OEt}}{\overset{|}{\text{P}}}}\text{..Et} \qquad (2.27)$$

The displacement reactions on 4-coordinate phosphorus compounds seen above have been described as occurring through an addition-elimination (AE) process, but in special cases this sequence may be reversed, the reaction occurring through elimination of a molecule to form a low-coordinate, highly reactive species that then adds the nucleophile (the EA process). The two processes may be distinguished by kinetics measurements: AE is first order in both the nucleophile and the phosphorus reactant as the trigonal bipyramid is created, and EA is first order only in the phosphorus reactant, because the attack of the nucleophile on the highly reactive low-coordinate intermediate would be extremely rapid. Further background on the EA process and methods for its recognition may be found in the book by Kirby and Warren,[31] it is also the subject of several reviews,[39,40] especially because it has been proposed to be involved in some biological phosphorylations. The EA mechanism with the hydrolysis of a monoaryl phosphate (**34**) is illustrated in Eq. 2.28. There is a rate maximum at about pH 4, which suggests that it is the monoanion **35**, at a maximum concentration at this pH, that undergoes the elimination of the aryloxy group as a phenol to form the reactive low-coordinate **36** (the metaphosphate ion).

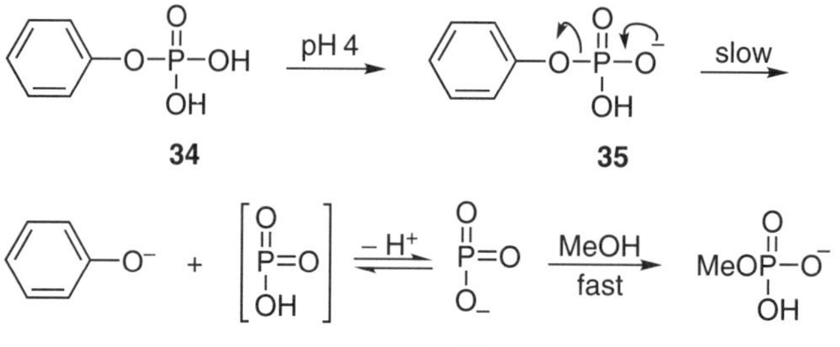

$$\underset{\textbf{34}}{\text{PhO-P(=O)(OH)}_2} \xrightarrow{\text{pH 4}} \underset{\textbf{35}}{\text{PhO-P(=O)(OH)(O}^-\text{)}} \xrightarrow{\text{slow}}$$

$$\text{PhO}^- + \left[\underset{\textbf{36}}{\overset{\text{O}}{\underset{\text{OH}}{\text{P=O}}}}\right] \underset{\text{}}{\overset{-\text{H}^+}{\rightleftharpoons}} \overset{\text{O}}{\underset{\text{O}^-}{\text{P=O}}} \xrightarrow[\text{fast}]{\text{MeOH}} \text{MeOP(=O)(OH)(O}^-\text{)} \qquad (2.28)$$

A more recent example, illustrating a stereochemical approach to distinguishing the EA and AE mechanisms, has been provided by Harger[41] who studied the nucleophilic substitution of a benzylphosphonamidothioic chloride (Scheme 2.2) by diethylamine. Here a chiral substituent was present on nitrogen, and because the phosphorus is also chiral, it was possible to obtain two diastereomeric forms (**37A** and **37B**) of the starting chloride. When reacted with diethylamine, both diastereoisomers gave nearly the same mixture of diastereomeric diamide products. This suggests that the reactions proceed through a common intermediate **38** with 3-coordinate phosphorus, formed by elimination of HCl. Because the intermediate is planar, the subsequent addition of diethylamine can occur nearly equally from the two faces. The stereochemical result is quite different from that expected of the more common AE mechanism, which as noted proceeds with inversion of the phosphorus configuration. (The subject of phosphorus stereochemistry will be developed more fully in chapter 9.)

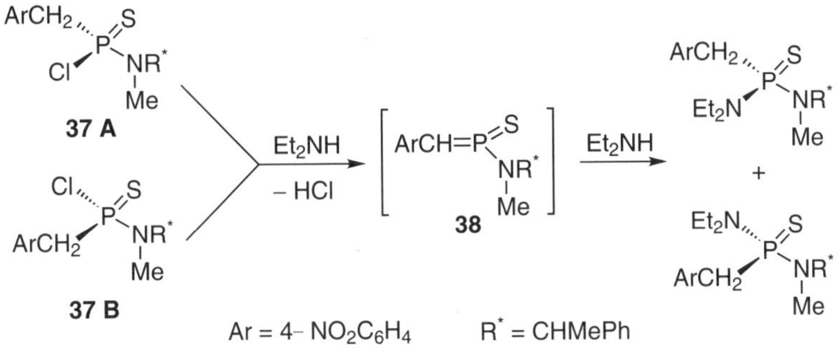

2.K. REACTIVE INTERMEDIATES AND REACTION MECHANISMS

Scheme 2.2

Ar = 4-NO$_2$C$_6$H$_4$ R* = CHMePh

Three-coordinate compounds may also undergo an elimination reaction of a different type to form 2-coordinate species, which can then add a nucleophile. This process is usually observed not as a mechanism but in two discrete synthetic steps. Examples will be seen in chapter 10.C. Some mention of P=C species as reactive intermediates does appear in early literature, but without experimental proof. An example of this will be seen in the discussion of phosphine reactions in chapter 3.C.

Cycloaddition reactions at 3-coordinate phosphorus constitute another type of mechanistic pathway. The best known case is that of the reaction of phosphorus halides with conjugated dienes, which is a major process for the synthesis of the five-membered ring containing phosphorus (Eq. 2.29). Here the 3-coordinate phosphorus uses its lone pair to form a new bond to carbon, while at the same time accepting an electron pair from the diene. This is referred to as biphilic behavior by 3-coordinate phosphorus. The initial product may be formulated as the 5-coordinate species **39**, but with chlorine derivatives, it is known that ionization to the more stable 4-coordinate state (**40**) takes place.

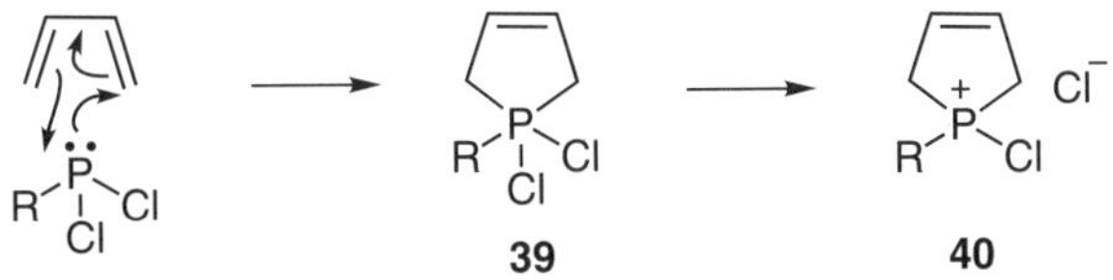

(2.29)

From stereochemical studies, it is known that the process has typical cheletropic properties and resembles the reaction of sulfur dioxide with dienes. Thus from *trans,trans*-2,4-hexadiene only the cis-substituted five-membered ring is formed, as expected of a disrotatory [4+2] cycloaddition (Eq. 2.30). Many examples are known of this cycloaddition reaction,[42] which will be seen elsewhere in this book.

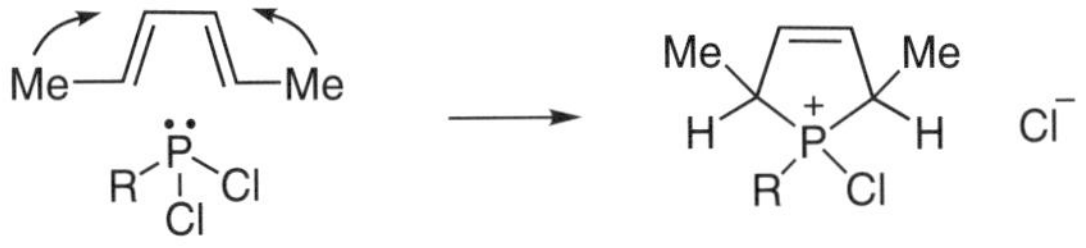

(2.30)

Another related addition reaction type for 3-coordinate phosphorus is that which occurs with α-dicarbonyl compounds, a process that is best known for trialkyl phosphites (Eq. 2.31). In this case, the cycloadduct is of a form well known to be stable, in which five oxygen atoms are attached to phosphorus (chapter 10.E).

$$\underset{\underset{\text{OR}}{\overset{\text{RO}\diagdown \text{P}\diagup \text{OR}}{}}}{\overset{\text{Me}\diagdown \diagup \text{Me}}{\underset{\text{O}\diagdown \diagup \text{O}}{}}}\quad \longrightarrow \quad \underset{\underset{\text{OR}}{\overset{\text{RO}\diagdown \text{P}\diagup \text{OR}}{}}}{\overset{\text{Me}\diagdown \diagup \text{Me}}{\underset{\text{O}\diagdown \diagup \text{O}}{}}} \tag{2.31}$$

The biphilic conversion of 3-coordinate to 5-coordinate phosphorus can occur in other reactions as well, as in the addition of peroxides seen in Eq. 2.32.

$$(RO)_3P\colon \; + \; R'{-}O{-}O{-}R' \; \longrightarrow \; RO{-}\underset{\underset{OR}{|}}{\overset{\overset{OR'}{|}}{P}}{\cdots}\underset{OR}{\overset{OR'}{}} \tag{2.32}$$

Compounds with double and triple bonds to phosphorus undergo many cycloaddition reactions, especially with 1,3-dipoles. These reactions are discussed in chapter 10.

Photochemical reactions of phosphorus compounds have been extensively studied, and some have practical value, especially among heterocyclic derivatives. Some examples are found in reference 42. The literature has been extensively reviewed on the photochemistry of phosphates,[43] phosphonium salts,[4h] ylides,[4h] and phosphoranes[4h].

To conclude this brief survey of reaction types open to phosphorus compounds, mention is needed of the existence of several skeletal rearrangement processes of phosphoryl compounds that bear a resemblance to well-known rearrangements of carbon compounds:

The Beckmann rearrangement (thermal) of oximes to amides:[44]

$$\underset{\underset{Ph}{|}}{\overset{HO\diagdown N \quad O}{\underset{Ph-C-P-OMe}{\overset{\parallel \quad \parallel}{}}}} \quad \overset{\Delta}{\longrightarrow} \quad \underset{\underset{Ph}{|}}{\overset{O \quad\quad O}{\underset{Ph-C-NH-P-OMe}{\overset{\parallel \quad\quad \parallel}{}}}} \tag{2.33}$$

The Lossen rearrangement (acid-catalyzed) of acyl hydroxamates to isocyanates:[45]

$$\overset{O\;\;\;O}{\underset{(RO)_2P-C-NHOH}{\overset{\parallel\;\;\parallel}{}}} \quad \xrightarrow{\text{pyr., pyr.}\cdot \text{HCl}} \quad \overset{O}{\underset{(RO)_2P-N=C=O}{\overset{\parallel}{}}} \tag{2.34}$$

2.K. REACTIVE INTERMEDIATES AND REACTION MECHANISMS

1,2-Shifts of a phosphinyl group to a carbocation:[46]

$$\text{Ph}_2\overset{O}{\overset{\|}{P}}-\overset{Me}{\underset{Me}{\overset{|}{C}}}-CH_2-OTs \xrightarrow[120°]{HCOOH} \text{Ph}_2\overset{O}{\overset{\|}{P}}-\overset{Me}{\underset{Me}{\overset{|}{C}}}-\overset{+}{CH}_2 \longrightarrow \overset{Me}{\underset{Me}{>}}\overset{+}{C}CH_2-\overset{O}{\overset{\|}{P}}Ph_2 \xrightarrow{-H^+} \overset{Me}{\underset{Me}{>}}C=CH-\overset{O}{\overset{\|}{P}}Ph_2 \quad (2.35)$$

Another type of rearrangement among phosphorus compounds, although quite rare, is found among enol or aryl phosphates. Strong base may induce in enol phosphates a migration of the phosphoryl moiety from oxygen to the β-carbon, thus producing a β-ketophosphonate. An example is shown in Eq. 2.36;[47a] the diastereoselectivity of the rearrangement has recently been studied.[47b]

$$\text{(camphor)} \xrightarrow[\text{2. (RO)}_2POCl]{\text{1. base}} \text{(enol phosphate)} \xrightarrow{LiN(i\text{-}Pr)_2} \text{(ketophosphonate)} \quad (2.36)$$

A related rearrangement is known among certain aryl phosphates,[48] wherein a 2-hydroxyarylphosphonate rather than a ketophosphonate is formed (Eq. 2.37).

$$X-C_6H_4-O-P(O)(OEt)_2 \xrightarrow{LiN(i\text{-}Pr)_2} X-C_6H_3(OH)-P(O)(OEt)_2 \quad (2.37)$$

Other types of rearrangements are known; for example, sigmatropic [1,5] rearrangements of groups from P (usually H or Ph) are common in phosphole chemistry (described in chapter 8.B), and migrations of P-substituents to nitrogen in certain structures can be observed. An example[49] is the rearrangement of N-phosphinoylhydroxylamine derivatives, as seen in Eq. 2.38. Another example[50] is the migration of a phosphoryl group from C to N when an α-lithioalkylphosphonate is reacted with a dialkylcyanamide (Eq. 2.39).

$$\underset{Ph}{\overset{PhMeCH}{>}}P\underset{NHOSO_2Me}{\overset{O}{<}} \xrightarrow{t\text{-}BuNH_2} \left[\underset{MeSO_2O}{\overset{PhMeCH}{>}}P\underset{NHPh}{\overset{O}{<}}\right] \xrightarrow{t\text{-}BuNH_2} \underset{t\text{-}BuNH}{\overset{PhMeCH}{>}}P\underset{NHPh}{\overset{O}{<}} \quad (2.38)$$

$$(EtO)_2\overset{O}{\overset{\|}{P}}-CH_2^- Li^+ + Me_2N-C\equiv N \longrightarrow (EtO)_2\overset{O}{\overset{\|}{P}}-CH_2-\overset{N^-}{\overset{\|}{C}}-NMe_2 \longrightarrow$$

$$(EtO)_2\overset{O}{\overset{\|}{P}}-\underset{\overset{\|}{\underset{CH_2}{C-NMe_2}}}{N} \xrightarrow{H^+} (EtO)_2\overset{O}{\overset{\|}{P}}-N=C\underset{CH_3}{\overset{NMe_2}{<}} \quad (2.39)$$

2.L. SAFETY IN THE HANDLING OF ORGANOPHOSPHORUS COMPOUNDS

We start with the statement that all operations with organophosphorus compounds should be conducted in efficient hoods, and the operator should be protected by safety glasses and rubber gloves. As always, safety shields or hood windows should be used to isolate vacuum distillations and reaction apparatus when there is a chance of excessive or sudden heat release. Toxicity is far less of a problem than is frequently thought to be true of organophosphorus compounds, but the fact remains that certain types of compounds have been discovered that have very high toxicity, and one should be extremely cautious with new types of compounds until their toxicity has been evaluated. Safety data sheets as supplied with commercial chemicals should always be consulted before using them.

The toxicity hazard is greatest with certain types of esters of phosphoric and phosphonic acids. Although the great majority of these chemicals are relatively harmless, some substitution patterns are associated with very high levels of anticholinesterase activity, such that they have been developed as agents for chemical warfare or insecticides. The combination of an ester group and a good leaving group—frequently fluorine, thioxy, or cyano—can lead to this toxicity in either phosphoryl or thiophosphoryl compounds. Examples of highly toxic compounds are given as structures **41** to **43**, and any work with substances having any resemblance to such structures should be avoided unless extra care is exerted, including the scrubbing of hood gases to prevent release of toxics to the atmosphere. This subject has been reviewed[4i] and receives some elaboration in chapter 11.D.

$$\underset{\textbf{41}}{\underset{F}{\overset{O}{\underset{\|}{CH_3P}}}-OPr\text{-}i} \qquad \underset{\textbf{42}}{(i\text{-}PrO)_2\overset{O}{\underset{\|}{P}}-F} \qquad \underset{\textbf{43}}{\underset{OEt}{\overset{O}{\underset{\|}{CH_3P}}}-SCH_2CH_2N(Pr\text{-}i)_2}$$

To illustrate the need for caution, in work on bicyclic esters of type **44** being made as ligands for metal coordination,[51] it was found unexpectedly that they possessed very high toxicity with γ-aminobutyric acid inhibitory properties. The compounds did not have the anticholinesterase activity found in certain phosphorus esters; they were metabolized with no change of the phosphorus function. Toxic properties are also known for the structure in which CH replaces the P group.

$$X=P{\begin{matrix}O-CH_2\\O-CH_2\\O-CH_2\end{matrix}}C-Bu\text{-}t$$

44   X = O or S

Phosphines have acquired a reputation for having high toxicity. The parent PH_3 certainly has this property; it is said to be extremely toxic at concentrations of 10 ppm. This does not prevent its use as a fumigant in agriculture, however. Few data are available on primary and secondary phosphines, but it is prudent to assume they will share in this toxicity. The lower members are gases or quite volatile, increasing the danger in their use. Tertiary phosphines have been more extensively studied, particularly as they are appearing more frequently as intermediates in the chemical industry. A thorough study of the data available on their toxicity has been published[4j] and should be consulted before work is initiated with them. In general, the higher molecular weight solid phosphines appear to present little hazard if they are handled properly. At the conclusion of any experiment with malodorous phosphines, it is good practice to immerse completely all glass apparatus in a bleach (or other oxidizing) solution to oxidize them to the harmless and odorless phosphine oxides or acids.

Any compound with a phosphorus-halogen bond should be considered hazardous to handle. Some of the simple reagents such as PCl_3 and $POCl_3$, and simple derivatives in which one or two chlorines are replaced, will undergo hydrolysis rapidly, releasing the highly irritating HCl.

Other types of phosphorus compounds, such as phosphine oxides and sulfides, phosphonic and phosphinic acids, and nonhalo derivatives, seem in general to have been little studied from a toxicological standpoint, but they do not appear to have hazards associated with them. Nevertheless, they should be handled by the general safety procedures mentioned at the outset of this section.

2.M. IMPORTANT LITERATURE RESOURCES IN PHOSPHORUS CHEMISTRY

The first book devoted entirely to organic derivatives of phosphorus did not appear until 1950, when G. M. Kosolapoff published his extremely valuable *Organophosphorus Chemistry*.[52] This book was revised with L. Maier in 1972 as *Organic Phosphorus Compounds*;[53] because of the tremendous growth of the field, seven volumes were needed for complete coverage. This outstanding work covers methods of synthesis and properties, with exhaustive tables of data on known compounds. It is the first place to go to determine if a particular compound had been synthesized before about 1970. R. S. Edmundson's *Dictionary of Organophosphorus Compounds*[54] published in 1988 is also an invaluable source of data on many known compounds. The well-known Houben–Weyl series on organic chemistry summarizes organophosphorus chemistry in two editions 1963[55] and of 1982[56]. They are outstanding sources of detailed information on synthesis and properties. Extensive coverage is provided in the four-volume, highly detailed treatment of phosphorus functional groups by E. R. Hartley, *The Chemistry of Organophosphorus Compounds*[57], in the Patai series, *The Chemistry of Functional Groups*. J. R. Van Wazer's *Phosporus and Its Compounds*,[58] Vol. 1 in 1958, Vol. 2 in 1961 was a highly influential series that provided pioneering insight into the nature of phosphorus chemistry and much information on compounds and their uses. Other books that treat the subject in a general or introductory way include Emsley and Hall (1976),[59] which is especially recommended for an introduction to some aspects of the field; Corbridge's broad-coverage reference work, *Phosphorus, an Outline of its Chemistry, Biochemistry and Technology* now in its fourth edition,[20] Walker's, *Organophosphorus Chemistry*,[60] and Goldwhite's *Introduction to Phosphorus Chemistry*,[61] the latter two though brief being of special help in beginning a study of phosphorus chemistry. An extensive collection of references to the review literature prior to 1981 is a valuable feature of Goldwhite's book.

As the field of phosphorus chemistry grew, many books were written on more specific aspects of the subject; their titles, given below, generally indicate their area of specialization.

- R. F. Hudson, *Structure and Mechanism in Organo-Phosphorus Chemistry*[62]
- A. J. Kirby and S. G. Warren, *The Organic Chemistry of Phosphorus*[31] (for reaction mechanisms).
- A. D. F. Toy and E. N. Walsh, *Phosphorus Chemistry in Everyday Living*[63]
- J. I. G. Cadogan, *Organophosphorus Reagents in Organic Synthesis*[64]
- F. G. Mann, *The Heterocyclic Derivatives of Phosphorus, Arsenic, Antimony and Bismuth*, second edition[65]
- L. D. Quin, *The Heterocyclic Chemistry of Phosphorus*[66]
- R. Luckenbach, *Dynamic Stereochemistry of Pentacoordinated Phosphorus and Related Elements*[67]
- W. E. McEwen and K. D. Berlin, *Organophosphorus Stereochemistry*[68]
- R. R. Holmes, *Pentacoordinated Phosphorus*[38]
- M. Regitz and O. J. Scherer, *Multiple Bonds and Low Coordination in Phosphorus Chemistry*[69]
- A. W. Johnson, *Ylides and Imines of Phosphorus*[16]
- M. Halmann, *Analytical Chemistry of Phosphorus*[70]
- K. S. Bruzik and W. J. Stec, *Biophosphates and Their Analogues*[71]
- T. Hori, M. Horiguchi and A. Hayashi, *Biochemistry of Natural C-P Compounds*[72]
- R. L. Hilderbrand, *The Role of Phosphonates in Living Systems*[73]
- K. B. Dillon, F. Mathey, and J. F. Nixon, *Phosphorus: The Carbon Copy*[29]
- D. Gorenstein, *Phosphorus-31 NMR: Principles and Applications*[74]
- J. C. Tebby, *Handbook of Phosphorus Nuclear Magnetic Resonance Data*[75]
- J. G. Verkade and L.D. Quin, *Phosphorus-31 NMR Spectroscopy in Stereochemical Analysis*[76]
- L. D. Quin and J. G. Verkade, *Phosphorus-31 NMR Spectral Properties in Compound Characterization and Structural Analysis*[77]

There are several invaluable collections of reviews on specific topics. The most extensive is that by M. Grayson and E. J. Griffith, *Topics in Phosphorus Chemistry*,[78] which consists of 11 volumes and covers many important aspects of the field. The on-going series *Organophosphorus Chemistry* of the Specialists Periodical Reports of The Royal Chemical Society[79] has provided annual coverage of the literature of the field since 1970. Other useful review collections are R. Engel, *Handbook of Organophosphorus Chemistry*[80] and A. N. Pudovik, *Chemistry of Organophosphorus Compounds*.[81] The second edition of *Comprehensive Heterocyclic Chemistry*[17] has several chapters devoted to heterocyclic phosphorus compounds, and *Comprehensive Organic Chemistry*[19] has provided some valuable introductory material. In addition to these collections, many reviews of specific topics, too numerous to mention here, have appeared in the major review journals. Many are cited at appropriate places in this book.

REFERENCES

1. L. D. Quin, J. Szewczyk, B. G. Marsi, X.-P. Wu, J. C. Kisalus, and B. Pete, *Phosphorus Sulfur* **30**, 249 (1987).
2. D. C. R. Hockless, Y. B. Kang, M. A. McDonald, M. Pabel, A. C. Willis, and S. B. Wild, *Organometallics* **15**, 1301 (1996).
3. J. Szewczyk, E.-Y. Yao, and L. D. Quin, *Phosphorus Sulfur Silicon* **54**, 135 (1990).
4. F. R. Hartley, ed., *The Chemistry of Organophosphorus Compounds*, Vol. 2, John Wiley & Sons, Inc., New York, 1990, (a) D. G. Gilheany, Vol. 2, Chapter 1, (b) R. S. Edmundson, Vol. 2, Chapter 7, (c) D. G. Gilheany, Vol. 3, Chapter 1, (d) S. Bachrach and C. I. Nitsche, Vol. 3, Chapter 4, (e) R. Burgada and R. Setton, Vol. 3, Chapter 7, (f) W. G. Bentrude, Vol. 1, Chapter 14, (g) O. Dahl, Vol. 4, Chapter 1, (h) M. Dankowski, Vol. 3, Chapter 6, (i) R. M. Black and J. M. Harrison, Vol. 4, Chapter 10, (j) N. R. Price and J. Chambers, Vol. 1, Chapter 16.
5. (a) R. F. Hudson, *Structure and Mechanism in Organo-Phosphorus Chemistry*, Academic Press, New York, 1965, Chapter 3, (b) H. Kwart and K. G. King, *d-Orbitals in the Chemistry of Silicon, Phosphorus, and Sulfur*, Springer-Verlag, Berlin, 1977.
6. E. Magnusson, *J. Am. Chem. Soc.* **112**, 7940 (1990).
7. W. Kutzelnigg, *Angew. Chem., Int. Ed. Engl.* **23**, 272 (1984).
8. D. G. Gilheany, *Chem. Rev.* **94**, 1339 (1994).
9. A. E. Reed and P. von R. Schleyer, *J. Am. Chem. Soc.* **112**, 1434 (1990).
10. D. B. Chesnut, *J. Am. Chem. Soc.* **120**, 10504 (1998).
11. J. A. Dobado, H. Martínez-García, J. M. Molina, and M. R. Sundberg, *J. Am. Chem. Soc.* **120**, 8461 (1998).
12. U. S. Rai and M. C. R. Symons, *J. Chem. Soc. Faraday Trans.* **90**, 2649 (1994).
13. D. B. Chesnut and A. Savin, *J. Am. Chem. Soc.* **121**, 2335 (1999).
14. L. D. Quin, J. C. Kisalus, J. J. Skolimowski, and N. S. Rao, *Phosphorus Sulfur Silicon* **54**, 1 (1990).
15. N. Weferling and S. Hoerold, paper presented at the International Conference on Phosphorus Chemistry, Cincinnati, Ohio, July 12–17, 1998.
16. A. W. Johnson, *Ylides and Imines of Phosphorus*, John Wiley & Sons, Inc., New York, 1993.
17. D. G. Hewitt in A. R. Katritzky, C. W. Rees, and E. F. V. Scriven, eds., *Comprehensive Heterocyclic Chemistry II*, Pergamon Press, Oxford, UK, 1996, Chapter 12.
18. G. Wittig and M. Rieber, *Liebigs Ann. Chem.* **562**, 187 (1948).

19. D. J. H. Smith in D. H. R. Barton and W. D. Ollis, eds., *Comprehensive Organic Chemistry*, Vol. 2, Pergamon Press, Oxford, UK, 1979. (a) Chapter 10.4, (b) Chapters 10.2, 10.6, (c) R. S. Edmundson, Chapters 10.3, 10.5.
20. D. E. C. Corbridge, *Phosphorus, An Outline of Its Chemistry, Biochemistry, and Technology*, 4th ed., Elsevier, Amsterdam, 1990, (a) p. 378, (b) p. 44.
21. R. R. Holmes, *Acc. Chem. Res.* **12**, 257 (1979); R. R. Holmes and J. A. Dieters, *J. Am. Chem. Soc.* **99**, 3318 (1977); R. R. Holmes, *J. Am. Chem. Soc.* **100**, 433 (1978).
22. R. J. Hach and R. E. Rundle, *J. Am. Chem. Soc.* **73**, 4321 (1951), J. I. Musher, *J. Am. Chem. Soc.* **94**, 1370 (1972).
23. E. L. Muetterties, W. Mahler, K. J. Packer, and R. Schmutzler, *Inorg. Chem.* **3**, 1298 (1964).
24. J. E. Griffiths and A. B. Burg, *J. Am. Chem. Soc.* **84**, 3442 (1962).
25. S. A. Mamedov, A. B. Kuliev, B. R. Gasanov, and A. M. Mirmovsumova, *Zh. Obshch. Khim.* 2669 (1991); *Chem. Abstr.* **117**, 69920 (1992).
26. G. O. Doak and L. D. Freedman, *Chem. Rev.* **61**, 31 (1961).
27. J. P. Rigaudy and S. P. Klesney, *Nomenclature of Organic Compounds*, Pergamon Press, Oxford, UK, 1979, pp. 382–408.
28. R. Appel in M. Regitz and O. J. Scherer, eds., *Multiple Bonds and Low Coordination in Phosphorus Chemistry*, Georg Thieme Verlag, Stuttgart, Germany, 1990, p. 157, (a) R. Appel, p. 157, (b) M. Sanchez, M. R. Mazières, L. Lamandé and R. Wolf, p.129.
29. K. B. Dillon, F. Mathey, and J. F. Nixon, *Phosphorus: The Carbon Copy*, John Wiley & Sons, Inc., New York, 1998.
30. G. Grüttner and M. Wiernik, *Ber.* **48**, 1473 (1915).
31. A. J. Kirby and S. G. Warren, *The Organic Chemistry of Phosphorus*, Elsevier, Amsterdam, 1967.
32. (a) S. Fleming, M. K. Lupton, and K. Jekot, *Inorg. Chem.* **11**, 2534 (1972); (b) B. E. Maryanoff and R. O. Hutchins, *J. Org. Chem.* **37**, 3475 (1972).
33. (a) A. Schmidpeter and M. Thiele, *Angew. Chem., Int. Ed. Engl.* **30**, 308 (1991), (b) S. E. Johnson and C. B. Knobler, *Phosphorus Sulfur Silicon* **115**, 227 (1996), (c) W.-W. du Mont, H.-J. Kroth, and H. Schumann, *Chem. Ber.* **109**, 3017 (1976).
34. S. Lochschmidt, F. Mathey, and A. Schmidpeter, *Tetrahedron Lett.* **27**, 2635 (1986).
35. E. Jungermann, J. J. McBride, Jr., R. Clutter, and A. Mais, *J. Org. Chem.* **27**, 606 (1962).
36. Z. Skrzypczynski, *J. Phys. Org. Chem.* **3**, 23, 35 (1990).
37. Ref. 30, Chap. 5.
38. R. R. Holmes, *Pentacoordinated Phosphorus*, Vols. 1–2, American Chemical Society, Washington, DC, 1980.
39. F. H. Westheimer, *Chem. Rev.* **81**, 313 (1981).
40. G. R. J. Thatcher and R. Kluger, *Adv. Phys. Org. Chem.* **25**, 99 (1989).
41. M. J. P. Harger, *Chem. Commun.*, 2339 (1998).
42. L. D. Quin, *The Heterocyclic Chemistry of Phosphorus*, John Wiley & Sons, Inc., New York, 1981, Chapter 2.
43. R. S. Givens and W. L. Kueper III, *Chem. Rev.* **93**, 55 (1993).
44. E. Breuer, A. Schlossman, M. Safadi, D. Gibson, M. Chorev, and H. Leader, *J. Chem. Soc. Perkin 1*, 3263 (1990).
45. C. J. Salomon and E. Breuer, *J. Org. Chem.* **62**, 3858 (1997).
46. S. G. Warren, *Acc. Chem. Res.* **11**, 401 (1978).
47. (a) T. Calogeropoulou, G. R. Hammond, and D. F. Wiemer, *J. Org. Chem.* **52**, 4185 (1987), (b) J. An, J. M. Wilson, Y.-Z. An, and D. F. Wiemer, *J. Org. Chem.* **61**, 4040 (1996).
48. B. Dhawan and D. Redmore, *J. Org. Chem.* **49**, 4018 (1984) and **51**, 179 (1986).

49. M. J. P. Harger and R. Sreedharan-Menon, *Chem. Commun.* 867 (1996).
50. W. B. Jang, K. Lee, C.-W. Lee, and D. Y. Oh, *Chem. Commun.* 609 (1998).
51. J. E. Casida, M. Eto, A. D. Moscioni, J. L. Engel, D. S. Milbrath, and J. G. Verkade, *Toxicol. Appl. Pharmacol.* **36**, 261 (1976); *Chem. Abstr.* **85**, 73049 (1976).
52. G. M. Kosolapoff, *Organophosphorus Chemistry*, John Wiley & Sons, Inc., New York, 1958.
53. G. M. Kosolapoff and L. Maier, eds., *Organic Phosphorus Compounds*, Vol. 1–7 Wiley-Interscience, New York, 1972.
54. R. S. Edmundson, *Dictionary of Organophosphorus Compounds*, Chapman & Hall, London, 1988.
55. K. Sasse in *Methoden der Organischen Chemie (Houben-Weyl), Band XII, Part I and II (Organischen Phosphorverbindungen)*, Georg Thieme Verlag, Stuttgart, Germany, 1963.
56. M. Regitz, ed., *Methoden der Organischen Chemie (Houben-Weyl), Band E(1) and E(2) (Organischen Phosphorverbindungen)*, Georg Thieme Verlag, Stuttgart, Germany, 1982.
57. F. R. Hartley, ed., *The Chemistry of Organophosphorus Compounds*, Vol. 1–4, John Wiley & Sons, Inc., New York, 1990.
58. J. R. Van Wazer, *Phosphorus and Its Compounds*, Vol. I–II, Interscience, New York, 1958.
59. J. Emsley and D. Hall, *The Chemistry of Phosphorus*, Harper & Row, London, 1976.
60. B. J. Walker, *Organophosphorus Chemistry*, Penguin, Harmondsworth, UK, 1972.
61. H. Goldwhite, *Introduction to Phosphorus Chemistry*, Cambridge University Press, Cambridge, UK, 1981.
62. R. F. Hudson, *Structure and Mechanism in Organo-phosphorus Chemistry*, Academic Press, New York, 1965.
63. A. D. F. Toy and E. N. Walsh, *Phosphorus Chemistry in Everyday Living*, 2nd ed., American Chemical Society, Washington, DC, 1987.
64. J. I. G. Cadogan, ed., *Organophosphorus Reagents in Organic Synthesis*, Academic Press, London, 1979.
65. F. G. Mann, *The Heterocyclic Derivatives of Phosphorus, Arsenic, and Antimony*, 2nd ed., Wiley-Interscience, New York, 1970.
66. L. D. Quin, *The Heterocyclic Chemistry of Phosphorus*, John Wiley and Sons, Inc., New York, 1981.
67. R. Luckenbach, *Dynamic Stereochemistry of Pentacoordinated Phosphorus and Related Elements*, Georg Thieme Verlag, Stuttgart, Germany, 1973.
68. W. E. McEwen and K. D. Berlin, eds., *Organophosphorus Stereochemistry*, Parts 1 and 2, Halsted Press, New York, 1975.
69. M. Regitz and O. J. Scherer, eds., *Multiple Bonds and Low Coordination in Phosphorus Chemistry*, Georg Thieme Verlag, Stuttgart, Germany, 1990.
70. M. Halmann, ed., *Analytical Chemistry of Phosphorus*, Wiley-Interscience, New York, 1972.
71. K. S. Bruzik and W. J. Stec, eds., *Biophosphates and Their Analogues*, Elsevier, Amsterdam, 1987.
72. T. Hori, M. Horiguchi, and A. Hayashi, eds., *Biochemistry of Natural C-P Compounds*, Japan Association for Research on the Biochemistry of C-P Compounds, Maruzen, Kyoto, Japan, 1984.
73. R. L. Hilderbrand, ed., *The Role of Phosphonates in Living Systems*, CRC Press, Boca Raton, FL, 1983.
74. D. G. Gorenstein, ed., *Phosphorus-31 NMR: Principles and Applications*, Academic Press, Orlando, FL, 1984.

75. J. C. Tebby, ed., *Handbook of Phosphorus-31 Nuclear Magnetic Resonance Data*, CRC Press, Boca Raton, FL, 1991.
76. J. G. Verkade and L. D. Quin, eds., *Phosphorus-31 NMR Spectroscopy in Stereochemical Analysis*, VCH Publishers, Deerfield Beach, FL, 1987.
77. L. D. Quin and J. G. Verkade, eds., *Phosphorus-31 NMR Spectral Properties in Compound Characterization and Structural Analysis*, VCH Publishers, New York, 1994.
78. M. Grayson and E. J. Griffith, eds., *Topics in Phosphorus Chemistry*, Vol. 1–11, John Wiley & Sons, Inc., New York, 1964–1983.
79. *Specialist Periodical Report, Organophosphorus Chemistry*, Royal Society of Chemistry, London, 1970.
80. R. Engel, ed., *Handbook of Organophosphorus Chemistry*, Marcel Dekker, New York, 1992.
81. A. N. Pudovik, ed., *Chemistry of Organophosphorus Compounds*, MIR Publishers, Moscow, 1989.

APPENDIX 2.1

SOME NOTES ON IUPAC NOMENCLATURE

Many proper names of compounds are used throughout this book, and in various places a brief explanation is provided for the names. Some of this material is summarized here for ready reference. The systems approved by IUPAC are used; the rules are stated in Rigaudy and Klesney,[27] and a concise summary is provided in Hartley's introduction.[57] There are many cases in which several names are acceptable for a particular structure. Here, the most common names are presented.

Phosphines and their chalcogenides are easily named, simply by stating the names (in alphabetical order) of the groups attached to phosphorus; they require no further comment. As prefixes, *dialkylphosphino* may be used for R_2P-, *dialkylphosphinoyl* for $R_2P(O)$-, *dialkylthiophosphinoyl* for $R_2P(S)$-, etc.

Phosphonium salts are also named simply by stating the names of substituents, in alphabetical order. The prefix *trialkylphosphonio* may be used. Phosphonium salts with heteroatoms are named in the same way.

The parent acids of phosphorus chemistry have names that express both the number of attached carbons (an infix *-on-* to indicate one C—P bond, *-in-* for two C—P bonds) and the oxidation state (a suffix *-ic* for the highest oxidation state, and *-ous* for the lower). In the latter compounds, as discussed in section 2.G, a proton shift occurs; only derivatives of these structures are known, but the parent names are used in nomenclature.

$MeP(O)(OH)_2$	methylphosphonic acid
$Me_2P(O)(OH)$	dimethylphosphinic acid
$MeP(OH)_2$	methylphosphonous acid
$Me_2P(OH)$	dimethylphosphinous acid

The acid name is retained in simple derivatives

MeP(O)Cl$_2$	methylphosphonic dichloride
MeP(O)(OR)$_2$	dialkyl methylphosphonate
MeP(O)(NHR)$_2$	methylphosphonic di(alkyl)amide
MePCl$_2$	methylphosphonous dichloride (dichloromethylphosphine is acceptable)
MeP(OR)$_2$	dialkyl methylphosphonite
Me$_2$P(OR)	alkyl dimethylphosphinite
MeP(NHR)$_2$	methylphosphonous bis(alkylamide) (methylbis(alkylamino)phosphine is acceptable)

Many derivatives of the acids are known with a mixture of substituents. Infixes have been created to handle these cases where necessary and are placed in alphabetical order. The common infixes are *amid* or *amido-*, *hydrazid(o)*, *fluorid(o)*, *chlorid(o)*, *bromid(o)*, *cyanid(o)*, *thio*, *seleno*, etc. These are inserted in the name just before the suffix for the oxidation state. Examples should make this clear.

MeP(O)Cl(OR)	alkyl methylphosphonochloridate
MeP(O)(OR)(NR$_2$)	*O*-alkyl(*N*,*N*-dialkyl)methylphosphonamidate
MeP(OR)(NHR)	*O*-alkyl(*N*,*N*-dialkyl)methylphosphonamidite; alkoxy(alkyl)(dialkylamino)phosphine is acceptable)
MePCl(SEt)	ethyl methylphosphonochloridothioite

Derivatives of phosphoric and phosphorous acid are named with similar rules, using the prefix *phosphoro* for both, with the suffixed *ate* and *ite*, respectively.

P(O)(OMe)$_2$Cl	dimethyl phosphorochloridate (also dimethyl chlorophosphate)
P(O)(OMe)Cl$_2$	methyl phosphorodichloridate
P(OMe)Cl$_2$	methyl phosphorodichloridite (also methoxydichlorophosphine)
P(O)(OMe)(RHN)$_2$	*O*-methyl(*N*,*N'*-dialkyl)phosphorodiamidate
P(O)Cl(OMe)(RNH)	*O*-methyl(alkylamino)phosphoramidochloridate
P(S)BrCl(OMe)	*O*-methyl thiophosphorobromidochloridate

Here we have illustrated only the basic ideas behind the naming systems for acid derivatives; many more complicated cases are known, and the rules in Rigaudy and Klesney[27] should be consulted. Occasionally, it may be necessary to indicate a phosphonic acid function or one of its derivatives as a prefix; *phosphono* is used for this purpose.

The 5-coordinate structures are named as derivatives of PH_5, phosphorane, and substituents are given in alphabetical order. Common names for low-coordination species will be given in chapter 10.

CHAPTER 3

THE COMMON 3-COORDINATE FUNCTIONS (σ^3, λ^3)

3.A. HALOPHOSPHINES

3.A.1. General

The halophosphines RPX_2 and R_2PX are an extremely important family of organophosphorus compounds and a fitting one with which to begin our study of phosphorus functional groups. Like other 3-coordinate halides, they are very reactive, especially to nucleophiles, and are widely used in synthetic operations. In fact, it is possible to synthesize most of the common phosphorus functional groups starting with a halophosphine; because they in turn can be prepared in various processes using phosphorus trichloride, they serve as relays in the conversion of inorganic to organic phosphorus compounds. As a reflection of the importance of halophosphines as intermediates, there are 12 practical preparations in *Organic Syntheses* and *Inorganic Syntheses* and many applications of their use as starting materials in the synthesis of other 3-coordinate compounds. Appendix 3.1 provides references to the halophosphines (compounds 1 to 12) and the other 3-coordinate compounds (compounds 13 to 57).

Halophosphines are generally more reactive to nucleophiles than are the corresponding 4-coordinate phosphoryl halides, a point that can be used to advantage in synthesis, because after a substitution reaction the 3-coordinate products can be easily oxidized to the phosphoryl form. Much more is known about the chloro derivatives than the other halides; although the bromophosphines do have some importance in certain applications, the chlorides will be stressed in this discussion. Chlorophosphines are of two types: the alkyldichlorophosphines or aryldichlorophosphines ($RPCl_2$, more properly known as phosphonous dichlorides, because they are considered the acid chlorides from the alkylphosphonous acids, albeit unknown), and the dialkylchlorophosphines or diarylchlorophosphines (R$_2$PCl, the phosphinous chlorides, from phosphinous acids, also unknown). We should remember here a point of nomenclature: the prefix *phosphon* always implies a single carbon or hydrogen attached to phosphorus, the prefix *phosphin* always indicates two attached carbons, a carbon and a hydrogen, or even two hydrogens. These prefixes are independent of coordination number; this is indicated by the suffixes *-ic* for 4-coordinate phosphoryl compounds and *-ous* for 3-coordinate structures. The phosphonous dichlorides are generally distillable liquids, unless of high molecular weight: methylphosphonous dichloride boils at 81°, and phenylphosphonous dichloride at 222°. The phosphinous chlorides Me$_2$PCl and Ph$_2$PCl have b.p. 72 to 75° and 320°, respectively. Like all 3-coordinate halides, the halophosphines are very sensitive to water and are also easily oxidized; they require protection from the atmosphere. Coupled with their notorious malodorous nature, they are difficult materials to handle. Nevertheless, methylphosphonous and phenylphosphonous dichlorides are manufactured on a commercial scale.

3.A.2. Major Synthetic Methods

Halophosphines were among the earliest phosphorus compounds to be synthesized, and they have remained of great importance ever since as building materials for other structures. There are many methods for the synthesis of halophosphines, and only the more important can be discussed here. Extensive literature coverage is provided by reviews in the Kosolapoff-Maier series,[1] the Hartley series[2] and in Houben–Weyl.[3,4]

Phosphorus trichloride, which is cheap and commercially available, is the starting material for several chlorophosphine syntheses. It was discovered as early as 1876 by Michaelis[5] that phenylphosphonous dichloride could be prepared by the aluminum chloride catalyzed reaction of benzene with phosphorus trichloride, in a variant of the Friedel-Crafts reaction (Eq. 3.1).

$$\text{C}_6\text{H}_6 + \text{PCl}_3 \xrightarrow{\text{AlCl}_3} \text{C}_6\text{H}_5\text{-PCl}_2 \cdot \text{AlCl}_3 \xrightarrow{\text{POCl}_3} \text{C}_6\text{H}_5\text{-PCl}_2 \quad (3.1)$$
                                                  1

Later many other aromatic compounds were subjected successfully to this reaction, and consequently the area of arylphosphonous dichloride chemistry developed much more rapidly than did that of the more difficultly prepared alkyl compounds. Early problems in the breakup of the intermediate AlCl$_3$ complex, presumed to be **1**, kept the yields low, but in later years, it was discovered that water or better phosphorus oxychloride[6] could replace the dichloride in the complex, and yields reached the 70 to 80% range. Pyridine and other Lewis bases have also been used for this purpose. This old process has been run on a commercial scale for many years and is also in current use for the preparation of tolylphosphonous dichlorides, although an alternative reaction of benzene with phosphorus trichloride at high temperatures also has been commercialized.[7] Another problem with the Michaelis approach is that only halogens and electron-releasing groups may be present on the benzene ring; as is typical of the Friedel-Crafts, the reaction fails if meta directors such as nitro or carbonyl groups are present. Also, a mixture of isomers will be formed from substituted benzenes. The reaction has also been extended to the synthesis of heterocyclic dichlorophosphines; thiophene is used as an example in Eq. 3.2, in which attack is seen to be as usual at the 2-position (**2**). Furan and *N*-methylpyrrole are so reactive that no catalyst is needed to form the phosphonous dichlorides.

$$\underset{S}{\bigcirc\!\!\!\!\!\diagup} \quad \xrightarrow[AlCl_3]{PCl_3} \quad \underset{\underset{\mathbf{2}}{S}}{\bigcirc\!\!\!\!\!\diagup}\!\!-PCl_2 \qquad (3.2)$$

The scope of the Friedel-Crafts synthesis is well presented in a detailed review by Kosolapoff.[8] Its mechanism has not been fully established; the early view was that PCl_3 first reacted with $AlCl_3$ to form the PCl_2^+ ion, which attacked the ring in the conventional manner for an electrophile. It is now known[9] that no detectable reaction or complex formation takes place between these compounds, and the role of the catalyst in this reaction remains uncertain. Monitoring of the reaction by ^{31}P NMR did lead to clarification of the structure of the $AlCl_3$ complexes of the product (see chapter 6.I), but not of the mechanism. Recently, it has been reported[10] that $BiCl_3$ is a useful catalyst for substitution on aromatic ethers and that $Bi(OSO_2CF_3)_2$ is useful for the less-reactive aromatic hydrocarbons. A major advantage is that the dichlorophosphine product can be distilled directly from the reaction mixture, and the catalyst recovered for later use.

Benzene may also be caused to react with PCl_3 at high temperatures (600°) without a catalyst, a process that is used commercially. A short contact time is used because phenylphosphonous dichloride undergoes further reaction when exposed to high temperature; with or without a catalyst ($AlCl_3$ or $ZnCl_2$), it can be converted to diphenylphosphinous chloride in a disproportionation process (Eq. 3.3). This reaction has made diphenylphosphinous chloride commercially available; as will be seen, this chloride plays an important role in the synthesis of phosphines.

$$2\;\underset{}{C_6H_5}\!-\!PCl_2 \quad \xrightarrow{600°} \quad [C_6H_5-PCl]_2 \;+\; PCl_3 \qquad (3.3)$$

There are a few examples of the reaction of alkenes with PCl_3 in the presence of $AlCl_3$ to form addition products, but the best way to conduct this addition is under free radical conditions, using such agents as diacetyl peroxide or azobisisobutyronitrile as radical initiators.[11] The addition can also be effected photochemically.[12] The first mechanistic studies[8] suggested that an alkyl radical abstracted Cl from PCl_3 to form the $Cl_2P\cdot$ radical, which then added to the double bond. A later view[13] is that the alkyl radical adds to PCl_3 to form a phosphoranyl radical ($Cl_3RP\cdot$) as an intermediate. The method is of value for the synthesis of β-chloroalkylphosphonous dichlorides as in the example of Eq. 3.4.

$$PCl_3 + C_6H_{13}CH=CH_2 \xrightarrow{AIBN} C_6H_{13}\underset{Cl}{CH}CH_2PCl_2 \quad (3.4)$$

More recent extensions include the photochemical 1,4-addition of PCl_3 to butadiene[14] (taking precedence over the much slower McCormack cycloaddition, vide infra) and to phenylacetylene.[15] Oxygen catalyzes the addition (trans) of PBr_3 to cyclohexene and norbornene,[16] and UV radiation or radical initiators effect the addition of PBr_3 to ethylene.[17] The mechanism for the free radical process is given in Eq. 3.5.

$$PBr_3 \xrightarrow{h\nu} Br\cdot + Br_2P\cdot$$

$$Br\cdot + CH_2=CH_2 \longrightarrow \dot{C}H_2-CH_2Br$$

$$PBr_3 + \dot{C}H_2-CH_2Br \longrightarrow Br_3\dot{P}-CH_2CH_2Br \quad (3.5)$$

$$Br_3\dot{P}-CH_2CH_2Br + CH_2=CH_2 \longrightarrow Br_2PCH_2CH_2Br + \dot{C}H_2CH_2Br$$

The $AlCl_3$ catalyzed addition to alkenes is more of interest in connection with the synthesis of heterocyclic systems. Thus, when highly branched alkenes are used, the phosphetane ring (as in **3**) can be constructed. This process (Eq. 3.6) will be discussed in more detail in chapter 8.

$$MeCH=CH-CMe_3 + PCl_3 \xrightarrow{AlCl_3} Me-\underset{Cl_2\overset{+}{P}:}{CH}-\overset{+}{C}HCMe_3 \longrightarrow$$

$$Me-\underset{Cl_2\overset{+}{P}:}{\overset{Me}{CH}}-\overset{+}{C}HCMe_2 \longrightarrow Me-\underset{\underset{Cl}{\overset{|}{P}}\diagdown Cl}{\overset{Me}{\diagup}\diagdown}Me_2 \quad (3.6)$$

$$\mathbf{3}$$

Many alkylphosphonous dichlorides have been prepared by replacement of one chlorine of phosphorus trichloride by reaction with an organometallic reagent. Grignard reagents and alkyl lithiums, the most common organometallics, however, are generally too reactive for this purpose and give large amounts of dialkyl and trialkyl products. Such reagents can be used successfully for alkylphosphonous dichloride synthesis only if they are of great steric bulk, which retards the overalkylation. Thus a useful synthesis has been devised for *tert*-butylphosphonous dichloride[18] (Eq. 3.7).

$$PCl_3 + t\text{-BuMgCl} \xrightarrow{-48°} t\text{-Bu}-PCl_2 \quad (3.7)$$

Reactions with aryl Grignard and lithium reagents are more successful than with alkyl reagents, especially with aryl groups of great steric demand. This reaction, for example, is used to prepare 2,4,6-tri-*t*-butylphenylphosphonous dichloride[19] (Eq. 3.8). The latter is of particular importance as a reagent for creating other phosphorus functional groups that require steric blockage to achieve stabilization, as in the case of the phosphorus–phosphorus double bond (chapter 10.C).

$$\text{ArBr} \xrightarrow{\text{Li}} \text{ArLi} \xrightarrow{\text{PCl}_3} \text{ArPCl}_2 \quad (3.8)$$

(Ar = 2,4,6-tri-*t*-butylphenyl)

Alkyl cadmium reagents are of moderate reactivity; because they are easily prepared from Grignard reagents, they have found use in the synthesis of some simple alkylphosphonous dichlorides (Eq. 3.9).

$$PCl_3 + n\text{-Bu}_2Cd \longrightarrow n\text{-BuPCl}_2 \quad (3.9)$$

Tetraethyl lead is of value in the synthesis of ethylphosphonous dichloride (Eq. 3.10).

$$Et_4Pb + PCl_3 \longrightarrow EtPCl_2 \quad (3.10)$$

Other metallic derivatives, especially of zinc and tin, have also found use; the latter are especially valuable, because they can be prepared with C-functional groups that can survive the reaction with phosphorus trichloride, as seen in Eq. 3.11.[20]

$$PCl_3 + Bu_3SnCH_2COOR \longrightarrow Cl_2P-CH_2COOR \quad (3.11)$$

The overalkylation of phosphorus trichloride can be avoided by first replacing one or two chlorines by amino groups. This is easily accomplished by reaction with a dialkylamine or by an exchange reaction with $(R_2N)_3P$. Under controlled conditions, the remaining chlorine can be replaced by alkyl or aryl using Grignard or lithium reagents. The P–N bonds in the product are easily cleaved by hydrogen halide, as will be described below. This sequence is illustrated by a recent synthesis of dicyclopropylphosphinous chloride[21] (Eq. 3.12) and of a cyclic phosphinous chloride[22] (Eq. 3.13).

$$PCl_3 \xrightarrow{Et_2NH} Et_2NPCl_2 \xrightarrow{\triangleright\text{-Li}} (\triangleright)_2PNEt_2 \xrightarrow{\text{dry HCl}} (\triangleright)_2PCl \quad (3.12)$$

$$Et_2NPCl_2 + BrMg(CH_2)_5MgBr \longrightarrow \text{(cyclo-C}_5\text{H}_{10}\text{)P-NEt}_2 \xrightarrow{\text{dry HCl}} \text{(cyclo-C}_5\text{H}_{10}\text{)P-Cl} \quad (3.13)$$

The first member of the series, methylphosphonous dichloride, remained unknown until the 1950s, because its high reactivity complicated the use of the organometallic reagent route with PCl$_3$. It was first reported in 1958 using the approach of Eq. 3.14, in which a complex **4** is formed that contains the methyltrichlorophosphonium ion.[23] Reduction then removed chlorine to provide the product. This process appears to be the most popular of the small-scale methods reported;[1a] it has been improved over the years, and the preferred reducing agent is metallic aluminum.

$$PCl_3 + CH_3Cl \xrightarrow{AlCl_3} CH_3\overset{+}{P}Cl_3\, Al\overset{-}{C}l_4 \xrightarrow{[H]} CH_3PCl_2 \quad (3.14)$$
$$\textbf{4}$$

For large-scale production of MePCl$_2$, the seemingly unlikely reaction of methane with phosphorus trichloride (Eq. 3.15) was developed by the FMC Corporation;[25] this free-radical process occurs at about 500° and is catalyzed by traces of oxygen. It is useful also for preparing ethylphosphonous dichloride but not higher derivatives. Between 15 and 20% conversion of PCl$_3$ to MePCl$_2$ can be achieved under the most favorable conditions. A form of this reaction is presently being carried out commercially by Clariant GmbH (formerly Hoechst) and is described further in reviews.[26,27] Methylphosphonous dichloride is a versatile starting material, although highly reactive and dangerous to handle; several uses have been developed for it.[28]

$$CH_4 + PCl_3 \xrightarrow[O_2\, cat.]{>500°} CH_3PCl_2 + HCl \quad (3.15)$$

There are several other processes by which a chlorophosphonium cation can be generated and then reduced to 3-coordinate form. The first of these[29] was introduced as a means of making isomer-specific arylphosphonous dichlorides, including those with meta-directing groups that had never before been accessible. The method consists of the reaction of diazonium fluoroborates in ethyl acetate with phosphorus trichloride, with a cuprous halide as catalyst, to give a chlorophosphonium fluoroborate (**5**). This type of salt was first prepared for hydrolysis to the phosphonic acid,[30] but it was found that addition of a metallic reducing agent, generally magnesium or aluminum, converts the intermediate **5** to the phosphonous dichloride in modest yields of 30 to 40%[29] (Eq. 3.16).

$$R\text{-}C_6H_4\text{-}NH_2 \xrightarrow[HBF_4]{NaNO_2} R\text{-}C_6H_4\text{-}\overset{+}{N_2}B\overset{-}{F_4} \xrightarrow[CuBr]{PCl_3} R\text{-}C_6H_4\text{-}\overset{+}{P}Cl_3\, B\overset{-}{F_4} \xrightarrow{Mg} R\text{-}C_6H_4\text{-}PCl_2 \quad (3.16)$$
$$\textbf{5}$$

This method has also been used to prepare diarylphosphinous chlorides, by substituting an arylphosphonous dichloride for phosphorus trichloride in the reaction with the diazonium salt[31] (Eq. 3.17).

$$Ar\overset{+}{N_2}B\overset{-}{F_4} + Ar'PCl_2 \xrightarrow{CuX} ArAr'\overset{+}{P}Cl_2\, B\overset{-}{F_4} \xrightarrow{Al} ArAr'PCl \quad (3.17)$$

Another synthetic method using a chlorophosphonium chloride intermediate has been used in the synthesis of trichloromethylphosphonous dichloride.[32,33] The compound CH_3PCl_4 can be made by addition of chlorine to methylphosphonous dichloride; the hydrogens can be replaced by chlorine to give the salt **6**, which then is reduced by CH_3O-PCl_2.[32] This compound acts by removing two chlorine atoms to form a new, but unstable, chlorophosphonium salt **7**, decomposing to methyl chloride and phosphorus oxychloride (Eq. 3.18). Halophosphonium ions have been generated in several other processes and reduced by CH_3OPCl_2 (Appendix 3.1, compound 11) and a variety of agents.[1a]

$$CH_3PCl_2 \xrightarrow{Cl_2} CH_3PCl_4 \xrightarrow{Cl_2} CCl_3PCl_4 \xrightarrow{CH_3OPCl_2} CCl_3PCl_2 + [CH_3OPCl_4]$$
$$\mathbf{6} \hspace{5cm} \mathbf{7}$$
$$CH_3Cl + POCl_3 \quad (3.18)$$

The monochloromethyl compound is more easily prepared, simply by reaction of diazomethane with phosphorus trichloride (Eq. 3.19). Higher diazoalkanes can also be used.

$$PCl_3 + CH_2N_2 \longrightarrow ClCH_2PCl_2 \quad (3.19)$$

There are a variety of reductive transformations of 4-coordinate phosphorus functions that give halophosphines; where the starting materials are readily available, these reactions can be of synthetic value. Eq. 3.20 illustrates a simple route to dimethylphosphinous chloride; it employs a novel Grignard reaction to prepare a diphosphine disulfide **8**,[28a] which is then cleaved with thionyl chloride and the product reduced with tributylphosphine.[34]

$$PSCl_3 + MeMgBr \longrightarrow Me_2\overset{S}{\overset{\|}{P}}-\overset{S}{\overset{\|}{P}}Me_2 \xrightarrow{SOCl_2} Me_2\overset{S}{\overset{\|}{P}}-Cl \xrightarrow{Bu_3P} Me_2PCl$$
$$\mathbf{8} \hspace{8cm} (3.20)$$

Generally, the corresponding reduction of phosphoryl chlorides to halophosphines cannot be realized, although a useful case has been discovered in which a cyclic phosphinic chloride can be reduced with trichlorosilane to a phosphinous chloride[35] (Eq. 3.21).

(3.21)

Another type of transformation involving phosphoryl compounds is illustrated in Eq. 3.22. Here an H-phosphinic acid is converted to a phosphonous dichloride with phosphorus trichloride,[36] possibly by attack of the phosphoryl oxygen on the trichloride to give an intermediate (9), which undergoes a displacement reaction with chloride ion (Eq. 3.22). A less likely alternative is that the enol form Ph_2P-OH is the reactive species.

$$Ph-\underset{OH}{\underset{|}{P}}=O \xrightarrow{PCl_3} Ph-\underset{Cl}{\underset{|}{P}}=O \xrightarrow{PCl_3} Ph-\underset{Cl}{\underset{|}{\overset{H}{\overset{|}{P}^+}}}-OPCl_2 \xrightarrow{Cl^-} Ph-\underset{Cl}{\underset{|}{\overset{H}{\overset{|}{P}^+}}}-Cl \xrightarrow{-H^+} Ph-\underset{Cl}{\underset{|}{P}}-Cl$$

$$PCl_3 \downarrow \qquad \xrightarrow{Cl^-}_{-[HOPCl_2]} \qquad 9 \qquad \qquad \qquad (3.22)$$

$$Ph-\underset{\underset{H}{\overset{|}{O-PCl_2}}}{\overset{H}{\overset{|}{P}}}=O$$

A related process is the conversion of a secondary phosphine oxide, $R_2P(O)H$, to a dialkylphosphinous chloride with PCl_3.[37] The required phosphine oxides are available from the reaction of Grignard reagents with the commercially available dialkyl phosphites (dialkyl H-phosphonates), so the process is of value for phosphinous chloride synthesis (Eq. 3.23).

$$ROH + PCl_3 \longrightarrow (RO)_2\overset{H}{\underset{|}{P}}=O \xrightarrow{R'MgX} R'_2\overset{H}{\underset{|}{P}}=O \xrightarrow{PCl_3} R'_2PCl \qquad (3.23)$$

Phosphonous dibromides and phosphinous bromides can be prepared by using phosphorus tribromide rather than the trichloride in these reactions (Eqs. 3.22 and 3.23), but a more general approach to their synthesis is to employ an exchange reaction (Appendix 3.1, compound 5) between phosphorus tribromide, or hydrogen bromide, and the chloro compound. Related exchanges can also be used to generate the fluorides and even the iodides.

A final process that represents a transformation of a phosphorus function to a halophosphine is that of the replacement of hydrogen of a primary phosphine with halogen, either by the free element or better by phosgene ($COCl_2$). A recent example is the sequence of Eq. 3.24, in which ortho-diphosphinobenzene is converted to the bis(phosphonous dichloride).[38] This reagent avoids a problem existing with use of the free halogens, because the halophosphines, after formation, can add a mole of halogen to form a halophosphonium halide, as described in the next section.

$$\underset{PH_2}{\underset{|}{C_6H_4}}\text{-}PH_2 \xrightarrow[-78°]{COCl_2} \underset{PCl_2}{\underset{|}{C_6H_4}}\text{-}PCl_2 \qquad (3.24)$$

3.A.3. Important Reactions

Four different reaction pathways can be identified for halophosphines: (a) they can act as powerful electrophiles to many nucleophilic reagents; (b) they can donate their lone electron pair to form 4-coordinate compounds; (c) with dienes and other α,β-unsaturated systems, they can both accept and donate an electron pair by participating in cheletropic cycloaddition and in related biphilic processes; (d) with base, an elimination of HCl from P—Cl and an α-CH bond can occur to form the (usually unstable) P=C bond.

3.A.3.a. Nucleophilic Substitutions.
Countless nucleophilic substitutions have been performed on the halophosphines; this is a principle route to many other 3-coordinate phosphorus compounds, as suggested by Eqs. 3.25 to 3.27, in which the products are dialkyl alkylphosphonites, alkylphosphonamidites, and dialkyl alkylphosphonodithioites, respectively.

$$RPCl_2 + R'OH \xrightarrow{base} RP(OR')_2 \qquad (3.25)$$

$$RPCl_2 + R_2'NH \xrightarrow{base} RP(NR_2')_2 \qquad (3.26)$$

$$RPCl_2 + R'SH \xrightarrow{base} RP(SR')_2 \qquad (3.27)$$

The phosphonous dichlorides are also of value in the synthesis of heterocyclic compounds; this takes place when the co-reactant is a diol (Eq. 3.28), diamine, or amino alcohol.

$$RPCl_2 + HO(CH_2)_3OH \xrightarrow{-HCl} RP\begin{pmatrix}O\\O\end{pmatrix} \qquad (3.28)$$

Similar substitutions are known for the dialkylphosphinous chlorides and, of course, for aryl derivatives of both types of halophosphines. The mechanism of these substitutions may be of two related types. As outlined in chapter 2.K, the most likely[39] appears to be the one in which the nucleophile forms a trigonal bipyramidal (TBP) transition state with four groups and the lone pair on phosphorus (**10**), which collapses to the substitution product. The process is the counterpart of the S_N2 reaction of alkyl halides and similarly occurs with inversion of configuration at phosphorus (Eq. 3.29). The nucleophile can also form a distinct TBP intermediate **11** (called a phosphoranide) with the halide, which then collapses.

$$\underset{R}{\overset{Cl}{\underset{|}{R'-P}}}: \;\; \xrightarrow{\bar{O}R} \;\; \left[\underset{OR}{\overset{Cl}{\underset{|}{R'-P}}}\overset{:}{\underset{R}{}}\right]^{-} \;\; \text{or} \;\; \left[\underset{OR}{\overset{Cl}{\underset{|}{R'-P}}}\overset{:}{\underset{R}{}}\right]^{-} \;\; \xrightarrow{-\bar{C}l} \;\; \underset{R}{\overset{:}{\underset{|}{R'-P}}}\text{''}OR \qquad (3.29)$$

$$\qquad\qquad\qquad\qquad\; \textbf{10} \qquad\qquad \textbf{11}$$

Unless the halophosphine has special steric crowding effects, these substitutions proceed very rapidly and exothermically and require care in controlling the reaction rate. Typically, a base such as triethylamine or pyridine is used to accept the hydrogen halide that is released. This is essential when an alcohol is the reactant, because the HCl can react with the initial ester and cleave the P–O–C bond (Eq. 3.30).

$$RPCl_2 + R'OH \xrightarrow{-HCl} RP(O-R')_2 \xrightarrow{HCl} RP^+(O-R')_2 H \bar{Cl} \xrightarrow{-R'Cl} \left[RP(OR')(OH) \right] \longrightarrow R-P(=O)(OR')H \qquad (3.30)$$

When water is used as the nucleophile (Eq. 3.31), the products are the alkylphosphonous (**12**, Eq. 3.31) or alkylphosphinous acid (**14**, Eq. 3.32), but these structures, it will be recalled from chapter 2.G, have no reality because they tautomerize to 4-coordinate phosphoryl compounds, as seen in Eqs. 3.31 and 3.32. These hydrolysis reactions constitute, respectively, major pathways to the alkyl H-phosphinates (**13**) and secondary phosphine oxides (**15**).

$$RPCl_2 + H_2O \longrightarrow \left[RP(OH)_2 \right] \longrightarrow R-\underset{OH}{\overset{H}{P}}=O \qquad (3.31)$$
$$\qquad\qquad\qquad\qquad \mathbf{12} \qquad\qquad \mathbf{13}$$

$$R_2PCl + H_2O \longrightarrow \left[R_2\ddot{P}-OH \right] \longrightarrow R_2\overset{H}{P}=O \qquad (3.32)$$
$$\qquad\qquad\qquad\qquad \mathbf{14} \qquad\qquad \mathbf{15}$$

It is difficult to control the reaction of a phosphonous dichloride to replace only one of the chlorine atoms; most synthetic operations are aimed at giving disubstituted products. Several cases are, however, known in which monosubstitution has been accomplished, as in the reaction of phenylphosphonous dichloride with ethanol (Eq. 3.33).

$$PhPCl_2 + EtOH \xrightarrow{R_3N} Ph-P(Cl)(OEt) \qquad (3.33)$$

Monosubstitution is more feasible if a sterically demanding nucleophile is used, because the second substitution is made more difficult. An example is seen in Eq. 3.34.[40]

$$PhPCl_2 + \text{(1-adamantyl)}-NH_2 \xrightarrow{R_3N} \text{(1-adamantyl)}-NHPPh(Cl) \qquad (3.34)$$

Many organometallic compounds, but especially Grignard and lithium reagents, can perform nucleophilic substitution on the halophosphines; this constitutes one of the

54 THE COMMON 3-COORDINATE FUNCTIONS (σ^3, λ^3)

best ways to synthesize tertiary phosphines, especially those with mixed substituents (Eq. 3.35).

$$RPCl_2 + R'MgX \longrightarrow RPR'_2 \qquad (3.35)$$

The first phosphorus heterocycle was prepared in 1915[41] by taking advantage of the difunctionality of phosphonous dichlorides in their reaction with Grignard reagents; thus phenylphosphorinane was prepared by reacting phenylphosphonous dichloride with the di-Grignard reagent from 1,4-dibromopentane (Eq. 3.36). Many heterocycles have since been prepared by this method.

$$PhPCl_2 + BrMg(CH_2)_5MgBr \longrightarrow PhP\langle\bigcirc\rangle \qquad (3.36)$$

Metallic hydrides may also react with halophosphines, replacing halogen with hydrogen and thus providing an excellent approach to the families of primary and secondary phosphines. The most commonly used reagent is lithium aluminum hydride at low temperatures;[42] silicon hydrides, especially $HSiCl_3$, are also useful.

3.A.3.b. Formation of 4-Coordinate Derivatives.
Oxidation must rank as the most important of these reactions involving the formation of 4-coordinate derivatives; several oxidizing agents have been used with the halophosphines to convert them to the phosphoryl derivatives. The reagents must, of course, be anhydrous and include oxygen, $N_2O_4-NO_2$, sulfur trioxide, sulfuryl chloride, etc. The oxychlorides resulting from phosphonous dichlorides are of great value as precursors of phosphonic acids and their ester and amide derivatives (Eq. 3.37), whereas the phosphinous chlorides are precursors of phosphinic acids and their derivatives (Eq. 3.38).

$$RPCl_2 \xrightarrow{(O)} \underset{Cl}{\overset{O}{\underset{\|}{RP}}-Cl} \xrightarrow{H_2O} \underset{OH}{\overset{O}{\underset{\|}{RP}}-OH} \qquad (3.37)$$

$$R_2PCl \xrightarrow{(O)} \underset{R}{\overset{O}{\underset{\|}{RP}}-Cl} \xrightarrow{H_2O} \underset{R}{\overset{O}{\underset{\|}{RP}}-OH} \qquad (3.38)$$

Thiophosphoryl and selenophosphoryl derivatives can be prepared by the addition of sulfur or selenium to the halophosphines, a process sometimes expedited by the presence of aluminum chloride (Eq. 3.39).

$$RPCl_2 + S_8 \xrightarrow{AlCl_3} \overset{S}{\underset{\|}{R-PCl_2}} \qquad (3.39)$$

The halophosphines can add one mole of a halogen to form halophosphonium halides (chapter 4.D), crystalline solids that are useful as precursors, on hydrolysis, of phosphonic or phosphinic acids. These halides are generally written in ionic form (e.g., **17**), for the product from a phosphonous dichloride), but as is true for phosphorus pentachloride, the 5-coordinate form (**16**) is probably in equilibrium with it in solution in some solvents (Eq. 3.40; see chapter 4.D).

$$RPCl_2 + Cl_2 \longrightarrow \left[R-\underset{\underset{Cl}{|}}{\overset{\overset{Cl}{|}}{P}}\diagdown_{Cl}^{Cl} \right] \longrightarrow R-\underset{\underset{Cl}{|}}{\overset{\overset{Cl}{|}}{\overset{+}{P}}}-Cl \ \ \bar{Cl} \qquad (3.40)$$

$$\qquad\qquad\qquad\qquad\quad \mathbf{16} \qquad\qquad\quad \mathbf{17}$$

The lone pair on phosphines is readily attacked by the electrophilic alkyl halides, but the electron density of chlorophosphines is so reduced by the powerful electron attraction of chlorine that this reaction is unknown for the chlorophosphines. The phosphorus in bromophosphines is not as deactivated, however, and cases are known in which their alkylation is possible. An example from the heterocyclic field is shown in Eq. 3.41; the cyclic phosphinous bromide **18** reacts with methyl or benzyl bromides[43] to form bromophosphonium bromides (**19**), useful in forming cyclic phosphines by reduction, or oxides by hydrolysis.

(3.41)

Tertiary phosphines form an incredible number of metallic coordination compounds using their lone electron pair, but this is a property of much less importance for the halophosphines, presumably because of the reduced electron density on phosphorus from the presence of electron-withdrawing halogens. Bonding of halophosphines is, however, favorable to certain metals in the "zero-valent" state. Some examples are shown as structures **20** to **22**, which are formed by displacements of CO or hydrocarbon ligands from metallic complexes.

$$\text{Ni(PhPCl}_2)_4 \qquad \text{Mo(CO)}_3(\text{PhPCl}_2)_3 \qquad \text{Ni(CF}_3\text{PF}_2)_4$$

$$\mathbf{20}^{44} \qquad\qquad \mathbf{21}^{45} \qquad\qquad \mathbf{22}^{46}$$

Methylphosphonous dichloride and dibromide have the unusual property of reacting directly with elemental nickel simply on refluxing, forming stable zero-valent complexes[47,48] (Eq. 3.42).

$$\text{Ni} + \text{MePX}_2 \longrightarrow \text{Ni(MePX}_2)_4 \qquad (3.42)$$

3.A.3.c. Cycloaddition Reactions.

Phosphonous dihalides participate in an electrocyclic reaction with dienes in which they both donate and accept an electron pair; the result is the formation of a 5-membered ring with phosphorus initially in the 5-coordinate state. This intermediate, however, can ionize into the more stable halophosphonium ion, which is the cycloadduct actually observed (Eq. 3.43).

$$\text{(3.43)}$$

In practice, the cycloadduct is usually hydrolyzed to the phosphine oxide, as shown in Eq. 3.41, although it can participate in other typical reactions of halophosphonium ions. Another characteristic of the process is that some phosphonous dichlorides give adducts that undergo double bond migration (Eq. 3.44), unless the bond in the initial adduct is stabilized by two substituents.

$$\text{(3.44)}$$

Phosphonous dibromides give adducts that do not undergo the double bond rearrangement. A great variety of phosphonous dihalides and even some phosphinous halides (which give quaternary phosphonium salts as in Eq. 3.45) have been used in this reaction. The cycloaddition, however, is retarded by large substituents on phosphorus; *tert*-butylphosphonous dichloride, for example, is incapable of reacting with a diene.

$$\text{(3.45)}$$

This cycloaddition is known as the McCormack reaction, first described in U.S. patents,[49] and is in very wide use in the synthesis of phosphorus heterocycles. It has been extensively reviewed,[50,51] and further discussion will be deferred to chapter 8.B. Looked at as a reaction of halophosphines, it represents a type of mechanistic behavior that can be described as "biphilic," as noted in chapter 2.K. Other examples of this biphilic behavior have been found, as in the cyclization with α,β-unsaturated carbonyl compounds (Eq. 3.44). This reaction occurs in the presence of acetic anhydride, which may act to stabilize the initial ring system (23) by conversion to phosphinate form (24)[52] (Eq. 3.46).

$$\text{(3.46)}$$

   23   24

3.A.3.d. The HX Elimination Reaction.

In principle, it might be expected that a mole of HX could be eliminated from a halophosphine on reaction with a base that could not perform simple displacement of the halogen. Although there may have been indications of this form of reactivity over the years, it was not until 1978 that it was conclusively demonstrated that the carbon-phosphorus double bond could be formed by reaction of a base with a phosphinous halide.[53] The synthesis is outlined in Eq. 3.47.

$$\text{Mesityl-PCHPh}_2\text{(Cl)} \xrightarrow{\text{DBU}} \text{Mesityl-P=CPh}_2 \quad (3.47)$$

This was, in fact, one of the first successful syntheses of the C=P bond; with the realization that this bond could exist, a massive amount of research followed in this area (chapter 10.C). It was also discovered that certain phosphonous dichlorides could undergo the elimination, seen, for example, in Eq. 3.48.[54]

$$\text{Me}_3\text{Si-CH(Ph)-PCl}_2 \xrightarrow{\text{DBU}} \text{Me}_3\text{Si-C(Ph)=PCl} \quad (3.48)$$

3.B. ALKOXY AND AMINO DERIVATIVES OF PHOSPHINES

3.B.1. General

Displacement reactions of halogen from halophosphines or phosphorus trihalides with alcohols or amines provide ready access to alkoxy and amino derivatives of 3-coordinate phosphorus, and they have been made by the thousands. Indeed, the phosphorus-halogen bond plays a role as the ultimate starting structure in every viable synthesis of these compounds (as in the syntheses of compounds 13 to 18, Appendix 3.1). It is convenient to organize the discussion around the structural possibilities that can arise from each of the three possible halide starting materials, which have the structures R_2PX, RPX_2, and PX_3. As will be seen, from the latter two, there can be some mixture of halogen, alkoxy, or amino substitution on phosphorus, which increases the number of structural possibilities greatly. The phosphinous chlorides are free of this complication; whereas they are less used than the other chlorides, they are a convenient starting point for our brief discussion. The simple alkoxy and amino products derived from them can be used as models for understanding the behavior of the more complicated forms. Extensive coverage of the literature is available, in chapters in the Kosolapoff–Maier series,[1b,1c] the Hartley series[2a] and Houben–Weyl.

3.B.2. Derivatives from Phosphinous Chlorides

To form an alkoxy derivative, the reaction with an alcohol is conducted in the presence of an equivalent of a base, usually triethylamine, to remove HCl (Eq. 3.49).

$$Me_2PCl + ROH \xrightarrow{Et_3N} Me_2P-OR \qquad (3.49)$$

Acid removal is critical to success, because the P–O–R bond first formed, here and in all 3-coordinate compounds, is easily cleaved with the formation of an alkyl chloride (Eq. 3.50). This leaves an –OH group on 3-coordinate phosphorus, which, of course, tautomerizes to the phosphoryl form (**25**).

$$Me_2P-O-R \xrightarrow{HCl} Me_2\overset{+}{P}\underset{H}{\overset{(Cl^-)}{-}O}-R \xrightarrow{-RCl} [Me_2P-OH] \longrightarrow \underset{H}{\overset{Me_2P=O}{|}} \qquad (3.50)$$
$$\hspace{11cm} \mathbf{25}$$

If a phenol is the reactant, cleavage by HCl cannot occur, because the aryl-O bond can be cleaved only by acids under much more forcing conditions (e.g., refluxing with concentrated HBr or HI); base is not a requirement in the phenol reaction, although it is frequently used. The ester product of Eq. 3.49 is properly named as an alkyl dialkylphosphinite, as if it were the ester of the dialkylphosphinous acid, R_2P-OH, which, of course, has no existence. As is common in inorganic nomenclature, the *-ous* ending in the acid compels the *-ite* ending in the ester. Reaction of a phosphinous chloride with a primary or secondary amine, usually 2 moles, converts the chloride to the amino derivative (Eq. 3.51). This product is properly known as an aminophosphine.

$$Me_2P-Cl + R_2NH \longrightarrow Me_2P-NR_2 \qquad (3.51)$$

Just as was seen in the chemistry of 3-coordinate phosphorus halides, the phosphinous esters and amides are sensitive to attack by both nucleophilic and electrophilic reagents. Among the important nucleophiles are aqueous NaOH, which accomplishes hydrolysis to the secondary phosphine oxide via the phosphinous acid (Eq. 3.52); Grignard and other organometallic reagents, useful for displacing the alkoxy or amino group with the formation of a tertiary phosphine (Eq. 3.53); and metallic hydrides, which provide access to secondary phosphines (Eq. 3.54).

$$Me_2P-OR \xrightarrow{NaOH} [Me_2P-OH] \longrightarrow \underset{H}{\overset{Me_2P=O}{|}} \qquad (3.52)$$

$$Me_2P-OR \xrightarrow{R'MgX} Me_2P-R' \qquad (3.53)$$

$$Me_2P-NR_2 \xrightarrow{LiAlH_4} Me_2P-H \qquad (3.54)$$

The hydrolysis of phosphinous acid derivatives requires further comment, because this reaction occurs at much faster rates than are known for corresponding

4-coordinate phosphoryl derivatives. Phosphinites and aminophosphines can be sensitive to water alone, although the reaction is catalyzed by both acids and bases, and this needs to be remembered in their handling. This is, in fact, a general property of 3-coordinate ester and amide derivatives, as is illustrated by the fact that trimethyl phosphite reacts with weakly acidic water 10^{12} times faster than does trimethyl phosphate.[55] A related characteristic is that phosphinites readily participate in transesterification reactions, which are much less common among 4-coordinate esters. Phosphinite transesterification can be a useful way to make esters with alcohols that may be sensitive to the conditions of the more common synthesis based on the phosphinous chloride.

The mechanism[39] of these nucleophilic displacements, as has already been noted, probably involves creation of a TBP transition state (**10**) or intermediate (**11**) from in-line attack behind the group to be displaced, which is supported by the observation of stereochemical inversion at phosphorus when chiral. There are occasions when stereochemical retention occurs, and this suggests that pseudorotation can precede the departure of the leaving group from the required apical position. An excellent example of this mechanism has recently been observed in the special case of intramolecular attack of a carbanionic center on a chiral phosphinite, which occurs with retention of the phosphorus configuration[56] (Eq. 3.55). Here the attacking nucleophile is geometrically constrained from performing the usual in-line displacement of the alkoxy group, so inversion cannot occur. The incoming anionic group must take up an apical position in the forming TBP, and the departing alkoxy group is forced to be equatorial. This leaving group can move to the required apical position only by a pseudorotation, which leads to a product with retained configuration at phosphorus.

(3.55)

The lone pair on phosphorus in both the ester and the amide is involved in many valuable reactions. Oxygen may be added to form the 4-coordinate phosphinate (**26**) and phosphinamide structures (**27**); anhydrous oxidizing agents are preferred, because both reactants undergo hydrolysis rather readily. Peroxy acids and *tert*-butyl peroxide are frequently used. Sulfur adds to phosphorus without the need for a catalyst to form the thiophosphinate derivatives (**28**).

$R_2\overset{O}{\overset{\|}{P}}-OR$ $R_2\overset{O}{\overset{\|}{P}}-NR_2$ $R_2\overset{S}{\overset{\|}{P}}-OR$

26   **27**   **28**

Another type of oxidized product is formed by the reaction with halogen; these are 1:1 addition products (**29**), which as noted (e.g., Eq. 3.43) are more properly described as the ionic halophosphonium halides rather than the 5-coordinate dihalophosphoranes (**30**). They are very rapidly hydrolyzed to give the corresponding phosphoryl compound.

$$\underset{\mathbf{29}}{\overset{+}{R_2P}-NR_2 \ \overset{-}{Cl}} \qquad \underset{\mathbf{30}}{\overset{Cl}{\underset{Cl}{|}}{R_2P-NR_2}}$$

Halophosphonium halides made from tertiary phosphines are reasonably stable, but those with an alkoxy group, as prepared from phosphinites, are quite unstable thermally and decompose by attack of the halide ion on carbon of the alkoxy group (Eq. 3.56). This dealkylation process plays a vital role in the famous Michaelis-Arbusov synthesis of 4-coordinate phosphoryl compounds from 3-coordinate esters, by which similar halophosphoniun ions are formed as intermediates, as will be seen shortly.

$$R_2P-O-R' \xrightarrow{Cl_2} \overset{+}{R_2P}-O-R' \ \overset{-}{Cl} \xrightarrow{-R'Cl} R_2P=O \qquad (3.56)$$

Anhydrous hydrogen halides readily add to both the ester and the amide. In both cases, protonation occurs on phosphorus, even though the lone pair on nitrogen would appear to present an alternative site in the latter. The protonated species are not stable and decompose readily. The secondary phosphine oxide is obtained from the ester by the dealkylation process already seen as Eq. 3.50. The P-protonated aminophosphine, however, follows a different route; halide cannot cleave the C−N bond, but it does cleave the P−N bond by attack on the positive phosphorus. The product is the phosphinous halide (Eq. 3.57).

$$Ph_2P-NR_2 \xrightarrow{HCl} Ph_2\overset{+}{P}-NR_2 \overset{Cl^-}{\underset{H}{}} \longrightarrow Ph_2P-Cl + R_2NH \qquad (3.57)$$

Another addition reaction to the lone pair that has synthetic value is that with azides in the well-known Staudinger-type reaction (Eq. 3.58). Here an adduct is formed initially that then decomposes by loss of nitrogen. The product is named a phosphinimidate.

$$R_2\ddot{P}-OR + PhN_3 \longrightarrow \underset{R_2P-OR}{\overset{N-N=N-Ph}{||}} \xrightarrow{-N_2} \underset{R_2P-OR}{\overset{N-Ph}{||}} \qquad (3.58)$$

As is true of the lone pair in halophosphines, that in the phosphinous ester and amide derivatives is available for complexation with metals, both in ionic and zerovalent form, and many complexes are known. Pertinent reviews of such complexes are those of Verkade[1d] and Pignolet.[57]

3.B. ALKOXY AND AMINO DERIVATIVES OF PHOSPHINES

The lone pair on the esters and amides is much more available for formation of a carbon–phosphorus bond with alkyl halides than is the lone pair of halophosphines. Such reactions are known in great number and proceed under mild conditions. Although the quaternary salts so formed from the phosphinous amide are relatively stable, those formed from the ester are readily decomposed, during the course of their preparation, by attack of the released halide ion on carbon of the alkoxy group. An alkyl halide is eliminated and the oxygen is converted to phosphoryl form in a tertiary phosphine oxide (Eq. 3.59).

$$PhP(Cy)-OMe \xrightarrow{MeI} PhP^+(Cy)(Me)-O-Me \; I^- \longrightarrow Ph-P(Cy)(Me)=O \; + \; MeI \qquad (3.59)$$

This is a variant of the highly important Michaelis–Arbusov reaction, which will be seen to be of general value in the conversion of 3-coordinate esters to 4-coordinate phosphoryl species. In the present case, the reaction provides access to tertiary phosphine oxides, which can be made in great variety with the substitution patterns $R_3P=O$, $R_2R'P=O$ and $RR'R''=O$ as determined by the selection of reagents. The reaction shown in Eq. 3.59 is an example of the synthesis of the latter type and is of special interest because the phosphorus is chiral. The Michaelis–Arbusov reaction may also be performed as a rearrangement (commonly referred to as simply the Arbusov rearrangement), by using a small amount of the same alkyl halide that will be eliminated in the dealkylation step. This is the case in Eq. 3.59. Upon elimination, this halide is available to alkylate the phosphorus of the phosphinite, thus continuing the reaction. We will see other examples of the Michaelis–Arbusov reaction, especially as being a major synthetic method for alkyl phosphonates, for which trialkyl phosphites are the starting material.

A unique property of the phosphorus lone pair is its ability to attack on halogen, rather than carbon, in certain polyhalo and α-halocarbonyl compounds. This will be seen to be an important property of tertiary phosphines, but it also occurs with phosphinites. In the reaction with polyhalides, phosphorus acquires the halophosphonium ion structure, and the halo compound is converted to a carbanion (Eq. 3.60). This reaction will occur only when the carbanion has some stability, as is provided by halogen substitution. Carbon tetrachloride is a particularly well known participant in this reaction, because the relatively stable trichloromethyl anion is released (which can nevertheless decompose to dichlorocarbene in the usual way). For this reason, the use of carbon tetrachloride and chloroform is not recommended for work with 3-coordinate phosphorus compounds. The reaction with α-halocarbonyl compounds is a form of the Perkow reaction, to be discussed in chapter 5.C.

$$R_2P(OR'): \; Cl-CCl_3 \longrightarrow R_2P^+(OR')-Cl \; + \; {}^-CCl_3 \longrightarrow \left[\begin{array}{c} OR' \\ | \\ R_2P-Cl \\ | \\ CCl_3 \end{array} \right] \xrightarrow{-Cl^-} R_2P^+(OR')(CCl_3) \; Cl^- \longrightarrow R_2P(=O)-CCl_3 \qquad (3.60)$$

3.B.3. Derivatives from Phosphonous Dichlorides

With two displaceable chlorine atoms, there is a large number of derived structural types. These are summarized below along with their correct names. For simplicity only alkyl groups will be used in the examples, but aryl groups are also commonly present.

RP(OR)$_2$	dialkyl alkylphosphonite
RP(OR′)Cl	alkyl alkylphosphonochloridite
RP(OR′)(NR′$_2$)	O-alkyl N,N-dialkyl alkylphosphonamidite
RP(NR′$_2$)$_2$	N,N,N′,N′-tetraalkyl alkylphosphonous diamide
RP(NR′$_2$)Cl	alkylchloro(N,N-dialkylamino)phosphine

In these structures, nitrogen may also have only one alkyl group, or it may only have hydrogen. Such structures, however, are generally unstable unless sterically bulky substituents are present to hinder decomposition reactions (elimination of HCl or ROH to form P=N bonds, undergoing further reaction). We will see many examples of the stabilization of phosphorus functions by this device of steric protection. An example of a sterically stabilized aminophosphine is compound **31**. Strongly electronegative groups are also useful for stabilization, as in compound **32**. A structure (**33**) is also shown that has hydrogen rather than alkyl on phosphorus, a type that is quite rare and has low stability when steric protection is not present.

$$
\begin{array}{ccc}
t\text{-BuP(NH}_2)_2 & \text{CF}_3\text{P(NH}_2)_2 & t\text{-Bu}-\overset{\overset{\text{H}}{|}}{\text{P}}-\text{NHBu-}t \\
\mathbf{31} & \mathbf{32} & \mathbf{33}
\end{array}
$$

Synthesis of the various phosphonous acid derivatives can be accomplished by the appropriate displacement reactions on the phosphonous dichloride with an alcohol or an amine, conducted in the presence of a tertiary amine. For products retaining one chlorine atom, it is necessary to employ a 1:1 ratio of reactants. Another technique uses silyl derivatives of the alcohol (Eq. 3.61) or amine (Eq. 3.62); no base is required in these reactions, because no acid is released. Silyl groups take the place of hydrogen in other important synthetic operations of phosphorus chemistry and are especially useful in creating some low coordination species (chapter 10).

$$\text{RPCl}_2 + \text{R'}-\text{O}-\text{SiMe}_3 \longrightarrow \text{RP(OR')}_2 + \text{Me}_3\text{SiCl} \quad (3.61)$$

$$\text{RPCl}_2 + \text{R'}_2\text{N}-\text{SiMe}_3 \longrightarrow \text{RP(NR'}_2)_2 + \text{Me}_3\text{SiCl} \quad (3.62)$$

The alkoxy and amino substituents on 3-coordinate phosphorus are of such a high order of reactivity that base- or acid-catalyzed transesterification with a higher alcohol, allowing the displaced lower boiling one to escape, is a simple synthetic procedure. Similarly, transamination is a feasible process.

The mixed chloro-amino derivatives can also be made readily by a group interchange reaction of the diamino and dichloro derivatives, as in Eq. 3.63. Also called a redistribution reaction, this process well illustrates the ease of displacement of substituents on 3-coordinate phosphorus;[58] such exchange processes are quite rare among 4-coordinate species.

$$\text{MePCl}_2 + \text{MeP(NMe}_2)_2 \longrightarrow 2\ \text{MeP}\begin{smallmatrix}\text{Cl}\\\text{NMe}_2\end{smallmatrix} \quad (3.63)$$

The chemical reactions of the phosphonite derivatives in general resemble those already presented for the phosphinite derivatives, but two new features deserve mention. With the possibility of mixed chloro-alkoxy, chloro-amino and alkoxy-amino groups on phosphorus, the question arises about the order of reactivity of the substituents. Although all are susceptible to nucleophilic displacement, it has been found that the ease of displacement is chloro > amino > alkoxy. Some examples illustrate this reactivity order (Eqs. 3.64,[59] 3.65[60] and 3.66[61]).

$$\text{MeP}\begin{smallmatrix}\text{Cl}\\\text{N(}i\text{-Pr)}_2\end{smallmatrix} + \text{HO-C}_6\text{H}_4\text{-NO}_2 \xrightarrow{\text{Et}_3\text{N}} \text{MeP}\begin{smallmatrix}\text{O-C}_6\text{H}_4\text{-NO}_2\\\text{N(}i\text{-Pr)}_2\end{smallmatrix} \quad (3.64)$$

$$\text{Ph-P}\begin{smallmatrix}\text{Cl}\\\text{OEt}\end{smallmatrix} + \text{Et}_2\text{NH} \longrightarrow \text{Ph-P}\begin{smallmatrix}\text{NEt}_2\\\text{OEt}\end{smallmatrix} \quad (3.65)$$

$$\text{Me-P}\begin{smallmatrix}\text{OR'}\\\text{NR''}_2\end{smallmatrix} \xrightarrow[\text{tetrazole}]{\text{ROH}} \text{Me-P}\begin{smallmatrix}\text{OR'}\\\text{OR}\end{smallmatrix} \quad (3.66)$$

The other new feature is that the difunctionality on phosphorus makes possible the creation of heterocyclic systems when the reactant is a difunctional nucleophile (Eqs. 3.67[62] and 3.68[63]).

$$\text{MeP(NEt}_2)_2 + \text{o-C}_6\text{H}_4(\text{NH}_2)_2 \xrightarrow{\Delta} \text{Me-P benzodiazaphosphole} \quad (3.67)$$

$$\text{PhP(NEt}_2)_2 + \text{HOCH}_2\text{CH}_2\text{NH}_2 \xrightarrow{\Delta} \text{Ph-P oxazaphospholidine} \quad (3.68)$$

The reaction of the phosphonous acid derivatives with organometallics resembles that already seen for the phosphinous acid derivatives but has more synthetic utility here. It is possible to replace any of the three substituents—chloro, amino, and alkoxy—with alkyl groups, in that order of reactivity. Thus to add one more alkyl group to a phosphonous acid derivative to synthesize a phosphinite, a chloro amino derivative would be preferable over a dichloride, because the amino group reacts less rapidly than chloro, allowing only monosubstitution under mild conditions (Eq. 69).

$$R-P(Cl)(NR'_2) \xrightarrow{R''MgX} R-P(R'')(NR'_2) \quad (3.69)$$

3.B.4. Derivatives from Phosphorus Trichloride

There are many structural possibilities from full or partial replacement of chlorine in PCl_3 by alkoxy or amino groups, as well as from various combinations of these substituents; all possibilities are known, and a very rich chemistry has developed around the 3-coordinate substitution products of phosphorus trichloride. No effort will be made to describe this enormous area in any detail; the review literature[1e] should be consulted for additional information.

On hydrolysis, phosphorus trichloride gives phosphorous acid, and this has led to the description of the various substitution products as derivatives of phosphorous acid. This description would be quite adequate if the structure of phosphorous acid were correctly represented by the formula $P(OH)_3$. It is not, however, because as we have seen before (chapter 2.G) a tautomeric shift occurs so that the proper structure is $HP(O)(OH)_2$. Substitution products from this structure are also known, and indeed there is a rich chemistry based on 4-coordinate derivatives of phosphorous acid. In earlier times (and to some extent even today), these derivatives were also referred to as phosphites. Thus $HP(O)(OCH_3)_2$ would have been called dimethyl phosphite, whereas $P(OCH_3)_3$ would be called trimethyl phosphite, a clearly confusing ambiguity for the uninitiated. This has been corrected by viewing $HP(O)(OH)_2$ as the parent of the family of phosphonic acids, with H rather than a carbon fragment on phosphorus. It has become known as H-phosphonic acid; therefore, the structure $HP(O)(OCH_3)_2$ may be named dimethyl H-phosphonate. The chemistry of the extremely important family of H-phosphonates will be discussed in chapter 5.

Replacement of chlorine in phosphorus trichloride by alkoxy can give the following structural types:

(RO)PCl$_2$ alkyl phosphorodichloridite or alkoxydichlorophosphine
(RO)$_2$PCl   dialkyl phosphorochloridite
(RO)$_3$P   trialkyl phosphite

Note that the syllable *-or-* indicates the state of phosphorus as having no H or carbon substituents, just as *-on-* has been seen to indicate one such substituent and *-in-* to indicate two as in phosphinous acids. The 3-coordinate state is indicated by the suffix *ite* in the case of esters; in other cases it may be simpler, and is acceptable, to name the compound as a substituted phosphine.

R$_2$NPCl$_2$	(*N*,*N*-dialkylamino)dichlorophosphine
(R$_2$N)$_2$PCl	chloro(*N*,*N*,*N'*,*N'*-tetraalkyldiamino)phosphine
(R$_2$N)$_3$P	hexaalkyl phosphorous triamide or *tris*(dialkylamino)phosphine
(R$_2$N)(OR)PCl	*O*-alkyl *N*,*N*-dialkyl phosphoroamidochloridite
(R$_2$N)(OR)$_2$P	*O*,*O*-dialkyl *N*,*N*-dialkyl phosphoramidite
(R$_2$N)$_2$(OR)P	*O*-alkyl *N*,*N*,*N'*,*N'*-tetraalkyl phosphorodiamidite

From the background already provided, the synthetic techniques for performing nucleophilic substitutions on phosphorus trichloride can easily be visualized. Care must be given to using the proper ratio of reactants, and an acceptor for the released HCl should be employed. This is critical in the case of reaction with alcohols, to avoid C—O cleavage by HCl. Derivatives can also be obtained by transesterification and transamination reactions. Silyl derivatives of alcohols and dialkylamines are frequently used in reactions with phosphorus chlorides, and have the advantage that no HCl is released (Eq. 3.70).

$$\text{PCl}_3 + 3\,\text{R}_2\text{N-SiMe}_3 \longrightarrow \text{P(NR}_2)_3 + 3\,\text{ClSiMe}_3 \qquad (3.70)$$

Taken individually, the three common P-substituents chloro, alkoxy and amino have basically the same chemical properties as already described for the phosphonous and phosphinous derivatives and show the same facile displacement reactions with nucleophilic reactants. Much more is known about the chemistry of the phosphite structures, because they are so easily obtained from phosphorus trichloride. This is especially true of the trialkyl phosphites, which have been used extensively as nucleophiles in many processes. This has given rise to a very large number of unique reactions and new structural types. An excellent summary of many of the fundamental reactions of trialkyl phosphites as nucleophiles is given by Kirby and Warren[64] and Emsley and Hall,[65] and we can consider only some generalities in the present discussion. Trialkyl phosphites played an important role in the early development of phosphorus chemistry; their alkyl halide-catalyzed rearrangement to dialkyl phosphonates (Eq. 3.71) and the related alkylation with a stoichiometric amount of alkyl halide bearing a different alkyl group (Eq. 3.72) provided an early, excellent synthetic method for obtaining phosphonic acid derivatives. We now recognize these reactions as examples of nucleophilic attack on saturated carbon, creating a phosphonium ion intermediate that undergoes cleavage of a C—O bond by attack of chloride. This is the original form of the Michaelis-Arbusov reaction. Its scope and limitations will be described in chapter 5.

$$(\text{MeO})_3\text{P}: + \text{MeX} \longrightarrow \begin{array}{c}\text{OMe}\\|\\\text{MeO-P}^+\text{-Me}\\|\\\text{O-Me}\\\ \ \downarrow\\\ \ X^-\end{array} \longrightarrow \begin{array}{c}\text{OMe}\\|\\\text{MeO-P-Me}\\\|\\\text{O}\end{array} \qquad (3.71)$$

$$(MeO)_3P + RX \longrightarrow \underset{O}{\overset{OMe}{MeO-P-R}} \quad (3.72)$$

Examples are also known of 1,4-addition of phosphites to α,β-unsaturated systems (Eq. 3.73[66]). A characteristic of this and related reactions is that the initial zwitterionic adduct is not stable, and a secondary reaction must take place to give a product.

$$(3.73)$$

This is true also of attack on carbonyl carbon of aldehydes (the Abramov reaction). This is a useful process only when the initial zwitterionic adduct **34** is formed in the presence of a proton donor (including its use as the solvent) to quench the negative charge; a secondary reaction of dealkylation on a C−O bond of the positive phosphorus takes place to give a stable 1-hydroxyphosphonate product (Eq. 3.74).

$$(RO)_3P + \underset{R'}{\overset{R'}{C=O}} \rightleftharpoons \underset{\mathbf{34}}{(RO)_3\overset{+}{P}-\overset{\overset{\bar{O}}{|}}{CR'_2}} \underset{R''\bar{O}}{\overset{R''OH}{\rightleftharpoons}} (RO)_2\overset{+}{P}-\overset{\overset{OH}{|}}{CR'_2} \longrightarrow (RO)_2\overset{\overset{O}{||}}{P}-\overset{\overset{OH}{|}}{CR'_2} + R-O-R'' \quad (3.74)$$

Phosphites containing one siloxy group are especially useful in this process, because the silyl group easily transfers to the negative oxygen of the initial adduct to give the required stabilization (Eq. 3.75[67]).

$$(EtO)_2\overset{O}{\overset{||}{P}}-H + Me_3SiCl \xrightarrow{Et_3N} (EtO)_2P-OSiMe_3 \xrightarrow{RCHO} \underset{OSiMe_3}{RCH-\overset{O}{\overset{||}{P}}(OEt)_2} \xrightarrow{H_2O} \underset{OH}{RCH-P(O)(OEt)_2} \quad (3.75)$$

The reagent (RO)$_2$P−OTMS is easily prepared in situ from (EtO)$_2$PHO and TMS−Cl; it is useful also in giving α-aminophosphonates by reaction with C=N groups (Eq. 3.76).

$$(EtO)_2P-OSiMe_3 \xrightarrow{RCH=NR'} \underset{R'N-SiMe_3}{RCH-P(O)(OEt)_2} \xrightarrow{H_2O} \underset{R'NH}{RCH-P(O)(OEt)_2} \quad (3.76)$$

Attack of trialkyl phosphites on unsaturated nitrogen—as in azo, diazo, and azido compounds—is also known. Biphilic attack of peroxides gives especially valuable products. With dialkyl peroxides at low temperatures, adducts are formed that have

3.B. ALKOXY AND AMINO DERIVATIVES OF PHOSPHINES

5-coordinate phosphorus (Eq. 3.77); pentaoxyphosphoranes were among the first compounds known to have this structural feature.

$$(RO)_3P \ + \ R'O-OR' \ \longrightarrow \ RO-\underset{\underset{OR'}{|}}{\overset{\overset{OR'}{|}}{P}}\!\!\begin{smallmatrix}OR\\OR\end{smallmatrix} \quad (3.77)$$

Pentacoordinate phosphorus is especially easily recognized by ^{31}P NMR spectroscopy. This type of phosphorus is characterized by strong shielding [e.g., (EtO)$_5$P, δ −70.9], distinguished from phosphite and phosphonate structures that have positive shifts (chapter 6). The simple adducts generally decompose to phosphates and dialkyl ethers, and the overall reaction is that of oxidation of phosphorus. With proper substitution, however, the 5-coordinate structures can be stabilized and some even can be distilled. Pentacoordinate phosphorus is also formed in the biphilic reaction with α-dicarbonyl compounds, a reaction discovered early in the development of 5-coordinate structures[68] (Eq. 3.78). Many structures have been prepared by this reaction.

$$(MeO)_3P \ + \ \begin{smallmatrix}O=\\O=\end{smallmatrix}\!\!\!\underset{Me}{\overset{Me}{\diagup}} \ \longrightarrow \ (MeO)_3P\!\!\underset{O}{\overset{O}{\diagup}}\!\!\underset{Me}{\overset{Me}{\diagdown}} \quad (3.78)$$

Trialkyl phosphites add to oxygen of 1,4-quinones to form tetraoxyphosphonium intermediates (**35**) that rearrange to para-alkoxyaryl phosphates by transfer of alkyl to the phenoxide group formed, or are converted by water if present to aryl phosphates[69] (Eq. 3.79).

$$(MeO)_3P \ + \ \text{[quinone]} \ \longrightarrow \ \left[\text{aryl-}O\overset{+}{P}(OMe)_3,\ O^-\right] \ \longrightarrow \ \begin{matrix}\text{aryl-OP(O)(OMe)}_2\\ \text{OMe}\end{matrix} \ \xrightarrow{H_2O} \ \begin{matrix}\text{aryl-OP(O)(OMe)}_2\\ \text{OH}\end{matrix} \quad (3.79)$$

35

In the above discussion, we have concentrated on reactions of trialkyl phosphites, but many of these are applicable to amino and thio phosphite derivatives as well. In closing, we will consider a process that is uniquely based on alkyl phosphoroamidite chemistry and is of great significance in biological chemistry. In very recent years, phosphite derivatives have found use in the synthesis of oligonucleotides.[70] As mentioned previously, the displacement of groups from 3-coordinate phosphorus takes place much more readily than with 4-coordinate phosphoryl compounds; this reaction is used to attach phosphorus to the nucleoside, and the resulting phosphite derivative can be oxidized to the phosphate. Two phosphorodichloridites (o-chlorophenyl and 2,2,2-trichloroethyl) at low temperatures (−80°) were first used for the phosphitylation of the open 3-OH group of 2-deoxyribose in suitably blocked nucleosides,[71] but a newer reagent that seems to be in general acceptance is MeOP(i-Pr$_2$N)$_2$, which is of value in avoiding the release of HCl.[70] As outlined in Eq. 3.80, the ribose 3-OH displaces one di-isopropylamino group from phosphorus to form **36**; the reaction is catalyzed by the weak acid tetrazole. The second amino group is then displaced by attack of the ribose 5-OH of a blocked nucleoside, thus forming the trialkyl phosphite **37**. The phosphite is oxidized to the phosphate by mild reagents such as I$_2$ in water. The methoxy group is then removed by attack on carbon by thiophenolate ion or ammonia; deblocking at other points in the molecule follows to give the nucleotide **38** (R = H). This entire procedure gives very high yields in every step, and it serves as excellent example of the application of the properties of organophosphorus compounds to solve a major problem in organic synthesis. It has been adapted for use in automated oligonucleotide synthesis by performing the reactions on a nucleoside first anchored at its 2-deoxyribose 3-position to the surface of silica gel, and it now is a major source of synthetic oligonucleotides.

$$(3.80)$$

Advances in nucleoside phosphoramidite technology continue to be made; as an example, deoxynucleosides with photolabile blocking groups and with the N-isopropyl 2-cyanoethylphosphoramidite function have been used in solid-phase DNA syntheses on microchips.[72] Phosphoramidites with O-allyl groups have been found useful because the allyl group can be removed by reaction with Pd(Ph$_3$P)$_4$.[73] This technology can also be applied to the synthesis of other biophosphates, such as phospholipids.[74]

3.C. PHOSPHINES

3.C.1. General

As early as the 1870s, simple phosphine derivatives were being prepared and characterized, mostly in the pioneering work of A. W. von Hofmann and K. A. A.

Michaelis. These compounds are difficult to handle, because they are sensitive to the atmosphere, highly malodorous, and perhaps toxic. It is a tribute to the skill (and bravery) of the early workers that so much information was gathered, and some of the methods of synthesis that we use today are based on their findings. Phosphine chemistry up to 1970 was summarized completely[1f] by Maier in the Kosolapoff–Maier series; thousands of compounds and their properties are listed there, along with the many methods used in their synthesis. Houben–Weyl[3,4] provides other coverage, and a 1990 review covering more recent work was published by Gilheany and Mitchell in the Hartley series.[2b] Heterocyclic phosphines are well treated in the monographs by F. G. Mann,[75] a pioneer in the field, and by Quin,[50] and others.[2c,76,77] Many new phosphines have been synthesized since Maier outlined the general principles of synthesis, but for the most part, they have been prepared by the methods already in place at the time of his review or by specialized methods of limited versatility. Indeed, there has been significant new activity in tertiary phosphine chemistry as a result of more recent discoveries of powerful catalytic activity of certain metal coordination compounds, and a review of this aspect of phosphine chemistry has been published.[2d] Some highlights of this commercially important application of phosphines will be described in chapter 11. Many other commercial applications of phosphines have been discovered, however; a lengthy summary prepared in 1972 by Maier[1f] is representative, but probably many more recent applications need to be added to this list. References for procedures for the synthesis of some 35 phosphines can be found in Appendix 3.1.

3.C.2. Synthesis of Phosphines

It is possible to summarize the most useful and versatile of the many methods for synthesis of phosphines according to the phosphorus-containing starting material, and this is the approach taken in the present concise survey of this very large field. More complete coverage will be found in references 1f and 2b.

3.C.2.a. From Phosphorus Halides. Phosphonous dihalides are excellent starting materials for the synthesis of tertiary phosphines of the form RR''_2P or R_3P. One simply reacts them with a small excess over 2:1 of an organometallic reagent to replace both halogens. These reactions are generally exothermic and are held at or below room temperature. Alkyl groups are most commonly introduced with Grignard reagents, and aryl groups with the easier formed lithium reagents, although there are many cases of the opposite choice for each, and other metals (Al, Sn, Zn, etc.) have also found use. An example of the process is seen in Eq. 3.81.[78]

$$PhPCl_2 \ + \ 2\ BrMgCH_2CH_2CH_2NMe_2 \longrightarrow PhP(CH_2CH_2CH_2NMe_2)_2 \quad (3.81)$$

Phosphinous halides are also used for the synthesis of the RR'_2P type by the reaction with organometallics. The most commonly used reactant in contemporary work is diphenylphosphinous chloride, which can be purchased commercially. The diphenylphosphino group consequently appears in several of the phosphine ligands used in metal complexes, as is illustrated by the synthesis[79] of the optically active ligand **39** (Eq. 3.82).

$$\text{[pinanyl]-CH}_2\text{MgBr} + \text{Ph}_2\text{PCl} \longrightarrow \text{[pinanyl]-CH}_2\text{PPh}_2 \quad \textbf{39} \tag{3.82}$$

Phosphinous halides with two different substituents can be used to make fully unsymmetrical phosphines, RR'R"P. Such phosphines are chiral and can be obtained in enantiomerically pure form (chapter 9); they are of great importance as ligands used in complexes employed in asymmetric synthesis. The organometallic method is useful in attaching the exocyclic substituent in cyclic tertiary phosphines, in cases in which the cyclic phosphinous halide is available (Eq. 3.83).

$$\text{(3-methyl-1-chlorophospholene)} + \text{PhCH}_2\text{MgBr} \longrightarrow \text{(3-methyl-1-benzylphospholene)} \tag{3.83}$$

For the synthesis of symmetrical tertiary phosphines R_3P, the reagent of choice is phosphorus trichloride, and countless compounds have been made from it. This method can be used to prepare commercially the best known of all tertiary phosphines, triphenylphosphine (Eq. 3.84). This is a solid with little or no odor and is widely used as a reagent in organic synthesis.

$$\text{PCl}_3 + 3\ \text{PhMgX} \longrightarrow \text{PPh}_3 \tag{3.84}$$

The phosphonous dihalides are useful in the construction of phosphorus heterocycles from the reaction with di-Grignard or lithium reagents. The first carbon-phosphorus heterocycle to be reported in (1915), 1-phenylphosphorinane, was prepared by this method[41] (Eq. 3.85), which remains of value today.

$$\text{PhPCl}_2 + \text{BrMg(CH}_2)_5\text{MgBr} \longrightarrow \text{Ph-P(cyclohexyl ring)} \tag{3.85}$$

Primary and secondary phosphines are also available from phosphonous dihalides (Eq. 3.86[80]) and phosphinous halides (Eq. 3.87) by reaction with anhydrous reducing agents. Of the many reagents that have been used, the most commonly

$$\text{(2,4,6-tri-tert-butylphenyl)PCl}_2 \xrightarrow{\text{LiAlH}_4} \text{(2,4,6-tri-tert-butylphenyl)PH}_2 \tag{3.86}$$

employed is lithium aluminum hydride, preferably at low temperatures. Also useful are the silanes $PhSiH_3$, Ph_2SiH_2, and $HSiCl_3$.

$$Ph_2PCl \xrightarrow{LiAlH_4} Ph_2PH \qquad (3.87)$$

3.C.2.b. From Metal Phosphides. The anions from PH_3 and from primary and secondary phosphines are excellent nucleophiles and are highly reactive to alkylating reagents. This has been put to advantage for the synthesis of phosphines of all types; old as the reaction is (the first report[81] is 150 years old!), it is one of the major synthetic procedures in use today, especially in the construction of phosphines as ligands. The required anions are generally prepared by reaction of the phosphine with the metal (usually sodium, potassium, or calcium) in liquid ammonia; a valuable alternative is to employ butyllithium, and other methods are also in use. The ratio of reactants must be controlled to give the desired degree of replacement of hydrogen. Thus, for primary phosphine synthesis from PH_3, the species MPH_2 is required, as in the recent example[82] seen in Eq. 3.88. Because the phosphine product also is a good nucleophile, in all of these syntheses overalkylation must be avoided with the proper 1 : 1 reactant ratio.

$$KPH_2 + ClCH_2CH_2NHCH_2CH=CH_2 \longrightarrow H_2PCH_2CH_2NHCH_2CH=CH_2 \quad (3.88)$$

To synthesize a secondary phosphine, the starting material must be the anion derived from a primary phosphine, as in Eq. 3.89.

$$PhPHNa + MeX \longrightarrow PhMePH \qquad (3.89)$$

From secondary phosphines may be prepared tertiary phosphines, and herein lies one of the real advantages of the metal phosphide synthetic method: to construct tertiary phosphines that have three different substituents and hence are chiral. As an example, methylphenylphosphine has been converted to the sodium salt and reacted with *l*-menthyl chloride to give diastereomeric tertiary phosphines; these were converted to oxides, separated, and reduced to the phosphines with phenylsilane[83] (Eq. 3.90).

As will be seen in chapter 9, to obtain optically active phosphines, it is sometimes the practice to convert the racemic phosphine to the quaternary salt or oxide derivative, by which the resolution is more conveniently performed, then return the derivative to the free phosphine form.

A useful alternative to the synthesis of secondary phosphide anions from secondary phosphines is to prepare them from phosphinous chlorides by metallation, as with lithium, or by halogen-metal exchange with butyllithium and phosphorus halides. This is of real practical value for the introduction of the common diphenylphosphino group.

Alkylene dihalides can react with phosphide ions to give either cyclic or bis(phosphino) derivatives. The former reaction was used in the synthesis of the first phosphorus heterocycle with three members, the phosphirane system[84] (Eq. 3.91) and has also given access to rings of larger size.[85]

$$ClCH_2CH_2Cl + NaPH_2 \longrightarrow \underset{H}{\overset{\triangle}{P}} \qquad (3.91)$$

With the proper ratio of reactants, however, alkylene dihalides can give bis(phosphino)alkane derivatives[86] as seen in Eq. 3.92; these compounds are of great value as bidentate ligands in coordination chemistry.

$$2\ Ph_2\bar{P} + BrCH_2CH_2Br \longrightarrow Ph_2P-CH_2CH_2-PPh_2 \qquad (3.92)$$

The versatility of the phosphide anion reaction is seen also in its valuable application to the synthesis of arylphosphines by halogen displacements from aryl halides (by nucleophilic substitution rather than by a benzyne intermediate), and to vinylic phosphines by reaction with monohaloalkenes and dihaloalkenes (the latter with stereochemical retention). In addition, acyl halides can be used to form 1-oxoalkylphosphines R_2PCOR, and epoxides can give 2-hydroxyethylphosphines (Eq. 3.93), to illustrate the value of the method to form C-functional phosphines.

$$Ph_2PNa + \underset{O}{\triangle} \longrightarrow Ph_2PCH_2CH_2OH \qquad (3.93)$$

3.C.2.c. From Phosphine or Organic Phosphines.

Phosphine and its primary and secondary derivatives can add to carbon-carbon multiple bonds by either a radical or an ionic mechanism, and this addition constitutes a third outstanding way to synthesize new phosphines. Phosphine itself, which can be purchased or generated as needed from commercially available metallic (generally calcium or magnesium) phosphides, is the most commonly used reactant; the tris-trimethylsilyl derivative is a very useful alternative, because it acts as a masked phosphine and is easy to handle. Phosphine adds to the double bond of alkenes in the presence of one equivalent of an acid (sulfonic acids, liquid HF, BF_3) to form the salt of a primary phosphine. The addition follows the Markovnikov rule, as seen in Eq. 3.94,[87] and appears to involve the initial formation of a carbocation from the alkene that then attacks the nucleophilic phosphine. The primary phosphine product is tied up as the acid salt, which prevents further alkylation from occurring.

$$(CH_3)_2C=CH_2 + PH_3 \xrightarrow{MeSO_2OH} (CH_3)_3CPH_2 \qquad (3.94)$$

Phosphine can add to α,β-unsaturated nitriles and carbonyl compounds under basic conditions, in which the attacking species is probably the phosphide anion. The reaction[88] is a variant of the well-known Michael reaction for conjugate addition. Depending on the reactant ratio, the reaction can be used to obtain monoalkylation, dialkylation, or trialkylation of phosphine. From acrylonitrile are formed valuable 2-cyanoethylphosphines (Eq. 3.95). They are quite useful as starting materials in other syntheses; the bis-cyanoethyl derivative from phenylphosphine, for example, can be cyclized in the Thorpe reaction to give derivatives of the 6-membered phosphorinane ring system[89] (chapter 8.B).

$$PhPH_2 + CH_2=CHCN \xrightarrow{conc.\ KOH} PhP(CH_2CH_2CN)_2 \xrightarrow{KOBu-t}$$

$$Ph-P\underset{CN}{\overset{}{\diagdown}}=NH \xrightarrow[-CO_2]{H_2O,\ HCl} Ph-P\diagdown=O \qquad (3.95)$$

Phosphine also can participate in free radical reactions with alkenes. Radical formation from PH_3 can be initiated in the usual ways by peroxides, azobisisobutyronitrile, and UV radiation. The $PH_2\cdot$ radical adds to the alkene to give the most stable carbon radical; this then abstracts a hydrogen atom from phosphine to give an alkylphosphine with the regeneration of $PH_2\cdot$ to propagate the chain. Usually this reaction is run with an excess of alkene to give the tertiary phosphine, although it is easy to control the process to generate the primary phosphine. Obtaining a good yield of the secondary phosphine is more difficult. The addition of phosphine to 1-butene is used in the synthesis of tri-n-butylphosphine (Eq. 3.96).

$$3\ CH_3CH_2CH=CH_2 + PH_3 \xrightarrow{AIBN} (CH_3CH_2CH_2CH_2)_3P \qquad (3.96)$$

Primary and secondary phosphines undergo many of the addition reactions described for phosphines, under the same variety of conditions. The products are predictable and will not be shown again. There are, however, some special applications that are important. Thus diphenylphosphine will add to acetylene under basic conditions to give the bisphosphine **40** (Eq. 3.97), valuable as a bidentate ligand.

$$Ph_2PH + HC\equiv CH \xrightarrow{base} Ph_2PCH=CH_2 \xrightarrow{Ph_2PH} Ph_2PCH_2CH_2PPh_2 \qquad (3.97)$$
$$\mathbf{40}$$

Another valuable application is in the synthesis of heterocycles; although this is discussed in chapter 8, we can note here examples of the formation of phospholes by addition (basic) to 1,3-diynes[90] (Eq. 3.98), of phosphorinanes by addition (also basic) to bis-(α,β-unsaturated) carbonyl compounds[91] (Eq. 3.99) and of bicyclic phosphines by addition (radical) to cyclic dienes[1g] (Eq. 3.100).

$$RC\equiv C-C\equiv CR + PhPH_2 \xrightarrow{PhLi} \text{[phosphole with R, R, Ph]} \quad (3.98)$$

$$O=C(CH=CH_2)_2 + PhPH_2 \xrightarrow{NaOEt} \text{[phosphorinanone with Ph]} \quad (3.99)$$

$$\text{[cyclooctadiene]} + PhPH_2 \xrightarrow{(t\text{-BuO})_2} \text{[bicyclic Ph]} + \text{[bicyclic Ph]} \quad (3.100)$$

Phosphine additions are, therefore, seen to be of considerable versatility and synthetic value. As noted, a handicap is the malodorous nature and easy oxidizability of phosphines as reactants, but in many cases this can be circumvented by the use of masked phosphines, which includes not only the silyl derivatives but also the hydroxymethyl derivatives made from reaction of phosphine PH groups with formaldehyde.

Finally, we note that it is possible to effect the controlled alkylation of monoalkyl phosphines with alkyl halides to prepare mixed secondary and tertiary phosphines in high yields.[92]

3.C.2.d. From Phosphoryl or Other 4-Coordinate Compounds.

Tertiary phosphine oxides are the most important of the 4-coordinate species used as precursors of phosphines, although phosphonic acid esters can be reduced by $LiAlH_4$ to primary phosphines. The availability of acyclic and cyclic forms is discussed in chapters 4 and 8, respectively. Until relatively recently, there were no general methods for the deoxygenation of these compounds and conversion to phosphines. The introduction of metallic hydrides, especially lithium aluminum hydride, into synthetic organic chemistry led to some useful syntheses, but other conventional reducing agents of organic chemistry, including catalytic hydrogenation, are without effect on phosphine oxides. In 1964, Fritzche and co-workers reported[93] that a number of silicon hydrides are extremely efficient reducing agents for phosphine oxides, and the use of these reagents is now standard practice in organophosphorus chemistry. The reductions occur in high yield under mild conditions and are generally without effect on other functional groups in the substrate. The specific compounds in use at this time include trichlorosilane

(Cl$_3$SiH) and the phenylsilanes (PhSiH$_3$, Ph$_2$SiH$_2$, and Ph$_3$SiH). Another silicon-containing reagent, Si$_2$Cl$_6$, has also been used in the reduction.[94] Trichlorosilane, which is commercially available, is probably the most popular of the reducing agents. It is frequently used in the presence of a tertiary amine, usually triethylamine or pyridine. The latter forms a complex with the silane, whereas triethylamine appears to act by proton abstraction to give SiCl$_3$$^-$ as the active species. Reductions are performed in an inert solvent such as benzene, at temperatures from ambient to reflux. An excellent discussion of the scope of this and other reduction reactions is provided by Engel,[95] who also gives some typical experimental procedures. Especially in the construction of heterocyclic phosphines is the silane reduction of phosphine oxides of great value, because there are several methods for the synthesis of heterocyclic phosphine oxides that cannot be modified for direct use to produce phosphines. An example is found in the synthesis of the five-membered phospholene series. It was noted earlier that the McCormack cycloaddition of 3-coordinate phosphorus halides to dienes can be used to give adducts that on hydrolysis are converted to phospholene oxides. These oxides can then be reduced to the corresponding phosphines, as in Eq. 3.101.

$$\text{diene} \xrightarrow[\text{2. H}_2\text{O}]{\text{1. RPX}_2} \underset{R}{\overset{O}{\underset{\|}{P}}} \text{(phospholene oxide)} \xrightarrow{\text{HSiCl}_3} \underset{R}{\overset{|}{P}} \text{(phospholene)} \quad (3.101)$$

The silane reduction method is also of great current importance in the synthesis of diphosphines as ligands for homogeneous catalyst systems. Thus the synthesis of the optically active BINAP (Eq. 3.102) is reported in *Organic Syntheses*.[96]

$$\text{binaphthyl-Br}_2 \xrightarrow[\text{2. Ph}_2\text{POCl}]{\text{1. 2Mg}} \text{binaphthyl-(P(O)Ph}_2)_2 \xrightarrow[\text{2. HSiCl}_3, \text{Et}_3\text{N}]{\text{1. separate}} \text{BINAP} \quad (3.102)$$

(±) → (±) → (+) or (−)

Because of the ease of interconversion of phosphine oxides and phosphines, it is possible to consider the oxo function as a blocking group on phosphorus; thus reactions can be performed on the carbon skeleton, such as oxidation, without effect on the unreactive phosphine oxide, but that would destroy the phosphine function.

A valuable feature of the silane reduction is that it can be used to effect a definite stereochemical outcome when applied to a chiral phosphine oxide. With acyclic phosphine oxides, trichlorosilane alone or better as its pyridine complex reduces the oxide with retention of configuration. Many examples are also known of the reduction of heterocyclic phosphine oxides with the trichlorosilane–pyridine complex, with the usual retention of configuration. Details of the mechanism are unclear, but one that is consistent with the observed reduction with retention is outlined in Eq. 3.103, in which it is seen that the first step is the complexation of phosphoryl oxygen by the silane. The complex collapses with transfer of the oxygen to silicon, leaving phosphorus in the protonated 3-coordinate form.

$$\begin{array}{c} \text{A} \\ \text{B} \\ \text{C} \end{array}\!\!P{=}O \ + \ HSiCl_3 \longrightarrow \begin{array}{c} \text{A} \\ \text{B} \\ \text{C} \end{array}\!\!\overset{+}{P}{-}O\underset{H}{\diagdown}SiCl_3 \longrightarrow \begin{array}{c} \text{A} \\ \text{B} \\ \text{C} \end{array}\!\!P{\cdots}O \atop H{\cdots}SiCl_3 \longrightarrow$$

$$\bar{O}SiCl_3 \ + \ \begin{array}{c} \text{A} \\ \text{B} \\ \text{C} \end{array}\!\!\overset{+}{P}{-}H \longrightarrow [\,HOSiCl_3\,] \ + \ \begin{array}{c} \text{A} \\ \text{B} \\ \text{C} \end{array}\!\!P\!: \qquad (3.103)$$

With a chiral phosphine oxide, the phosphine formed has primarily, but seldom exclusively, the retained configuration. With amines of base strength greater than about pK_b 5, such as triethylamine, the stereochemical result is predominantly inversion, possibly by the operation of the mechanism of Eq. 3.104.

$$[\,HSiCl_3 \ + \ R_3N \ \rightleftharpoons \ \bar{S}iCl_3 \ + \ R_3NH^+\,]$$

$$\begin{array}{c} R^1 \\ R^2{\cdots}P{=}O \\ R^3 \end{array} \ + \ \overset{+}{H}NR_3 \longrightarrow \begin{array}{c} R^1 \\ R^2{\cdots}\overset{+}{P}{-}OH \\ R^3 \end{array} \xrightarrow[-OH^-]{-SiCl_3} Cl_3Si{-}\overset{R^1}{\underset{OH\ R^3}{\overset{+}{P}}}{\diagdown}R^2 \longrightarrow :\!\!P\overset{R^1}{\underset{R^3}{\diagdown R^2}}$$

(3.104)

In the 1990s the list of reagents useful for deoxygenating phosphine oxides grew. The new reagents appear to be as effective as the silicon hydrides, although less is known about them at this time. These reducing agents include the Schwartz reagent,[97] $[Cp_2ZrHCl]_n$, a combination of polymethylhydrosiloxaxe (PMHS; $Me_3Si(MeHSiO)_nOSiMe_3$) and titanium tetra(isopropoxide),[98] and alane, AlH_3.[99]

The reduction of the phosphoryl group in esters of phosphonic and phosphinic acids is most commonly accomplished with lithium aluminum hydride or $AlHCl_2$, which usually simultaneously cleave the P—O single bonds and produce a phosphine. This is, in fact, a quite useful synthesis of primary and secondary phosphines (Eq. 3.105). Although the phosphorus acids can be reduced similarly to phosphines, it is more common to accomplish this transformation by first converting the acid to the acid chloride, which is reduced with greater ease.

$$R\overset{O}{\overset{\|}{P}}(OH)_2 \xrightarrow{SOCl_2} R\overset{O}{\overset{\|}{P}}Cl_2 \xrightarrow{LiAlH_4} R\!\cdot\! PH_2 \qquad (3.105)$$

Phosphines can also be synthesized by the reduction of other 4-coordinate species, including quaternary phosphonium ions, halophosphonium ions, oxyphosphonium ions, and thiophosphoryl comounds (Eq. 3.106).

$$Bu_3P{=}S \xrightarrow{LiAlH_4} Bu_3P\!: \qquad (3.106)$$

3.C.3. Important Properties of Phosphines

The chemistry of the phosphines, although extensive and of great practical importance, is centered on the lone pair and its availability for the formation of new bonds to phosphorus. Exceptions to this generality do exist, and we will touch on some of these, but it is the donation of the lone pair to electrophilic centers that accounts for much of the high reactivity of the phosphines. Many reactions are known, and only the most important can be presented here, in a manner as to convey the general characteristics of the family. The reviews already cited provide excellent coverage with full referencing.

Phosphines have a major structural difference from their nitrogen relatives the amines: in both families, the geometry is pyramidal, but whereas pyramidal inversion is rapid in amines at room temperature, it is so slow in phosphines as to give them fixed pyramidal structure. As will be discussed further in chapter 9, this has important consequences in stereochemistry. With three different substituents, phosphines can be optically active, and in cyclic systems, substituents on P are held in fixed positions, such as axial and equatorial in six-membered rings. An excellent summary of our understanding of structure and bonding in phosphines is provided by Gilheany.[2e] It is known that the C−P−C bond angles in tertiary phosphines expand as group size increases; the smallest value is 98.6° in trimethylphosphine, whereas the largest so far reported are 109.7° for trimesitylphosphine and 109.9° for tri-*tert*-butylphosphine.[100] The H−P−C angle in primary and secondary phosphines is in the range of 95 to 97°. It is convenient in practical work with phosphines to consider the lone pair as lying in an orbital directed from the apex of the pyramid, having mixed s and p character, but this is a great simplification of the situation. With this view, however, we can understand how increases in steric size of P-substituents can reduce reactivity by hindering the approach of reagents to the lone pair or by changing the hybridization to give more s-character in the C−P bonds with an increase in p-character in the lone pair orbital. In reactions in which a fourth bond is formed to phosphorus, without passing through a 5-coordinate transition state or intermediate, the new group takes the position of the lone pair, with the important consequence that the stereochemistry of the phosphine is retained.

Another major difference found on comparing lone pair characteristics of amines to phosphines is the relative importance of delocalization phenomena on interaction with unsaturated groups. Although delocalization is extensive and of critical importance when the amino group is conjugated with double or triple bonds, aromatic rings, carbonyl groups and others, it is of little significance in phosphine chemistry. The point has been examined extensively in various structural types, and the general conclusion is that overlap is much less efficient (but not absent) between the lone pair orbital on the large phosphorus atom with that of the adjacent p-orbital on the smaller carbon atom. Furthermore, for efficient overlap, the lone pair should acquire more p-character, meaning the phosphorus pyramid should be flattened and preferably phosphorus should reach planarity. There is an energy barrier to this flattening process that would offset the benefit of delocalization. This leads to some profound differences between amine and phosphine chemistry, viz., (1) Phosphino groups have only small electron donating effects on the benzene ring, showing up mostly in spectroscopic properties (chapter 7); electrophilic substitution of aromatic phosphines is virtually unknown, because electron density remains so high on phosphorus that it is the site of attack rather than a ring carbon. (2) Phosphino groups attached to C=O remain pyramidal, as in the case of tribenzoylphosphine (with bond angles[101] of 95.0°) and do not have the restricted rotation of amides.[102] (3) The phosphine lone pair in pyramidal phosphorus makes only a small contribution to establishing an aromatic sextet, thus the 5-membered phosphole ring has low aromaticity unless the phosphorus pyramid is flattened (chapter 8.B). (4) Enephosphines have no significant buildup of negativity on the β-carbon as is so important in enamines, etc. Indeed, there is even some evidence that phosphino groups can act as weak electron acceptors. In *para*-dimethylaminophenylphosphines, it has been proposed[1b,1c] that the resonance structure **41** makes a contribution.[103,104]

41

The phosphide group R_2P^-, in contrast, shows a pronounced tendency to donate electrons to π systems; thus aryl phosphides are colored, due to resonance of form **42**, and phosphole anions are delocalized and aromatic.[105]

42

One important property that *is* in common with the amines is the basicity of tertiary phosphines, which has been much studied.[2f] Phosphines are just slightly weaker bases than their amine counterparts. Thus triethylphosphine has pK_b 5.4, and other simple tertiary phosphines have values in the range pK_b 4.5 to 6. For comparison, triethylamine has pK_b 3.2. Aryl substitution reduces the basicity of phosphines, but the effect is weaker than that in amines; $PhMe_2P$ has pK_b 7.7, compared to 8.9 for $PhMe_2N$, and Ph_3P has pK_b 11.2 with Ph_3N at 19. This is consistent with the notion of greater resonance interaction in the amine series. In a practical sense, the basicity of trialkyl and dialkylarylphosphines is sufficient for them to form stable salts with acids, useful as a protective device from oxidation, and to allow their easy extraction from organic solvents with dilute aqueous acids, a property of great value in the isolation of phosphines.

As hydrogen replaces alkyl in phosphines, the basicity is progressively decreased, so that secondary alkylphosphines have pK_b values around 9.5 to 10.5, and primary phosphines of around 13.5 to 14. Here the departure from amine basicity effects is considerable, because there is relatively little effect on basicity (actually a small increase) as one replaces alkyl on nitrogen by hydrogen, and indeed with no additivity effect. Thus diethylamine has pK_b 2.9, and ethylamine has pK_b 3.2. The P—H bond has little tendency to form hydrogen bonds, either in the phosphine as a proton acceptor or in the protonated form as a proton donor; this is another striking difference from amines and thus hydration effects are less important. Furthermore, with the greater size of the phosphorus atom steric interaction among the substituents is of less importance. The net effect is to make phosphine basicity largely determined by the inductive effect of the substituents; as the number of electron-releasing alkyl groups increases, so does the basicity. Phenyl substitution decreases the basicity by inductive electron attraction, with little contribution from a resonance effect.

In the first detailed study of the basicity of phosphines[106] (where most of the cited pK_b values can be found), it was noted that the basicity of each of the three types of phosphines could be correlated with the use of Taft σ^*-constants, emphasizing the unique importance of inductive effects in phosphines, and equations for predicting basicity were developed. A later study showed that the use of Kabachnik σ^{ph} constants[107] (derived from dissociation constants for phosphorus acids) gave a better correlation.

Phosphines have boiling points[1f] that are in the same general range as their amine counterparts, despite the unimportance of H-bonding in the former and the greater polarity of the N—C bonds. Thus $MePH_2$ boils at $-14°$, somewhat below the H-bonded $MeNH_2$, b.p. $-6.7°$, and similarly $PhPH_2$ boils below $PhNH_2$ (160° and 184°, respectively). When H-bonding is absent, the values can be quite similar or the heavier P counterpart may have a higher value: $PhPMe_2$, b.p. 192°, and $PhNMe_2$, b.p. 194°; Et_3P, b.p. 126°, and Et_3N, b.p. 89.4°. (Amine boiling points were taken from reference 108). The absence of significant H-bonding among phosphines is responsible for their generally low solubility in water.

Tertiary phosphines act as powerful nucleophiles in many reactions. Nucleophilicity is strongest in the trialkylphosphines and decreases in the secondary and then in the primary phosphines as the electron-release by alkyl diminishes. Phenyl substitution also decreases the nucleophilicity; nevertheless, triphenylphosphine, unlike triphenylamine, is still a useful nucleophile. The nucleophilicity of phosphines in their reaction with alkyl halides has been studied extensively and placed on a quantitative basis by Henderson and Buckler,[109] and certainly quaternization is one of their most characteristic reactions (Eq. 3.107).

$$R_3P + R'X \longrightarrow R_3\overset{+}{P}-R'\ \ \overset{-}{X} \tag{3.107}$$

It has already been noted that phosphine and primary and secondary phosphines participate in Michael-type additions to activated double bonds, as with acrylonitrile.

It is of synthetic value in the synthesis of diphosphino compounds that certain phosphines, notably diphenylphosphine, add readily under basic conditions to the double bond of vinylphosphines[110] (Eq. 3.108). It may be that the addition is aided by electron delocalization, provided by the phosphine group, of the initially formed carbanion. Resonance of this form was used earlier (e.g., **41**) to explain other phosphine properties.

$$Ph_2P-CH=CH_2 \xrightarrow[(PhPH_2, PhLi)]{Ph_2P^-} [Ph_2P-\overset{..}{C}H-CH_2PPh_2 \leftrightarrow Ph_2\bar{P}=CH-CH_2PPh_2] \xrightarrow{H^+}$$

$$Ph_2PCH_2CH_2PPh_2 \qquad (3.108)$$

Many examples are also known of addition of trialkylphosphines to activated double bonds. The initial adduct would be a zwitterion, which in some cases can be stabilized by electron-withdrawing groups as in Eq. 3.109.[111]

$$PhCH=C\begin{matrix}CN\\CN\end{matrix} + Et_3P \longrightarrow \underset{Et_3P^+}{PhCH-\bar{C}(CN)_2} \qquad (3.109)$$

Stable adducts also can be formed with quinones (Eq. 3.110) and other systems.[69]

[quinone] + Ph₃P ⟶ [adduct with PPh₃⁺ and O[−]]   (3.110)

In the case of the highly reactive tetracyanoethylene,[112] another path is followed, giving the unusual adduct **43**.

[structure of 43: cyclic compound with Ph₃P=N and multiple CN groups]

43

A valuable application of this type of phosphine reactivity is in the polymerization of acrylonitrile.[113] Tributylphosphine is used for this purpose; the initial zwitterionic adduct adds to a second molecule of acrylonitrile, and so on, to give a high polymer (Eq. 3.111).

$$Bu_3P + CH_2=CHCN \longrightarrow Bu_3\overset{+}{P}CH_2\overset{|}{\underset{CN}{C}}-\bar{H} \xrightarrow{CH_2=CHCN} Bu_3\overset{+}{P}CH_2\overset{|}{\underset{CN}{C}}H\left[CH_2\overset{|}{\underset{CN}{C}}H\right]_n^-$$

$$(3.111)$$

Tertiary phosphines can also react with activated alkynes to give vinylphosphonium ions, but frequently secondary reactions follow. Particularly interesting is the reaction of tertiary phosphines with acetylene monocarboxylates or dicarboxylates, which gives a variety of complex products. These reactions are the subject of detailed reviews,[114,115] and also have recently been succinctly summarized.[2j] In the presence of aqueous acid, the reaction is simplified and the vinylphosphonium salts can be isolated (Eq. 3.112).

$$Ph_3P + HC \equiv CCOOH \xrightarrow{HX} Ph_3\overset{+}{P}-\underset{\underset{H}{|}}{C}=\underset{\underset{H}{|}}{C}-COOH \quad X^- \qquad (3.112)$$

As might be expected, phosphines can attack the electrophilic carbon in carbonyl and imino compounds. With primary and secondary phosphines, the products that form initially are α-hydroxyphosphines. They are, however, subject to other reactions, and the final products depend on such factors as acidity, ratio of reactants, and electronic effects in both the phosphine and the electrophile. Ketones are less reactive than aldehydes, and aliphatic aldehydes are more reactive than aromatic aldehydes. Probably the cleanest and most useful of the carbonyl condensation reactions is that between phosphine and formaldehyde. In aqueous solution with metallic salt catalysts, the reaction provides tris(hydroxymethyl)phosphine, which is useful in synthesis as a phosphine generator. When this reaction is conducted in strongly acidic media, a fourth hydroxymethyl group is added; tetrakis(hydroxymethyl)phosphonium ion is formed, a material that is made commercially as a flame retardant (Eq. 3.113).

$$\underset{H}{\overset{H}{O=C}}:PH_3 \xrightarrow{HCl} \left[\underset{OH}{\overset{H_2C-PH_2}{|}}\right] \xrightarrow[HCl]{2\,H_2CO} (HOCH_2)_3P \xrightarrow[HCl]{H_2CO} (HOCH_2)_4\overset{+}{P}\ \bar{Cl} \qquad (3.113)$$

The most common secondary reactions of the initial α-hydroxyphosphines from higher aldehydes and ketones are the formation of cyclic products from further reaction of the hydroxy group with the carbonyl group and a P–H group if present, as is seen in the reaction of phenylphosphine with excess benzaldehyde (Eq. 3.114), and a rearrangement (in strongly acidic media) of oxygen from carbon to phosphorus to form phosphine oxides (Eq. 3.115).

$$PhPH_2 + PhCHO \longrightarrow \underset{Ph}{\overset{Ph}{\underset{|}{\overset{|}{\underset{O\diagdown\;\diagup O}{Ph\diagdown P \diagup Ph}}}}} \qquad (3.114)$$

$$Ph_2PH + ArCHO \xrightarrow{H^+} Ph_2\overset{+}{P}-\underset{\underset{}{|}}{\overset{OH}{C}HAr} \xrightarrow{H^+} [\,Ph_2\overset{..}{\underset{}{P}}-\overset{+}{C}HAr \longleftrightarrow Ph_2\overset{+}{P}=CHAr\,] \xrightarrow{H_2O}$$

$$[\,Ph_2\overset{+}{P}=\underset{\underset{}{|}}{\overset{OH}{C}}HAr\,] \longrightarrow Ph_2\overset{O}{\overset{\|}{P}}-CH_2Ar \qquad (3.115)$$

The literature on these complex phosphine-carbonyl condensations is extensive,[1f,2g] but only in special cases have the reactions proved to have synthetic value.

Tertiary phosphines give unstable zwitterionic adducts with carbonyl groups. However, if excess carbonyl compound is present, the adduct can be trapped as a cyclic product, the structure of which is determined by conditions and reactant structure. An example of one such reaction is seen in Eq. 3.116, where the product is a 1,4,2-dioxaphospholane with phosphorus in the 5-coordinate state.[116]

$$R_3P: \ C=O \longrightarrow R_3\overset{+}{P}-\underset{O^-}{\overset{R'}{\underset{|}{C}}}-R' \xrightarrow{O=CR'_2} \cdots \longrightarrow \cdots \quad (3.116)$$

One might ask if the initially formed α-hydroxyphosphines from primary phosphines with carbonyl compounds might undergo intramolecular dehydration to form the C=P bond, as is so well known in primary amine chemistry. This process had never been observed until recently, but with the use of sterically hindered phosphines, it can in fact take place (Eq. 3.117). As will be seen in chapter 10.C, the C=P bond is now well known and constructed by various methods.

$$\text{2,4,6-tri-}t\text{-Bu-C}_6H_2\text{-PH}_2 + \text{ArCHO} \longrightarrow \text{2,4,6-tri-}t\text{-Bu-C}_6H_2\text{-P=C(Ar)(H)} \quad (3.117)$$

Of no less importance in phosphine chemistry than their reaction with electrophilic reagents is their very great reactivity to oxidizing agents, of practically every sort. Trialkylphosphines are the most reactive and require protection from atmospheric oxygen in their handling. They can be cleanly converted to tertiary phosphine oxides, $R_3P=O$, by the conventional oxidizing agents of organic chemistry, but the most popular reagents in current use are peroxide derivatives, the preferred one being t-butyl hydroperoxide. The process is easily performed at room temperature and gives oxides with retained stereochemistry. With trialkylphosphines, oxygen itself should be avoided as the oxidizing agent, because it has the potential of cleaving the P—C bond. Thus in the O_2-oxidation of tributylphosphine, significant amounts of O-insertion products accompanied the main product, $Bu_3P=O$[117] (Eq. 3.118).

$$Bu_3P \xrightarrow{O_2} Bu_3P=O + Bu_2\overset{O}{\underset{\|}{P}}-OBu + Bu-\overset{O}{\underset{\|}{P}}(OBu)_2 \quad (3.118)$$

Triarylphosphines are also easily oxidized by chemical reagents to give phosphine oxides, but less readily by the atmosphere, so that they can be handled without special precautions. Primary and secondary phosphines are more rapidly oxidized and do require care in handling. They have the potential of undergoing additional change after the attachment of oxygen to phosphorus; although the phosphine oxides can be obtained in some cases by proper control, especially of the ratio of reactants, the more common result (of synthetic value), is the formation of phosphonic and phosphinic acids (Eqs. 3.119 and 3.120, respectively).

$$RPH_2 \xrightarrow{O_2} R-\overset{O}{\underset{\|}{P}}H_2 \xrightarrow{O_2} R-\overset{O}{\underset{\|}{P}}(OH)_2 \qquad (3.119)$$

$$R_2PH \xrightarrow{O_2} R_2\overset{O}{\underset{\|}{P}}-H \xrightarrow{O_2} R_2\overset{O}{\underset{\|}{P}}-OH \qquad (3.120)$$

In another context, the great oxidizability of phosphines allows them to be used as reducing agents in organic chemistry. Among other applications may be cited the deoxygenation of epoxides to alkenes, and of amine oxides to amines.

Sulfur, selenium, and halogens add readily to tertiary phosphines to give $R_3P=$(S or Se) and R_3PX_2. These reactions and products will be considered in more detail in chapter 4.B. With primary and secondary phosphines, the P—H bond can be attacked also by these reagents. As seen in Eq. 3.121, sulfur will first add to secondary phosphines to form phosphine sulfides, which can be isolated, and then just as in the case of oxidation, sulfur proceeds to insert into the P—H bond to form thiophosphinic acids.

$$R_2PH \xrightarrow{S_8} R_2\overset{S}{\underset{\|}{P}}-H \xrightarrow{S_8} R_2\overset{S}{\underset{\|}{P}}-SH \qquad (3.121)$$

Primary phosphines also undergo S-insertion reactions, but a mixture of products with cyclic structures is formed, and the reaction is of little value.

With halogens, substitution of hydrogen on phosphorus appears to take place before addition to phosphorus. Thus dibutylphosphine is known[118] to give a 49% yield of Bu_2PBr on bromination at 15 to 20° with one equivalent of bromine, and butylphosphine under the same conditions gives $BuPBr_2$ in 44% yield. It is more common, however, to use sufficient halogen as to give the species R_2PX_3 and RPX_4. An alternative to halogen for controlled substitution of hydrogen in both types of phosphine without oxidation is the use of phosgene in inert solvents, which proceeds smoothly in high yield. This reaction was presented in Eq. 3.23 as a synthesis of chlorophosphines.

Tertiary phosphines participate in a number of reactions with nitrogen compounds. Probably the best known is that with azides, which is referred to as the Staudinger reaction (reviewed in reference 119). The reaction proceeds through a triazo intermediate, which decomposes with loss of nitrogen to form iminophosphoranes (Eq. 3.122). This is an excellent way of forming this functional group, which is of great value as a starting material for the synthesis of C=N compounds from C=O compounds. This application has become known as the Aza-Wittig reaction.

$$R_3P + N_3Ar \longrightarrow R_3P=N-N=N-Ar \longrightarrow \underset{Ar}{\overset{R_3P-N}{\underset{N-N}{\|}}} \xrightarrow{-N_2} R_3P=NAr \quad (3.122)$$

Another useful reaction occurs with azo compounds, most commonly dialkyl azodicarboxylates (the Mitsunobu reaction). Here an initial adduct is formed from addition of a tertiary phosphine to the double bond; the adduct has dipolar character and is widely used as a coupling agent for acids of various types with alcohols to form esters (Eq. 3.123). It is discussed further in chapter 6.I.

$$Ph_3P + EtO_2C-N=N-CO_2Et \longrightarrow \underset{EtO_2C}{\overset{Ph_3\overset{+}{P}}{\diagdown}}N-\bar{N}CO_2Et \xrightarrow{RCOOH}$$

$$\left[\underset{EtO_2C}{\overset{Ph_3\overset{+}{P}}{\diagdown}}N-\overset{H}{\underset{|}{N}}CO_2Et\right] \xrightarrow{R'OH} \underset{Ph_3\overset{+}{P}}{\overset{OR'}{|}} \xrightarrow{RCOO^-} Ph_3P=O + R'OOCR \quad (3.123)$$

Phosphines can be classed as Lewis bases and are known to form adducts with several Lewis acids, chiefly BF_3 and other boron derivatives and $AlCl_3$. But by far the most important type of adduct based on electron donation by phosphines is that formed with metals, both in the ionic and the zero-valent state. Thousands of phosphine complexes are known in great structural variety, and many are of major significance in catalysis. Further discussion is not possible here.

Primary and secondary phosphines are weak acids but somewhat stronger than amines. This makes it possible to form metallic derivatives by reaction with metallic amides, especially of sodium and potassium. These phosphines also react with Grignard reagents to form magnesium derivatives and with alkyl lithiums to form lithio derivatives. It is also possible to effect metallation by direct reaction of the phosphines with sodium or potassium. From primary phosphines, either monometallation or dimetallation can be effected. Phosphide anions are powerful nucleophiles and have found value in various synthetic operations.[1f] For example, phosphide ions can add to conjugated unsaturated systems in a Michael-type manner (Eq. 3.124) to form functionalized phosphines and can be acylated with acid chlorides to form α-ketophosphines.

$$Ph_2PNa + PhHC=CHPh \xrightarrow{\text{then } H^+} Ph_2P-CHPh-CH_2Ph \quad (3.124)$$

Metallic phosphides can also be obtained from tertiary phosphines containing an aryl group, by cleavage of aryl with alkali metals in solvents such as tetrahydrofuran and dioxane (Eq. 3.125) or with potassium amide in liquid ammonia. As one might expect from the ready availability of triphenylphosphine, the diphenylphosphide ion is quite commonly used as a nucleophile in various processes, and it is also a source on hydrolysis of diphenylphosphine.[120] The phenyl anion formed in the process is usually quenched by reaction with *tert*-butyl chloride, which forms benzene and isobutylene.

$$Li\cdot + Ph_3\overset{..}{P} \xrightarrow{THF} Ph\overset{..}{:}{}^- Li^+ + Ph_2\overset{..}{P}\cdot \xrightarrow{Li\cdot} Ph_2\overset{..}{\underset{..}{P}}{}^- Li^+ \xrightarrow{H_2O} Ph_2PH$$

(3.125)

Another valuable application of the metal cleavage process is in heterocyclic chemistry; an exocyclic P-phenyl substituent can be cleaved, and the resulting heterocyclic anion can be used as a nucleophile for the attachment of various substituents by alkylation or conjugate addition reactions[121] (Eq. 3.126).

$$\text{Ph-phosphole(Ph)-Ph} \xrightarrow{Li} \text{Ph-phospholide}^- \text{-Ph} \xrightarrow{RX} \text{Ph-phosphole(R)-Ph}$$

(3.126)

In both alkyl and aryl phosphines, a number of functional groups are known to be compatible with the phosphino group. Obviously, groups that are oxidizing in character, or that are among the electrophiles that react with phosphines, should be avoided, but this leaves many common functional groups that are compatible with the phosphino group, and there are a large number of known functionalized phosphines. For the most part, the normal organic chemical reactions can be performed on such groups. To mention a few, the nitrile group may be reduced or hydrolyzed, the C=O group may be reduced to C–OH with metal hydrides, aromatic phosphines with halo substituents may form Grignard reagents, alkoxy groups on alkyl or aryl substituents may be cleaved to –OH with hydrogen halides, the triple bond in phosphinoalkynes may be reduced to the double- or singly-bonded structure, primary amino groups can be converted to imino groups, and hydroxy groups can be acylated or in the case of hydroxymethyl groups reacted with amines to give aminomethylphosphines in a Mannich-like reaction. Only in rare cases does the C–P bond undergo cleavage in phosphine reactions. A notable case is that of sodium hydroxide cleavage of primary or secondary phosphines with polyfluoro substitution. Thus $(CF_3)_2PH$ is sensitive to aqueous NaOH and releases CF_3H. A particularly interesting example is that of base acting on $CHFClCF_2PH_2$ (Eq. 3.127); this and related cleavage reactions were proposed[122] to involve a phosphaalkene intermediate from base abstraction of HF, a postulate made years before true phosphaalkenes became known.

$$HCFCl-CF_2-PH_2 \xrightarrow{aq.NaOH} [HCFCl-CF=PH] \xrightarrow{H_2O} [HCFCl-CF_2-P(H)(OH)] \xrightarrow{OH^-}$$

$$H_3PO_2 + [HCFCl-CF_2^-] \longrightarrow Cl^- + CHF=CHF$$

(3.127)

Finally, we must point out that both free radical and photochemical processes are known for the phosphines. The phosphoranyl radical, $R_3(OR)P\cdot$, is easily formed from tertiary phosphines, and its chemistry has been extensively explored.[2h] Many examples are known of the use of photochemical reactions for the synthesis of heterocyclic phosphorus compounds.[2c,2i]

In concluding this section, it needs to be mentioned that the purpose of this section is to give an appreciation of the wide scope of phosphine reactivity. Many other reactions are known of phosphines that could not be included, and indeed it is fair to say that phosphines are perhaps the most reactive, the most useful, and the most studied of any family of phosphorus compounds. The reviews cited must be examined to gain a greater appreciation of phosphine reactivity.

3.D. BIPHOSPHINES

The phosphorus-phosphorus bond has remarkable thermal stability (86 kcal/mol in Et_2P-PEt_2,[123] and many compounds based on this bond type are known. It is an area in which future growth can be expected, as pointed out in two reviews that envisioned numerous structures that remain to be synthesized.[124,125] Here we will treat only the family of biphosphines, but polyphosphines are also well known, especially with cyclic structure.[126] These can be of simple monocyclic structure, typically $(R-P)_{4-6}$ (e.g., **44**) but some surprising bridged structures are possible, for example, the norbornane skeleton where all carbons are replaced by phosphorus (**45**). Such cyclic polyphosphines are fascinating substances from the standpoint of their ^{31}P NMR spectra, which are quite complex and have various coupling patterns.[127]

The parent molecule, H_2P-PH_2, is oxidized very rapidly and indeed is spontaneously inflammable in the atmosphere, as are derivatives with small alkyl groups. Substitutions do lessen the oxidizability and provide greater stability in the atmosphere. Phosphines without P–H bonds are in fact the best known type, and this discussion will be limited to them.

The P–P bond in biphosphines is long (2.19 Å in Me_2P-PMe_2[2k]), and phosphorus retains its stable pyramidal character, with bond angles typified by those in Me_2P-PMe_2, 99.6° for C–P–C and 101.1° for C–P–P. The barrier to inversion, however, is somewhat less than that in tertiary phosphines; these have activation energies for inversion around 35 kcal/mol, whereas biphosphines have values about 10 kcal/mol lower (e.g., $MePhCH_2P-PMeCH_2Ph$, 24 kcal/mol[1h]). This effect has been attributed to some stabilization of the planar transition state in the inversion by lone pair donation to a d-orbital of the attached phosphorus. The high thermal stability of the pyramid has also been attributed to the presence of some multiple bonding of this type. The importance of d-orbitals in such an interaction is, of course, a matter of present debate. The pyramidal stability is sufficient to allow the existence and separation of diastereoisomeric forms when each of the two phosphorus atoms bears different substituents. The diastereoisomers are conveniently written in the conventional Fischer projection formulas of organic chemistry, by treating the lone pair as a substituent on a tetrahedral, chiral phosphorus atom. Thus P,P'-dimethyl-P,P'-diphenylbiphosphine is known to exist in meso (**46**) and dl (**47**) forms.

$$\begin{array}{c} \overset{..}{Ph-P-Me} \\ \underset{..}{Ph-P-Me} \end{array} \qquad \begin{array}{c} \overset{..}{Me-P-Ph} \\ \underset{..}{Ph-P-Me} \end{array} \bigg| \begin{array}{c} \overset{..}{Ph-P-Me} \\ \underset{..}{Me-P-Ph} \end{array}$$

$$\textbf{46} \qquad\qquad\qquad \textbf{47}$$

Although several methods can be used to construct the P—P bond, we will mention only three here; others may be found in the reviews cited. An obvious method, and one that is in fact quite satisfactory, is the base-promoted reaction between a phosphinous chloride and a secondary phosphine This method allows the construction of unsymmetrical biphosphines, as outlined in Eq. 3.128. The dialkylamino group may also act as the leaving group on phosphorus in the reaction with the secondary phosphine.

$$Ph_2PCl \ + \ Ph_2PH \ \xrightarrow{\text{base}} \ Ph_2P-PPh_2 \qquad (3.128)$$

Another common process is the Wurtz-Fittig type coupling of two moles of a phosphinous chloride. Mercury was used in early work as the coupling agent, but alkali or alkaline earth metals are now in common use. Very likely the metal first reacts with the phosphinous chloride to form a metallic phosphide, as we have seen earlier. This then acts as a nucleophile toward a second molecule of the phosphinous chloride (Eq. 3.129).

$$Ph_2PCl \ \xrightarrow{Na} \ Ph_2PNa \ \xrightarrow{Ph_2PCl} \ Ph_2P-PPh_2 \qquad (3.129)$$

Biphosphine disulfides are available from the unusual reaction of Grignard reagents with $PSCl_3$ (Eq. 3.20), and the desulfurization of these compounds constitutes the third approach to biphosphines. The reaction to form biphosphines from $PSCl_3$ is at its best when simple alkyl Grignard reagents are used; Grignards with long alkyl chains, aryl groups, and some unsaturated groups give the more expected tertiary phosphine sulfides. Tetramethylbiphosphine disulfide and tetraethylbiphosphine disulfide are common compounds because of the ease of this reaction. The coupling reaction also occurs with the phosphonothioic dihalides, giving it a broader scope; the halide reactants may contain either alkyl or aryl substituents, and this is true also of the Grignard reagent. The desulfurization proceeds smoothly with a variety of metals, as well as with tertiary phosphines and phosphites. Tributylphosphine is used in the example of Eq. 3.130.[128] This process constitutes a major pathway to simple biphosphines.

$$Cl_3P=S + EtMgBr \longrightarrow Et_2\overset{\overset{S}{\|}}{P}-\overset{\overset{S}{\|}}{P}Et_2 \xrightarrow{Bu_3P} Et_2\ddot{P}-\ddot{P}Et_2 \qquad (3.130)$$

The chemistry of the biphosphines is to some extent that of tertiary phosphines. Controlled oxidation can add oxygen to both P atoms to produce biphosphine dioxides, and reaction with proper amount of sulfur can give either the biphosphine disulfides or the monosulfides. Both oxides and sulfides have good stability and are well known. Quaternization with alkyl halides, on the other hand, occurs on only one P atom. The P–P bond is more susceptible to cleavage in these salts. This is made use of for introducing the Ph_2P^+ group into heterocyclic systems, which can be accomplished by reacting the biphosphine with certain dihalides, as in the example[129] of Eq. 3.131.

$$\text{(o-C}_6\text{H}_4\text{(CH}_2\text{Br)}_2\text{)} + Ph_2P-PPh_2 \longrightarrow \text{cyclic } [o\text{-C}_6\text{H}_4(CH_2)_2\overset{+}{P}Ph_2] \; Br^- \qquad (3.131)$$

Biphosphines serve as ligands in forming complexes with transition metals and also act in conventional manner with Lewis acids. In both processes, the P–P bond is retained, but in many other reactions it undergoes cleavage, as with halogens (to form phosphinous halides) and metals (to form metallic phosphides). A more useful cleavage occurs with certain alkenes and alkynes; in Eq. 3.132, it is seen that tetraphenyldiphosphine when heated in the presence of an alkyne undergoes cleavage with addition to the double bond, constituting a synthesis of 1,2-bis(diphenylphosphino)alkenes.[130]

$$RC\equiv CR + Ph_2P-PPh_2 \longrightarrow Ph_2P-\underset{R}{C}=\underset{R}{C}-PPh_2 \qquad (3.132)$$

There is much highly specialized chemistry[1i,124,125] on the synthesis and properties of the P–P bond in 3-coordinate form, not only in biphosphines but also in linear triphosphine and tetraphosphine structures and particularly, as mentioned, in many different cyclic systems.

REFERENCES

1. G. M. Kosolapoff and L. Maier, eds., *Organic Phosphorus Compounds*, John Wiley & Sons, Inc., New York, 1972, (a) M. Fild and R. Schmutzler, Vol. 4, Chapter 8, (b) A. W. Frank, Vol. 4, Chapter 10, (c) L. A. Hamilton and P. S. Landis, Vol. 4, Chapter 11, (d) J. G.

Verkade, Vol. 2, Chapter 3, (e) W. Gerrard and H. R. Hudson, Vol. 5, Chapter 13, (f) L. Maier, Vol. 1, Chapter 1, (g) L. Maier, Vol. 1, p. 64, (h) L. Maier, Vol. 1, p. 325, (i) L. Maier, Vol. 1, Chapter 2.
2. F. R. Hartley, ed., *The Chemistry of Organic Phosphorus Compounds*, John Wiley & Sons, Inc., New York, 1990, (a) O. Dahl, Vol. 1, Chapter 1, (b) D. G. Gilheany and C. M. Mitchell, Vol. 1, Chapter 7, (c) L. D. Quin and A. N. Hughes, Vol. 1, Chapter 10, (d) H. B. Kagan and M. Sasaki, Vol. 1, Chapter 3, (e) D. G. Gilheany, Vol. 1, Chapter 2, (f) H. R. Hudson, Vol. 1, Chapter 12, (g) H. R. Hudson, Vol. 1, p. 410, (h) W. G. Bentrude, Vol. 1, Chapter 14, (i) M. Dankowski, Vol. 1, Chapter 13, (j) H. R. Hudson, Vol. 1, pp. 419–426, (k) D. G. Gilheany, Vol. 1, p. 42.
3. K. Sasse, *Methoden der Organischen-Chemie (Houben-Weyl)*, Band XII, Part 1 (Organischen Phosphorverbindungen), Georg Thieme Verlag, Stuttgart, Germany, 1963.
4. M. Regitz, ed., *Methoden der Organischen-Chemie (Houben-Weyl)*, Band E1 (Organischen Phosphorverbindungen), Georg Thieme Verlag, Stuttgart, Germany, 1982.
5. A. Michaelis, *Ber.* **12**, 1009 (1879).
6. *Organic Syntheses*, IV, 784 (1963).
7. A. D. F. Toy and E. N. Walsh, *Phosphorus Chemistry in Everyday Living*, 2nd ed., American Chemical Society, Washington, D.C., 1987, p. 228.
8. G. M. Kosalopoff in G. A. Olah, ed., *Friedel-Crafts and Related Reactions*, Vol. IV, Interscience Publishers, New York, 1965, p. 213.
9. C. Symmes, Jr. and L. D. Quin, *J. Org. Chem.* **43**, 1250 (1978).
10. F. Siméon, P.-A. Jaffrès, and D. Villemin, *Tetrahedron* **54**, 10111 (1998) and references cited therein.
11. M. S. Kharash, E. V. Jensen, and W. H. Urry, *J. Am. Chem. Soc.* **67**, 1864 (1945).
12. R. Little and P. F. Hartmann, *J. Am. Chem. Soc.* **88**, 96 (1966).
13. L. A. Errede and W. A. Pearson, *J. Am. Chem. Soc.* **83**, 954 (1961).
14. M. N. Krivchun, M. V. Sendyurev, B. I. Ionin, and A. A. Petrov, *Zhur. Obshch. Khim.* **60**, 2395 (1990); *Chem. Abstr.* **115**, 136184 (1991).
15. T. N. Belyaeva, M. N. Krivchun, M. V. Sendyurev, A. V. Dogadina, V. V. Sokolov, B. I. Ionin, and A. A. Petrov, *Zhur. Obshch. Khim.* **56**, 1184 (1986); *Chem. Abstr.* **106**, 50303 (1987).
16. N. S. Zefirov, N. Y. Zyk, A. A. Borisenko, and M. Y. Krysin, *Tetrahedron* **39**, 3145 (1983).
17. B. Fontal and H. Goldwhite, *Chem. Commun.*, 111 (1965); see also reference 14.
18. *Inorganic Syntheses* **14**, 4 (1973).
19. *Inorganic Syntheses* **27**, 236 (1990).
20. L. Maier, *Phosphorus Sulfur* **11**, 149 (1981).
21. H. Schmidbaur and A. Schier, *Synthesis*, 372 (1983).
22. H. Schmidbaur and A. Mörtl, *Z. Chem.* **23**, 249 (1983).
23. I. P. Komkov, K. V. Karavanov, and S. Z. Ivin, *Zhur. Obshch. Khim.* **28**, 2963 (1958); *Chem. Abstr.* **53**, 9035 (1959).
24. (a) M. Sokora, *Synthesis*, 450 (1970), (b) J. A. Sikorski and E. W. Logusch in R. Engel, ed., *Handbook of Organophosphorus Chemistry*, Marcel Dekker, Inc., New York, 1992, p. 779.
25. J. A. Pianfetti and L. D. Quin, *J. Am. Chem. Soc.* **84**, 851 (1962); J. A. Pianfetti (to FMC Corp.), U.S. Patent 3,210,418 (Oct. 5, 1965) *Chem. Abstr.* **64**, 2128 (1966).
26. N. Weferling and S. Hoerold, paper presented at the *International Conference on Phosphorus Chemistry*, Cincinnati, Ohio, July 12–17, 1998.
27. K. Weissermel, H.-J. Kleiner, and M. Finke, *Angew. Chem., Int. Ed. Engl.* **20**, 223 (1981).

28. B. Cornils, *Phosphorus Sulfur* **30**, 623 (1987).
29. L. D. Quin and J. S. Humphrey, Jr., *J. Am. Chem. Soc.* **83**, 4124 (1961).
30. G. O. Doak and L. D. Freedman, *J. Am. Chem. Soc.* **73**, 5658 (1951).
31. L. D. Quin and R. E. Montgomery, *J. Org. Chem.* **27**, 4120 (1963).
32. L. D. Quin and C. H. Rolston, *J. Org. Chem.* **23**, 1693 (1958).
33. V. A. Ginsburg and A. Ya. Yakubovich, *Zhur. Obshch. Khim.* **28**, 728 (1958); *Chem. Abstr.* **52**, 17092 (1958).
34. (a) *Organic Syntheses*, Coll. Vol. V, 1016, (1973), (b) *Inorganic Syntheses* **15**, 191 (1974).
35. L. D. Quin and J. Szewczyk, *Phosphorus Sulfur* **21**, 161 (1984).
36. A. W. Frank, *J. Org. Chem.* **26**, 850 (1961).
37. L. D. Quin and R. E. Montgomery, *J. Org. Chem.* **30**, 2393 (1965).
38. E. P. Kyba, M. C. Kerby, and S. P. Rines, *Organometallics* **5**, 1189 (1986).
39. J. Nielsen and O. Dahl, *J. Chem. Soc., Perkin Trans.* 2, 553 (1984).
40. G. S. Quin, S. Jankowski, and L. D. Quin, *Phosphorus Sulfur Silicon* **115**, 93 (1996).
41. G. Grüttner and M. Wiernik, *Ber.* **48**, 1473 (1915).
42. R. C. Taylor, R. Kolodny, and D. B. Walters, *Synth. React. Inorg. Metal-Org. Chem.* **3**, 175 (1973).
43. L. D. Quin, S. G. Borleske, and J. F. Engel, *J. Org. Chem.* **38**, 1858 (1973).
44. L. Malatesta and A. Sacco, *Ann. Chim.* **44**, 134 (1954).
45. E. W. Abel, M. A. Bennett, and G. Wilkinson, *J. Chem. Soc.*, 2323 (1959).
46. A. B. Burg and G. B. Street, *Inorg. Chem.* **5**, 1532 (1966).
47. L. D. Quin, *J. Am. Chem. Soc.* **79**, 3681 (1957).
48. L. Maier, *Angew. Chem.* **71**, 574 (1959).
49. W. B. McCormack (to E. I. DuPont de Nemours and Co.) U.S. Patents 2,663,736 and 2,663,737 (Dec. 22, 1953), *Chem. Abstr.* **49**, 7601 (1955).
50. L. D. Quin, *The Heterocyclic Chemistry of Phosphorus*, John Wiley & Sons, Inc., New York, 1981.
51. L. D. Quin in A. R. Katritzky, C. W. Rees, and E. F. V. Scriven, eds., *Comprehensive Heterocyclic Chemistry II*, Vol. 1, Pergamon Press, Oxford, UK, 1996, Chapter 15.
52. K. Bergesen, *Acta Chem. Scand.* **19**, 1784 (1965).
53. Th. C. Klebach, R. Lourens, and F. Bickelhaupt, *J. Am. Chem. Soc.* **100**, 4886 (1978).
54. R. Appel and A. Westerhaus, *Tetrahedron Lett.*, 2159 (1981).
55. F. H. Westheimer, S. Huang, and F. Covitz, *J. Am. Chem. Soc.* **110**, 181 (1988).
56. M. B. Tollefson, J. J. Li, and P. Beak, *J. Am. Chem. Soc.* **118**, 9052 (1996).
57. L. H. Pignolet, ed., *Homogeneous Catalysis with Metal Phosphine Complexes*, Plenum Press, New York, 1983.
58. L. Maier, *Helv. Chim. Acta* **46**, 2667 (1963).
59. J. Helinski, W. Dabkowski, and J. Michalski, *Tetrahedron Lett.* **32**, 4981 (1991).
60. E. Steininger, *Chem. Ber.* **95**, 2993 (1962).
61. A. Jäger and J. Engels, *Tetrahedron Lett.* **25**, 1437 (1984).
62. E. G. Bent, R. Schaeffer, R. C. Haltiwanger, and A. D. Norman, *J. Organomet. Chem.* **364**, C25 (1989).
63. M. A. Pudovik, S. A. Terent'eva, and A. N. Pudovik, *Zhur. Obshch. Khim.* **52**, 491 (1982); *Chem. Abstr.* **97**, 39134 (1982).
64. A. J. Kirby and S. G. Warren, *The Organic Chemistry of Phosphorus*, Elsevier Publishing, Amsterdam, 1967, Chapter 3.
65. J. Emsley and D. Hall, *The Chemistry of Phosphorus*, Harper & Row, London, 1976, Chapter 4.

66. F. Ramirez, C. P. Madan, and S. R. Heller, *J. Am. Chem. Soc.* **87**, 731 (1965); D. Gorenstein and F. H. Westheimer, *J. Am. Chem. Soc.* **92**, 634 (1970).
67. K. Afarinkia, C. W. Rees, and J. I. G. Cadogan, *Tetrahedron* **46**, 7175 (1990).
68. G. H. Birum and J. L. Dever (to Monsanto Chemical Co.) U.S. Patent 2,961,455 (Nov. 22, 1960); *Chem. Abstr.* **55**, 8292 (1961).
69. F. Ramirez and S. Dershowitz, *J. Am. Chem. Soc.* **78**, 5614 (1956).
70. M. H. Caruthers, *Science* **230**, 281 (1985).
71. R. L. Letsinger and W. B. Lunsford, *J. Am. Chem. Soc.* **98**, 3655 (1976).
72. M. C. Pirrung, L. Fallon, and G. McGall, *J. Org. Chem.* **63**, 242 (1998).
73. W. Bannwarth and E. Küng, *Tetrahedron Lett.* **30**, 4219 (1989).
74. N. Hébert and G. Just, *J. Chem. Soc., Chem. Commun.*, 1497 (1990).
75. F. G. Mann, *The Heterocyclic Derivatives of Phosphorus, Arsenic, Antimony, and Bismuth*, Wiley-Interscience, New York, 1970.
76. K. D. Berlin and D. M. Hellwege, *Chem. Rev.* **77**, 121 (1977).
77. A. R. Katritzky, C. W. Rees, and E. F. V. Scriven, eds., *Comprehensive Heterocyclic Chemistry II*, Pergamon Press, Oxford, UK, 1996.
78. *Organic Syntheses*, Coll. Vol. VI, 776 (1988).
79. H. Brunner and H. Leyerer, *J. Organomet. Chem.* **334**, 369 (1987).
80. *Inorganic Syntheses* **27**, 237 (1990).
81. P. E. Thénard, *C. R. Acad. Sci. Paris* **25**, 892 (1847).
82. K. Issleib, U. Kühne, and F. Krech, *Z. Anorg. Allg. Chem.* **523**, 7 (1985).
83. C. Fischer and H. S. Mosher, *Tetrahedron Lett.*, 2487 (1977).
84. R. I. Wagner, L. V. D. Freeman, H. Goldwhite, and D. G. Rowsell, *J. Am. Chem. Soc.* **89**, 1102 (1967).
85. S. Chan, H. Goldwhite, H. Keyzer, D. G. Rowsell, and R. Tang, *Tetrahedron* **25**, 1097 (1969).
86. J. Chatt, *J. Chem. Soc.*, 1378 (1960).
87. M. C. Hoff and P. Hill, *J. Org. Chem.* **24**, 356 (1959).
88. M. M. Rauhut, I. Hechenbleikner, H. A. Currier, F. C. Schaefer, and V. P. Wystrach, *J. Am. Chem. Soc.* **81**, 1103 (1959).
89. *Organic Syntheses*, Coll. Vol. VI, 932 (1988).
90. G. Märkl and R. Potthast, *Angew. Chem., Int. Ed. Engl.* **6**, 861 (1967).
91. R. P. Welcher and N. E. Day, *J. Org. Chem.* **27**, 1824 (1962).
92. H. R. Hayes, *J. Org. Chem.* **31**, 3817 (1966).
93. H. Fritzsche, U. Hasserodt, and F. Korte, *Chem. Ber.* **97**, 1988 (1964).
94. K. Naumann, G. Zon, and K. Mislow, *J. Am. Chem. Soc.* **91**, 7012 (1969).
95. R. Engel in R. Engel, ed., *Handbook of Organophosphorus Chemistry*, Marcel Dekker, Inc., New York, 1992, Chapter 5.
96. *Organic Syntheses*, Coll. Vol. VIII, 57 (1993).
97. M. Zablocka, B. Delest, A. Igau, A. Skowronska, and J.-P. Majoral, *Tetrahedron Lett.* **38**, 5997 (1997).
98. N. J. Lawrence and F. Muhammad, *Tetrahedron* **54**, 15361 (1998).
99. S. Griffin, L. Heath, and P. Wyatt, *Tetrahedron Lett.* **39**, 4405 (1998).
100. H. Oberhammer, R. Schmutzler, and O. Stelzer, *Inorg. Chem.* **17**, 1254 (1978).
101. A. Cogne, L. Wiesenfeld, J. B. Tyka, and J. B. Robert, *Org. Magn. Reson.* **13**, 72 (1980).
102. D. Kost, F. Cozzi, and K. Mislow, *Tetrahedron Lett.*, 1983 (1979).
103. G. P. Schiemenz, *Angew. Chem., Int. Ed. Engl.* **4**, 603, (1965).
104. G. P. Schiemenz, *Tetrahedron Lett.*, 2729 (1964).

105. (a) R. F. Hudson, *Structure and Mechanism in Organo-Phosphorus Chemistry*, Academic Press, New York, 1965, p. 26, (b) L. D. Quin and W. L. Orton, *J. Chem. Soc., Chem. Commun.*, 401 (1979).
106. W. A. Henderson, Jr. and C. A. Streuli, *J. Am. Chem. Soc.* **82**, 5791 (1960).
107. M. I. Kabachnik, *Dokl. Akad. Nauk SSSR* **110**, 393 (1956); *Chem. Abstr.* **51**, 5513 (1957).
108. F. A. Carey, *Organic Chemistry*, 3rd ed., McGraw-Hill Co., New York, 1996, Chapter 22.
109. W. A. Henderson, Jr. and S. A. Buckler, *J. Am. Chem. Soc.* **82**, 5794 (1960).
110. R. B. King and P. N. Kapoor, *J. Am. Chem. Soc.* **93**, 4158 (1971).
111. (a) L. Horner and W. Klüpfel, *Liebigs Ann. Chem.* **591**, 69 (1955); (b) J. A. Ford, Jr. and C. V. Wilson, *J. Org. Chem.* **26**, 1433 (1961).
112. H. Fritz and C. D. Weis, *J. Org. Chem.* **43**, 4900 (1978).
113. H. Takashina and C. C. Price, *J. Am. Chem. Soc.* **84**, 489 (1962).
114. A. N. Hughes, *Heterocycles* **15**, 637 (1981).
115. M. A. S. Howard and S. R. Ward, *Top. Phosphorus Chem.* **7**, 1 (1972).
116. F. Ramirez, A. V. Patwardhan, and S. R. Heller, *J. Am. Chem. Soc.* **86**, 514 (1964).
117. (a) S. A. Buckler, *J. Am. Chem. Soc.* **84**, 3093 (1962); (b) M. B. Floyd and C. E. Boozer, *J. Am. Chem. Soc.* **85**, 984 (1963).
118. C. Walling (to E. I. DuPont de Nemours Co.), U.S. Patent 2,437,795 (Mar. 16, 1948); *Chem. Abstr.* **42**, 4199 (1948).
119. (a) Y. G. Gololobov, I. N. Zhmurova, and L. F. Kasukhin, *Tetrahedron* **37**, 437 (1981); (b) A. W. Johnson, *Ylides and Imines of Phosphorus*, John Wiley & Sons, Inc., New York, 1993, Chapter 13.
120. *Inorganic Syntheses* **16**, 161 (1976).
121. E. H. Braye, I. Caplier, and R. Saussez, *Tetrahedron* **27**, 5523 (1971).
122. G. M. Burch, H. Goldwhite, and R. N. Haszeldine, *J. Chem. Soc.*, 572 (1964).
123. N. N. Grishin, G. M. Bogolyubov, and A. A. Petrov, *Zhur. Obshch. Khim.* **38**, 2683 (1968); *Chem. Abstr.* **70**, 96045 (1969).
124. R. R. Holmes, *Phosphorus Sulfur Silicon* **109**, 1 (1996).
125. L. Lamonde, K. Dillon, and R. Wolf, *Phosphorus Sulfur Silicon* **103**, 1 (1995).
126. M. Baudler, *Angew. Chem., Int. Ed. Engl.* **21**, 492 (1982); *ibid.* **26**, 419 (1987).
127. J. Hahn in J. G. Verkade and L. D. Quin, eds., *Phosphorus-31 NMR Spectroscopy in Stereochemical Analysis*, VCH Publishers, Deerfield Beach, FL, 1987, Chapter 10.
128. L. Maier, *J. Inorg. Nucl. Chem.* **24**, 275 (1962).
129. G. Märkl, *Angew. Chem., Int. Ed. Engl.* **2**, 620 (1963).
130. W. R. Cullen and D. S. Dawson, *Can. J. Chem.* **45**, 2887 (1967).

APPENDIX 3.1

SYNTHESES OF PERTINENT COMPOUNDS IN *ORGANIC SYNTHESES* (OS) AND *INORGANIC SYNTHESES* (IS)

[1] i-Pr$_2$PCl
OS **V**, 211

[2] PhPCl$_2$ 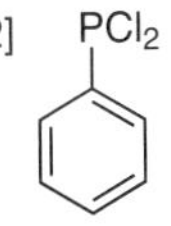
OS **IV**, 784

[3] (2,4,6-tri-t-Bu-C$_6$H$_2$)PCl$_2$
IS **27**, 236

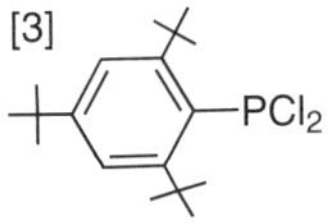

[4] (Me$_3$Si)$_3$CPCl$_2$
IS **27**, 239

[5] PhPBr$_2$
IS **9**, 73

[6] t-BuPCl$_2$
IS **14**, 4

[7] Me$_2$PCl
IS **15**, 191

[8] t-Bu$_2$PCl
IS **14**, 4

[9] Cl$_2$PCH$_2$PCl$_2$
IS **25**, 121

[10] Cl$_2$PCH$_2$CH$_2$PCl$_2$
IS **23**, 141

[11] CCl$_3$PCl$_2$
IS **12**, 290

[12] Me$_3$C–PF$_2$
IS **18**, 174

[13] ROPCl$_2$
R = Me, Et
IS **4**, 63

[14] ClCH$_2$CH$_2$–OPCl$_2$
IS **4**, 66

[15] (EtO)$_3$P
OS **IV**, 955

[16] (Me$_2$N)$_3$P
OS **V**, 602

[17] (PhO)$_2$PCl
IS **8**, 68

[18] Me$_2$NPCl$_2$
IS **10**, 147

[19] (2,4,6-tri-t-Bu-C$_6$H$_2$)PH$_2$
IS **27**, 237

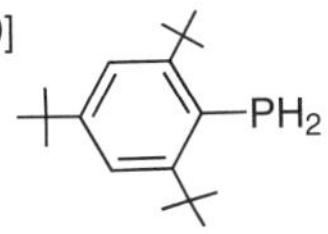

[20] (2,4,6-tri-t-Bu-C$_6$H$_2$)PHSiMe$_3$
IS **27**, 238

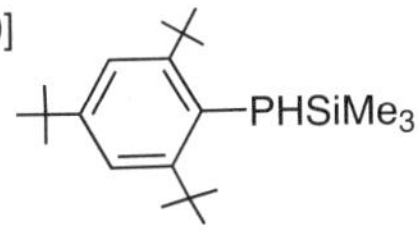

[21] Ph$_2$PH
IS **9**, 19; **16**, 161

[22] Me$_2$PH
IS **11**, 126, 157

[23] H$_2$PCH$_2$CH$_2$PH$_2$
IS **14**, 10

[24] MePH$_2$
IS **11**, 124

[25] MePPh$_2$
IS **16**, 128

[26] P[(CH$_2$)$_3$NMe$_2$]$_2$

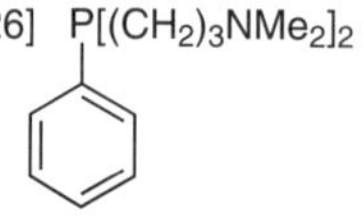

OS **VI**, 776

[27] Ph$_2$P$^-$ Li$^+$

IS **18**, 155

[28] PhEtP(CH$_2$)$_4$OH

IS **18**, 190

[29] S(−) with biphenyl bearing two –PPh$_2$ groups

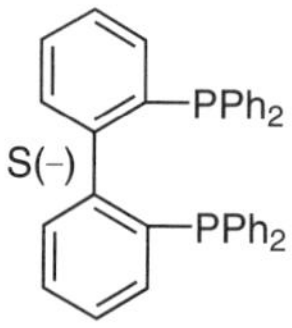

OS **VIII**, 57

[30] (S,S) Ph$_2$PCHCHPPh$_2$ with Me, Me substituents

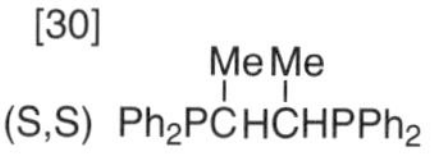

IS **31**, 134

[31] Ph$_2$PCH$_2$COPh

IS **31**, 140

[32] Me$_3$P

IS **18**, 153

[33] (Me$_3$Si)$_2$PLi

IS **27**, 248

[34] (Me$_3$Si)$_3$P

IS **27**, 243

[35] n-Bu$_3$P

IS **6**, 87

[36] PhCH$_2$PPh

IS **16**, 159

[37] 4-Br-C$_6$H$_4$-PPh$_2$

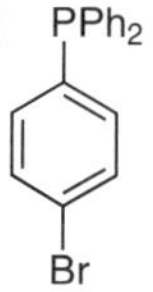

OS **IV**, 496

[38] n-BuPPh$_2$

IS **16**, 158

[39] Cy-PPh$_2$

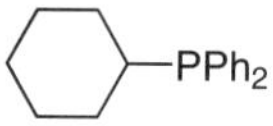

IS **16**, 159

[40] EtPh$_2$P

IS **16**, 158

[41] n-BuPh$_2$P

IS **16**, 158

[42] (PhCH$_2$)$_2$PPh

IS **18**, 171

[43] (Cy)$_2$PPh

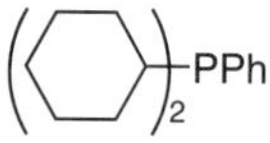

IS **18**, 171

[44] Bu$_2$PPh

IS **18**, 171

[45] Et$_2$PPh

IS **18**, 170

[46] MePPh$_2$

IS **16**, 157

[47] HO(CH$_2$)$_4$PEtPh

IS **18**, 189

[48] (PhCH$_2$)$_2$PPh

IS **18**, 171

[49] Me$_2$PCH$_2$CH$_2$CH$_2$NMe$_2$

IS **14**, 21

[50] Me$_2$PCH$_2$CH$_2$CH$_2$PMe$_2$

IS **14**, 17

[51] Me$_2$PCH$_2$CH$_2$PMe$_2$

IS **15**, 188

[52] [—C$_6$H$_2$(t-Bu)$_2$—CP(Cl or H or Me$_3$Si)$_2$—]$_3$

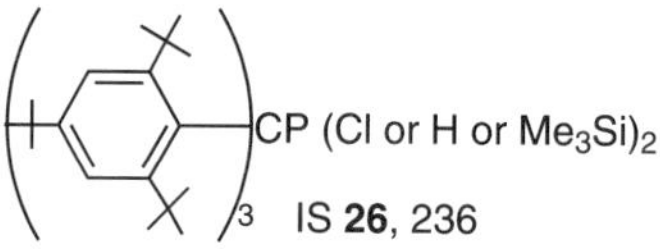

IS **26**, 236

[53] HP(CH$_2$CH$_2$CN)$_2$

OS **VI**, 932

[54] 1-phenyl-4-oxo-phosphinane

OS **VI**, 932

[55] Me$_2$P—PMe$_2$

IS **13**, 30; **14**, 15; **15**, 187

[56] (MeP)$_4$ ring: MeP—PMe / MeP—PMe

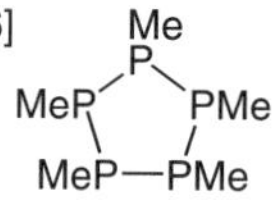

IS **25**, 4

[57] (t-BuP)$_3$ triangle

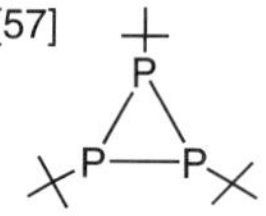

IS **25**, 2

CHAPTER 4

THE 4-COORDINATE PHOSPHINE OXIDES, OTHER CHALCOGENIDES, AND PHOSPHONIUM SALTS

4.A. PHOSPHINE OXIDES

4.A.1. General

Tertiary phosphine oxides are the most stable and least reactive of all families of phosphorus compounds. Many thousands are known, with both acyclic and cyclic carbon frameworks. With very little exception, they are crystalline, nontoxic solids, and except for occasional instances of hygroscopicity (which is severe with the lower molecular weight compounds), they are easily handled. Their thermal stability is also very high.[1a] Trimethylphosphine and triphenylphosphine oxides are stable up to 700°, while dimethylethylphosphine oxide decomposes above 330° with elimination of ethylene, an event probably characteristic of other alkyl derivatives. Phosphine oxides are highly polar, and some have surprisingly high solubility in water. As we shall see, the phosphoryl group generally resists chemical change (a major exception being its reducibility), and most of the chemistry known for phosphine oxides occurs at the carbon substituents. It is important in the beginning to recognize the total absence of similarity between the P=O bond and the C=O bond, even though drawn similarly. Thus the multitude of nucleophilic addition products known for carbonyl compounds are not found in phosphine oxide chemistry. The strength of the P=O bond is responsible for this very high stability; as discussed in chapter 2.D, a typical P=O bond has a bond dissociation energy of about 128 to 139 kcal/mol, and indeed the formation of this bond is a common driving force in several reactions of organophosphorus chemistry.

Primary and secondary phosphine oxides are also known, but in rather small numbers. In these structures, tautomeric equilibria with the −OH form must be considered, although these must make only a tiny contribution to the system, as discussed in chapter 2.G (Eqs. 4.1 and 4.2).

$$\underset{H}{\overset{O}{\underset{\|}{R-P-H}}} \rightleftharpoons R-\overset{OH}{\underset{..}{P}}-H \quad (4.1)$$

$$\underset{R}{\overset{O}{\underset{\|}{R-P-H}}} \rightleftharpoons R-\overset{OH}{\underset{..}{P}}-R \quad (4.2)$$

The presence of a P—H bond gives these oxides a chemistry that is quite different from that of the the unreactive tertiary phosphine oxides. For example, they are easily oxidized to acids and can be converted to anions with strong bases. The phosphoryl group, however, remains intact in the reactions of the oxides. Their thermal stability is much lower than that of the tertiaries, and they undergo a form of disproportionation, to be discussed.

The chemistry of tertiary phosphine oxides through about 1970 is covered completely in a review by Hayes.[1b] In 1992, an entire book was published on the subject of phosphine oxides, sulfides and selenides.[2] Secondary phosphine oxides are covered there as well as by Hamilton and Landis.[3] The references to phosphine oxides (compounds 1 to 4) in *Organic Syntheses* and *Inorganic Syntheses* are given in Appendix 4.1.

4.A.2. Synthesis of Phosphine Oxides

One of the most useful syntheses of tertiary phosphine oxides is to construct the desired carbon framework in the phosphine series, making use of the many versatile methods leading to that functionality, and then simply to oxidize the phosphine group to the oxide. We have already seen that a great variety of oxidizing agents can be used for this conversion. There are, however, two oxidizing agents with which problems might develop. We noted in chapter 3 that air is not recommended for oxidation of alkyl-substituted phosphines, because by-products with P—O—C bonds might be formed.[3,4] Another problem can occur with angle-constrained cyclic phosphines on oxidation with peroxy acids; O-insertion in the product phosphine oxide can occur, in a process reminiscent of the Baeyer–Villiger reaction of ketones. This is, in fact, a useful way to make certain cyclic phosphonates and phosphinates, but the scope of the reaction is quite limited. It is known to occur when the internal C—P—C angle of a cyclic system is contracted well below the normal tetrahedral angle of unstrained systems. Thus it occurs easily with the four-membered ring of the phosphetane oxides[5] (Eq. 4.3) but not with the five-membered phospholanes and never with noncyclic oxides.

Bridged cyclic structures, such as the well-known 7-phosphanorbornenes in which the C—P—C angle is around 83°, also smoothly undergo O-insertion.[6,7] The example of Eq. 4.4 is particularly instructive, because the O-insertion is seen to

4.A. PHOSPHINE OXIDES

[Structure: methylated phosphetane oxide] → ArCO₃H → [epoxidized product] (4.3)

be specific for the bridging P, while neither the five-membered ring component nor the double bonds are affected. The latter can be epoxidized only with excess peroxy acid. We will comment on the proposed mechanism of this insertion reaction in section 4.A.3.

[Bicyclic bis-phosphine oxide structure] → ArCO₃H → [epoxidized bicyclic product] (4.4)

The principle of approaching tertiary phosphine oxides by first synthesizing the phosphine for oxidation is an old one in organophosphorus chemistry, yet continues to be of contemporary importance. One example, of interest in employing modern techniques, is that used in the synthesis of *para*-aminotriphenylphosphine oxide and related compounds of possible value as second-order nonlinear optical materials.[8] Bromoaniline was protected by reacting the amino group with 1,2-bis(chlorodimethylsilyl)ethane to form the cyclic product **1**; the lithium reagent was prepared and reacted with diphenylphosphinous chloride. The protecting group was removed by hydrolysis, and the phosphine oxidized with hydrogen peroxide (Eq. 4.5).

[Reaction scheme: 4-bromoaniline + ClMe₂SiCH₂CH₂SiMe₂Cl → cyclic silyl-protected bromoaniline **1**; then 1. Mg, 2. Ph₂PCl → silyl-protected aryl-PPh₂ → H₂O → 4-aminophenyl-PPh₂ → H₂O₂ → 4-aminophenyl-P(=O)Ph₂] (4.5)

Another synthesis of the phosphine involved the preparation of *para*-bromotriphenylphosphine, reaction with azidomethyl phenyl sulfide to form the triazene **2**, and finally hydrolysis to the amine (Eq. 4.6).

98 THE 4-COORDINATE PHOSPHINE OXIDES

$$\text{(4.6)}$$

Phosphonic and phosphinic acids can be synthesized in great variety by methods to be discussed in chapter 5. Conversion to the acid chlorides or esters, followed by reaction with organometallic compounds, constitutes another widely used synthesis of tertiary phosphine oxides (Eq. 4.7). Phosphorus oxychloride similarly can be used to prepare symmetrical phosphine oxides.

$$R-\overset{O}{\underset{\|}{P}}(OH)_2 \longrightarrow R-\overset{O}{\underset{\|}{P}}Cl_2 \xrightarrow{R'MgX} R-\overset{O}{\underset{\|}{P}}R'_2 \quad (4.7)$$

However, the reactions with organometallics can be sluggish and do not always give high yields. The synthesis of trimethylphosphine oxide proceeds in only 52% yield from $POCl_3$ and of triphenylphosphine oxide in 65% yield.[1c] If they are available, the corresponding 3-coordinate chlorides are better partners with organometallics, and indeed it is a general observation that nucleophilic displacements are faster and cleaner with 3-coordinate halides and esters than with phosphoryl compounds. The desired carbon framework thus established, the phosphine is oxidized to the oxide. Nevertheless, many phosphine oxides have been produced directly from the phosphoryl chlorides or esters.

Another way to use tertiary phosphines as precursors of phosphine oxides is to quaternize them with alkyl halides and then decompose the quaternary ion by action of 20 to 40% NaOH. One of the P-substituents is eliminated as a hydrocarbon, and the tertiary phosphine oxide is formed. The method is illustrated in Eq. 4.8 with the use of an optically active quaternary ion, which preferentially loses the benzyl group with inversion of configuration.[9] The mechanism of this decomposition will be considered later in this chapter as a property of phosphonium ions.

$$\underset{\underset{CH_2Ph}{|}}{\overset{\overset{Ph}{|+}}{Me-P-Et}} \xrightarrow{HO^-} \underset{\underset{Ph}{|}}{\overset{\overset{O}{\|}}{Me-P-Et}} + PhC\bar{H}_2 \quad (4.8)$$

In this process, the P-substituent is eliminated as a carbanion, and in an unsymmetrical ion, as in the case of Eq. 4.8, it is the group that forms the most stable carbanion that is eliminated. For good results, one group should form a clearly more stable carbanion than the others to avoid mixture formation. The experimentally established priority of leaving groups is allyl > benzyl > phenyl > methyl > ethyl > higher alkyls.[1d] A useful procedure is to quaternize the readily available triphenylphosphine with an alkyl halide; a phenyl group is lost in the basic decomposition, and thus the product is an alkyldiphenylphosphine oxide. Many oxides have been prepared in this fashion.

Secondary phosphine oxides can also be used as precursors of tertiary derivatives by making use of the fact that the P–H bond exhibits acidity to very strong bases. A particularly useful reaction is that with aldehydes and ketones, which gives rise to α-hydroxyalkyl derivatives. The reaction can be conducted in neutral or basic media. Chloral is commonly used to form stable derivatives of the secondary oxides (Eq. 4.9).

$$Ph_2\overset{O}{\overset{\|}{P}}-H \;+\; Cl_3C\overset{O}{\overset{\|}{C}}-H \;\longrightarrow\; Ph_2\overset{O}{\overset{\|}{P}}-\overset{OH}{\overset{|}{C}H}CCl_3 \qquad (4.9)$$

Another useful reaction is Michael addition to α,β-unsaturated systems to give alkyl substituents with a functional group, as in the use of acrylonitrile (Eq. 4.10). Primary phosphine oxides have similar reactivity, but they are not as accessible or as stable and are not widely used in synthesis.

$$Ph_2\overset{O}{\overset{\|}{P}}-H \;+\; CH_2=CHCN \;\xrightarrow{NaOH}\; Ph_2\overset{O}{\overset{\|}{P}}-CH_2CH_2CN \qquad (4.10)$$

As was seen in the discussion of the properties of phosphines in chapter 3, halogen addition products are formed readily, and here phosphorus appears in the same oxidation state as that of a phosphine oxide. These adducts are rapidly hydrolyzed to phosphine oxides. Although there is little value in converting phosphines to their oxides in this two-step manner when one-step oxidations can be performed, there are several cases in which a product of a reaction is found to have the same structure as the phosphine–halogen adducts (properly known as halophosphonium halides, emphasizing their ionic character); these compounds are extremely useful as precursors on hydrolysis to phosphine oxides. An excellent example is the case of the cycloadducts from reaction of phosphonous dihalides with dienes (see section 3.A.3 and Appendix 4.1, compound 4); these adducts are halophosphonium ions and are easily hydrolyzed to give the corresponding phosphine oxides (Eq. 4.11). This will be seen to be an excellent method for entry into the 5-membered phospholene oxide system (chapter 8.B).

$$\diagup\!\!\!\diagdown \quad \xrightarrow{RPX_2} \quad \underset{R\;\;X}{\overset{+}{P}}\;X^- \quad \xrightarrow{H_2O} \quad \underset{R\;\;O}{P\!\!=\!\!O} \qquad (4.11)$$

Halophosphonium ions are formed as intermediates in other cyclization procedures, for example in the formation of the four-membered phosphetane ring from reaction of phosphonous dihalides with highly branched alkenes in the presence of AlCl$_3$ (see chapter 8.B). These adducts are hydrolyzed to give phosphetane oxides (Eq. 4.12).

$$Me_3CCH=CH_2 \xrightarrow{PhPCl_2 \atop AlCl_3} \underset{Ph}{\overset{Me}{\underset{Cl}{P^+}}}\hspace{-0.2em}Me_2 \ AlCl_4^- \xrightarrow{H_2O} \underset{Ph}{\overset{Me}{\underset{O}{P}}}\hspace{-0.2em}Me_2 \quad (4.12)$$

Finally, there is a version of the Michaelis–Arbusov reaction, introduced in chapter 2.K, that can lead to tertiary phosphine oxides. Thus esters of phosphinous acids, R$_2$P–OR', can either be rearranged by action of a catalytic amount of an alkyl halide corresponding to the alkoxy group or can be reacted with an equivalent of a halide containing a different alkyl substituent than that in the alkoxy group. In the illustration of Eq. 4.13, cyclic phosphinite is converted to the P-benzyl phosphine oxide, a procedure that avoids the sometimes troublesome reaction of a P-halide with benzyl Grignard reagent. Many other processes involving the Michaelis–Arbusov reaction have been described and are well covered by Bhattacharya and Roy.[12]

$$\text{(cyclic P-Br, Me)} \xrightarrow[\text{base}]{PhCH_2OH} \text{(cyclic P-OCH}_2\text{Ph, Me)} \longrightarrow \text{(cyclic P(=O)CH}_2\text{Ph, Me)} \quad (4.13)$$

There are, in addition to the major synthetic routes mentioned above, a number of more specialized processes for obtaining tertiary phosphine oxides; procedures will be found in the main references on phosphine oxide chemistry.[1b,2] Primary and secondary phosphine oxides, by comparison, are synthesized by a relatively small number of methods.

Secondary phosphine oxides, it will be recalled, are tautomeric forms of phosphinous acids, and a method of great generality is to hydrolyze phosphinous halides, esters, or amides so as to generate, presumably, the phosphinous acid, which then rearranges. Probably the best known of the series is diphenylphosphine oxide (Eq. 4.14), easily prepared from diphenylphosphinous chloride.

$$Ph_2P-Cl \xrightarrow{H_2O} [Ph_2P-OH] \longrightarrow Ph_2\overset{O}{\underset{\|}{P}}-H \quad (4.14)$$

Another valuable approach to the secondary oxides is the alkylation of dialkyl H-phosphonates (HP(O)(OR)$_2$) (dialkyl phosphites), with Grignard reagents. The P–H bond is weakly acidic and reacts first with one mole of the Grignard reagent to form a magnesium derivative; both alkoxy groups are then displaced by further reaction with the Grignard reagent. Hydrolysis restores the hydrogen to phosphorus, giving the secondary phosphine oxide. Dimethylphosphine oxide, a crystalline low-melting solid, was first prepared by this method[10] (Eq. 4.15), and it has been used to synthesize a number of other compounds.

$$(MeO)_2\overset{O}{\underset{\|}{P}}-H \xrightarrow{MeMgX} [(MeO)_2\overset{O}{\underset{\|}{P}}-MgX] \xrightarrow{MeMgX}$$

$$Me_2\overset{O}{\underset{\|}{P}}-MgX \xrightarrow[H^+]{H_2O} Me_2\overset{O}{\underset{\|}{P}}-H \qquad (4.15)$$

The most obvious way to prepare a secondary oxide is to oxidize a secondary phosphine; this can in fact be accomplished, but because the secondary phosphine oxide can be oxidized to the phosphinic acid, most chemical oxidizing agents must be avoided. In fact, the preferred procedure employs air as the oxidant. The reaction is conducted in an aprotic solvent at 50 to 70° simply by bubbling the gas through the solution. An example is the synthesis of bis(2-cyanoethyl)phosphine oxide from the phosphine[11] (Eq. 4.16).

$$(NCCH_2CH_2)_2P-H \xrightarrow{O_2} (NCCH_2CH_2)_2\overset{O}{\underset{\|}{P}}-H \qquad (4.16)$$

Primary phosphine oxides, seldom encountered in practical organophosphorus chemistry, were unknown until 1962, when two methods of synthesis were reported by the same group. These oxides may be prepared[12] by the careful direct oxidation of primary phosphines, employing hydrogen peroxide at 0°. n-Octylphosphine oxide so prepared (Eq. 4.17) is a crystalline solid.

$$n\text{-}C_8H_{17}PH_2 \xrightarrow{H_2O_2} n\text{-}C_8H_{17}-\overset{O}{\underset{\|}{P}}H_2 \qquad (4.17)$$

Phosphine itself is the source of phosphorus in the second process,[13] which is quite limited in scope. Phosphine reacts with ketones in strongly acidic media, typically concentrated HCl, presumably to give an adduct of structure 3. Rearrangement to the phosphine oxide follows (Eq. 4.18). The mechanism of this rearrangement is not clear, and several possibilities have been suggested.[14a] Note that the product necessarily has a secondary carbon attached to P. Ketones with moderate steric hindrance give the best results; with less hindered ketones, a further reaction of the primary phosphine oxide can occur, giving an adduct with a second mole of the ketone, as shown in Eq. 4.18.

$$(CH_3CH_2CH_2)_2C=O + PH_3 \xrightarrow{conc.\ HCl} \left[(CH_3CH_2CH_2)_2\overset{OH}{\underset{|}{C}}-PH_2\right] \longrightarrow$$
$$\mathbf{3}$$

$$(CH_3CH_2CH_2)_2CH-\overset{O}{\underset{\|}{P}}H_2 \xrightarrow{(CH_3CH_2CH_2)_2C=O} (CH_3CH_2CH_2)_2CH-\overset{O}{\underset{\|}{P}}H-\overset{OH}{\underset{|}{C}}(CH_2CH_2CH_3)_2$$

$$(4.18)$$

4.A.3. Properties of Phosphine Oxides

Tertiary phosphine oxides are much less reactive than are other classes of organophosphorus compounds but are certainly not devoid of chemical activity. Probably their most important chemical property is their reducibility to phosphines, shown in chapter 3.C to be a most valuable method of obtaining tertiary phosphines. It was noted there that the reduction is best performed with silicon hydrides or metallic hydrides; phosphine oxides are inert to the other common reducing systems, such as metal-acid combinations or catalytic hydrogenation. Phosphine oxides do react with metals such as sodium or potassium but generally undergo cleavage of a C–P bond along with deoxygenation. Bond cleavage, with release of the most stable carbanion if the phosphine oxide has different carbon substituents, can also be effected by heating in very strong aqueous base or on fusion with metallic hydroxides. Neither cleavage reaction has any practical value. There is, however, a cleavage reaction of monoacylphosphine oxides and diacylphosphine oxides that has attained some practical importance; on irradiation, the P–C bond is broken homolytically to form free radicals (Eq. 4.19), which can be used to initiate polymerization processes. A recent study of the cleavage reaction has been reported.[15]

$$\text{ArC(O)-PPh}_2 \xrightarrow{h\nu} \text{ArC(O)} \cdot + \text{Ph}_2\text{P(O)} \cdot \qquad (4.19)$$

The oxygen atom of phosphine oxides can be exchanged for sulfur in a process useful for forming tertiary phosphine sulfides (Eq. 4.20). The most common agent for this purpose is P_4S_{10}, although the Lawesson reagent (**4**, used to replace oxygen of carbonyl groups with sulfur) has also been employed.[16]

$$R_3P=O + \mathbf{4} \longrightarrow R_3P=S \qquad (4.20)$$

Another form of reactivity is found at the negative phosphoryl oxygen, which can act as an electron pair donor to protic species and strong electrophiles. Hydrogen bonding is pronounced with various proton donors, and the solubility of phosphine oxides in water can be quite high. Alcohols, amines, phenols, and carboxylic acids also form H-bonds to the oxygen, and to a lesser extent, bonding occurs to chloroform. In some cases, as with triphenylphosphine oxide and water, definite H-bonded complexes are formed. These hydrogen-bonding effects can have important consequences in spectroscopy; for example, the ^{31}P NMR chemical shift of phosphine oxides can vary considerably when measured in solvents of different H-donating ability. In water, the shift can be several ppm downfield of that in aprotic solvents, and the solvent should always be specified in reporting NMR or other properties.

Phosphine oxides can also act as weak bases and undergo full protonation to form salts with strong nonaqueous acids such as hydrogen bromide or concentrated sulfuric acid. The pK_a for dimethylphenylphosphine oxide, for example, is -2.4,[1e] and it is a stronger base than a ketone (pK_a -6.5 for acetophenone) but weaker than an amine oxide (pK_a 4.7 for trimethylamine oxide). The protonated species can be classed as a hydroxyphosphonium ion; the ion from triphenylphosphine oxide (Eq. 4.21) has nearly the same ^{31}P NMR shift (δ +58) as that of an alkoxyphosphonium ion (e.g., triphenylethoxyphosphonium tetrafluoroborate, δ +62). The latter species is itself formed as a result of the nucleophilicity of the phosphoryl group to potent electrophiles, in this case triethyloxonium tetrafluoroborate[17] (Eq. 4.22).

$$Ph_3P=O \xrightarrow{H_2SO_4} Ph_3\overset{+}{P}-OH \quad HSO_4^- \qquad (4.21)$$

$$R_3P=O + Et_3\overset{+}{O} \; BF_4^- \longrightarrow R_3\overset{+}{P}-OEt \; BF_4^- \qquad (4.22)$$

A more recent case is that of the silylation of certain cyclic phosphine oxides with various silylating agents.[18] Two examples are shown in Eqs. 4.23 and 4.24. In each, the initial siloxy phosphonium salt derives stabilization by loss of a proton to form a delocalized system. The structures were confirmed by ^{31}P and ^{13}C NMR, but the compounds were too hydrolytically unstable to allow isolation. In fact, it is a common property of all species made from attack on phosphoryl oxygen—be they protonated, alkylated, silylated, or other—that they return readily to the oxide condition.

$$\text{(4.23)}$$

$$\text{(4.24)}$$

With the recognition in recent years that the phosphoryl group is not as inactive as formerly believed[19a] have come some interesting applications as nucleophilic catalysts. For example, the catalytic effect is useful in the rearrangement of epoxides to carbonyl compounds[20] (Eq. 4.25). Another application is in the synthesis of carbodiimides from isocyanates (Eq. 4.26). Phospholene oxides are especially active in this reaction and have been prepared on a large scale to accomplish this

$$R'_3P=O \; +RCH-CHR \longrightarrow R'_3\overset{+}{P}-O-CHR-\underset{H}{\overset{\overset{\displaystyle O^-}{|}}{C}}-R \xrightarrow[-H^+]{-R'_3PO}$$

$$RCH=\underset{|}{\overset{\overset{\displaystyle O^-}{|}}{C}}-R \xrightarrow{H^+} RCH_2-\overset{\overset{\displaystyle O}{\|}}{C}-R \qquad (4.25)$$

industrially important process (Eq. 4.26).[21] The carbodiimides are used in the production of thermoplastic polyurethanes and polyester fibers. The reaction mechanism is described[22] as starting with the formation of an iminophosphorane (**6**) via a 5-coordinate intermediate (**5**); the iminophosphorane then forms another 5-coordinate intermediate (**7**) with a second molecule of the isocyanate. Collapse of **7** then produces the carbodiimide. The process provides an excellent example of the importance of 5-coordinate reaction intermediates in phosphine oxide chemistry, a point to which we will return shortly.

$$(4.26)$$

Tertiary phosphine oxides participate in various reactions with Lewis acids, and complexes are especially well known with $AlCl_3$ and BF_3. Of great importance also are the complexes formed with transition metal compounds. Many metals have been used in this work; the extensive literature has been reviewed.[2c] A practical application that has arisen from the metal-coordinating ability of phosphine oxides is their use as extractants for metal ions from aqueous media. Many metals have been extracted in this manner; a prominent case is that of uranium extraction. The most commonly used phosphine oxide for metal extraction is the water-insoluble tri-*n*-octylphosphine oxide, used in an inert solvent such as benzene, carbon tetrachloride, and cyclohexane; also useful are alkylenediphosphine dioxides, such as the chelating $Ph_2P(O)CH_2CH_2CH_2CH_2P(O)Ph_2$. Complexes of phosphine oxides have also been found to catalyze a number of reactions.

Only recently has it been discovered that the phosphorus in certain phosphine oxides can be converted to the 5-coordinate state of the oxyphosphoranes. This is by no means a general reaction and is best known in structures in which an intramolecular ring closure is involved in creating the oxyphosphorane. As an example,[23] the dibenzophosphole oxide **8**, with nearby hydroxymethyl groups that can interact with the phosphoryl group, gives a cyclic oxyphosphorane on being heated at 200°. Presumably, a 5-coordinate adduct (**9**) is formed first with one hydroxymethyl group, and water elimination with the second hydroxymethyl follows (Eq. 4.27).

$$\text{8} \xrightarrow{200°} \text{9} \xrightarrow{-H_2O} \quad (4.27)$$

Oxyphosphoranes can also be formed spontaneously, as in Eq. 4.28, a process again facilitated by the proximity of nucleophilic groups.[24]

$$\xrightarrow{H^+} \quad (4.28)$$

A case is also known[25] in which a phosphine oxide has been converted to a noncyclic oxyphosphorane, detected spectroscopically; in this case, the powerfully electron-attracting trifluoromethyl groups in **10** increase the electrophilicity of phosphorus, facilitating the addition process (Eq. 4.29). This new appreciation of the oxyphosphorane-forming ability of phosphine oxides may lead to fresh discoveries in this area.

$$(CF_3)_3P=O \; + \; (Me_3Si)_2O \longrightarrow (CF_3)_3P\begin{matrix}OSiMe_3\\OSiMe_3\end{matrix} \quad (4.29)$$
10

Five-coordinate intermediates or transition states probably are involved in the few reactions in which the carbon-phosphorus bond is broken in a phosphine oxide. Thus the attack of hydroxide ion to displace the most stable carbanion may proceed through a species such as **11** in Eq. 4.30.

It has also been postulated[6,7] that a 5-coordinate intermediate plays a role in the insertion by peroxides of oxygen into the C–P bond of certain strained cyclic phosphine oxides. The process is illustrated with a phosphetane oxide[7] (Eq. 4.31). It

$$Ph_2\overset{\underset{\displaystyle CH_2Ph}{|}}{P}=O \xrightarrow{HO^-} Ph-\overset{\underset{\displaystyle Ph}{|}}{\underset{}{P}}\overset{\underset{\displaystyle }{OH}}{(}CH_2Ph \longrightarrow$$

$$\qquad \qquad \qquad \qquad \qquad \overset{}{O^-}$$

$$\qquad \qquad \qquad \qquad \qquad \quad \mathbf{11}$$

$$Ph_2\overset{\underset{\displaystyle }{|}}{\overset{OH}{P}}=O + PhCH_2^- \longrightarrow Ph_2PO_2^- + PhCH_3 \qquad (4.30)$$

is proposed that the peroxy acid adds to the electrophilic phosphorus to form the species **12**, which then rearranges with migration of a ring carbon to the attached peroxy oxygen.

$$(4.31)$$

Equally as important as the direct reactions on the phosphoryl group are the reactions occurring at carbon substituents that are promoted by the strongly electron-withdrawing phosphoryl group. There is a form of conjugation that prevails in vinyl and alkynyl phosphine oxides that leads to significant polarization of the unsaturated group and to stabilization of a developing negative charge when nucleophiles add to such groups. The nature of the electronic interaction between P=O and C–C multiple bonds must be treated as uncertain at this time. For many years, it was expressed with the resonance hybrid shown in Eq. 4.32, in which a phosphorus d-orbital was assumed to accept the π-electrons of a C=C group.

$$RCH=CH-\overset{\underset{\displaystyle R}{|}}{\overset{\overset{\displaystyle O}{\|}}{P}}-R \quad \longleftrightarrow \quad R\overset{+}{C}H-CH=\overset{\underset{\displaystyle R}{|}}{\overset{\overset{\displaystyle O^-}{|}}{P}}-R \qquad (4.32)$$

The literature makes frequent use of this representation. With the view now taking hold that d-orbitals play no primary role in the bonding of the P=O group, the once-popular concept of d-orbital resonance in general seems best considered as simply an empirical aid that does not have a theoretical basis. We can avoid an explicit representation of the bonding by shifting the emphasis away from phosphorus and onto the unsaturated substituent. Thus in structure **13** we simply show the $R_2P=O$

group as a whole in causing the polarization and withdrawal of electron density from the double bond.

$$R^{\delta+}CH\text{----}CH\text{----}(POR_2)^{\delta-}$$
$$\mathbf{13}$$

The reality of the polarization is evident from several spectroscopic properties (discussed in more detail in chapter 7), which are exactly those so familiar from α,β-unsaturated carbonyl compounds: (1) in the infrared spectrum, the C=C stretching absorption shifts to lower frequency; (2) in the proton NMR spectrum, the protons on the β-carbon are strongly shifted downfield; and (3) in the ^{13}C NMR spectrum, the β-carbon is far downfield of the α-carbon.

The most important chemical consequence of the phosphoryl interaction is that the double bond is activated to cause the addition of nucleophilic reagents, making possible the occurrence of a number of reactions that resemble the classical Michael type. Many additions of this sort are known and have provided a great variety of C-functional phosphine oxides. An example is shown in Eq. 4.33, in which an amine acts as the nucleophile toward diphenylvinylphosphine oxide.[26]

$$\underset{\text{O}}{\overset{\text{O}}{\|}}Ph_2PCH=CH_2 + RNH_2 \longrightarrow Ph_2\overset{\text{O}}{\overset{\|}{P}}-\bar{C}H-CH_2NHR \xrightarrow{H^+} Ph_2\overset{\text{O}}{\overset{\|}{P}}-CH_2CH_2NHR \quad (4.33)$$

Among other agents adding to activated double or triple bonds are alkoxide ions, amines, thiolates, dialkyl copper lithiums, Grignard reagents, primary and secondary phosphines, secondary phosphine oxides, and dialkyl hydrogenphosphonates. The reactions with phosphorus anions have special synthetic significance as they provide access to valuable diphosphorus species. If in this reaction an unsymmetrical secondary phosphine oxide adds to the double bond, both phosphorus atoms in the product are chiral, and the product is obtained as a mixture of diastereoisomers. In one published example,[27] the S-stereoisomer of the vinylphosphine oxide was used in a reaction with an R,S secondary phosphine oxide, giving a mixture of optically active diastereomers.

$$t\text{-Bu}\overset{\text{O}}{\overset{\|}{\underset{\underset{\text{Ph}}{|}}{P_1}}}-H + H_2C=HC-\overset{\text{O}}{\overset{\|}{\underset{\underset{\text{Me}}{|}}{P_2}}}-Ph \xrightarrow[110°]{\text{toluene}} t\text{-Bu}\overset{\text{O}}{\overset{\|}{\underset{\underset{\text{Ph}}{|}}{P_1}}}-CH_2CH_2-\overset{\text{O}}{\overset{\|}{\underset{\underset{\text{Me}}{|}}{P_2}}}-Ph \quad (4.34)$$
$$(R,S) \qquad (S) \qquad\qquad (R_{P\text{-}1},S_{P\text{-}2}) \text{ and } (S_{P\text{-}1},S_{P\text{-}2})$$

Another valuable consequence of the conjugative effect is that the multiple bonds are activated to cycloadditions with both 1,3-dienes in the Diels–Alder reaction[28] (Eq. 4.35) and with 1,3-dipoles[29] (Eq. 4.36).

Many cycloadditions are known that produce quite valuable structures. Among the numerous examples are some in which the phosphorus atom is included in a ring, as in the 2-phospholene oxide or the phosphole oxide systems; the former has been used as a precursor of the phosphindoline ring system[30] (Eq. 4.37).

$$\text{//}\diagdown + H_2C=CH-\overset{\overset{O}{\|}}{\underset{Me}{P}}-Ph \longrightarrow \underset{\underset{Me}{O=PPh}}{\bigodot} \quad (4.35)$$

$$R_2\overset{\overset{O}{\|}}{P}-C\equiv CMe + PhN_3 \longrightarrow \underset{Me}{\underset{\underset{Ar}{N}}{\overset{R_2P}{\diagdown}}}\overset{O}{\underset{N}{\diagdown}}_N \quad (4.36)$$

$$\text{(diene-OAc)} + \underset{O}{\overset{}{\diagdown}}\underset{Ph}{P} \xrightarrow{150°} \left[\text{intermediate}\right] \longrightarrow \text{product} \quad (4.37)$$

Another chemical manifestation of the strong electron-withdrawing effect of the P=O group in phosphine oxides is found in aromatic substitution; the group ranks as a moderately strong meta director and deactivator, and this has been used in the synthesis of substituted triarylphosphine oxides, as in Eq. 4.38.

$$Ph_3P=O \xrightarrow{HNO_3} \left(\text{m-NO}_2\text{-C}_6\text{H}_4\text{-P=O}\right)_3 \quad (4.38)$$

There is a considerable literature on attempts to separate the inductive effect of the phosphinyl group ($R_2P(O)-$) on the benzene ring from the resonance effect.[1f] The inductive effect in a saturated system is considerable, as was evaluated by comparing the pK_a values for diphenylphosphinyl acetic acid ($Ph_2P(O)CH_2COOH$) to that of acetic acid. This led to a σ^* value (the polar substituent constant) for Ph_2PO of +0.68, revealing an inductive effect comparable in size to that of the carbomethoxy group. The resonance effect can be seen by comparing the σ-meta (about 0.4) and σ-para (about 0.5 to 0.6) that were determined for Ph_2PO as a substituent influencing pK_a values for benzoic acids and phenols. The size of the effect at the para position indicates the operation of moderate electron withdrawal by resonance, which can be pictured, in the case of the para-OH substituent, by structure **14**, for which, as before, no attempt is made to define the effect of the electron release to phosphorus in terms of bonding at the phosphoryl group.

4.A. PHOSPHINE OXIDES

[Structure **14**: resonance structures of para-hydroxyphenyl phosphine oxide]

Other studies have given similar results; reviews cover much of the fundamental work in this field.[31] NMR spectroscopy in various forms, including studies based on ^{1}H, ^{13}C, and ^{19}F NMR, has also been used in the effort to separate inductive from resonance effects; comparable conclusions to those of the pK$_a$ measurements were obtained. The NMR shifts at the para position of phenyl-substituted phosphine oxides can be considered as especially indicative of the resonance effect; for triphenylphosphine oxide, the ^{1}H shift[32a] is δ 7.5 and the ^{13}C shift is δ 131.8;[32b] both show some deshielding relative to benzene (δ 7.3 and 128.5, respectively). The deshielding is stronger than that at the meta position, which is not controlled by the resonance effect (^{1}H, δ 7.4; ^{13}C, δ 128.5).

One consequence of the electron withdrawal at the ortho and para positions is the activation of substituents to attack by nucleophiles. Although this effect has indeed been observed, it has received only limited study. An example[33] is seen in Eq. 4.39.

[Equation 4.39: p-Cl-C$_6$H$_4$-P(=O)Ph$_2$ + NaOMe → p-MeO-C$_6$H$_4$-P(=O)Ph$_2$] (4.39)

The strong electron-withdrawing effect of phosphinoyl substituents also has important consequences on attached saturated carbon and leads to pronounced resonance stabilization of carbanionic centers, as in **15**.

[Structure **15**: RCH⁻–P(=O)R$_2$ ⇌ R$^{δ-}$CH=(POR$_2$)$^{δ-}$]

This stabilization makes it possible for protons to be easily removed from the α-carbon of alkyl substituents by reaction with strong bases, such as metallic amides, alkyl lithiums, and Grignard reagents. These carbanions are then available for condensation with electrophilic centers of various types, including alkyl halides to lengthen the chain, carbon dioxide to form carboxylic acids, and acyl halides and esters to form ketone derivatives. Phosphinoyl carbanions also undergo Michael addition to α,β-unsaturated carbonyl compounds, as well as to vinylphosphine oxides[34] (Eq. 4.40), a process of value in the synthesis of diphosphine oxides (and hence of diphosphines by reduction).

$$\text{Ph}-\overset{\text{O}}{\underset{\text{Ar}}{\text{P}}}-\text{CH}_2\text{COOBu-t} + \text{Ph}-\overset{\text{O}}{\underset{\text{Ar}}{\text{P}}}-\text{CH}=\text{CH}_2 \xrightarrow{\text{NaH (catalyst)}} \text{Ph}-\overset{\text{O}}{\underset{\text{Ar}}{\text{P}}}-\underset{\text{COOBu-t}}{\text{CHCH}_2\text{CH}_2}-\overset{\text{O}}{\underset{\text{Ar}}{\text{P}}}\cdots\text{Ph} \longrightarrow$$

$$\xrightarrow[\Delta]{\text{Tos-OH}} \text{Ph}-\overset{\text{O}}{\underset{\text{Ar}}{\text{P}}}-\text{CH}_2\text{CH}_2\text{CH}_2-\overset{\text{O}}{\underset{\text{Ar}}{\text{P}}}\cdots\text{Ph} \xrightarrow[\substack{\text{Et}_2\text{N}(c\text{-C}_6\text{H}_{11}) \\ \text{(inversion)}}]{\text{HSiCl}_3} \text{Ar}-\overset{\cdot\cdot}{\underset{\text{Ph}}{\text{P}}}-\text{CH}_2\text{CH}_2\text{CH}_2-\overset{\cdot\cdot}{\underset{\text{Ph}}{\text{P}}}\cdots\text{Ar} \quad \text{(S,S)}$$

(4.40)

The reaction that has attracted the most attention is that with aldehydes and ketones (Eq. 4.41). The initial adducts, after protonation, would have the β-hydroxyalkyl structure that would be expected from analogy to the addition of conventional carbonyl-stabilized carbanions to carbonyl groups, as in the aldol condensation. Some cases are known with aromatic aldehydes in which the initial adduct does indeed decompose to form the expected α,β-unsaturated system so familiar from aldol condensations.

$$\text{R}_2\overset{\text{O}}{\text{P}}-\bar{\text{C}}\text{H}_2 + \text{RCHO} \longrightarrow \text{R}_2\overset{\text{O}}{\text{P}}-\text{CH}_2-\overset{\bar{\text{O}}}{\text{CHR}} \xrightarrow{\text{H}^+} \text{R}_2\overset{\text{O}}{\text{P}}-\text{CH}_2-\overset{\text{OH}}{\text{CHR}} \xrightarrow{-\text{H}_2\text{O}} \text{R}_2\overset{\text{O}}{\text{P}}-\text{CH}=\text{CHR} \quad (4.41)$$

However, it is more common, especially with aliphatic aldehydes and ketones, for the initial oxyanion product to follow a different course of elimination, involving cleavage of the carbon-phosphorus bond with loss of a phosphinate anion (Eq. 4.42). It is the other product that is more interesting, because it is an alkene. We have here an excellent method of alkene synthesis; known as the Horner reaction, this stands as a rival to the widely used Wittig and Wadsworth–Emmons olefination procedures that employ phosphonium ions and phosphonates as the starting materials, respectively.

$$\underset{\underset{\text{R}}{\text{CH}}}{\overset{\text{R}_2\text{P}}{\underset{\text{H}_2\text{C}}{\diagup}}}\overset{\text{O}}{\underset{\diagdown}{\diagdown}}\text{O}^- \longrightarrow \text{R}_2\overset{\text{O}}{\text{P}}-\text{O}^- + \text{CH}_2=\text{CHR} \quad (4.42)$$

The extensive chemistry that has developed since the initial development of the olefination process by Horner[35] in 1966 has been recently reviewed.[2a] Many applications of the method have been reported, and much is known about its stereochemistry and regiochemistry. In practice, the usual Horner reagent is an alkyldiphenylphosphine oxide, so that the leaving group is the diphenylphosphinate ion. The elimination takes place from the original basic medium with heat. In Eq. 4.43, the mechanism of the olefination is considered with a simple case that avoids the important question of E vs. Z olefin formation. In this mechanism, we must deal with the possibility that the initial anion **16** may be in equilibrium with a ring-closed 5-coordinate structure (**17**), an oxaphosphetane as has been demonstrated for the Wittig reaction. The cleavage of the P—C bond, with transfer of oxygen from carbon to phosphorus, can be envisioned as occurring with either form. We thus have another probable instance of the role of a 5-coordinate intermediate being formed from a phosphine oxide, followed by P—C cleavage.

$$Ph_2P(O)-CH_2-CHR-O^- \rightleftharpoons [\text{17}] \longrightarrow Ph_2P=O\ + CH_2=CHR \quad (4.43)$$

16   **17**

One of the more interesting reactions of phosphine oxides to be discovered in recent years is that of the migration of the diphenylphosphinoyl group, with its pair of electrons, to a carbocationic center at the β-carbon.[36] This reaction occurs readily when the carbocation is generated by acid treatment of a β-hydroxyalkyl phosphine oxide, or solvolysis of a β-tosylate with TsOH or trifluoroacetic acid. An excellent example, illustrating the greater migratory aptitude of the phosphinoyl group relative to that of methyl, is seen in Eq. 4.44. Here the solvolysis of the tosylate produces a primary carbocation (**18**), which can stabilize itself by the migration of methyl or phosphinoyl to form a tertiary carbocation. It is, in fact, the phosphinoyl group that migrates to give ion **19**, which loses a proton to give the alkene **20**.

$$Ph_2P(O)-CMe_2-CH_2-OTos \longrightarrow Ph_2P(O)-CMe_2-CH_2^+ \longrightarrow$$

18

$$Me_2C^+-CH_2-PPh_2(O) \xrightarrow{-H^+} Me_2C=CH-PPh_2(O) \quad (4.44)$$

19   **20**

Many examples of this rearrangement in other structural types have been studied, and other migratory aptitude situations have been considered; a review of this work is available.[36] In a more refined view of the mechanism, it is proposed that a planar transition state, such as **21**, is involved (Eq. 4.45). This is supported by the retention of the configuration at the phosphorus when a single stereoisomer of the phosphine

$$Ph(Me)P(O)-CMe_2-CHMe-OTos \longrightarrow [\text{21}] \xrightarrow{-OTs^-} Ph(Me)P(O)-CMe(H)-CMe_2^+ \xrightarrow{-H^+} Ph(Me)P(O)-CMe=CMe_2$$

21

$$(4.45)$$

oxide is used and inversion at the migration terminus. Very little epimerization occurs in the rearrangement. The Ph$_2$P(O) group can also exert an activating effect at a β-substituent; the nature of this effect is still under study.[37]

The chemistry of secondary[3a] and primary[2a] phosphine oxides is far less extensive, and is dominated by the behavior of the P—H bond. Because of the greater difficulty in their synthesis, the primary oxides have received very little attention, and this discussion is directed to the more common secondary oxides. Even here, scope is lacking, and most of our knowledge has been derived from studies with diphenylphosphine oxide. The secondary oxides are weak acids; the proton can be removed with the usual strong bases (metals, metallic amides, organometallics), and the resulting anion is available for many valuable reactions. We must first note that the anion may be expressed by a resonance hybrid (Eq. 4.46), which places electron density on both phosphorus and oxygen.

$$R_2\overset{\overset{O}{\|}}{P}-H \longrightarrow R_2\overset{\overset{O}{\|}}{P}:^- \longleftrightarrow R_2\overset{\overset{O^-}{|}}{P}: \qquad (4.46)$$

It is generally the phosphorus atom that is the site of reaction with electrophiles, thus preserving the phosphoryl group in the product, but cases are known in which oxygen is attacked to give a 3-coordinate product (vide infra). The most common reactions of the anion are those of addition to an unsaturated system, in typical Michael fashion. A variety of derivatives of α,β-unsaturated acids have been used in such condensations, e.g., with ethyl acrylate[38] (Eq. 4.47).

$$(PhCH_2)_2\overset{\overset{O}{\|}}{P}-H + CH_2=CHCOOEt \longrightarrow (PhCH_2)_2\overset{\overset{O}{\|}}{P}-CH_2CH_2COOEt \qquad (4.47)$$

The acrylate condensation products are of special importance when one or two P-benzyl groups are present, for here the phosphoryl group activates the benzylic methylene group, making it subject to an intramolecular Claisen-like condensation with the ester group to form the five-membered phospholane ring system[39] (Eq. 4.48). Michael addition also occurs with vinylphosphine oxides, already shown in Eq. 4.33.

$$\begin{array}{c}\text{PhCH}_2 \\ \text{PhCH}_2\end{array}\!\!\overset{\overset{O}{\|}}{P}\!-\!CH_2\!-\!CH_2\!-\!\underset{O}{\overset{|}{C}}\!\!-\!OEt \xrightarrow{\text{base}} PhCH_2\overset{\overset{O}{\|}}{P}\!-\!CH_2\!-\!CH_2 \quad \longrightarrow \quad \text{(phospholane)} \qquad (4.48)$$

Secondary phosphine oxides also form adducts with aldehydes, ketones, nitriles, C=N, and C=S bonds. These additions occur quite readily, and a basic catalyst is not always required. Chloral is especially reactive and is used to make crystalline derivatives, as already noted in Eq. 4.9.

An example of the attack of an electrophile on oxygen rather than on phosphorus is afforded by the reaction of diarylphosphine oxides with PCl$_3$ or PBr$_3$. The nucleophilic oxygen displaces a halogen atom to form a species such as **22**, which is unstable in the presence of the displaced halide ion and suffers P-O bond cleavage. The final product is a phosphinous halide[40] (Eq. 4.49). The reaction is conducted at room temperature in high yield. The secondary phosphine oxides are frequently made by hydrolysis of phosphinous chlorides; when PBr_3 is the reactant, the process amounts to an exchange of chlorine for bromine.

$$Ph_2P-Cl \xrightarrow{H_2O} Ph_2\overset{O}{\underset{\|}{P}}-H \xrightarrow{PBr_3} \underset{22}{Ph_2\overset{H}{\underset{+}{P}}-O-PBr_2 \cdots Br^-}$$

$$Ph_2\overset{H}{\underset{+}{P}}-Br + {}^-OPBr_2 \xrightarrow{-H^+} Ph_2P-Br \qquad (4.49)$$

Another example of 3-coordinate product formation may be found in the reaction of diphenylphosphine oxide with acetic anhydride.[41] The initial product **23** is acetyldiphenylphosphine oxide, which then isomerizes by migration of the acetyl group to the phosphoryl oxygen until an equilibrium position is reached (Eq. 4.50). This is a rare example of the conversion of 4-coordinate to 3-coordinate phosphorus; it is the reverse of the Michaelis–Arbusov process, although the mechansm would be quite different and may well involve free radicals (as suggested by Eq. 4.19). The acetoxyphosphine combines with the oxide **23** to form **24** as the final product. The acetoxyphosphine can also be prepared by the reaction of acetyl chloride with MeO-PPh_2, and the same equilibration with **23** is observed.

$$Ph_2\overset{O}{\underset{\|}{P}}-H + (CH_3CO)_2O \xrightarrow{55°} \underset{23}{Ph_2\overset{O}{\underset{\|}{P}}-\overset{O}{\underset{\|}{C}}CH_3} \rightleftharpoons Ph_2\overset{\cdot\cdot}{P}-O-\overset{O}{\underset{\|}{C}}CH_3 \xrightarrow{23} \underset{24}{\overset{O\quad O}{\underset{\underset{Me}{\overset{|}{C}}}{\underset{\|}{Ph_2P}\diagdown\underset{C}{\diagup}PPh_2}}\diagdown Me}$$

(4.50)

4.B. PHOSPHINE SULFIDES, SELENIDES, AND TELLURIDES

4.B.1. General

The other members of the Chalcogen family, sulfur, selenium, and tellurium, all form bonds to phosphorus. The number of known compounds with these bonds is, however, far less than that with the P=O bond and decreases in the order S > Se > Te. Only the tertiary derivatives of the series have received much attention, and much less is known about primary and secondary phosphine chalcogenides. The secondary phosphine sulfide structure, $R_2P(S)H$, is the best known of these and is of interest in pointing out that the tautomeric behavior that greatly favors the form $R_2P(O)H$ over R_2P-OH is at work in the corresponding sulfur compounds as well. The main sources of information on the phosphine chalcogenides are again the review by Edmundson[2a] in the Hartley series and also by Maier[3b] in the Kosolapoff–Maier series.

Bond energy data are incomplete for the family $R_3P=X$, but some comparisons to the better known oxides can nevertheless be made. Gilheany[2d] has provided a summary of some of these data. The bond-dissociation energy of the sulfides, while quite high, is significantly less than that of the oxides: $Pr_3P=O$, 139 kcal/mol; $Pr_3P=S$, 91.6 kcal/mol. Selenides are stable in most chemical operations, but tellurides show instability that interferes with their purification. Thus small amounts of elemental tellurium are released on recrystallization from hot solvents, and longer heating periods can produce a mirror on the walls.[42] The P=X bonds have large dipole moments: for the series $Ph_3P=X$, O is 4.51D, S is 4.88, and Se is 5.17. It has been noted, however,[2d] that when charge separation across the P–X bond is considered, the polarity order becomes O > S ~ Se, which accords with practical observations made while working with the compounds. Just as true for the oxides, the higher chalcogenides are generally crystalline solids. Because hydrogen bonding will be of importance only to the oxides, the S, Se, and Te members of the family in general lack the water solubility so characteristic of the lower phosphine oxides. All members of the series retain similar tetrahedral geometry, as can be seen from the C–P–C bond angles[2d]: for $Me_3P=X$, O is 104.1°, S is 104.5°, and Se is 104.8°; for t-$Bu_3P=XO$ is 112.9° and Te is 110.2°. As expected, the P=X bond lengths increase as the atomic size increases: for $Me_3P=X$, O is 1.476 Å, S is 1.940 Å, and Se is 2.091 Å. The same difficulties seen in arriving at a satisfactory theoretical description of the nature of the multiple bonding in the phosphine oxides prevail in the higher members of the series. Although Gilheany states that the phosphoryl bond has a bond order greater than two, a value less than two seems to apply for the S and Se compounds, implying a reduced level of backbonding to P. Again the backbonding is unlikely to involve the participation of d-orbitals on P. As for the oxides, the best working formula to express the bonding is simply the familiar $R_3P=X$ representation, but the difference in polarity and bond strength from the oxides must be kept in mind. Several practical applications have been reported for phosphine sulfides, and a summary of some of these is available.[3b]

4.B.2. Synthesis

The most valuable method for the synthesis of the tertiary sulfides, selenides, and tellurides is the most obvious one: the direct reaction of the elements with tertiary phosphines. With S_8, the reaction is generally rapid and sometimes exothermic. It is usually conducted in a solvent (ether, arenes, alcohol, etc.) and, if incomplete, can be finished at reflux. A wide variety of phosphines, including those with cyclic structure and with multiple bonds and some other functionalities can be used. These reactions proceed with stereochemical retention. Selenium also reacts well with phosphines, although it may require more forcing conditions. Tellurium is still less reactive, and such reactions are conducted in refluxing toluene.

Another way to convert phosphines to the phosphine sulfides involves two steps, first the addition of halogen to form the species R_3PX_2, and then the displacement of halogen with hydrogen sulfide (Eq. 4.51). It is interesting that the latter reaction does not require the presence of a base and is indeed hindered by its presence; from Ph_3PBr_2 the sulfide was obtained in 82% yield without a base, but the yield was reduced to 53% when triethylamine was present.[43]

$$Ph_3P \ + \ Br_2 \longrightarrow Ph_3PBr_2 \xrightarrow{H_2S} Ph_3P=S \qquad (4.51)$$

Many types of organic and inorganic sulfur compounds can act as sulfur-transfer agents to phosphines. The list is long, and the older work is covered by reviews.[2a,3b] To illustrate the variety, we mention HSCN, ammonium polysulfides, episulfides, and various organic disulfides as useful reagents. The same type of reaction is useful in synthesizing phosphine selenides; here the choice of reagents is limited, and the most commonly used is potassium selenocyanate in acetonitrile (Eq. 4.52).

$$Ph_3P \ + \ KSeCN \longrightarrow Ph_3P=Se \qquad (4.52)$$

Another widely used method is that of the displacement of halogen from 4-coordinate thiophosphoryl halides. Thus the reaction of $PSCl_3$ with alkyl lithium or aluminum reagents is used to synthesize tertiary phosphine sulfides. As noted in chapter 3.C, simple alkyl Grignard reagents tend to give a biphosphine disulfide from a coupling reaction at phosphorus, although aryl and a variety of higher molecular weight and unsaturated alkyl Grignards give good yields of the tertiary phosphine sulfides. The product from methyl Grignard reagent is of some interest as an easily obtained biphosphine disulfide derivative (Eq. 4.53); the reaction appears in *Organic Syntheses*.[44]

$$Cl_3P=S \ + \ MeMgBr \longrightarrow \underset{Me_2P-PMe_2}{\overset{S \ \ S}{\overset{\| \ \ \|}{}}} \qquad (4.53)$$

The various factors influencing tertiary phosphine sulfide vs. biphosphine formation are discussed by Maier[3b] in some detail. The same problem can appear in the reaction of phosphonothioic dichlorides ($RP(S)Cl_2$) and phosphinothioic chlorides ($R_2P(S)Cl$) with organometallics. The preferred organometallic reagent to use for simple chlorine displacement is based on lithium; thus methyllithium reacts with $PSCl_3$ at $-30°$ to give a 49% yield of $Me_3P=S$.

Phosphine oxides can be used as precursors of phosphine sulfides, but surprisingly few applications of this reaction have been reported. Three reagents are available for this purpose. The most commonly used is phosphorus "pentasulfide", actually P_4S_{10}; the reaction is generally conducted in a solvent such as methylene chloride or toluene and has been especially used in the synthesis of heterocyclic phosphine sulfides, as seen in Eq. 4.54. Lawesson's reagent (**4**) has also been used in the heterocyclic field, as in the replacement of both oxygen atoms of the phosphole oxide dimer **25**[45] (Eq. 4.55).

$$\text{(structure: phospholene Ph, P=O)} \xrightarrow{P_4S_{10}} \text{(structure: phospholene Ph, P=S)} \quad (4.54)$$

$$\underset{\mathbf{25}}{\text{(bicyclic structure with two P=O groups, Me substituents)}} \xrightarrow{4} \text{(bicyclic structure with two P=S groups, Me substituents)} \quad (4.55)$$

In this case, retention of configuration occurred at both reaction sites. Retention has also been observed in reactions with P_4S_{10} and B_2S_3.[46]

A process that is useful for preparing unsymmetrical tertiary phosphine sulfides also is of interest in answering a question: Does the Michaelis–Arbusov rearrangement apply also to sulfur derivatives? The answer is positive. In this process, phosphinous halides are converted to phosphinothioites with mercaptans, and these then are rearranged to the phosphine sulfides The rearrangement occurs thermally and does not require a catalyst; it is most efficient when the S-substituent is allyl, benzyl or phenyl, as seen in the synthesis of benzyldiphenylphosphine sulfide[47] (Eq. 4.56).

$$Ph_2PCl \xrightarrow{PhCH_2SH} Ph_2P-SCH_2Ph \xrightarrow{\Delta} Ph_2\overset{S}{\underset{\|}{P}}-CH_2Ph \quad (4.56)$$

The Arbusov rearrangement also occurs with phosphinoselenoites but has been little studied. The reaction of Ph_2PCl with NaSePh directly gives the rearranged product Ph_3PSe, apparently passing through the phosphinoselenoite as an intermediate.[48]

There are a variety of other methods that can be used to form the phosphine chalcogenides, but they have been employed less frequently. These include the alkylation of P_4S_{10} with Grignard reagents, the Friedel–Crafts reaction of aromatic hydrocarbons with $PSCl_3$ and $AlCl_3$ (a process of little value for $POCl_3$), and the addition of secondary phosphine sulfides to carbonyl groups and to activated alkenes.

4.B.3. Properties

The phosphine chalcogenides respond to both oxidizing and reducing agents. With the former, phosphine oxides are the products, whereas reduction converts the phosphorus back to the phosphine stage. Many reagents have been used for both types of reactions. The sulfides are much easier reduced than are the oxides, and in addition to the metallic hydrides and silane derivatives required for the oxides, their reduction may be effected with agents such as sodium in naphthalene, metallic iron,

Raney nickel, and even tributylphosphine in the case of triaryl derivatives. Oxidation of the selenides and tellurides is so facile that even air can effect the reaction, and they are best preserved in an inert atmosphere.

The sulfides and selenides are known to act as nucleophiles and form addition products with the halogens and complexes with various metallic compounds. The sulfides and selenides are known to react with alkylating agents, including alkyl halides (Eq. 4.57), trialkyloxonium salts, and dimethyl sulfate, to form a type of phosphonium salt (**26**).

$$R_3P=S + R'X \longrightarrow R_3\overset{+}{P}SR'\ \bar{X} \qquad (4.57)$$
$$\mathbf{26}$$

An interesting example of this reaction occurs with a 3-bromopropylphosphine sulfide; intramolecular alkylation gives a cyclic salt[49] (Eq. 4.58). The solid product has the cyclic structure, but in solution some reversal of the reaction occurs.

$$\underset{Ph_2\overset{\overset{S}{\|}}{P}-CH_2CH_2CH_2Br}{} \xrightarrow{\Delta} \underset{Ph\ \ Ph}{\overset{+}{P}}\!\!\diagdown\!\!S\ \ Br^- \qquad (4.58)$$

Phosphorus in the chalcogenides is electron withdrawing when attached to unsaturated functions, although little study has been made of this effect. In electrophilic substitution of the phenyl group in tertiary phosphine sulfides, incoming groups are oriented to the meta position, and Hammett σ-constants are positive. The sensitivity of sulfur to oxidizing or electrophilic agents makes such substitutions have little versatility.

4.C. QUATERNARY PHOSPHONIUM SALTS

4.C.1. General

Quaternary phosphonium salts, which have tetrahedral geometry, are almost invariably stable crystalline solids; they are frequently made as derivatives for analysis of the air-sensitive tertiary phosphines and are known in the thousands. A more important contemporary use for them is as precursors of ylides, the reagents developed by Georg Wittig for the synthesis of alkenes (Eq. 4.59).

$$Ph_3\overset{+}{P}CH_2R \xrightarrow{base} Ph_3\overset{+}{P}\overset{-}{C}HR \xrightarrow{R_2CO} Ph_3P=O + R_2C=CHR \qquad (4.59)$$

The Wittig reaction will be considered more fully in chapter 11.F, but we should recognize even here the profound impact it has had on organophosphorus chemistry.

It brought the attention of many organic chemists to this field, and it was one of the first cases of the use of phosphorus reagents in organic synthesis, to be followed by many others. The extensive literature of quaternary phosphonium salts has been reviewed in the Kosolapoff–Maier series[50] and in the Hartley series.[51a]

4.C.2. Synthesis

Quaternary phosphonium salts are primarily synthesized by the direct reaction of tertiary phosphines with alkyl halides or other alkylating agents (Eq. 4.60).

$$R_3P + R'X \longrightarrow R_3\overset{+}{P}R' \ \bar{X} \qquad (4.60)$$

Much is known about structure-reactivity relations for this important reaction, which was presented in chapter 3 as a major reaction of phosphines. It is, of course, to be viewed simply as nucleophilic substitution of an alkyl halide (or other), and it follows the expected second-order kinetics. The reaction rates are significantly greater for phosphines than for the corresponding amines, possibly because steric requirements in passing from the pyramidal to the tetrahedral state are smaller for the larger phosphorus atom, which also forms longer bonds. Phosphorus is also more polarizable than is nitrogen.[14b] All types of phosphines, including phosphine itself, can be alkylated, but the rates diminish with the degree of substitution, so that alkylations of primary phosphines and phosphine are quite slow. Another point of difference from amines is that aryl substitution does not greatly diminish the nucleophilicity of a tertiary phosphine, and indeed one of the most common phosphines used in quaternization is triphenylphosphine. Triarylamines, it will be remembered, are quite difficult to protonate or alkylate. In part this difference can be explained by the fact that electron release to the ring by resonance is weak in phosphines but quite pronounced in amines. Phosphines can even be arylated with aryl halides, and this is an effective way to synthesize tetra-aryl phosphonium salts.[52] The reaction with an aryl halide is performed in the presence of a Friedel–Crafts catalyst; nickel and cobalt halides, as well as $AlCl_3$, are effective in this process (Eq. 4.61).

$$Ph_3P + PhBr \xrightarrow{NiCl_2} Ph_4\overset{+}{P} \ \bar{Br} \qquad (4.61)$$

Two methods employing free radical chemistry are also useful in the arylation of tertiary phosphines. One of these is accomplished photochemically, by ultraviolet irradiation of a solution of an aryl iodide and a tertiary phosphine[53] (Eq. 4.62).

$$Ph_3P + PhI \xrightarrow[C_6H_5Br]{h\nu} Ph_4\overset{+}{P} \ \bar{I} \qquad (4.62)$$

Aryl radicals can also be produced in the decomposition of aryl azoacetates and will attack tertiary phosphines present in the medium. The azoacetates are generated by adding acetate ion to diazonium ions formed in water in the usual way and extracting them into an organic solvent containing the phosphine; the azoacetate decomposes with release of nitrogen and formation of acetate and aryl radicals, which then add to the phosphine to produce the phosphoranyl radical (**27**). The process is outlined in Eq. 4.63.

$$ArN_2^+ + CH_3COO^- \longrightarrow ArN=N-OCOCH_3 \xrightarrow{-N_2}$$

$$Ar\cdot + CH_3COO\cdot \xrightarrow{R_3P} R_3\overset{\cdot}{P}Ar \xrightarrow{CH_3COO\cdot} R_3\overset{+}{P}Ar \; ^-Ac \qquad (4.63)$$
$$\mathbf{27}$$

Quaternary phosphonium ions are formed in several other reactions that are rather specialized and not of wide use. We mention some of these only in outline form (Eqs. 64–67).

1. The reaction of halophosphonium ions with Grignard reagents (Eq. 4.64).

$$R_3\overset{+}{P}-X + R'MgX \longrightarrow R_3\overset{+}{P}R' \qquad (4.64)$$

2. Ring opening of epoxides in acid media (Eq. 4.65).

$$R_3P + \overset{\triangle}{O} \xrightarrow{HCl} R_3\overset{+}{P}CH_2CH_2OH \; \overline{Cl} \qquad (4.65)$$

3. Conjugate addition of tertiary phosphines to α,β-unsaturated carbonyl compounds in acid media (Eq. 4.66).

$$R_3P + RCH=CHCO_2H \xrightarrow{HBr} R_3\overset{+}{P}CHRCH_2CO_2H \; \overline{Br} \qquad (4.66)$$

4. Condensation of phosphines with carbonyl compounds to form α-hydroxyalkyl phosphonium ion; a special application and one of commercial importance is the reaction of phosphine itself with formaldehyde in HCl, producing the valuable flame-retarding agent tetrahydroxymethylphosphonium chloride (THPC).

$$H_2CO + PH_3 \xrightarrow{HCl} (HOCH_2)_4\overset{+}{P} \; \overline{Cl} \qquad (4.67)$$

Some other specialized processes are useful in the synthesis of heterocyclic phosphonium salts. Thus there is an adaptation of the McCormack reaction that gives the 5-membered phospholenium ions. This is accomplished by using phosphinous halides, rather than phosphonous halides, in the cycloaddition with dienes[54] (Eq. 4.68).

$$\text{butadiene} + Me_2PCl \longrightarrow \text{[cyclic phospholenium]}^+ Cl^- \qquad (4.68)$$

An adaptation of phosphine alkylation can be used to form cyclic phosphonium salts. Here a tertiary monoarylphosphine (**28**) is synthesized in which the aryl ring contains a β-methoxyalkyl side chain in the ortho position; the methoxy group is cleaved with HBr, and cyclization follows[55] (Eq. 4.69).

$$\text{(Eq. 4.69)}$$

The monocyclic phospholanium ion can be obtained in the form of the 2,5-dihydroxy derivative by reacting a solution of succinaldehyde and a secondary phosphine in MeOH-THF with concentrated HCl at room temperature, followed by a short reflux period. The crystalline phosphonium salt is obtained on concentration[56] (Eq. 4.70).

$$\text{(Eq. 4.70)}$$

Other special procedures for preparing heterocyclic phosphonium ions may be found in the monograph by Mann on heterocyclic phosphorus compounds.[57]

4.C.3. Reactions

The reactions of quaternary phosphonium ions are not numerous but include some very important processes. There is little similarity to the reactions of quaternary ammonium salts. Phosphorus has two points of vulnerability that are lacking in nitrogen, and these differences dominate its reactivity: the ability to achieve pentacovalency and the ability of phosphorus to stabilize a negative charge on an α-carbon.

The most notable reaction of phosphonium ions is that with bases, of many types, whereupon an α-proton is abstracted and an alkylenephosphorane (or phosphine alkylene), an ylide, is formed. The ylides, of course, are employed in the Wittig olefin synthesis (Eq. 4.59) and have been made in great variety, some of considerable framework complexity in the synthesis of natural products.

With tetraarylphosphonium ions, the reaction with alkyllithiums cannot lead to ylides as it does when an alkyl substituent is present but instead gives an addition product with 5-coordinate phosphorus. The pentaarylphosphoranes (Eq. 4.71) were formed for the first time by this process,[58] and were of great importance in opening up the field of 5-coordinate phosphorus.

$$Ph_4\overset{+}{P} + PhLi \longrightarrow Ph_5P \qquad (4.71)$$

There are several methods that can effect the cleavage of the C—P bond. Simply heating an alkyl phosphonium salt at about 200° will cause dealkylation, but more useful methods include the following.

1. The reaction with concentrated NaOH, which gives a phosphine oxide and an alkane. The mechanism of this reaction has been carefully studied and well described.[14c] It follows third-order kinetics (first in the salt, second in OH⁻), and when optically active phosphonium salts are used, the reaction proceeds with inversion. It is also relevant that the group that is cleaved is the most basic of the possible carbanions. McEwen[9] and co-workers proposed the following mechanism (Eq. 4.72), which accounts for these facts.

$$\underset{Me}{\overset{Ph}{Et-\overset{|}{P}:}} \xrightarrow{PhCH_2Br} \underset{Me}{\overset{Ph}{Et-\overset{|+}{P}-CH_2Ph}} \xrightarrow{OH^-} \underset{Et \quad Me}{\overset{Ph}{HO-\overset{|}{P}-CH_2Ph}} \xrightarrow{OH^-}$$

$$\mathbf{29}$$

$$\underset{Et \quad Me}{\overset{Ph}{^-O-\overset{|}{P}-CH_2Ph}} \longrightarrow \underset{Me}{\overset{Ph}{O=\overset{|}{P}--Et}} \qquad (4.72)$$

$$\mathbf{30}$$

Note that the phosphine oxide product **30** has the opposite configuration of structure **31** from the oxidation of the phosphine (Eq. 4.73). Oxidation occurs with retention, thus revealing the occurrence of inversion in the phosphonium ion degradation. The inversion comes about from the addition of OH⁻ to phosphorus to form a

$$\underset{Me}{\overset{Ph}{Et--\overset{|}{P}:}} \xrightarrow{(O)} \underset{Me}{\overset{Ph}{Et--\overset{|}{P}=O}} \qquad (4.73)$$

$$\mathbf{31}$$

5-coordinate intermediate (**29**), with the incoming group in the apical position, followed by departure of the carbanion from an apical position. These apical-addition and apical-elimination processes are in accord with general principles of 5-coordinate intermediates in phosphorus mechanisms, as discussed in chapter 2.

2. Reductive cleavage with metals or metallic hydrides, or by electrolysis, which leads to the formation of a phosphine and a hydrocarbon. In practice, C–P cleavage is usually effected with lithium aluminum hydride and is most efficacious when performed on phosphonium salts that contain benzyl groups. Phenyl groups can also be cleaved, but less readily than benzyl; alkyl groups are the most difficult to cleave. This order of reactivity is consistent with the departing group being in the form of an anion, and a mechanism that involves formation of a 5-coordinate intermediate by attachment of hydride ion to phosphorus seems to be in effect (Eq. 4.74).

$$Ph_3\overset{+}{P}CH_2Ph \xrightarrow{LiAlH_4} Ph\text{-}\underset{H}{\overset{CH_2Ph}{\underset{|}{P}}}\underset{Ph}{\overset{Ph}{\diagdown}} \longrightarrow Ph_3\overset{+}{P}H \xrightarrow{base} Ph_3P \qquad (4.74)$$

This mechanism has received support from the actual isolation of a stable 5-coordinate hydride product (**32**) in the reaction of a spirocyclic phosphonium ion with LiAlH$_4$.[59]

32

Electrolysis is also a useful technique for the reductive cleavage of a group from a phosphonium ion. The process, first described in 1961,[60] is conducted in water solution at moderate temperatures (70 to 90°), generally with lead, mercury, or carbon anodes. Benzyl groups are especially easily cleaved, although many other groups have also been cleaved. The reduction may proceed through the addition of electrons to positive phosphorus, first to create a phosphoranyl radical, R$_4$P·, then a phosphoranide anion R$_4$P$^-$, which fragments to form the tertiary phosphine. The cleaved group acquires a proton from the solvent and appears as a hydrocarbon. The electrochemical conditions and scope of the reduction may especially be appreciated by examining reference 61; some of the many other reductions that have been performed over the years have been reviewed.[51c] Electrolytic reduction was of special importance in the classical first studies on optically active phosphines by Horner,[62] where it was discovered that the cleavage of a benzyl group occurs with retention of configuration at phosphorus. Phosphonium ions with four different substituents, one of which was benzyl, were prepared and resolved as their dibenzoyltartrate salts to provide the optically active forms, which were then submitted to electrolytic reduction to give the optically active phosphines (Eq. 4.75). Optical resolutions of many other phosphorus compounds have been performed and are discussed further in chapter 9.

$$\underset{\underset{Ph}{|}}{\overset{Me}{\underset{+}{\overset{\diagdown}{Pr\text{----}P}}}-CH_2Ph} \;\; Br^- \longrightarrow \underset{\underset{Ph}{|}}{\overset{Me}{\overset{\diagdown}{Pr\text{----}P}}}: \qquad (4.75)$$

4.D. HETEROPHOSPHONIUM (QUASIPHOSPHONIUM) SALTS

4.D.1. General

Phosphonium salts are also known where one or more of the carbon substituents of a quaternary ion is replaced by halogen or by groups based on oxygen or nitrogen; sulfur or selenium groups can also be attached to positive phosphorus, but less is known about them, and they will not be included in this discussion. Although hetero-substituted phosphonium ions have been known since the early days of phosphorus chemistry, there has been much activity in recent years, and the literature is now of such a considerable size that this discussion must concentrate on only the most important characteristics and synthetic approaches. Further discussions are found in various reviews.[19b,51c,63,64]

Hetero-substituted phosphonium ions, although structurally similar to the quaternaries, have some quite different properties. The electrophilic character of phosphorus is greatly enhanced by the presence of electronegative substituents, and these substituents can also function as efficient leaving groups. The result is that this type of phosphonium salt is highly reactive to nucleophiles and undergoes displacement reactions with great ease, probably through 5-coordinate intermediates (see reviews cited). The number of possible structural types is large, although, as will be seen, not all types are stable, and they frequently appear as transient reaction intermediates. The situation is complicated by the possibility that in some solvents combination of the cation and anion can occur to give an equilibrium with the neutral 5-coordinate species. This is well illustrated by the behavior of chlorotriphenylphosphonium chloride; this compound is ionic in the solid state[65] and ionizes in acetonitrile solution,[66] but has largely 5-coordinate character in benzene[67] and nitrobenzene[68] (Eq. 4.76).

$$Ph_3\overset{+}{P}Cl \;+\; \bar{C}l \;\rightleftharpoons\; Ph-\underset{\underset{Cl}{|}}{\overset{\overset{Cl}{|}}{P}}\diagup\overset{Ph}{\diagdown Ph} \qquad (4.76)$$

On the other hand, the bromide and iodide are known to adopt only the ionic form, as is true of halotrialkylphosphonium halides. It has been suggested that the greater electronegativity of phenyl relative to alkyl is responsible for the increased amount of 5-coordinate character.[69]

The early phosphorus literature referred to such compounds as "quasiphosphonium" salts,[19a] a name still in use[63a] that recognizes their ionic character but seems to indicate some reservation about the extent of this ionic structure relative to the true quaternary ions and the possible contribution of 5-coordinate forms. When ionic character is present, however, there are no differences at phosphorus in regard to the quaternary ions, and all may be considered to have 4-coordinate phosphorus with tetrahedral sp^3 structure. For this reason, the term *heterophosphonium* is preferable to *quasiphosphonium* for these ions and is the one used in this discussion. Even when the equilibrium with the 5-coordinate isomer is present, it is the tetrahedral form that one should consider as the reactive electrophilic species. Although exceptions abound (as already noted for Ph_3PCl_2), we will use the generalization that the ionic form is strongly, or even exclusively, favored when the counterion is derived from a strong acid, especially the halogen (Cl, Br, I) acids, and, of course, when the counterion has no nucleophilic character, as in BF_4^-, $AlCl_4^-$, PF_6^-, ClO_4^-, etc., there is no possibility for the 5-coordinate species to exist. This structural point can be explored by various techniques, but of most value is ^{31}P NMR spectroscopy, because 5-coordinate species would be recognized by chemical shifts that are well upfield of the heterophosphonium ion and are relatively independent of the counterion in the ionic form. Of course, X-ray crystallography is of great value when crystals can be obtained. The generalization has been made that with oxy-substituted compounds, only the phosphonium ion structure prevails.[63b] Five-coordination (chapter 2.F and 10.E) is generally restricted to cases in which the counterion, if free, is fluorine or a strongly nucleophilic group, such as alkoxide. Thus the 5-coordinate structure $(RO)_5P$, not the heterophosphonium structure $(RO)_4P^+RO^-$, is correct, and RPF_4 prevails over $RPF_3^+F^-$. Five-coordination is also promoted when an ethylenedioxy group is bonded to phosphorus; thus compound **33** is reported[70] to have the 5-coordinate rather than the phosphonium ion structure.

$$(EtO)_3P\begin{matrix}O-CH_2\\|\\O-CH_2\end{matrix}$$

33

The nature of the hetero atom has great influence on the characteristics of the heterophosphonium ions, and this calls for separate consideration of the halo-, oxy-, and amino-derivatives.

4.D.2. Halophosphonium Ions

The best known and most easily prepared type of heterophosphonium salt has one or more halogen atoms replacing carbon, as in the forms $RPX_3^+X^-$, $R_2PX_2^+X^-$, and

$R_3PX^+X^-$, where X is chlorine or bromine. These salts are not very stable thermally and decompose to alkyl halides and 3-coordinate phosphorus species on heating. They are easily formed simply by adding the free halogen to a halophosphine (Eq. 4.77) or to a tertiary phosphine (Eq. 4.78) in an aprotic solvent. Both reactions have been noted previously as reactions of these 3-coordinate compounds (chapter 3).

$$RPCl_2 \xrightarrow{Cl_2} RPCl_4 \qquad (4.77)$$

$$R_3P \xrightarrow{X_2} R_3PX_2 \qquad (4.78)$$

If the solvent is nonpolar (an alkane being of frequent use), the salt precipitates generally as a crystalline solid. The salts have some solubility in the more polar solvents; chloroform and methylene chloride are useful, and this has allowed structural determination by the powerful technique of ^{31}P NMR spectroscopy. The chemical shifts (chapter 6) are generally found to be consistent with the heterophosphonium ion structure rather than the 5-coordinate structure.

Halophosphonium ions are highly reactive to water, a property in common with the inorganic analogues, the phosphorus pentahalides. The reaction proceeds by nucleophilic attack on positive phosphorus, probably to form a 5-coordinate intermediate (e.g., **34** from a tertiary phosphine-halogen adduct), which decomposes by loss of a proton and HCl with formation of the phosphoryl bond (Eq. 4.79). Of course, if additional halogen is present on phosphorus, it too will be hydrolyzed.

$$R_3\overset{+}{P}Cl \;\; \bar{Cl} \xrightarrow{H_2O} \underset{\mathbf{34}}{R_3\overset{+}{P}Cl} \xrightarrow{-H^+} \underset{R_3P-Cl}{\overset{O-H}{|}} \longrightarrow R_3P=O \qquad (4.79)$$

This reaction (Eq. 4.79) has frequently been used as a source of phosphine oxides, and when hydrogen sulfide is the reactant, phosphine sulfides can be formed. Another valuable reaction is that with alcohols; the initial P-alkoxy product from the displacement is unstable and undergoes cleavage (a point developed below) to form alkyl halides in good yields (Eq. 4.80). This is, therefore, a reaction of synthetic value.

$$R_3\overset{+}{P}Br \;\; \bar{Br} \xrightarrow[-HCl]{ROH} \underset{Ph_3P-Br}{\overset{O\diagdown R}{|}} \; \bar{Br} \longrightarrow Ph_3P=O + R\text{-}Br \qquad (4.80)$$

Salts of structure $RPX_3^+X^-$, formed readily from phosphonous dihalides with halogen, are also of considerable synthetic value. They are converted to phosphonic acids on hydrolysis (Eq. 4.81) and to phosphonic dihalides with sulfur dioxide (Eq. 4.82). These transformations were discovered early in the development of phosphorus chemistry and remain of great practical value.

$$\text{RPCl}_2 + \text{Cl}_2 \longrightarrow \text{RPCl}_4 \xrightarrow{\text{H}_2\text{O}} \text{R}-\overset{\overset{\text{O}}{\|}}{\text{P}}(\text{OH})_2 \qquad (4.81)$$

$$\text{RPCl}_4 \xrightarrow{\text{SO}_2} \text{R}-\overset{\overset{\text{O}}{\|}}{\text{P}}\text{Cl}_2 + \text{SOCl}_2 \qquad (4.82)$$

Halophosphonium ions can be formed by other reactions than the halogen addition process, some of which were seen in chapter 3 (e.g., the McCormack cycloaddition of dienes and 3-coordinate halophosphines, and the reaction of PCl_3 with diazonium fluoroborates). A notable process is to conduct reactions of tertiary phosphines with "positive" halogen compounds, in which halogen can be considered to be bonded to a displaceable negative group (e.g., $-\text{OR}$, $-\text{NR}_2$, $-\text{CCl}_3$, etc.). The phosphine attacks as a nucleophile on this halogen to form the halophosphonium ion; this species can itself undergo nucleophilic substitution by the anion released from the halogen compound, in the absence of other nucleophiles. The well-known reaction with CCl_4 or CBr_4 is of this type; the initially formed halophosphonium ion undergoes attack by the released CX_3^- ion (Eq. 4.83) to form a trihalomethyl phosphonium salt.

$$\text{R}_3\text{P} + \text{CBr}_4 \longrightarrow \text{R}_3\overset{+}{\text{P}}\text{Br} + \text{C}\bar{\text{B}}\text{r}_3 \longrightarrow \text{R}_3\overset{+}{\text{P}}-\text{CBr}_3 \qquad (4.83)$$

If, however, a nucleophile such as an alcohol is present, it will trap the intermediate halophosphonium ion as already seen in Eq. 4.80, by displacing halide ion. This is, in fact, a valuable and widely used method for alkyl halide synthesis.[71,72]

The phosphine-CX_4 adducts are also highly effective as dehydrating agents, and have been employed in coupling reactions as in the synthesis of peptides,[72] as have halophosphonium ions from other sources (e.g., phospholenium ions from McCormack reactions[73]). They also serve as a source of dihalomethylene ylides on further reaction with a phosphine (Eq. 4.84); these ylides are used to synthesize dihaloalkene derivatives.

$$\text{Ph}_3\overset{+}{\text{P}}\text{CX}_3\ \bar{\text{X}} + \text{Ph}_3\text{P} \longrightarrow \text{Ph}_3\overset{+}{\text{P}}-\bar{\text{C}}\text{X}_2 + \text{Ph}_3\overset{+}{\text{P}}\text{X}\ \bar{\text{X}} \qquad (4.84)$$

In addition to the nucleophilic displacement reactions of halophosphonium ions that occur with simple alcohols or amines, these ions can form bonds with more complex nucleophiles, such as enolates and carbanions.[64b] Another reaction of halophosphonium halides of practical importance is that an equivalent of a halogen molecule can be removed with a variety of reducing agents, commonly metals, to return phosphorus to the 3-coordinate state. Applications of this reaction type were noted in chapter 3.

4.D.3. Alkoxyphosphonium Ions

When one or more of the P-substituents of a heterophosphonium ion is alkoxy and the counterion is halide, a new form of reactivity is present in addition to the easy displaceability of alkoxy by nucleophiles: a halide counterion can attack on carbon of the alkoxy group in S_N2 manner with elimination of alkyl halide and formation of the phosphoryl group, as exemplified by the reaction of a tertiary phosphite with chlorine (Eq. 4.85).

$$(EtO)_3P \ + \ Cl_2 \longrightarrow (EtO)_3\overset{+}{P}Cl \ \overset{-}{Cl} \longrightarrow (EtO)_2\overset{O}{\underset{\|}{P}}-Cl \ + \ EtCl \qquad (4.85)$$

This dealkylation can take place rapidly at room temperature or below, and as a general rule in phosphorus chemistry, one should consider any structure with an alkoxy group on positive phosphorus with a nucleophilic counterion, from whatever route, to represent an unstable combination. Aryloxy groups are not subject to such easy cleavage by halide, however, and consequently compounds such as the adducts of triaryl phosphites with halogen, $(ArO)_3PX_2$, among other types, are known to be stable compounds at room temperature. However, even the aryloxy compounds decompose at higher temperatures (typically approaching 200°), with the elimination of aryl halide.

The usual precursors of alkoxyphosphonium ions are 3-coordinate alkoxy species (tertiary phosphites, phosphonites, and phosphinites). The ions can be generated by addition of free halogen or by alkylation with alkyl halides. In both cases, dealkylation follows, with the creation of the phosphoryl group. It will be remembered from chapter 3 that the phosphite-alkyl halide reaction results in the formation of dialkyl phosphonates, in the classical Michaelis–Arbusov reaction. The mechanism of this reaction has been extensively studied, and the intermediacy of the alkoxyphosphonium ion firmly established.[63a] The intermediate (e.g., **35**) has actually been isolated from the reaction of a phosphinite with an alkyl halide at low temperature; here the two alkyl P-substituents act to moderate the reactivity of the alkoxyphosphonium intermediate, which was stable for 2 to 3 weeks at room temperature before complete conversion to the tertiary phosphine oxide[74] (Eq. 4.86).

$$Et_2P-OEt \ + \ MeI \longrightarrow \underset{OEt}{\overset{Me}{Et_2\overset{+}{P}}} \ I^- \longrightarrow Et_2\overset{Me}{\underset{\|}{P}}=O \ + \ EtI \qquad (4.86)$$

35

Phosphonium ion intermediates have also been observed with sterically hindered phosphonites[75] and phosphites (e.g., with trineopentyl phosphite[76]) and have varying lifetimes until conversion to the phosphoryl products. The salts from the reaction of methyl bromide with neopentyl diphenylphosphinite and with dineopentyl phenylphosphonite are crystalline solids whose tetrahedral structures were confirmed by X-ray crystallography.[77]

Several other techniques are available for the generation of alkoxyphosphonium ions.[63a] Among these is the O-alkylation of tertiary phosphine oxides that was mentioned earlier in this chapter (Eqs. 4.21 to 4.23); the analogous reaction occurs with phosphine sulfides to form thiophosphonium salts.

4.D.4. Aryloxyphosphonium Ions

As noted, aryloxy groups are less readily cleaved by halide from positive phosphorus, and structures such as the tertiary phosphite-halogen adducts, $(ArO)_3PX^+X^-$, and the tertiary phosphite-alkyl halide reaction products, $(ArO)_3PR^+X^-$, have been known for many years. Thus, the methiodide of triphenyl phosphite was reported as early as 1898 to be a crystalline solid[78] with m.p. 75°. Its ^{31}P NMR shift of δ +39 confirms the phosphonium ion structure.[79]

The phosphite-halogen adducts, especially those from triaryl phosphites, are quite electrophilic and highly reactive to nucleophiles, sometimes getting involved in complex group exchange (disproportionation) phenomena[69] with nucleophiles that proceed through 5-coordinate intermediates. An example of this type of disproportionation is found in Eq. 4.87, in which exchange of phenoxide and chloride in the triphenyl phosphite-chlorine adduct takes place.[80]

$$(PhO)_3\overset{+}{P}Cl \; \bar{Cl} \rightleftharpoons (PhO)_3PCl_2 \rightleftharpoons (PhO)_2\overset{+}{P}Cl_2 \; ^-OPh$$
$$\mathbf{36}$$
$$\mathbf{36} + {}^-OPh \rightleftharpoons (PhO)_4\overset{+}{P} \; \bar{Cl} \quad (4.87)$$

4.D.5. Aminophosphonium Ions

Relative to halo- and the oxy-substituents, amino groups on positive phosphorus are quite stable and less readily displaced. The tetraamino derivatives are said to resemble true quaternary phosphonium ions,[19b] even forming ionic hydroxides by metathesis with moist silver oxide. Hydrolysis can, however, be effected with aqueous acid or base. The cleavage reaction with halide ion so common with P-alkoxy groups on positive phosphorus does not occur in the amino series. Thus a simple method of preparation is to react tris(dialkylamino)phosphines (phosphorous acid triamides) with alkyl halides. The resulting salts are quite stable ionic compounds; the methiodide from the tripiperidide (Eq. 4.88) is a crystalline solid[81] with m.p. 251–255°.

$$\left(\bigcirc N-\right)_3 P + MeI \longrightarrow \left(\bigcirc N-\right)_3 \overset{+}{P} Me \; \bar{I} \quad (4.88)$$

Another common method of preparation, which provides tetraamino derivatives, is to react primary or secondary amines with phosphorus pentachloride. The reaction product with aniline (Eq. 4.89) is a solid[82] with m.p. 275°.

$$PhNH_2 + PCl_5 \longrightarrow (PhNH)_4\overset{+}{P} \; \bar{Cl} \quad (4.89)$$

In a related process, aryltrichlorophosphonium ions may be reacted with amines to form aryltriaminophosphonium ions.

Aminophosphonium ions can also be usefully created by reacting aminophosphines with alkyl halides; alkylation occurs on phosphorus, not on nitrogen, because of the greater nucleophilicity of the more polarizable heavier element[83] (Eq. 4.90). Much of the fundamental research on aminophosphonium ions was performed

$$R_2P-NR_2 + R'X \longrightarrow R_2\overset{+}{\underset{R'}{P}}-NR_2\ \overset{-}{X} \qquad (4.90)$$

years ago. A few scattered references to more recent work may be found in Reference 51a.

REFERENCES

1. H. R. Hayes and D. J. Peterson in G. M. Kosolapoff and L. Maier, eds., *Organic Phosphorus Compounds*, Vol. 3, John Wiley & Sons, Inc., New York, 1972, (a) p. 416, (b) Chapter 6, (c) p. 355, (d) p. 350, (e) p. 389, (f) p. 391.
2. H. R. Hartley, ed., *The Chemistry of Organophosphorus Compounds*, Vol. 2, John Wiley & Sons, Inc., New York, 1992, (a) R. S. Edmundson, Chapter 7, (b) A. K. Bhattacharya and N. K. Roy, Chapter 6, (c) T. S. Lobana, Chapter 8, (d) D. G. Gilheany, Chapter 1.
3. G. M. Kosalopoff and L. Maier, eds., *Organic Phosphorus Compounds*, Vol. 4, John Wiley & Sons, Inc., New York, 1972, (a) L. A. Hamilton and P. S. Landis, Chapter 11, (b) L. Maier, Chapter 7.
4. S. A. Buckler, *J. Am. Chem. Soc.* **84**, 3093 (1962); M. B. Floyd and C. E. Boozer, *J. Am. Chem. Soc.* **85**, 984 (1963).
5. J. Szewczyk, E.-Y. Yao and L. D. Quin, *Phosphorus Sulfur Silicon* **54**, 135 (1991).
6. Y. Kashman and O. Awerbouch, *Tetrahedron* **31**, 53 (1975).
7. L. D. Quin, *Rev. Heteroatom Chem.* **3**, 39 (1990).
8. C. M. Whitaker, K. L. Kott, and R. J. McMahon, *J. Org. Chem.* **60**, 3499 (1995).
9. W. E. McEwen, K. F. Kumli, A. Bláde-Font, M. Zanger, and C. A. Vander Werf, *J. Am. Chem. Soc.* **86**, 2378 (1964).
10. H. R. Hayes, *J. Org. Chem.* **33**, 3690 (1968).
11. M. M. Rauhut, I. Hechenbleikner, H. A. Currier, and V. P. Wystrach, *J. Am. Chem. Soc.* **80**, 6690 (1958).
12. S. A. Buckler and M. Epstein, *Tetrahedron* **18**, 1221 (1962).
13. S. A. Buckler and M. Epstein, *Tetrahedron* **18**, 1211 (1962).
14. A. J. Kirby and S. G. Warren, *The Organic Chemistry of Phosphorus*, Elsevier, Amsterdam, 1967, (a) p. 54, (b) p. 15, (c) p. 254.
15. S. Jockusch, I. V. Koptyug, P. F. McGarry, G. W. Sluggett, N. J. Turro, and D. M. Watkins, *J. Am. Chem. Soc.* **119**, 11495 (1997); S. Jockusch and N. J. Turro, *J. Am. Chem. Soc.* **120**, 11773 (1998).
16. C. C. Santini, J. Fischer, F. Mathey, and A. Mitschler, *J. Am. Chem. Soc.* **102**, 5809 (1980).
17. D. B. Denney, D. Z. Denney, and L. A. Wilson, *Tetrahedron Lett.*, 85 (1968).
18. L. D. Quin, J. C. Kisalus, J. J. Skolimowski, and N. S. Rao, *Phosphorus Sulfur Silicon* **54**, 1 (1990).
19. G. M. Kosolapoff, *Organophosphorus Compounds*, John Wiley & Sons, Inc., New York, 1950, (a) p. 111, (b) Chapter 11, (c) p. 328.

20. D. E. Bissing and A. J. Speziale, *J. Am. Chem. Soc.* **87**, 1405 (1965).
21. N. Weferling and S. Hoerold, paper presented at the International Conference on Phosphorus Chemistry, Cincinnati, Ohio, July 12–17, 1998.
22. J. Monagle, *J. Org. Chem.* **27**, 3851 (1962).
23. D. Hellwinkel and W. Krapp, *Chem. Ber.* **111**, 13 (1978).
24. Y. Segall and I. Granoth, *J. Am. Chem. Soc.* **100**, 5130 (1978).
25. R. G. Cavell and R. D. Leary, *J. Chem. Soc., Chem. Commun.*, 1520 (1970).
26. G. Märkl and B. Merkl, *Tetrahedron Lett.* **22**, 4459 (1981).
27. (a) K. M. Pietrusiewicz and M. Zablocka, *Tetrahedron Lett.* **29**, 1987 (1988), (b) K. M. Pietrusiewicz, M. Zablocka, and W. Wieczorek, *Phosphorus Sulfur Silicon* **42**, 183 (1989).
28. K. M. Pietrusiewicz, M. Zablocka, and J. Monkiewicz, *J. Org. Chem.* **49**, 1522 (1984).
29. A. N. Pudovik, N. G. Khusainova, E. A. Berdnikov, and Z. A. Nasybullina, *Zhur. Obshch. Khim.* **44**, 222 (1974); *Chem. Abstr.* **80**, 107635 (1974).
30. T. H. Chan and L. T. Wong, *Can. J. Chem.* **49**, 530 (1971).
31. A. W. Johnson and H. L. Jones, *J. Am. Chem. Soc.* **90**, 5232 (1968).
32. (a) C. J. Pouchert and J. Behnke, *Aldrich ^{13}C and FT NMR Spectra*, Aldrich Chemical Co., Milwaukee, 1993, Spectrum 1691, (b) T. A. Albright, W. J. Freeman, and E. E. Schweizer, *J. Org. Chem.* **40**, 3437 (1975).
33. (a) B. Klabuhn, H. Goetz, P. Steirl, and D. Alscher, *Tetrahedron* **32**, 603 (1976).
34. C. R. Johnson and T. Imamoto, *J. Org. Chem.* **52**, 2170 (1987).
35. L. Horner, *Fortschr. Chem. Forsch.* **7**, 1 (1966).
36. S. G. Warren, *Accts. Chem. Res.* **11**, 401 (1978).
37. J. B. Lambert and Y. Zhao, *J. Am. Chem. Soc.* **118**, 3156 (1996).
38. R. C. Miller, J. S. Bradley, and L. A. Hamilton, *J. Am. Chem. Soc.* **78**, 5299 (1956).
39. R. Bodalski and K. M. Pietrusiewicz, *Tetrahedron Lett.*, 4209 (1972).
40. L. D. Quin and R. E. Montgomery, *J. Org. Chem.* **30**, 2393 (1965).
41. J. A. Miller and D. Stewart, *J. Chem. Soc., Perkin Trans.* 1, 1898 (1977).
42. R. A. Zingaro, B. H. Stever, and K. Irgolic, *J. Organomet. Chem.* **4**, 320 (1965).
43. L. Horner and H. Oediger, *Liebigs Ann. Chem.* **627**, 142 (1959).
44. *Organic Syntheses Coll.* Vol. V, 1016 (1973).
45. K. Moedritzer, *Syn. React. Inorg. Metal-Org. Chem.* **4**, 119 (1974).
46. B. E. Maryanoff, R. Tang, and K. Mislow, *J. Chem. Soc., Chem. Commun.*, 273 (1973).
47. A. E. Arbusov and K. V. Nikonorov, *Zhur. Obshch. Khim.* **18**, 2008 (1948); *Chem. Abstr.* **43**, 3801 (1949).
48. R. A. N. McLean, *Inorg. Nucl. Chem. Lett.* **5**, 745 (1969).
49. T. A. Mastryukova, I. M. Aladzheva, D. I. Lobanov, O. V. Bykhovskaya, P. V. Petrovskii, K. A. Lyssenko, and M. I. Kabachnik, paper presented at the International Conference on Phosphorus Chemistry, Cincinnati, Ohio, July 12–17, 1998.
50. P. Beck in G. M. Kosolapoff and L. Maier, eds., *Organic Phosphorus Compounds*, Vol. 2, John Wiley & Sons, Inc., New York, 1972, Chapter 4.
51. F. R. Hartley, ed., *The Chemistry of Organophosphorus Compounds*, Vol. 3, John Wiley & Sons, Inc., New York, 1994, (a) H.-J. Cristau and F. Pléna, Chapter 2, (b) H.-J. Cristau and F. Pléna, pp. 96–105, (c) K. S. V. Santhanam, Chapter 5.
52. L. Horner, G. Mummenthey, H. Moser, and P. Beck, *Chem. Ber.* **99**, 2782 (1966).
53. J. B. Plumb and C. E. Griffin, *J. Org. Chem.* **27**, 4711 (1962).
54. A. Bond, M. Green, and S. C. Pearson, *J. Chem. Soc. B*, 929 (1968).
55. F. G. Mann and I. T. Millar, *J. Chem. Soc.*, 2205 (1951).
56. S. A. Buckler and M. Epstein, *J. Org. Chem.* **27**, 1090 (1962).

57. F. G. Mann, *Heterocyclic Derivatives of Phosphorus, Arsenic, Antimony and Bismuth*, Wiley-Interscience, New York, 1970, Chapter 1.
58. G. Wittig and M. Rieber, *Naturwissenschaften* **35**, 345 (1948).
59. D. Hellwinkel, *Angew. Chem., Int. Ed. Engl.* **5**, 968 (1966).
60. L. Horner and A. Mentrup, *Liebigs Ann. Chem.* **646**, 65 (1961).
61. L. Horner and J. Haufe, *J. Electroanal. Chem.* **20**, 245 (1969).
62. L. Horner, H. Winkler, A. Rapp, A. Mentrup, H. Hoffmann, and P. Beck, *Tetrahedron Lett.*, 161 (1961).
63. H. R. Hudson, *Topics Phos. Chem.* **11**, (a) pp. 339–435, (b) p. 346 (1983).
64. R. F. Hudson, *Structure and Mechanism in Organo-Phosphorus Chemistry*, Academic Press, New York, 1965, (a) p. 213, (b) p. 218, (c) p. 346.
65. K. B. Dillon, R. J. Lynch, R. N. Reeve, and T. C. Waddington, *J. Chem. Soc., Dalton Trans.*, 1243 (1976).
66. B. V. Timokhin, V. P. Feshin, V. I. Dmitriev, V. I. Glukhikh, G. V. Dolgushin, and M. G. Voronkov, *Dokl. Akad. Nauk SSSR* **236**, 938 (1977); *Chem. Abstr.* **88**, 5761 (1978).
67. B. V. Timokhin, V. I. Dmitriev, B. I. Istomin, and V. I. Donskikh, *Zhur. Obshch. Khim.* **51**, 1989 (1981); *Chem. Abstr.* **95**, 202820 (1981).
68. G. A. Wiley and W. R. Stine, *Tetrahedron Lett.*, 2321 (1967); D. B. Denney, D. Z. Denney, and B. C. Chang, *J. Am. Chem. Soc.* **90**, 6332 (1968).
69. H. R. Hudson in F. R. Hartley, ed., *The Chemistry of Organophosphorus Compounds*, Vol. 1, John Wiley & Sons, Inc., New York, 1990, pp. 450–451.
70. D. B. Denney and D. H. Jones, *J. Am. Chem. Soc.* **91**, 5821 (1969).
71. H. Teichmann, *Z. Chem.* **14**, 216 (1974).
72. R. Appel, *Angew. Chem., Int. Ed. Engl.* **14**, 801 (1975).
73. E. Vilkas, M. Vilkas, and J. Sainton, *Nouv. J. Chem.* **2**, 307 (1978).
74. A. I. Razumov and N. N. Bankovskaya, *Zhur. Obshch. Khim.* **34**, 1859 (1964); *Chem. Abstr.* **61**, 8335 (1964).
75. A. I. Razumov and N. N. Bankovskaya, *Dokl. Akad. Nauk SSSR* **116**, 241 (1957); *Chem. Abstr.* **52**, 6164 (1958).
76. (a) H. R. Hudson, R. G. Rees, and J. E. Weekes, *J. Chem. Soc., Chem Commun.*, 1297 (1971); (b) H. R. Hudson, R. G. Rees, and J. E. Weekes, *J. Chem. Soc., Perkin Trans.* 1, 982 (1974).
77. K. Henrick, H. R. Hudson, and A. Kow, *J. Chem. Soc., Chem. Commun.*, 226 (1980).
78. A. Michaelis and R. Kaehne, *Ber.* **31**, 1048 (1898).
79. L. V. Nesterov, A. Ya. Kessel, Yu. Yu. Samitov, and A. A. Musina, *Dokl. Akad. Nauk SSSR* **180**, 116 (1968); *Chem. Abstr.* **69**, 77368 (1968).
80. H. N. Rydon and B. L. Tonge, *J. Chem. Soc.*, 3043 (1956).
81. A. Michaelis and K. Luxemburg, *Ber.* **28**, 2205 (1897).
82. J. E. Gilpin, *Am. Chem. J.* **19**, 352 (1897).
83. N. L. Smith and H. H. Sisler, *J. Org. Chem.* **28**, 272 (1963).

APPENDIX 4.1

SYNTHESES OF PERTINENT COMPOUNDS IN *ORGANIC SYNTHESES* (OS) AND *INORGANIC SYNTHESES* (IS)

[1] MePPh$_2$=O

IS **17**, 184

[2] Me$_2$PPh=O

IS **17**, 185

[3] 2,2'-biphenylene-bis(diphenylphosphine oxide)

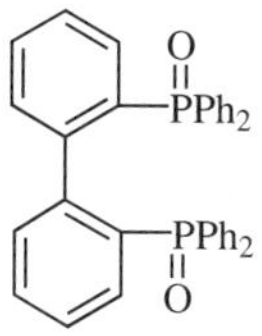

OS **VIII**, 57

[4] 3-methyl-1-phenyl-2-phospholene 1-oxide

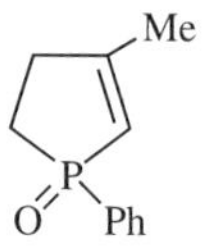

OS **V**, 787

[5] Me$_2$P(S)-H

IS **26**, 162

[6] Me$_2$P(S)-P(S)Me$_2$

OS **V**, 1016

[7] Ph$_3$P=Se

IS **10**, 157

[8] PhOCH$_2$CH$_2$PPh$_3^+$ Br$^-$

OS **V**, 1145

[9] CH$_2$=CH-PPh$_3^+$ Br$^-$

OS **V**, 1145

[10] PhCH$_2$PPh$_3^+$ Cl-Au-C$_6$F$_5^-$

IS **26**, 88

[11] 1-ethyl-1-phenylphospholanium ClO$_4^-$

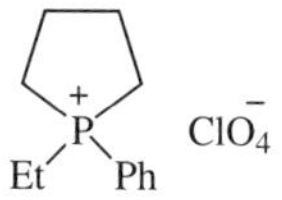

IS **18**, 189, 191

[12] PhCH=CHPCl$_3^+$ PCl$_6^-$

OS **VI**, 1005

CHAPTER 5

THE ACIDS OF ORGANOPHOSPHORUS CHEMISTRY AND THEIR DERIVATIVES

5.A. STRUCTURES AND GENERAL CHARACTERISTICS

The requisite structure for a phosphorus acid contains the grouping P(O)OH with 4-coordinate phosphorus or, much less commonly, this function in which sulfur atoms replace one or both oxygen atoms. A variety of groups can be present to complete the valence requirements of phosphorus, including alkyl, aryl, hydrogen, OH and OR groups, and numerous other heteroatom groups. For classification purposes, we use only the C, H, and OH substituents to provide parent structures; other forms are considered as derivatives of these parents. This leads to the following collection of parent acids (using *R* to represent either alkyl or aryl groups).

1. Phosphonic Acids, $RP(O)(OH)_2$. These are by far the best known of the acids with a C–P bond; they are highly stable and easily made by a number of methods, to be discussed, and have found use in the free acid or in derivatized forms in many practical applications. In recent years, they have achieved great importance in agricultural and medicinal chemistry (see chapter 11), and much current research is directed to the development of these applications. As is true of almost all phosphorus acids, phosphonic acids are classed as moderately strong acids. The first ionization (pK_a) for methylphosphonic acid is 2.38. The ionization is increased by electronegative substituents on the α-carbon as would be expected (e.g., trichloromethyl, 1.63; hydroxymethyl, 1.91; styrylmethyl, 2.00). The available data indicate that most alkylphosphonic acids can be considered to have K_a values in the range 2.4 to 2.9, with the second ionization 7.7 to 9.0. Aromatic phosphonic acids respond to substituent effects in much the same way as do carboxylic acids; the acidity of phenylphosphonic acid (pK_a 1.86) is increased by electron-withdrawing agents and decreased by electron-releasing substituents. These acids have Hammett reaction constants of 0.755 for the first ionization and 0.949 for the second (in water).[1] Generally they can be assigned a first-ionization range of pKa 1 to 2. Thus simple alkyl and phenyl phosphonic acids are seen to be stronger than carboxylic acids (with a general K_a range of 5 to 7) but weaker than sulfonic acids which have quite negative pKa values. In fact, this order of acidity holds for all of the parent types of phosphorus acids; the ionization of P—OH is surely influenced by the nature of the substituents directly on P, but not to change their relative position between the carboxylic and sulfonic acids. With this level of acidity for phosphonic and all other phosphorus parent acids, salts are easily formed with metal hydroxides, ammonia, and amines. In the resulting phosphonate anion, the negative charge is shared equally between the two oxygen atoms, as is expressed by the resonance hybrid of Eq. 5.1.

$$\begin{array}{c} O \\ \| \\ R-P-O^- \\ | \\ OH \end{array} \quad \longleftrightarrow \quad \begin{array}{c} O^- \\ | \\ R-P=O \\ | \\ OH \end{array} \qquad (5.1)$$

The phosphonic acid group appears in thousands of compounds, with every conceivable type of attached carbon (including C—C double and triple bonds, alicyclic and bridged multicyclic rings, heterocyclic systems) and with carbon groups bearing all of the common substituents of organic chemistry. The phosphonic group is quite polar and makes many of these compounds, especially those with short carbon chains, readily soluble in water. In organic solvents, intermolecular H-bonding can be pronounced.

The literature of phosphonic acids is vast, but valuable reviews are available. As for other phosphorus families, the best source of information on the literature to about 1970 is found in the Kosolapoff–Maier series.[2a] The literature is carried forward up to about 1994 in the Hartley series.[3a,4] Another recent review that surveys methods for forming the carbon-phosphorus bond, that in acids in particular, has been published by Engel.[5] An older but still valuable review was prepared by Freedman and Doak in 1957.[6] The Houben–Weyl series also has extensive summaries of the chemistry of phosphonic acids in the editions of 1963[7] and 1982.[8] More recent advances in phosphonate synthesis are outlined in a review by Mitchell and Kee.[9]

Chemists are not alone in synthesizing phosphonic acids; Nature has found ways to do this also, and a small number of phosphonic acids have been isolated from living systems, especially in the marine environment and from bacterial fermentation processes. This intriguing aspect of organophosphorus chemistry, completely unknown and unexpected until 1959,[10] is discussed in chapter 11.B. Illustrative of the naturally occurring compounds, which now number about a dozen, are 2-aminoethylphosphonic acid (**1**), the major form found in marine animals, and 1,2-epoxypropylphosphonic acid (**2**), a clinically useful antibiotic isolated from a *Streptomyces* strain.

$$\text{NH}_2\text{CH}_2\text{CH}_2-\overset{\overset{\text{O}}{\|}}{\text{P}}(\text{OH})_2 \qquad \text{H}_3\text{CHC}-\overset{\text{O}}{\underset{\diagdown\diagup}{\text{CH}}}-\overset{\overset{\text{O}}{\|}}{\text{P}}(\text{OH})_2$$

$$\mathbf{1} \qquad\qquad\qquad\qquad \mathbf{2}$$

Very recently, phosphonic acids have even been detected in meteoritic material,[11] adding a fascinating new dimension to their chemistry. The ramifications of these discoveries remain to be appreciated, but one can expect new developments as a role for organophosphorus compounds in extraterrestrial chemistry is considered.

2. H-Phosphonic Acid Esters, H—P(O)(OR)$_2$. The first member of the phosphonic acid family in which H rather than carbon is present on P is, strictly speaking, an inorganic compound. It is known, however, in the form of many ester and amide derivatives, of great importance in organophosphorus chemistry. We will use the term *H-phosphonate* to describe the ester derivatives, but the parent acid is, of course, phosphorous acid, and the diesters can also be called dialkyl phosphites, as is done in much of the older literature.

The starting material generally used in the synthesis of simple H-phosphonates is PCl$_3$ (Appendix 5.1, compounds 1 and 2). This compound is reacted with at least 3 moles of an alcohol in the absence of a base, generally with vacuum removal of the released HCl (in chapter 3.B we saw that in the presence of a base tertiary phosphites are formed). The mechanism of this process, reviewed by Gerrard and Hudson,[2b] involves the initial formation of the tertiary phosphite from stepwise displacement of the three chlorines, but the HCl released in the final step proceeds to cleave one P-alkoxy group with the creation of the phosphoryl group and release of alkyl chloride (Eq. 5.2).

$$\text{PCl}_3 + 3\,\text{ROH} \longrightarrow (\text{RO})_2\text{P-O-R}\underset{\text{Cl}}{\overset{}{\diagdown}} \xrightarrow{-\text{RCl}} (\text{RO})_2\bar{\text{P}}{=}\text{O} \xrightarrow{\overset{+}{\text{H}}} (\text{RO})_2\overset{\text{H}}{\underset{|}{\text{P}}}{=}\text{O} \quad (5.2)$$

In support of this mechanism, HCl has been found to cleave pre-formed tertiary phosphites to H-phosphonates. The PCl$_3$-alcohol reaction has been conducted on a commercial scale, using appropriate variations of the reaction conditions.[2b] H-phosphonates have a multitude of practical uses, including additives in various plastics formulations, lubricants and fuel oils.

The chemistry of the H-phosphonates, most recently reviewed by Stawinski[12] and earlier by others,[2b,13] is complicated by the presence of the tautomeric equilibrium with the P—OH form (HO—P(OR)$_2$), which, although present in only tiny amounts, can play a critical role as a reactive intermediate. For example, dialkyl H-phosphonates can be oxidized to dialkyl phosphates, and when this is done by the halogens (Cl$_2$, Br$_2$, I$_2$), the same reaction rate, independent of the halogen identity and concentration, is observed, indicating that the 3-coordinate P—OH structure is formed in a rate-determining step and that its oxidation is relatively fast. Many studies over the years have delved into this equilibrium; the equilibrium constant for diethyl H-phosphonate, for example, is 10^7 in favor of the H-phosphonate form,[14] and the P—OH form has never been directly detected by any spectroscopic technique. Thus the ^{31}P NMR spectrum for this compound shows only the signal from the P—H form, easily recognized from the very large ^{1}H—^{31}P coupling constant (e.g., 691 Hz for diethyl H-phosphonate[15]). The early literature on the tautomeric equilibrium has been reviewed;[16] references 2b, 12, and 13 bring the coverage up to 1992.

The H-phosphonates possess a most valuable property: the hydrogen can be removed by alkali metals and strong bases, and the resulting salts (some of which are soluble in nonpolar organic solvents) can be used as a building block in phosphonic acid synthesis (vide infra). Thus alkyl halides attack on phosphorus of the anion to form phosphonates; the anion also adds at phosphorus to α, β-unsaturated systems in the Michael reaction. Certain other reagents, however, notably trimethylsilyl chloride, attack on oxygen to form phosphites, and thus the anion can be described as being ambident (Scheme 5.1). Tris(trimethylsilyl) phosphite, a valuable synthetic intermediate, can be made simply by silylation of phosphorous acid.

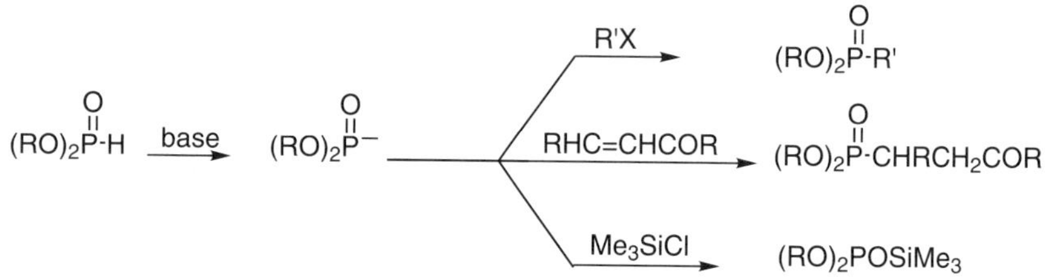

Scheme 5.1

H-Phosphonates have recently proved to be of value in nucleotide synthesis. A characteristic of H-phosphonates is that they undergo nucleophilic substitution by alcohols much faster than do phosphates, and after a substitution, they can be oxidized to the phosphates, either directly or after conversion to 3-coordinate form, e.g., by silylation. These procedures (and others, as seen in a recent review[12] of the extensive literature) have been put to advantage in bonding phosphate groups so as to form nucleotides. A recent approach,[17] which embodies this chemistry and introduces some new modifications, is outlined in Eq. 5.3.

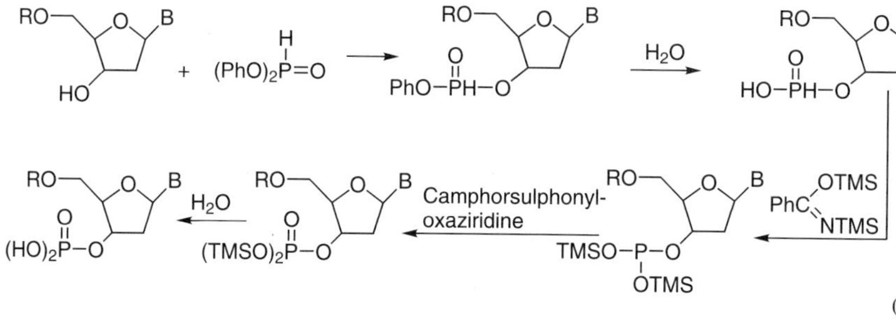

(5.3)

The H-phosphonate method can be extended to the synthesis of other biophosphates, such as the glycerophospholipids.[18]

3. *Phosphinic Acids, $R_2P(O)OH$*. Replacement of one OH group of a phosphonic acid by alkyl or aryl results in another stable and well-known family of phosphorus acids. The phosphinic acids are less readily prepared and not as numerous as the phosphonic acids, however, and their acidity is diminished by the replacement of the electronegative OH on phosphorus ($Me_2P(O)OH$, pK_a 3.08; $Ph_2P(O)OH$, pK_a 5.80 in 95% ethanol). The diaryl acids are generally rather insoluble in water. The chemistry of the phosphinic acids has been reviewed,[2c,3a,7,8] and a typical synthesis is referenced in Appendix 5.1, compound 3.

4. *H-Phosphinic Acids, RHP(O)OH*. These compounds are also known in the earlier literature as phosphonous acids. That name is now reserved for the tautomeric structure $RP(OH)_2$; there are, in fact, no compounds known with this structure, but there are many examples of esters, $RP(OR)_2$, and amides, $RP(NR_2)_2$, of this structure, and these are known as dialkyl alkylphosphonites and alkylphosphonamidites. Hydrolysis of these derivatives simply gives the H-phosphinic acids (the subject of reviews[2d,3a,7,8]). Just as in the case of the H-phosphonates, there appears to be an equilibrium with the phosphonous acid form, heavily dominated by the H-phosphinic acid form, probably at about the same level as seen for the H-phosphonates. As for the H-phosphonates, the structure is easily proved by various methods, notably ^{31}P NMR, which reveals the presence of H directly bonded to P from the very large $^1H-^{31}P$ coupling constant, typically 500 to 525 Hz.

Phenylhydrogenphosphinic acid (more simply phenylphosphinic acid, an item of commerce especially as an additive in the polymer industry) is a crystalline solid, with m.p. 70–71° and K_a 1.75, of reasonable stability at mild temperatures. It and related acids can easily be oxidized to phosphonic acids. The lower alkylhydrogenphosphinic acids are much less stable and on being heated undergo an oxidation-reduction disproportionation process that results in the formation of the phosphonic acid and the alkyl phosphine (Eq. 5.4). Impure alkylhydrogenphosphinic acids frequently have a phosphine odor associated with them from this form of instability. The aryl acids also undergo this disproportionation but at higher temperatures.

$$3 \; R\overset{\overset{O}{\|}}{\underset{\underset{OH}{|}}{P}}-H \quad \xrightarrow{\Delta} \quad RPH_2 \; + \; 2\,R\overset{\overset{O}{\|}}{P}(OH)_2 \qquad (5.4)$$

5. *Phosphoric Acid Esters, $(RO)_2P(O)OH$ and $ROP(O)(OH)_2$*. These compounds are of wide occurrence in organophosphorus chemistry, natural and synthetic, and can be considered as parents for the classification of the many known derivatives. Triesters also are important in synthetic, but not natural, phosphorus chemistry. The dialkyl phosphates (also called phosphodiesters) are the better known of the two structures; there are many complex compounds with this structural feature that play vital roles in biochemical processes (see chapter 11.A), and there is much current activity in this field, especially as related to the synthesis of

nucleotides and modified nucleotides. These phosphoric acid derivatives are somewhat stronger acids than are those with C−P bonds; thus diethyl phosphate has pK_a 1.39. Simple dialkyl phosphates are highly water soluble, thermally stable substances; they can be hydrolyzed on heating in acid or base to phosphoric acid. The chemistry and mechanistic aspects of phosphoric acid derivatives have been reviewed.[20,21]

6. *Hypophosphorous Acid Esters, ROP(O)H₂.* The acid structure can be visualized as the dihydrogen parent of the phosphinic acids, and in fact a more systematic name for the acid is *phosphinic acid*. Very few esters with this structure are known, as is evident from a review.[2f] The methyl and ethyl derivatives are the most common and are easily derived from the reaction of hypophosphorous acid with a trialkyl orthoformate. As is true of H-phosphonates, displacements of the alkoxy group are quite easy; with organometallic reagents, the C−P bond can be formed to give primary phosphine oxides, which can be oxidized to phosphonic acids. Another useful property is the weak acidity of the P−H bonds; anions can be generated and used in various reactions, as in the two-fold Michael addition[19] of Eq. 5.5 for the generation of heterocycles. The dicarboxylate product participates in the Dieckmann condensation (Eq. 5.5) to give a six-membered ring.

$$\text{MeO}\overset{\text{O}}{\underset{\text{H}}{\overset{\|}{\text{P}}}}\text{-H} + \text{H}_2\text{C=CHCO}_2\text{Me} \xrightarrow{\text{R}_3\text{N}} \left[\text{MeO}\overset{\text{O}}{\underset{\text{H}}{\overset{\|}{\text{P}}}}\text{-CH}_2\text{CH}_2\text{CO}_2\text{Me}\right] \xrightarrow[\text{2. CH}_2\text{=CHCO}_2\text{Me}]{\text{1. MeONa}}$$

$$\text{MeO}\overset{\text{O}}{\overset{\|}{\text{P}}}(\text{CH}_2\text{CH}_2\text{CO}_2\text{Me})_2 \xrightarrow{\text{NaOEt}} \text{(cyclohexene with OH, CO}_2\text{Me, P(=O)OMe)} \xrightarrow[-\text{CO}_2]{\text{H}^+} \text{(cyclohexanone with P(=O)OMe)} \quad (5.5)$$

5.B. DERIVATIVES OF PHOSPHORUS ACIDS

Phosphorus acids of all types are frequently prepared and used in the form of ester, amide, or thioester derivatives. Anhydrides with the P−O−P link also are encountered, especially with phosphoric acid derivatives. All such derivatives can ultimately be converted to the free acids by conventional acid or base hydrolysis, and many undergo interchanges with other nucleophiles to create new possibilities. For the preparation of derivatives from the acids, it is the practice to convert the OH groups to Cl or to some other reactive leaving group, since in general the direct displacement of OH on phosphorus by simple nucleophiles is too slow to be practical. With the chlorides (or other) in hand, a multitude of displacement reactions can be performed, and huge numbers of derivatives are known for the parent acids with C−P bonds and for phosphoric acid. In this section, we will see just a bare outline (Scheme 5.2) of the many possibilities that can be prepared, using only the phosphorus chloride bond as the reactant. Many of these reagents can be applied to the displacement of the other P-functions, especially the alkoxy and aryloxy

5.B. DERIVATIVES OF PHOSPHORUS ACIDS

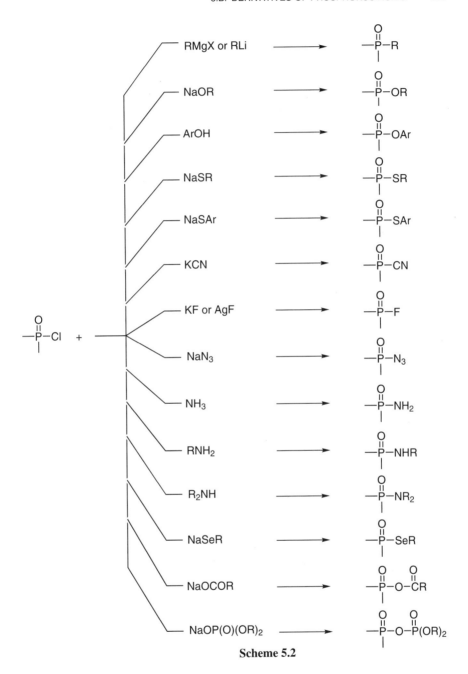

Scheme 5.2

groups. In the phosphonic series, it is possible to have two different substituents on P, and three in the phosphates. The review literature[2–8,13] is available for greater depth in this subject. In Appendix 5.1, compounds 5 to 17 were all prepared by some form of nucleophilic displacement on phosphorus.

Much attention has been paid to the physical chemistry and mechanism of substitution at phosphoryl compounds and reviews[3b,20,21] can be consulted for a coverage of some of the fundamental experimental work in this area. For simple, noncyclic compounds, the reaction can be visualized as occurring through a 5-coordinate transition state or intermediate (the latter is preferred by Holmes[21]) with the usual trigonal bipyramidal geometry. This mechanism was described in chapter 2.F. The incoming nucleophile adopts an apical position in the bipyramid (structure 3 in Eq. 5.6 for the hydrolysis of a phosphinate). The leaving group is believed to require the apical position for departure. As this group departs, phosphorus returns to the tetrahedral shape. In the case in which three different substituents are present in the phosphoryl compound and a nucleophile other than $^-$OH is used, stereochemical inversion is the general result.

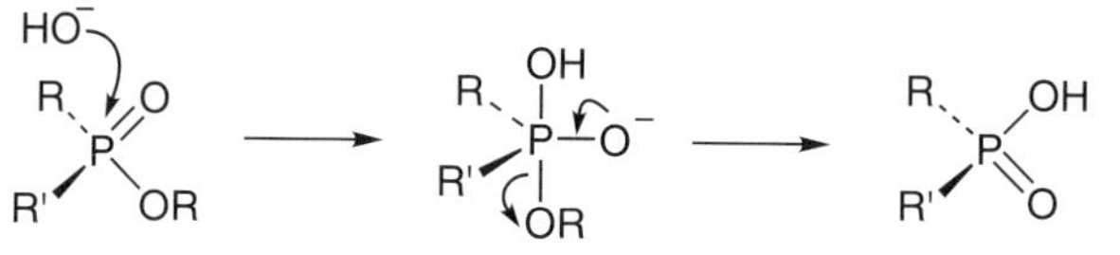

$$\text{(structures shown)} \qquad (5.6)$$

An early demonstration of inversion was accomplished by the equilibration of an optically active form of a methyl ester of a phosphinic acid, in which the methoxy carbon is labelled with ^{14}C, by the attack of unlabelled methoxide[22] (Eq. 5.7). If inversion occurs, the optical activity will be lost after one-half of the optically active ester has reacted, thus racemization would occur at half the rate of the loss of the ^{14}C label. This is in fact what was observed experimentally. Inversion has been demonstrated for several other types of phosphoryl compounds over later years of research.

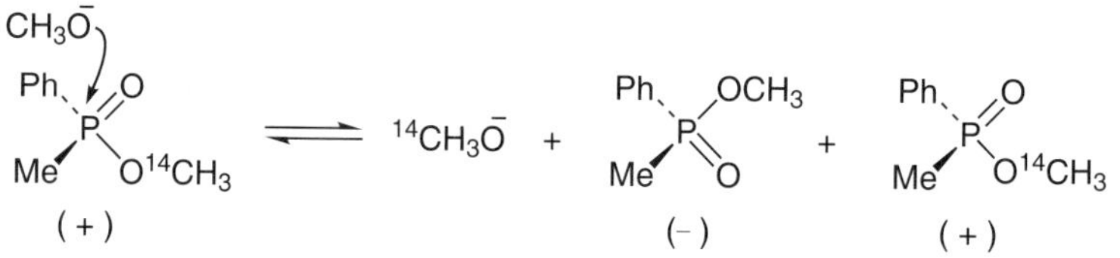

$$\text{(structures shown)} \qquad (5.7)$$

Each structural type of phosphoryl derivative will, of course, have its own rate of substitution, and much effort has gone into sorting out the factors leading to these rates. For a given structural type, it can be demonstrated that increasing the size of the substituents on the phosphorus tetrahedron will reduce the reaction rate as steric hindrance develops in the intermediate TBP. Thus for phosphonates, it has been demonstrated that increasing the size of both the P-alkyl substituent, and of the O-alkyl substituent reduces the rate of basic hydrolysis.[23] Many examples are known of such effects. Another practical effect to consider is the influence of induction within a given P-substituent type; as might be expected, increased electron attraction in the P-substituent causes an increase in the rate of nucleophilic substitution. Thus for the diethyl phosphonates in which the P-alkyl substituent is varied from methyl to chloromethyl to dichloromethyl, the rate of basic hydrolysis is in the order 1:15.6:108, even though steric hindrance increases in the same direction.[24a] When phosphorus is included in a cyclic structure, pronounced influences on reaction rates and stereochemistry can be observed. In general, the displacement of a group from a 5-membered ring, such as the cyclic phosphinate **4**, will be faster than will the displacement from a 6-membered ring (**5**) or a noncyclic relative (**6**).[24b] Cyclic 5-coordinate intermediates[21] continue to receive attention in the literature.[25] Indeed, the entire subject of nucleophilic substitution at 4-coordinate phosphorus remains under consideration.

4   5   6

Finally, we note that an entirely different mechanism than that of the addition-elimination discussed above can operate; in certain structures, a 4-coordinate phosphoryl compound can first undergo an elimination reaction to produce a transient 3-coordinate species (σ^3, λ^5) which then adds the nucleophile to give a 4-coordinate product (Eq. 5.8). Discussion of the process, and the nature of the 3-coordinate intermediates, appears in chapters 2.K and 10.D.

$$\text{RO–P(=O)(OH)–OH} \xrightarrow{-H_2O} [\text{RO–P(=O)=O}] \xrightarrow{\text{R'OH}} (RO)(R'O)\text{P(=O)–OH} \quad (5.8)$$

5.C. THE FORMATION OF PHOSPHORUS ACIDS WITH C–P BONDS

The numerous methods that have been developed for the synthesis of phosphonic, phosphinic, and H-phosphinic acids are well described in the reviews already cited.[2–9] Only the most widely used methods will be presented here. The material is primarily organized around the type of phosphorus compound used as the starting material.

Phosphonous dihalides are excellent starting materials for the synthesis of both phosphonic and H-phosphinic acids. These halides are available from a variety of synthetic methods, both with alkyl and aryl substituents, as was discussed in chapter 3. The hydrolysis of phosphonous dihalides gives rise to H-phosphinic acids. This process is easier to perform with the aryl derivatives and indeed has been conducted for many years on a commercial scale with phenylphosphonous dichloride (Eq. 5.9). Phenylphosphinic acid is one of the most readily available organophosphorus compounds; the overall commercial synthesis simply involves the two basic steps of reaction of benzene with PCl_3, by either the $AlCl_3$-catalyzed or the thermal process, followed by the hydrolysis of $PhPCl_2$. It has recently been announced that tolylphosphinic acid (as an isomer mixture) is now also commercially available (Ferro Corporation).

$$Ph-PCl_2 \xrightarrow{H_2O} Ph-\underset{\underset{H}{|}}{\overset{\overset{O}{\|}}{P}}-OH \qquad (5.9)$$

The hydrolysis reaction can be quite violent with the lower alkylphosphonous dihalides, and must be conducted with great care and control of conditions. With methylphosphonous dichloride, it is not uncommon to have inflammation take place if the compound is added carelessly to water. This is probably due to the heat of the reaction causing some disproportionation of the methylphosphinic acid, releasing methylphosphine that ignites when hot. A more controlled conversion to the simple alkylphosphinic acids can be accomplished by first reacting the phosphonous dichloride with ethanol to give the diethyl alkylphosphonite, which then can be hydrolyzed to the phosphinic acid (Eq. 5.10).

$$RPCl_2 \xrightarrow[\text{base}]{EtOH} RP(OEt)_2 \xrightarrow{H_2O} R\underset{\underset{H}{|}}{\overset{\overset{O}{\|}}{P}}-OH \qquad (5.10)$$

Because of the instability of the acids, pure samples of the simple alkylphosphinic acids are difficult to prepare, and most are reported as oils.[2d] Cleavage of the C—P bond has been found to occur in a few special cases on hydrolysis of substituted-alkyl phosphonous dichloride derivatives en route to phosphinic acids, as with the CX_3[26] (Eq. 5.11) or Ph_3C group.[27]

$$CCl_3PCl_2 \xrightarrow[\Delta]{H_2O} CHCl_3 + H_3PO_3 + HCl \qquad (5.11)$$

With the H-phosphinic acids in hand, ready access to the phosphonic acids is assured, since many oxidizing agents can accomplish this conversion in excellent yield. However, there are ways to bypass the phosphinic acid intermediates in phosphonic acid synthesis. Thus the phosphonous dichlorides can be oxidized directly in inert solvents by such anhydrous reagents as sulfuryl chloride, SO_3, Me_2SO, NO_2, and oxygen to give the phosphonic dichlorides (phosphonyl dichlorides); these are stable, distillable liquids that are easily hydrolyzed without complicating side reactions to the phosphonic acids (Eq. 5.12).

5.C. THE FORMATION OF PHOSPHORUS ACIDS WITH C–P BONDS

$$\text{RPCl}_2 \xrightarrow{(O)} \text{R-}\overset{O}{\overset{\|}{P}}\text{Cl}_2 \xrightarrow{H_2O} \text{R-}\overset{O}{\overset{\|}{P}}(OH)_2 \quad (5.12)$$

Another common procedure for utilizing phosphonous dihalides in phosphonic acid synthesis is to add halogen, usually chlorine, to form the species RPX_4. As noted in chapter 4, these compounds are examples of heterophosphonium salts and are rapidly hydrolyzed with complete loss of halogen to give phosphonic acids (Eq. 5.13). The process is generally restricted to the synthesis of arylphosphonic acids; the chlorination of simple alkylphosphonous dichlorides can be accompanied by substitution in the alkyl group[28]

$$\text{RPCl}_2 \xrightarrow{Cl_2} \text{RPCl}_4 \xrightarrow{H_2O} \text{R-}\overset{O}{\overset{\|}{P}}(OH)_2 \quad (5.13)$$

Phosphinous halides are also useful as sources of phosphinic acids by the hydrolysis-oxidation approach. The intermediate in this case is the secondary phosphine oxide (Eq. 5.14).

$$\text{R}_2\text{PCl} \xrightarrow{H_2O} \text{R}_2\overset{O}{\overset{\|}{P}}\text{H} \xrightarrow{(O)} \text{R}_2\overset{O}{\overset{\|}{P}}\text{OH} \quad (5.14)$$

Several methods exist for the direct formation of the C–P bond in phosphonic acids or derived forms (usually esters). Among these is the highly versatile and widely used Michaelis–Arbusov method (Eq. 5.15), to which reference was made earlier (chapters 3 and 4). The extensive literature has been well reviewed.[2a,5a,29] In its most common form, this method consists of the alkylation of a trialkyl phosphite with a primary alkyl halide; secondary halides can be used, but frequently the product is contaminated with byproducts from HX-elimination, which is, of course, much more extensive for tertiary halides.

$$(\text{RO})_3\text{P} \xrightarrow{R'X} \underset{\underset{OR}{|}}{\overset{\overset{OR}{|}}{R'-\overset{+}{P}-O-R}} \; (X^-) \longrightarrow \underset{\underset{OR}{|}}{\overset{\overset{OR}{|}}{R'-P=O}} + RX \quad (5.15)$$

The Michaelis–Arbusov reaction often requires prolonged heating at moderately high temperatures, but a recent advance has been made that greatly reduces the severity of the reaction conditions; this should be a valuable aid in the synthesis of phosphonates with fragile organic fragments, as is often the case with the present use of the reaction to synthesize molecules resembling biophosphates. In this modification, the reaction is effected under microwave irradiation for only 5 min when simple alkyl halides are reacted with triethyl phosphite.[30]

The mechanism of this reaction has been widely discussed;[31] and the reality of the alkoxyphosphonium ion intermediate has been confirmed, as was noted in chapter 4.D. The scope of the reaction is very wide and well described,[5a] and it has been used to make countless numbers of phosphonates and phosphonic acids. The latter are usually generated by simple acid or base hydrolysis of the esters. In addition to the simple alkyl halides, many complex alkyl groups have been placed on phosphorus by this route. Other alkylating reagents, such as alkyl sulfates, also can be used in the Michaelis–Arbusov process. Dihalides can also be employed; the methylene halides give rise to the valuable *gem*-diphosphonates. Acyl halides are quite useful in the Michaelis–Arbusov reaction and lead to the important family of acylphosphonates, $RC(O)P(O)(OR)_2$.[4] Vinyl and aryl halides, of course, are not as reactive to nucleophiles, although cases are known in which the latter, when bearing activating groups, do give modest yields of phosphonates. The arylation or vinylation of phosphites, however, can be catalyzed by certain metals or metal halides. Probably the most useful catalysts are the nickel(II) halides[32] (5 to 10 mol %) at temperatures around 150°. Under these conditions, a number of aryl halides have been found to react in useful yields with trialkyl phosphites. Valuable catalytic effects have been observed with other metallic compounds, among them $NiCl_2$, $CuCl$, $Ni(CO)_4$ and $PdCl_2$; cases are known in which metal catalysts make possible the reaction of heteroaromatic halides, such as the halothiophenes,[33] with phosphites. Aryl iodides react with sodium dialkyl phosphonates under irradiation to form phosphonates.[34]

An important departure from the normal Michaelis–Arbusov pathway takes place when certain α-halocarbonyl compounds are used in the alkylation step; the products are mainly enol phosphates, in addition to some of the expected ketophosphonate. Enol phosphate formation is referred to as the Perkow reaction and has been developed into a useful synthetic method for this class of phosphates. It is the basis for the formation of important insecticidal materials. Thus the reaction of trimethyl phosphite with chloral (Eq. 5.16) gives dichlorovinyl dimethyl phosphate (DDVP).

$$(MeO)_3P \;+\; Cl_3CCHO \;\longrightarrow\; Cl_2C=CHO-\overset{\overset{O}{\|}}{P}(OMe)_2 \tag{5.16}$$

The mechanism of the Perkow reaction has been studied widely, and although it is possible that two mechanisms are operative, the most common mechanism, as emphasized in a recent review,[35a] involves the attack of the phosphite lone pair on carbonyl oxygen to form an intermediate (**7**) that rearranges to an enolphosphonium (**8**) ion, followed by its rapid dealkylation, as in the second stage of a typical Michaelis–Arbusov reaction (Eq. 5.17).

$$(RO)_3P \;+\; Y-\overset{\overset{O}{\|}}{C}-CHR'Br \;\longrightarrow\; \underset{\underset{\textbf{7}}{}}{Y-\overset{\overset{O^-}{|}}{\underset{\underset{CHR'Br}{|}}{C}}-\overset{+}{P}(OR)_3} \;\longrightarrow\;$$

$$\underset{\underset{R'HC-Br}{|}}{Y\overset{\overset{O}{\triangle}}{C}-P(OR)_3} \;\longrightarrow\; \underset{\underset{\textbf{8}}{}}{Y-\underset{\underset{CHR'}{\|}}{C}O\overset{+}{P}(OR)_3} \;\xrightarrow{Br^-}\; Y-\underset{\underset{CHR'}{\|}}{C}-O-\overset{\overset{O}{\|}}{P}(OR)_2 \;+\; RBr \tag{5.17}$$

In the other mechanism, the phosphite lone pair attacks on halogen to form a halophosphonium ion, which then undergoes nucleophilic attack by the enolate of the ketone (Eq. 5.18).

$$PhC(O)CH_2Br + (EtO)_3P \longrightarrow (EtO)_3\overset{+}{P}Br + PhC(O^-)=CH_2 \longrightarrow$$

$$Ph\underset{CH_2}{\overset{+}{C}}-O-\overset{+}{P}(OEt)_3 \xrightarrow{Br^-} Ph\underset{CH_2}{\overset{\parallel}{C}}-O-\overset{O}{\underset{\parallel}{P}}(OEt)_2 + EtBr \quad (5.18)$$

Both phosphonites and phosphinites participate in the Michaelis–Arbusov reaction, the latter to give tertiary phosphine oxides, the former to give phosphinates, which makes it of interest in the present discussion. This reaction is particularly useful for the synthesis of phosphinic esters with two different P-substituents (Eq. 5.19).

$$MeP(OEt)_2 + EtBr \longrightarrow MeEt\overset{O}{\underset{\parallel}{P}}-OEt \quad (5.19)$$

In the special case in which the alkyl group of the alkyl halide is the same as that of the trialkyl phosphite, the Michaelis–Arbusov reaction simply amounts to an isomerization of the phosphite to the phosphonate. To effect this process, one needs only to use a catalytic amount of the alkyl halide, because it is continually regenerated in the reaction. Dimethyl methylphosphonate and other derivatives are easily made by this process, which has been conducted on a commercial scale (Eq. 5.20).

$$(MeO)_3P \xrightarrow{MeX} Me-\overset{O}{\underset{\parallel}{P}}(OMe)_2 \quad (5.20)$$

The rearrangement can also be effected in some cases just on heating (to about 200°) the phosphite without added alkyl halide, possibly a result of impurities in the phosphite sample. It has been shown that photochemically induced rearrangement is also possible;[36,37] yields are especially high with benzyl and allyl phosphites (Eq. 5.21).

$$(MeO)_2P-O-CH_2C(Me)=CH_2 \xrightarrow{254\ nm,\ benzene} (MeO)_2\overset{O}{\underset{\parallel}{P}}-CH_2C(Me)=CH_2 \quad (5.21)$$

The dialkyl esters of phosphorous acid, having the H-phosphonate structure, are also widely used in phosphonate synthesis. This is accomplished by removing the proton on phosphorus with active metals (most commonly sodium) or strong bases (NaOEt, NaH, NaNH$_2$, BuLi), as already noted, to create an anion that then acts as a nucleophile in alkylation reactions. With alkyl halides (primary are preferred), the attack is virtually exclusively on phosphorus (Eq. 5.22).

$$(RO)_2\overset{O}{\underset{\|}{P}}-H \xrightarrow{\text{base}} (RO)_2\overset{O}{\underset{\|}{P}}^- \xrightarrow{R'X} (RO)_2\overset{O}{\underset{\|}{P}}-R' \quad (5.22)$$

The process is known as the Michaelis–Becker reaction (frequently just the Michaelis reaction) and was reported as early as 1897; it is well described in the various reviews on phosphonic acids already cited. The method offers some advantages over the Michaelis–Arbusov reaction, because the anion is a stronger nucleophile than is the tertiary phosphite and undergoes alkylation generally at or near room temperature. Conventional Michaelis–Arbusov reactions, as already noted, require strong heating.

H-phosphonate anions can also be used in conjugate addition reactions (versions of the Michael reaction) with α,β-unsaturated systems, including the usual functions such as ketones, esters, nitriles, and amides, and has been reviewed.[3a,38,39] The process is sometimes referred to as the Pudovik reaction,[3a] but others[5a,9] use this term for the addition of H-phosphonates to carbonyl groups. With the ketone or aldehyde groups, base-catalyzed 1,2-addition can also occur in a kinetically controlled process, but 1,4-addition, which is under thermodynamic control, is the more common final result. An example of 1,4-addition is shown in Eq. 5.23;[40] many other cases of this type are known.

$$(RO)_2\overset{O}{\underset{\|}{P}}-H + PhHC=CH\overset{O}{\underset{\|}{C}}-Ph \xrightarrow{\text{base}} (RO)_2\overset{O}{\underset{\|}{P}}CHPhCH_2\overset{O}{\underset{\|}{C}}-Ph \quad (5.23)$$

H-phosphonates also add to carbonyl groups (and to C=N as well) in saturated compounds, giving rise to α-hydroxyphosphonates in a process sometimes called the Abramov reaction[3a] (again, there is variation in the term; to others,[5a,9] the reaction of trialkyl phosphites with carbonyl compounds constitutes an Abramov reaction). The hydroxy group then permits access to a number of other C-functional phosphonates. Probably the enol tautomer is the reacting species in this reaction, and indeed the 3-coordinate silyl phosphites and phosphorodiamidites are useful mimics of the enol form of the H-phosphonate. Some can be synthesized in chiral form and added to carbonyl groups with high enantioselectivity (retention[41]) at phosphorus as seen in Eq. 5.24. The reaction is proving to be of great value in constructing complex phosphonates; an example[42] is the synthesis of 2′-nucleoside phosphates, analogous to natural phosphates (Eq. 5.25). Other examples have been reviewed.[9]

(5.24)

5.C. THE FORMATION OF PHOSPHORUS ACIDS WITH C–P BONDS

$$\text{[structure: RO-furanose with HO and C=O, B=thymine]} + [(EtO)_2\overset{..}{P}-OH] \text{ or } (EtO)_2\overset{..}{P}-OSiMe_3 \longrightarrow \text{[structure: RO-furanose with HO, B, -(OH or OSiMe}_3\text{), P(O)(OEt)}_2\text{]} \quad (5.25)$$

(R= trityl; B= thymine)

We have already seen that a related carbonyl condensation occurs between another type of P–H compound, the secondary phosphine oxides, and the condensation appears to be a general reaction of P–H compounds. Of use in phosphinic acid synthesis is the condensation with H-phosphinates (Eq. 5.26)

$$\underset{\underset{OEt}{|}}{\overset{\overset{O}{\|}}{R-P-H}} + R'_2CO \longrightarrow \underset{\underset{OEt}{|}}{\overset{\overset{O}{\|}}{R-P-\underset{|}{\overset{OH}{C}R'_2}}} \quad (5.26)$$

The P–H bonds of hypophosphorous acid esters can also participate in the carbonyl condensation; one or both of the P–H bonds may react (Eq. 5.27).

$$\underset{\underset{H}{|}}{\overset{\overset{O}{\|}}{RO-P-H}} + R'CHO \longrightarrow \underset{\underset{H}{|}}{\overset{\overset{O}{\|}}{RO-P-\underset{|}{\overset{OH}{C}R'H}}} \xrightarrow{R'CHO} \underset{}{\overset{\overset{O}{\|}}{RO-P(\underset{|}{\overset{OH}{C}R'H})_2}} \quad (5.27)$$

Another useful synthesis starting with H-phosphonates is based on their free radical reactions with alkenes in what is known as the hydrophosphonation process. Radicals are formed by abstracting H from P in the usual way with peroxides, azo compounds, or by irradiation; the radical then adds to the carbon-carbon double bond in a chain reaction. The mechanism is illustrated with the addition to 1-methylcyclohexene (Eq. 5.28), which reveals the regioselectivity of the addition.[43]

$$\text{[1-methylcyclohexene with Me]} + (RO)_2\overset{\overset{O}{\|}}{P}-H \xrightarrow{R'O-OR'} \text{[methylcyclohexane with Me and P(O)(OR)}_2\text{]} \quad (5.28)$$

H-Phosphonates have other uses in the synthesis of C–P acids. It was noted earlier (chapter 4.A) that they react with organometallic compounds to give secondary phosphine oxides, and these on oxidation are converted to phosphinic acids. This amounts to an excellent way to obtain the simple symmetrical phosphinic acids (Eq. 5.29).

$$\underset{\underset{H}{|}}{\overset{\overset{O}{\|}}{RO-P-OR}} + R'MgX \longrightarrow \underset{\underset{H}{|}}{\overset{\overset{O}{\|}}{R'-P-R'}} \xrightarrow{(O)} R'_2P(O)(OH) \quad (5.29)$$

In another approach,[44] the H-phosphinate **9**, which contains a labile acetal group, is used as a synthon for hypophosphorous acid; a P—H group is converted to P-aryl by reaction with an aryl bromide in the presence of a Pd(0) complex, and the acetal group is hydrolyzed and eliminated (Eq. 5.30).

$$(EtO)_2CH-\underset{H}{\underset{|}{\overset{O}{\overset{\|}{P}}}}-OEt \xrightarrow[Pd]{O_2N-\text{C}_6H_4-Br} (EtO)_2CH-\underset{OEt}{\overset{O}{\overset{\|}{P}}}-\text{C}_6H_4-NO_2 \xrightarrow{H_2O, H^+}$$

9

$$O_2N-\text{C}_6H_4-\underset{OH}{\overset{H}{\underset{|}{P}}}=O \qquad (5.30)$$

The inorganic phosphorus halides have played several roles in the synthesis of C—P acids. The most obvious way to prepare phosphonic acids would appear to be the reaction of an organometallic reagent with $POCl_3$ to obtain the phosphonic dichloride, which can then be hydrolyzed to the acid. This method, however, has never proved to be satisfactory, because overalkylation is inevitable. The use of partially blocked derivatives, or of trialkyl esters, has been much more fruitful. Thus replacement of one chlorine by a dialkylamino or ethoxy group has given derivatives that limit the reaction to the formation of two C—P bonds, and this approach has been used in the preparation of cyclic phosphinic acid derivatives, as in Eq. 5.31.[45]

$$Et_2N-\overset{O}{\overset{\|}{P}}Cl_2 + BrMg(CH_2)_5MgBr \longrightarrow \underset{O \quad NEt_2}{\text{cyclo-P}} \xrightarrow{H_2O, H^+} \underset{O \quad OH}{\text{cyclo-P}} \qquad (5.31)$$

A particularly useful method for the synthesis of unsaturated phosphonic acid derivatives depends on the addition of PCl_5 to the double bond. The initial adduct loses HCl to form a vinyl derivative.

There are a number of synthetic methods that have been developed in recent years that make use of blocked phosphoric acid chlorides. Particularly intriguing are reactions based on the use of phosphorochloridates where two ester bonds are derived from carbohydrate OH groups as in structure **10**.[46] Reaction with one mole of MeMgX gave the methylphosphonic ester (Eq. 5.32), in which the isomer with axial methyl predominated over that with equatorial methyl by 5:1. The reaction,

(5.32)

10

therefore, proceeds with retention, but inversion is the more common result of nucleophilic displacement on the 6-membered ring.[47] Related reactions and their stereochemical implications are discussed in a review[48] and also receive attention in chapter 9.

Other chiral templates with P—Cl bonds have also been employed in stereochemical studies, an excellent example being the synthesis of cyclic phosphonate **11** in which P has the S-configuration[49] (Eq. 5.33). Again the reaction proceeds with retention, but this is the expected result for displacements on 5-membered rings.[47]

$$\underset{\underset{\mathrm{CH(CH_3)_2}}{|}}{\overset{\mathrm{Ph}}{\underset{\mathrm{Me}}{\diagdown}}\mathrm{N}\overset{\mathrm{O}}{\diagup}\overset{\mathrm{Cl}}{\underset{\mathrm{O}}{\diagdown}}\mathrm{P}\overset{\mathrm{}}{\diagup}} \xrightarrow{\mathrm{H_2C=CHCH_2MgBr}} \underset{\underset{\underset{\mathbf{11}}{\mathrm{CH(CH_3)_2}}}{|}}{\overset{\mathrm{Ph}}{\underset{\mathrm{Me}}{\diagdown}}\mathrm{N}\overset{\mathrm{O}}{\diagup}\overset{\mathrm{CH_2CH=CH_2}}{\underset{\mathrm{O}}{\diagdown}}\mathrm{P}\overset{\mathrm{}}{\diagup}} \qquad (5.33)$$

Triesters of phosphoric acid are less reactive to organometallics, but a method has been developed for phosphonate synthesis by the reaction of two moles of an alkyl lithium with simple phosphates, as seen in Eq. 5.34.[50] Here the second mole of RLi acts to abstract a proton from the initially formed mono-alkyl product **12**, giving a carbanion (**13**) that is of reduced reactivity at phosphorus and causing the reaction to stop at this stage. The carbanion can be protonated at this point or subjected to alkylation with an alkyl halide. Phosphonate carbanions are of considerable importance in synthesis and are discussed further in section 5.F.

$$(RO)_2\overset{O}{\overset{\|}{P}}-OR \xrightarrow{R'CH_2Li} \underset{\mathbf{12}}{(RO)_2\overset{O}{\overset{\|}{P}}-CH_2R'} \xrightarrow{R'Li} \underset{\mathbf{13}}{(RO)_2\overset{O}{\overset{\|}{P}}-\overset{-}{C}HR'} \underset{Li^+}{} \xrightarrow{R''X} (RO)_2\overset{O}{\overset{\|}{P}}-CHR'R'' \qquad (5.34)$$

Other useful reactions are known based on alkylations of 4-coordinate phosphoryl compounds; the earlier work has been reviewed,[51] but new results continue to appear.

Phosphorus trichloride has been much more useful in the synthesis of acids with P—C bonds. In addition to its value in phosphonous and phosphinous halide synthesis, PCl_3 can be used, for example, in the direct synthesis of phosphonic dichlorides by reaction with saturated hydrocarbons in the presence of oxygen (the Clayton–Jensen oxyphosphonation reaction[52]). Cyclohexane is a particularly useful substrate because no isomer formation can occur (Eq. 5.35). The reaction is conducted simply by bubbling in oxygen to a mixture of the hydrocarbon and PCl_3. Much of the PCl_3 is oxidized to $POCl_3$ in the process.

$$\bigcirc \xrightarrow[O_2]{PCl_3} \bigcirc\!\!-\!\overset{O}{\overset{\|}{P}}Cl_2 \qquad (5.35)$$

The phosphonation procedure has been applied to alkyl halides, as well as to alkenes, which give β-chloroalkylphosphonic dichlorides by addition of the released HCl to the double bond. The phosphonation reaction has been widely studied[3d,2g] and appears to proceed through a free radical mechanism.[53]

Another approach to phosphonic acids employs PCl_3 in an alkylation reaction with alkyl halides promoted by $AlCl_3$. These three reactants, in equimolar amounts, combine to form a complex that can be described as a chlorophosphonium salt (**14**), hydrolyzable to the phosphonic dichloride and then to the phosphonic acid (Eq. 5.36).

$$PCl_3 + RX \xrightarrow{AlCl_3} R\overset{+}{P}Cl_3\ \overset{-}{AlCl_4} \xrightarrow{H_2O} RPO(OH)_2 \qquad (5.36)$$
$$\mathbf{14}$$

The reaction can be used with primary, secondary, and tertiary alkyl halides, including cyclic and benzylic derivatives. The details of the reaction mechanism leading to the chlorophosphonium salt (whose structure is not in doubt) remain somewhat obscure. There appears to be considerable development of carbocation character from the alkyl halide, because the order of reactivity is $3° > 2° > 1°$. Furthermore, typical carbocation rearrangements occur with primary and secondary halides. A possible expression of the mechanism is shown in Eq. 5.37. The literature speaks against the formation of a true carbocation intermediate, and this mechanism leaves open the question of how much carbocation character does develop. Phosphorus trichloride has no detectable reactivity to either alkyl halides or to $AlCl_3$, and it must require carbocationic character in the co-reactant halide for it to participate in the process.

$$R\cdots Cl \cdots AlCl_3 \longrightarrow R\overset{+}{P}Cl_3\ \overset{-}{AlCl_4} \qquad (5.37)$$
$$PCl_3$$

Regardless of the mechanism, the process is quite useful for phosphonic acid synthesis, and has been widely employed. An example in *Organic Synthesis* is the synthesis of CCl_3POCl_2, using carbon tetrachloride as the alkyl halide.[54]

Phosphorus trichloride is also used in an outstanding method for the synthesis of aromatic phosphonic acids. In this method, devised by Doak and Freedman,[55] stable anhydrous diazonium fluoroborates are suspended in dry ethyl acetate containing a cuprous halide as catalyst; addition of PCl_3 leads (generally after an induction period) to evolution of nitrogen and formation of an intermediate presumed to have halophosphonium ion structure **15**. Hydrolysis of the mixture gives the phosphonic acid (Eq. 5.38). This process was the first to allow the synthesis of a number of isomer-free phosphonic acids, because the phosphorus group specifically takes the place of the amino group used in the diazotization reaction. Furthermore, amines with electon-withdrawing groups, especially nitro, were especially useful, and this

$$\text{O}_2\text{N}-\text{C}_6\text{H}_4-\text{NH}_2 \xrightarrow[\text{HBF}_4,\ 0\text{-}5°]{\text{NaNO}_2} \text{O}_2\text{N}-\text{C}_6\text{H}_4-\text{N}_2^+\text{BF}_4^- \xrightarrow[\text{CuCl}]{\text{PCl}_3}$$

$$\text{O}_2\text{N}-\text{C}_6\text{H}_4-\overset{+}{\text{P}}\text{Cl}_3 \xrightarrow{\text{H}_2\text{O}} \text{O}_2\text{N}-\text{C}_6\text{H}_4-\text{PO(OH)}_2 \qquad (5.38)$$
15

led to the synthesis of several new phosphonic acids unavailable at the time from other methods (e.g., *p*-nitrophenylphosphonic acid in Eq. 5.38). A valuable extension to phosphinic acid synthesis was made later[56] by replacing PCl$_3$ with an arylphosphonous dichloride, again resulting in the first syntheses of many previously unattainable diarylphosphinic acids (Eq. 5.39).

$$\text{O}_2\text{N}-\text{C}_6\text{H}_4-\text{N}_2\text{BF}_4 \xrightarrow[\text{CuCl}]{\text{ArPCl}_2} \text{O}_2\text{N}-\text{C}_6\text{H}_4-\overset{+}{\text{P}}\text{ArCl}_2 \xrightarrow{\text{H}_2\text{O}} \text{O}_2\text{N}-\text{C}_6\text{H}_4-\text{P(O)(OH)Ar} \qquad (5.39)$$

Derivatives of methylenebisphosphonic acid have become of great importance in recent years, because certain of them are clinically useful in the treatment of bone disease. This valuable new aspect of phosphonic acid chemistry is discussed in chapter 11.C. The most effective compounds have an OH group and an alkyl substituent (frequently aminoalkyl) on the methylene carbon; such compounds are readily made in one step by a remarkable reaction of a carboxylic acid with a mixture of phosphorous acid and PCl$_3$. One may understand the process as being initiated by addition of the P–H bond of H$_3$PO$_3$ to the carbonyl group, followed by regeneration of the carbonyl for a second addition of a P–H bond (Eq. 5.40). In a valuable modification of this process used to synthesize the drug Alendronate from γ-aminobutyric acid, methanesulfonic acid replaced PCl$_3$ and gave a more tractable reaction medium.[57]

$$\text{NH}_2(\text{CH}_2)_3\text{COOH} + \text{H}-\overset{\text{O}}{\underset{}{\text{P}}}(\text{OH})_2 \xrightarrow{\text{MeSO}_3\text{H}} \left[\text{R}-\underset{\text{OH}}{\overset{\text{PO(OH)}_2}{\text{C}}}-\text{OH} \right] \xrightarrow{-\text{H}_2\text{O}}$$

$$\text{NH}_2(\text{CH}_2)_3-\underset{}{\overset{\text{PO(OH)}_2}{\text{C}}}=\text{O} \xrightarrow{\text{H}-\overset{\text{O}}{\text{P}}(\text{OH})_2} \text{NH}_2(\text{CH}_2)_3-\underset{\text{PO(OH)}_2}{\overset{\text{PO(OH)}_2}{\text{C}}}-\text{OH} \qquad (5.40)$$

Methylenebisphosphonates can be prepared by the Michaelis–Arbusov reaction with methylene halides as reactants; the anion may be generated with butyllithium, and alkylated with alkyl halides (Eq. 5.41). Several other methods, summarized in reference 57, are available for bisphosphonate synthesis.

$$CH_2X_2 + 2(EtO)_3P \longrightarrow CH_2[P(O)(OEt)_2]_2 \xrightarrow{BuLi} \bar{C}H[P(O)(OEt)_2]_2 \xrightarrow{RX} RCH[P(O)(OEt)_2]_2$$

(5.41)

Finally, to show that innovation in phosphonic acid synthesis is still possible, the use of white phosphorus (P_4) as a starting material in a free radical synthesis, initiated by radiation with "white light" can be cited.[58] Here, a Barton ester (**16** in Eq. 5.42) and P_4 are reacted in $CH_2Cl_2-CS_2$[58a] or THF;[58b] the mixture is oxidized with H_2O_2 to give excellent yields of phosphonic acids. The example in Equation 5.42 (DCC = dicyclohexylcarbodiimide) was selected to show the value of this new method in the synthesis of phosphonic acids with complex structures.[58b]

(5.42)

Several of the synthetic methods presented above have led directly to simple esters of phosphonic and phosphinic acids. When more complicated esters are required, an approach is to form the acid by hydrolysis (acidic or basic), convert it to the acid chloride with, e.g., $SOCl_2$, and then react the chloride with an alcohol to give the desired new ester. Amides may be made similarly from the reaction of the chlorides with amines. Phosphonic acid esters with mixed alkoxy substituents may be prepared by a recently announced method[59] that involves transesterification with a carboxylic acid ester, with catalysis by small amounts of KOBu-*t* or NaOBu-*t*. With care in the choice of conditions and reactant ratios, the transesterification may be stopped after replacement of one group on phosphorus (Eq. 5.43). If desired, both groups can be replaced. The readily available $CH_3P(O)(OMe)_2$ is an excellent starting material for the preparation of mixed esters.

$$MeP(OMe)_2 + MeCOR \xrightarrow[KOBu-t]{1-5 \text{ mol \%}} MeP(O)(OR)(OMe) + MeC-OMe \quad (5.43)$$

This discussion has centered primarily on ways that provide simple alkyl and aryl phosphinic and phosphonic acids. Other methods, of more limited scope, are available, and descriptions can be found in the review literature cited. To appreciate fully the scope of the C—P acid literature, one should also consider the synthesis of C-functional acids, a topic that cannot be included here. Many of the methods cited here are applicable to such compounds, but a number of specialized processes are also known. The important subject of α-ketophosphonic acids has been admirably reviewed by Breuer,[4] and Edmundson has provided an excellent coverage of many other types of C-functional acids.[3a]

5.D. SYNTHESIS OF PHOSPHORIC ACID DERIVATIVES

Although thousands of organic derivatives of phosphoric acid have been reported, there are in fact relatively few methods that have been employed in their synthesis. Phosphoric acid esters are of great commercial importance; a valuable summary has been prepared by Toy and Walsh[60] and the comments on commercial applications that follow were taken from this source. Other valuable reviews are references 2 and 13.

Phosphoric acid itself is never used as a starting material; as is typical of other P—OH groups, direct catalyzed esterification with alcohols, or amidation with amines, is far too slow to be practical. However, phosphorus oxychloride is in quite wide use for the synthesis of triaryl and trialkyl phosphates. The reaction with alcohols is performed in the presence of a base to accept the HCl produced. Alkyl chlorides can be formed as byproducts, but nevertheless the process is feasible and has been used to produce phosphates on a commercial scale. Tri-*n*-butyl phosphate is used as a selective extractant for metal compounds, particularly in the purification of uranium compounds, and as a hydraulic fluid in aircraft, as well as in other applications. Tris(2-ethylhexyl) phosphate is used as a plasticizer for vinyl polymers and is valuable also in imparting some flame resistance to the material. Triaryl phosphates are also manufactured commercially (Eq. 5.44) and have various uses, including additives for gasoline, polymer plasticizers, lubricants, coolants, hydraulic fluids, and flame retardants. The reactions require heat for completion, but the products are quite stable to the HCl produced, and no base is used to trap this acid.

$$3ArOH + POCl_3 \xrightarrow{\Delta} (ArO)_3P=O \quad (5.44)$$

Another valuable approach to tri-substituted phosphates is to prepare the corresponding phosphites and then perform an oxidation reaction. Similarly, dialkyl H-phosphonates can be oxidized to phosphates, in this case giving the dialkyl phosphates (Eq. 5.45).

$$PCl_3 \xrightarrow{ROH} (RO)_2\overset{O}{\overset{\|}{P}}H \xrightarrow{(O)} (RO)_2\overset{O}{\overset{\|}{P}}OH \quad (5.45)$$

Dialkyl phosphates can also be prepared by other means, including the easily controlled partial hydrolysis of tri-alkyl esters with NaOH (Eq. 5.46), and by the reaction of alcohols (or phenols) with "phosphorus pentoxide," P_4O_{10} (Eq. 5.47).

$$(RO)_3P=O + aq.NaOH \longrightarrow (RO)_2P(O)ONa \quad (5.46)$$

$$P_4O_{10} + ROH \longrightarrow (RO)_2P(O)OH + ROP(O)(OH)_2 \quad (5.47)$$

In the latter reaction, the mono ester is also formed; the identity of the alcohol and the conditions lead to variable ratios, but the mixed acids as such are commercially valuable for many applications (metal extractants, organic-soluble detergents, acidic catalysts in various processes, etc.) as are their ammonium salts (textile lubricants) or amine salts (corrosion inhibitors). It is feasible to separate the mixed acids by the different solubilities of their alkaline earth salts. The commercial mixed acids may also contain some trialkyl phosphates and pyrophosphates.

For the preparation of mixed-substituent phosphates, it is customary to prepare a chloro derivative of a di-substituted phosphate and react it with an alcohol. Base is used to absorb the HCl, although if pyridine is the base, evidence suggests[61a] that it may first interact to displace the chlorine and form a highly reactive species of formula $(RO)_2P(O)(NC_5H_5)^+$. A similar species has been detected on mixing phosphorochloridates with Et_3N[61b] These intermediates were detected by ^{31}P NMR and will be described in chapter 6.I. Dialkyl (and aryl) chlorophosphates (more properly, dialkyl phosphorochloridates) can be prepared from dialkyl phosphates with thionyl chloride or PCl_5, or by the chlorination of dialkyl H-phosphonates (Eq. 5.48), a particularly valuable procedure for preparing pure samples of the chlorophosphate.

$$(RO)_2\overset{O}{\underset{\|}{P}}-H \xrightarrow{Cl_2} (RO)_2\overset{O}{\underset{\|}{P}}-Cl \xrightarrow{R'OH} (RO)_2\overset{O}{\underset{\|}{P}}-OR' \qquad (5.48)$$

For the synthesis of complex phosphates of biological importance, several methods are in use. Phosphotriesters may be synthesized by the carbodiimide coupling method, as in Eq. 5.49, or by the Mitsunobu method (Eq. 5.50), which was described in chapter 3.C.

$$(RO)_2\overset{O}{\underset{\|}{P}}-OH + R'N=C=NR' \longrightarrow (RO)_2\overset{O}{\underset{\|}{P}}-\underset{NHR'}{\overset{|}{C}}=NR' \xrightarrow{R''OH} (RO)_2\overset{O}{\underset{\|}{P}}-OR'' \qquad (5.49)$$

$$Ph_3P + EtOOCN=NCOOEt \longrightarrow EtOOCN-\underset{Ph_3\overset{+}{P}}{\overset{|}{N}}-COOEt \xrightarrow[(RO)_2PO(OH)]{R'OH} (RO)_2\overset{O}{\underset{\|}{P}}-OR' \qquad (5.50)$$

Two specialized syntheses of nucleotides, also applicable to the synthesis of other types of biophosphates, are the phosphoramidite method and the H-phosphonate method, both discussed previously (chapter 3).

Nitrogen derivatives of phosphoric acid constitute another important class of compounds. The many reviews on this subject include those by Fluck[62] and Corbridge.[63a] The chemistry of the phosphazene bond (as seen in **17**) is of special current importance, because this bond can be found in valuable polymeric materials. Perhaps the best known compound containing this bond is the cyclic trimer phosphonitrilic chloride (**17**), easily formed (along with some tetramer) by refluxing a mixture of PCl_5 and ammonium chloride in an inert solvent such as tetrachloroethane (Eq. 5.51). Many organic derivatives have been formed by nucleophilic substitution with alcohols, amines, thiols, etc., and arylation to form the C—C bond is possible under Friedel–Crafts conditions. Cycic phosphazenes with exocyclic P—C bonds can also be formed by replacing PCl_5 in the reaction with ammonium chloride by the species Ar_2PX_3, and by other methods.

$$PCl_5 + NH_4Cl \xrightarrow{\Delta} \underset{\mathbf{17}}{\text{Cl}_2P\overset{N}{=}\overset{\parallel}{\underset{N}{\text{N}}}\overset{PCl_2}{\underset{\text{Cl}_2}{\text{P}}}N} \qquad (5.51)$$

Phosphazene chemistry is a well-developed part of phosphorus chemistry, treated variably as inorganic or organic, and is too broad to receive further consideration here. The subject is well treated in the reviews already cited.

5.E. SULFUR DERIVATIVES OF PHOSPHORUS ACIDS

In principle, sulfur (or selenium) can replace oxygen, either singly or multiply bound, in any of the phosphorus acids considered in this chapter. Many such compounds are known, again constituting an area we can touch on only briefly. Useful reviews can be found in various chapters in the Kosolapoff–Maier series,[2b,2g,2h,2i] in the Hartley series,[3a] and in the Houben–Weyl series.[7,8] Thioacids of various types are extremely important materials of commerce. Applications as pesticides are numerous and will receive consideration in chapter 11.D. Thio derivatives also are used as flotation agents, oil and gasoline additives, and rubber vulcanization accelerators. More recently, they have become important in studies on the chemical modification of natural phosphates, especially nucleotides.

The acidity of acids with P—SH groups differs little from that of the corresponding acids with P—OH groups, as seen in the phosphate series[63b] (($EtO)_2P(O)OH$, pK_a 1.37; $(EtO)_2P(O)SH$, 1.49; $(EtO)_2P(S)SH$, 1.62). This is also true of acids with P—C bonds. The situation is made complicated, however, by a form of tautomeric equilibrium involving the O=P—SH moiety. This is shown in Eq. 5.52 for dialkyl phosphorothiolates; here the equilibrium constant is about unity,[64] but it can vary considerably in other structures. Thus the diphenyl ester is about 80% in the P=S (thiono) form, and dialkylthiophosphinic acids are completely in the thiono form. The effect of structure on the equilibrium has been studied in some detail by Kabachnik, et al.,[65] who have used a type of Hammett free energy relationship to correlate the effects.

$$(RO)_2\overset{O}{\underset{\parallel}{P}}-SH \rightleftharpoons (RO)_2\overset{OH}{\underset{|}{P}}=S \qquad (5.52)$$

A related problem with thioacids is the site of attack of electrophiles on their anions, which are ambident. Such reactions are generally under kinetic control. This problem has been studied extensively and is the subject of a review.[66] As a general rule, alkyl halides (soft electrophiles) react with the anion of a monothiophosphate at the soft sulfur atom, and this constitutes a reliable method for preparing thiolo esters (Eq. 5.53).

$$(RO)_2\overset{O^-}{\underset{|}{P}}=S \longleftrightarrow (RO)_2\overset{O}{\underset{\|}{P}}-S^- \xrightarrow{R'X} (RO)_2\overset{O}{\underset{\|}{P}}-SR' \qquad (5.53)$$

The solvent, however, can influence the result, and in hexamethylphosphoramide (HMPA), O-alkylation is favored. Esterification with diazoalkanes in ether gives primarily S-alkyl esters;[67] in this reaction a proton is transferred initially from the acid to carbon to form the methyldiazonium ion as the alkylating reagent. Thus, again, it is the ambident ion that is being alkylated (Eq. 5.54). In this case, the soft diazonium ion attacks the soft sulfur atom. Acyl halides on the other hand are considered to be hard and indeed do give primarily O-acyl derivatives with salts of thioacids.

$$(RO)_2\overset{O}{\underset{\|}{P}}-SH + CH_2N_2 \xrightarrow{-N_2} \left[(RO)_2\overset{O}{\underset{\|}{P}}-S^- \longleftrightarrow (RO)_2\overset{O^-}{\underset{|}{P}}=S\right] + CH_3N_2^+ \longrightarrow$$

$$(RO)_2\overset{O}{\underset{\|}{P}}-SCH_3 + N_2 \qquad (5.54)$$

A common procedure for preparing thiolate esters of various phosphorus acids is to react the phosphoryl chloride derivative with a sodium thiolate. Similarly, thio-acids are formed when the chlorides are reacted with NaSH. The P=O group in esters, as in phosphine oxides (chapter 4.B) can be replaced with sulfur by reaction with phosphorus pentasulfide (P_4S_{10}) or the Lawesson reagent (**18**) (Eq. 5.55).

$$R-\overset{O}{\underset{\|}{P}}(OR)_2 + \underset{\mathbf{18}}{\overset{Ar\diagdown \underset{S}{\overset{S}{P}}\diagup S\diagdown \underset{S}{\overset{S}{P}}\diagup Ar}{}} \longrightarrow R-\overset{S}{\underset{\|}{P}}(OR)_2 \qquad (5.55)$$

$$Ar = MeO-\!\!\bigcirc\!\!-$$

Phosphorus pentasulfide, a cheap commercial chemical, is also used for the formation of dithio acids according to Eq. 5.56. This reaction with alcohols is of major importance in the construction of intermediates used in the synthesis of important commercial pesticides (chapter 11.D).

$$P_4S_{10} + ROH \longrightarrow (RO)_2\overset{S}{\underset{\|}{P}}-SH \qquad (5.56)$$

Thiophosphoryl chlorides can also be used to prepare monothio or dithio phosphates (or phosphonates), as in Eq. 5.57.

$$(RO)_2\overset{S}{\underset{\|}{P}}-Cl + ROH \longrightarrow (RO)_2\overset{S}{\underset{\|}{P}}-OR \qquad (5.57)$$

Such reactions are conducted commercially; for example, the insecticide Parathion is prepared from $(EtO)_2PSCl$ and the sodium derivative of *p*-nitrophenol (Eq. 5.58).

$$(EtO)_2\overset{\overset{S}{\|}}{P}-Cl + NaO-\!\!\left\langle\!\!\bigcirc\!\!\right\rangle\!\!-NO_2 \longrightarrow (EtO)_2\overset{\overset{S}{\|}}{P}-O-\!\!\left\langle\!\!\bigcirc\!\!\right\rangle\!\!-NO_2 \quad (5.58)$$

Numerous other reactions have been used in the synthesis of the various types of thiophosphorus acids and can be found in the several reviews already cited. Much of the synthetic work on thiophosphorus acids has been done in connection with the pesticidal activity found in many derivatives. Thio acids and esters are reasonably stable substances, but with strong oxidizing agents, both SH and P=S groups may be converted to the corresponding oxy structures. Vigorous hydrolytic conditions will also convert P=S to P=O, with release of H_2S. The P—SH group and its anionic form are quite nucleophilic; in addition to the alkyl halide alkylation already mentioned, many reactions take place at sulfur. One that has proved of great commercial value is the base-catalyzed addition of esters of the type $(RO)_2P(S)SH$ to activated double bonds in the Michael reaction; this reaction is the basis of the commercial synthesis of the insecticide Malathion (Eq. 5.59).

$$(MeO)_2\overset{\overset{S}{\|}}{P}-SH + \begin{matrix} HC-COOEt \\ \| \\ HC-COOEt \end{matrix} \xrightarrow{\text{base}} (MeO)_2\overset{\overset{S}{\|}}{P}-SCH-COOEt \quad (5.59)$$
$$\phantom{(MeO)_2\overset{\overset{S}{\|}}{P}-SCH-COOEt xxxxxxxxxxxxxx} CH_2COOEt$$

Thiol acids of various types have also been reported to react directly with alkenes to give Markovnikoff addition products, as shown in Eq. 5.60.

$$(MeO)_2\overset{\overset{S}{\|}}{P}-SH + Me_2C=CH_2 \longrightarrow (MeO)_2\overset{\overset{S}{\|}}{P}-SCMe_3 \quad (5.60)$$

The well-known conversion of thiol groups to disulfides by mild oxidizing agents (hydrogen peroxide or iodine water) is also known in the P—SH families, making possible the linking of two thiophosphorus groups (Eq. 5.61). The reaction occurs with SH derivatives of phosphoric acid as well as of phosphonic acids.[68,69]

$$(RO)_2\overset{\overset{S}{\|}}{P}-SH \xrightarrow{H_2O_2} (RO)_2\overset{\overset{S}{\|}}{P}-S-S-\overset{\overset{S}{\|}}{P}(OR)_2 \quad (5.61)$$

O,O,O-Trialkyl phosphorothionates are well known to undergo a rearrangement reaction whereby an alkyl group is transferred from O to S, thus forming a P=S group. The thione to thiolate rearrangement may be catalyzed by acids or by alkyl halides; in the latter case it is known as the Pischimuka reaction and has been used in pesticide manufacture as in the synthesis of Methamidophos (Eq. 5.62).

158   THE ACIDS OF ORGANOPHOSPHORUS CHEMISTRY AND THEIR DERIVATIVES

$$\underset{NH_2}{\overset{\overset{S}{\|}}{MeO-P-OMe}} \xrightarrow{MeX} \underset{NH_2}{\overset{\overset{SMe}{|}}{MeO-P=O}} \quad (5.62)$$

5.F. CARBANION FORMATION AND REACTIVITY AT THE α-CARBON OF PHOSPHONATES

Although it is not the intent of this chapter to describe the many reactions and derivatives of phosphonic acids, a topic admirably covered recently by Edmundson,[3a] there have been developments of such great importance in contemporary phosphorus chemistry as to compel a brief consideration here. This has to do with the creation and reactions of anions on carbon attached to phosphonate groups. Phosphoryl groups are known to be electron attracting groups; for example, the acidity of acetic acid is increased by a factor of 10 in the compound $Ph_2P(O)CH_2COOH$ to pK_a 4.45,[3f] and the possibility of some sort of electron attraction by a resonance interaction with attached double bonds is revealed by their activation to nucleophilic additions of the Michael type (Eq. 5.63).

$$H_2C=CH-P(O)(OR)_2 \longleftrightarrow [H_2\overset{+}{C}-CH=(P(O)(OR)_2)^-] \xrightarrow{R_2NH} R_2NCH_2CH_2P(O)(OR)_2$$
(5.63)

Therefore, the possibility of resonance stabilization of a negative charge α to phosphorus is suggested by these facts, and indeed the early literature records the formation of carbanions at the α-position of phosphonates by abstraction of hydrogen. In his 1950 monograph, Kosolapoff[70] mentions that sodium or potassium form metallic derivatives with phosphonoacetic esters that undergo C-alkylation, but the exploitation of phosphonate carbanions as synthetic intermediates had to await the introduction of such powerful bases as NaH, $NaNH_2$ and lithium reagents into organic chemistry. Two highly important processes are the result of studies over the last 3–4 decades.

The first of these is the olefin synthesis method attributed to Wadsworth and Emmons, which is the subject of extensive reviews.[71,72] In its original form, a carbanion is generated from a phosphonoacetic ester with one of the strong bases, and then reacted with an aldehyde or ketone. Addition of the carbanion to the carbonyl group occurs, but the addition product collapses by elimination of the phosphorus group as the dialkyl phosphate anion, leaving an olefin (E is generally

$$(RO)_2\overset{\overset{O}{\|}}{P}CH_2CO_2Et \xrightarrow{base} (RO)_2\overset{\overset{O}{\|}}{P}CHCOO_2Et \xrightarrow{R'CHO} (RO)_2\overset{\overset{O}{\|}}{P}\underset{\underset{CO_2Et}{|}}{\overset{O-}{\underset{CH}{\diagdown}}}CHR' \longrightarrow$$

$$(RO)_2\overset{-}{P}O_2 + \underset{EtO_2C}{\overset{H}{\diagdown}}C=C\underset{H}{\overset{R'}{\diagdown}} \qquad (5.64)$$

favored) as the other product (Eq. 5.64). This widely used olefin synthesis, and the related Wittig reaction of carbanions α to phosphonium groups (thus ylides), will receive further attention in chapter 11.F.

In 1966, Corey and Kwiatkowski[73a] reported that simple phosphonamides could be converted to α-carbanions with butyllithium, and that the usual condensation with carbonyl compounds could be effected. The initial β-hydroxy phosphonamide adducts (Eq. 5.65) were quite stable, unlike those from phosphonates, and diastereomeric mixtures proved to be separable. The hydroxy compounds could then be decomposed on heating to provide the olefin, in E or Z forms where structurally possible.[73b] The phosphonamides, therefore, offer a distinct advantage over phosphonates as precursors of alkenes, because use of the latter usually leads to E-dominated mixtures of alkenes. Carbanions can also be generated from thiophosphonates, $RCH_2P(S)(OR)_2$ and used in condensations with aliphatic carbonyl compounds to form alkenes.[74]

$$H_3C-\overset{O}{\underset{}{P}}(NMe_2)_2 \xrightarrow{BuLi} Li^+ \ H_2\bar{C}-\overset{O}{\underset{}{P}}(NMe_2)_2 \xrightarrow[2.H^+]{1.RCHO} R\overset{OH}{\underset{}{C}}H-CH_2-\overset{O}{\underset{}{P}}(NMe_2)_2$$

(5.65)

The carbanions from the phosphonamides, phosphonates, and thiophosphonates can be easily alkylated, and in the 1990s, carbanion alkylation became of great value in the synthesis of complex phosphonates of biological importance. The goal of this work was generally to prepare phosphonate analogs of natural phosphates where the α-CH_2 group can be viewed as an isosteric replacement for the O atom of the phosphate. An early example was the synthesis of the phosphonate analog **19** of geranyl phosphate **20**[75] by the alkylation of the lithium derivative of dimethyl methylphosphonate with geranyl bromide.

19 [CH$_2$]–P(O)(OMe)$_2$

20 [O]–P(O)(OH)$_2$

Compound **19** was converted to the phosphonyl-phosphonate **21** which can be viewed as an analogue of the natural geranyl pyrophosphate (**22**) in which the O–P(O)–O–P(O)–O linkage is replaced by C–P(O)–C–P(O)–O. This compound, and similar analogues of isopentenyl and farnesyl pyrophosphates, were effective inhibitors of squalene and kaurene biosynthesis.

21 –CH$_2$–P(O)(OH)–CH$_2$–P(O)(OH)$_2$

22 –O–P(O)(OH)–O–P(O)(OH)$_2$

The lithium carbanions (now frequently formed with lithium diisopropylamide) are highly reactive, but it has been found that other metallic derivatives can be prepared from them that are less reactive. Savignac and Mathey[76] reported Cu(I) derivatives for this purpose, which reacted with acyl chlorides at $-35°$ to give the acyl derivatives. Zinc compounds have also been used as derivatives of lower reactivity.[77] In addition to the common CN, COOR, and RCO substituents on the α-carbon, fluorine and chlorine can be tolerated and, indeed, as electronegative groups, assist in stabilizing the carbanion. Thus the carbanion can be prepared from diisopropyl fluoromethylphosphonate and phosphorylated by diethyl phosphorochloridate[78] (Eq. 5.66).

$$\text{FCH}_2-\overset{\overset{\text{O}}{\|}}{\text{P}}(\text{OPr-i})_2 \xrightarrow{\text{base}} \text{F}\overset{-}{\text{C}}\text{H}-\overset{\overset{\text{O}}{\|}}{\text{P}}(\text{OPr-i})_2 \xrightarrow{(\text{EtO})_2\text{POCl}} \underset{\underset{\text{PO(OEt)}_2}{|}}{\text{FCHPO(OPr-i)}_2} \quad (5.66)$$

Even the difluoromethylphosphonate group can be converted to a carbanion (used as the zinc derivative[79]) and this makes possible the use of the CF_2 group as a replacement for O in biophosphates; Blackburn and co-workers have pointed out this would be a better isostere for O than CH_2.[80] The carbanion from diethyl chloromethylphosphonate has been formylated to give the chloroaldehyde **23**, and that from diethyl 1-chloroethylphosphonate reacts with benzaldehyde to give the stable lithium salt **24**.

$$\underset{\underset{\text{CHO}}{|}}{\text{Cl}-\text{CHPO(OEt)}_2} \qquad \underset{\underset{\text{Ph}-\text{CHOLi}}{|}}{\text{H}_3\text{C}-\text{CClPO(OEt)}_2}$$

23   **24**

Other substituents can be tolerated on the α-carbon such as the trimethylsilyl[81] and the pyrrolidino groups.[82]

We noted the work of Corey and Kwiatkowski, who produced phosphonyl phosphonate analogues of pyrophosphates (e.g., **21**) using phosphonate carbanion chemistry. In later years, other approaches to these important variants were developed and have been reviewed recently.[35b] One approach that introduces a new aspect of phosphonate carbanion chemistry for the construction of P–C–P–C mimics of the natural pyrophosphate linkage P–O–P–O involves a novel dimerization (resembling a Claisen condensation) of lithio diisopropyl methylphosphonate to give phosphonyl-phosphonate **25**, a method introduced by Teulade, et al.[83] Compound **25** was then converted to the dicarbanion **26**, which was found to undergo alkylation first at the terminal, least-stabilized carbanionic center to give structure **27** (Eq. 5.67). This is an application of a general principle of organic chemistry developed years ago by Hauser[84] in studies of dicarbanions of dicarbonyl compounds (and others), in which alkylation was shown to occur at the most basic carbon. Hydrolysis of protonated **27** gave the free acid **28**, which exhibited competitive inhibition of farnesyl diphosphate synthase involved in the biosynthesis of squalene.[85] Phosphonate analogues of farnesyl pyrophosphate have also been prepared from farnesyl halides using $\text{LiCH}_2\text{P(O)(OEt)(morpholinyl)}$ to form the P–C bond, followed by development of the P–O–P linkage by conventional means.[86]

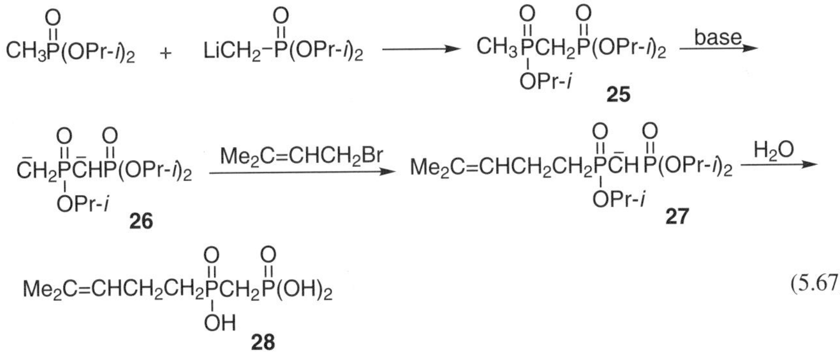

5.F. CARBANION FORMATION AND REACTIVITY AT THE α-CARBON OF PHOSPHONATES

$$CH_3\overset{O}{\overset{\|}{P}}(OPr\text{-}i)_2 + LiCH_2\text{-}\overset{O}{\overset{\|}{P}}(OPr\text{-}i)_2 \longrightarrow CH_3\overset{O}{\overset{\|}{\underset{OPr\text{-}i}{P}}}CH_2\overset{O}{\overset{\|}{P}}(OPr\text{-}i)_2 \xrightarrow{\text{base}}$$

25

$$\overset{-}{C}H_2\overset{O}{\overset{\|}{\underset{OPr\text{-}i}{P}}}CHP(OPr\text{-}i)_2 \xrightarrow{Me_2C=CHCH_2Br} Me_2C=CHCH_2CH_2\overset{O}{\overset{\|}{\underset{OPr\text{-}i}{P}}}\overset{-}{C}H\,\overset{O}{\overset{\|}{P}}(OPr\text{-}i)_2 \xrightarrow{H_2O}$$

26   **27**

$$Me_2C=CHCH_2CH_2\overset{O}{\overset{\|}{\underset{OH}{P}}}CH_2\overset{O}{\overset{\|}{P}}(OH)_2 \qquad\qquad (5.67)$$

28

In addition to their use in the synthesis of biophosphate analogues, phosphonate carbanions based on cyclic structures are useful in obtaining stereoselective or regioselective α-alkylations. To illustrate the former, the optically active phosphonamide **29** was converted to the anion with LiNPr-i_2 and alkylated with an alkyl halide; the product **30** was a separable mixture of diastereoisomers, which could then be hydrolyzed to the optically active phosphonic acids[87] (Eq. 5.68).

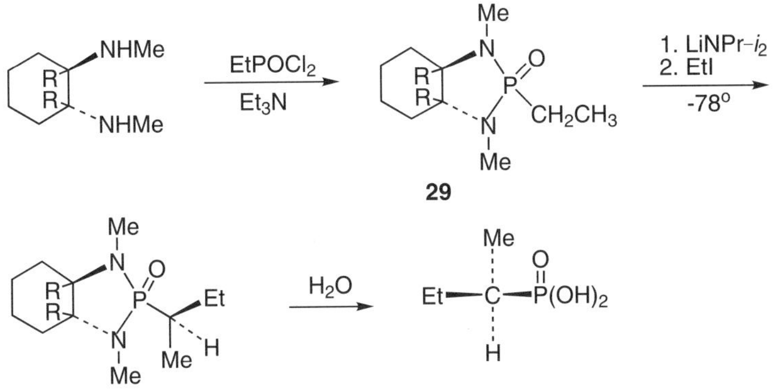

<chemical structures for Eq. 5.68 showing conversion of diamine via EtPOCl₂/Et₃N to compound 29, then 1. LiNPr-i₂, 2. EtI at -78°, to give 30 (major isomer), then H₂O to give Et-C(Me)(H)-P(OH)₂>

(5.68)

30 (major isomer)

The 1,3,2-oxazaphosphorinane ring system as in **31** has also been used in stereoselective alkylation studies[88] (Eq. 5.69).

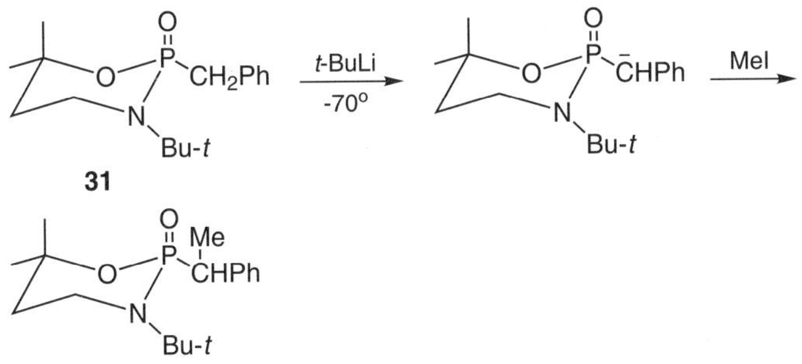

<chemical structures for Eq. 5.69: compound 31 with CH₂Ph, treated with t-BuLi at -70°, giving anion ⁻CHPh, then MeI, giving product with Me-CHPh>

(5.69)

Equally important are structural studies on the anions; one report[89] provides some references to earlier work and describes results on lithiated cyclic phosphonates employing solution NMR and single-crystal X-ray crystallography. The phosphonates **32** and **33** in solution were found to form freely rotating sp^2-hybridized anions, with lithium only associated with oxygen and not carbon.

32   **33**

This brief discussion can only introduce the reader to the basics of the subject; other examples of the use of phosphonate carbanions in the synthesis of biophosphate analogues (including other synthetic approaches) may be found in the review by McClard and Witte.[35b] Additional references to the chemistry of phosphonate carbanions are given in the discussions by Edmundson[3g] and Mastalerz.[35c]

REFERENCES

1. H. H. Jaffé, L. D. Freedman, and G. O. Doak, *J. Am. Chem. Soc.* **75**, 2209 (1953).
2. G. M. Kosolapoff and L. Maier, eds., *Organic Phosphorus Compounds*, Vol. 7, John Wiley & Sons, Inc., New York, 1972. (a) K. H. Worms and M. Schmidt-Dunker, Vol. 7, Chapter 18, (b) W. Gerrard and H. R. Hudson, Vol. 5, Chapter 13, (c) P. C. Crofts Chapter 14, (d) A. W. Frank, Vol. 4, Chapter 10, (e) E. Fluck and W. Hanbold, Vol. 6, Chapter 15, (f) M. Baudler, Vol. 5, Chapter 12, (g) M. Fild and R. Schmutzler, Vol. 4, Chapter 9, (h) A. W. Frank, Vol. 4, Chapter 10, (i) D. E. Ailman and R. J. Magee, Vol. 7, Chapter 19.
3. R. S. Edmundson in F. R. Hartley, ed., *The Chemistry of Organophosphorus Compounds*, Vol. 4, John Wiley & Sons, Inc., New York, 1996, (a) Chapters 2–6, (b) p. 598, (c) pp. 254–257, (d) pp. 74–76, (e) Chapter 5, (f) p. 499, (g) pp. 112–116.
4. E. Breuer in F. R. Hartley, ed., *The Chemistry of Organophosphorus Compounds*, John Wiley and Sons, Inc., 1996, Vol. 4, Chapter 7.
5. R. Engel, *Synthesis of Carbon-Phosphorus Bonds*, CRC Press, Boca Raton, Fla., 1988. (a) pp. 21–34, (b) pp. 101–107, (c) pp. 78–87.
6. L. D. Freedman and G. O. Doak, *Chem. Rev.* **57**, 479 (1957).
7. K. Sasse, *Methoden der Organischen-Chemie (Houben-Weyl), Band XII, Part 1 (Organischen Phosphor Verbindungen)*, Georg Thieme, Stuttgart, Germany, 1963, pp. 338–619.
8. B. Gallenkamp, W. Hofer, B.-W. Krüger, F. Maurer, and T. Pfister in M. Regitz, ed., *Methoden der Organischen-Chemie (Houben-Weyl), Band E(2), (Organischen Phosphor Verbindungen)* Georg Thieme, Stuttgart, Germany, 1982, pp. 300–476.
9. M. C. Mitchell and T. P. Kee, *Coordination Chem. Rev.* **158**, 359 (1997).
10. M. Horiguchi and M. Kandatsu, *Nature* **184**, 901 (1959).
11. G. W. Cooper, M. H. Thiemens, T. L. Jackson, and S. Chang, *Science* **277**, 1072 (1997).
12. J. Stawinski in R. Engel, ed., *Handbook of Organophosphorus Chemistry*, Marcel Dekker, Inc., New York, 1992. Chapter 8.

13. R. S. Edmundson in D. H. R. Barton and W. D. Ollis, eds., *Comprehensive Organic Chemistry*, Pergamon Press, Oxford, UK, 1979, Chapter 10.5.
14. J. P. Guthrie, *Can. J. Chem.* **57**, 236 (1979).
15. V. Mark and J. R. Van Wazer, *J. Org. Chem.* **32**, 1187 (1967).
16. G. O. Doak and L. D. Freedman, *Chem. Rev.* **61**, 31 (1961).
17. T. Wada, A. Mochizuki, Y. Sato, and M. Sekine, *Tetrahedron Lett.* **39**, 7123 (1998).
18. I. Lindh and J. Stawinski, *J. Org. Chem.* **54**, 1338 (1989).
19. (a) M. J. Gallagher and J. Sussman, *Phosphorus* **5**, 91 (1975), (b) A. E. Wróblewski and J. G. Verkade, *J. Am. Chem. Soc.* **118**, 10168 (1996).
20. A. J. Kirby and S. G. Warren, *The Organic Chemistry of Phosphorus*, Elsevier, Amsterdam, 1967, Chapter 10.
21. R. R. Holmes, *Pentacoordinated Phosphorus*, Vols. 1 and 2, American Chemical Society, Washington, DC, 1980.
22. M. Green and R. F. Hudson, *Proc. Chem. Soc.* 307 (1962).
23. R. F. Hudson and L. Keay, *J. Chem. Soc.*, 2463 (1956).
24. (a) G. Aksnes and J. Songstad, *Acta Chem. Scand.* **19**, 893 (1965), (b) G. Aksnes and K. Bergesen, *Acta Chem. Scand.* **20**, 2508 (1966).
25. R. R. Holmes, *Accts. Chem. Res.* **31**, 535 (1998); R. R. Holmes, *Chem. Rev.* **96**, 927 (1996).
26. V. A. Ginsburg and A. Ya. Yakubovich, *Zhur. Obshch. Khim.* **28**, 728 (1958); *Chem. Abstr.* **52**, 17092 (1958).
27. H. H. Hatt, *J. Chem. Soc.*, 776 (1953).
28. L. D. Quin and C. H. Rolston, *J. Org. Chem.* **23**, 1693 (1958).
29. R. G. Harvey and E. R. De Sombre, *Top. Phosphorus Chem.* **1**, 57 (1964).
30. D. Villemin, F. Simeon, H. Decreus, and P.-A. Jaffres, *Phosphorus Sulfur Silicon* **33**, 209 (1998).
31. H. R. Hudson, *Top. Phosphorus Chem.* **11**, 339 (1983).
32. P. Tavs, *Chem. Ber.* **103**, 2428 (1970).
33. E. L. Gavrilova, E. A. Krasil'nikova, V. V. Sentemov, T. V. Zykova, and R. A. Salakhutdinov, *Zhur. Obshch. Khim.* **62**, 1421 (1992); *Chem. Abstr.* **118**, 102078 (1993).
34. *Organic Synthesis, Coll.* Vol. VI, 451 (1988).
35. J. Engel, ed., *Handbook of Organophosphorus Chemistry*, Marcel Dekker Inc., New York 1992. (a) G. B. Borowitz and I. J. Borowitz, Chapter 3, (b) R. McClard and J. F. Witte, Chapter 13, (c) P. Mastalerz, Chapter 7.
36. R. B. Lacount and C. E. Griffin, *Tetrahedron Lett.* 3071 (1965).
37. (a) S. Ganapathy, K. P. Dockery, A. E. Sopchick, and W. G. Bentrude, *J. Am. Chem. Soc.* **115**, 8663 (1993), (b) W. G. Bentrude in E. N. Walsh, E. J. Griffith, R. W. Parry, and L. D. Quin, eds., *Phosphorus Chemistry*, American Chemical Society, Washington, DC, 1992. Chapter 11.
38. B. A. Arbusov and A. V. Fuzhenkova in A. N. Pudovik, ed., *Chemistry of Organophosphorus Compounds*, MIR Publishers, Moscow, 1989, Chapter 1.
39. A. N. Pudovik in Ref. 38, Chap. 2.
40. A. N. Pudovik, M. G. Zimin, A. A. Sobanov, and A. A. Musina, *Zhur. Obshch. Khim.* **46**, 1455 (1976); *Chem. Abstr.* **85**, 143203 (1976).
41. V. Sum, A. J. Davies, and T. P. Kee, *J. Chem. Soc., Chem. Commun.*, 1771 (1992), V. Sum and T. P. Kee, *J. Chem. Soc. Perkin Trans.* 1, 2701 (1993).
42. W. L. McEldoon, K. Lee, and D. E. Weimer, *Tetrahedron Lett.* **34**, 5843 (1993).
43. E. E. Nifant'ev, R. K. Magdeeva, A. V. Dolidze, K. V. Ingorokva, and L. K. Vasyanina, *Zhur. Obshch. Khim.* **63**, 1748 (1993); *Chem. Abstr.* **120**, 270572 (1994).

44. S. N. L. Bennett and R. G. Hall, *J. Chem. Soc., Perkin Trans.* 1, 1145 (1995).
45. G. Hilgetag, H.-G. Henning, and D. Gloyna, *Z. Chem.* **4**, 347 (1964).
46. D. B. Cooper, T. D. Inch, and G. J. Lewis, *J. Chem. Soc. Perkin Trans.* 1, 1043 (1974); J. M. Harrison, T. D. Inch, and G. J. Lewis, *J. Chem. Soc. Perkin Trans.* 1, 1053 (1974).
47. M. J. Gallagher in W. L. F. Armarego, ed., *Stereochemistry of Heterocyclic Compounds*, Part 2, Wiley-Interscience, New York, 1977, pp. 391–397.
48. C. R. Hall and T. D. Inch, *Tetrahedron* **36**, 2059 (1980).
49. D. H. Hua, J. S. Chen, S. Saha, H. Wang, D. Roche, S. N. Bharathi, R. Chan-Yu-King, P. D. Robinson, and S. Iguchi, *Synlett* 817 (1992).
50. M. P. Teulade and P. Savignac, *Tetrahedron Lett.* **28**, 405 (1987).
51. K. D. Berlin, T. H. Austin, M. Peterson, and M. Nagabhushanam, *Top. Phosphorus Chem.* **1**, 17 (1964).
52. J. O. Clayton and W. L. Jensen, *J. Am. Chem. Soc.* **70**, 3880 (1948).
53. R. E. Flurry and C. E. Boozer, *J. Org. Chem.* **31**, 2076 (1966).
54. *Organic Syntheses, Coll.* Vol. IV, 784 (1963).
55. G. O. Doak, and L. D. Freedman, *J. Am. Chem. Soc.* **73**, 5658 (1951).
56. L. D. Freedman, H. Tauber, G. O. Doak, and H. J. Magnuson, *J. Am. Chem. Soc.* **75**, 1379 (1953).
57. G. R. Kieczykowski, R. B. Jobson, D. G. Melillo, D. F. Reinhold, V. J. Grenda, and I. Shinkai, *J. Org. Chem.* **60**, 8310 (1995).
58. (a) D. H. R. Barton and J. Zhu, *J. Am. Chem. Soc.* **115**, 2071 (1993), (b) D. H. R. Barton and R. A. Vonder Embse, *Tetrahedron* **54**, 12475 (1998).
59. R. M. Kissling and M. R. Gagné, *J. Org. Chem.* **64**, 1585 (1999).
60. A. D. F. Toy and E. N. Walsh, *Phosphorus Chemistry in Everday Living*, 2nd ed., American Chemical Society, Washington, DC, 1987.
61. (a) J. W. Perich, P. F. Alewood and R. B. Johns, *Phosphorus Sulfur Silicon* **105**, 1 (1995), (b) R. Hirschmann, K. M. Yager, C. M. Taylor, J. Witherington, P. A. Sprengeler, B. W. Phillips, W. Moore, and A. B. Smith, III, *J. Am. Chem. Soc.* **119**, 8177 (1997).
62. E. Fluck, *Top. Phosphorus Chem.* **4**, 291 (1967).
63. D. E. C. Corbridge, *Phosphorus, an Outline of Its Chemistry, Biochemistry, and Technology*, 4th ed., Elsevier, Amsterdam, 1990, (a) Chapter 5, (b) p. 560.
64. R. F. Hudson, *Structure and Mechanism in Organo-Phosphorus Chemistry*, Academic Press, New York, 1965, p. 120.
65. M. I. Kabachnik, T. A. Mastryukova, A. E. Shipov, and T. A. Melentyeva, *Tetrahedron* **9**, 10 (1960).
66. T. A. Mastryukova in A. N. Pudovik, ed., *Chemistry of Organophosphorus Compounds*, MIR Publishers, Moscow, 1989. Chapter 7.
67. T. A. Mastryukova, *Phosphorus Sulfur* **1**, 211 (1976).
68. G. R. Norman, W. M. LeSuer, and T. V. Mastin, *J. Am. Chem. Soc.* **74**, 161 (1952).
69. W. E. Bacon and W. M. LeSuer, *J. Am. Chem. Soc.* **76**, 670 (1954).
70. G. M. Kosolapoff, *Organic Phosphorus Compounds*, John Wiley & Sons, Inc., New York, 1950, p. 143.
71. B. J. Walker in J. I. G. Cadogan, ed., *Organophosphorus Reagents in Organic Synthesis*, Academic Press, London, 1979, Chapter 3.
72. A. W. Johnson, *Ylides and Imines of Phosphorus*, John Wiley & Sons, Inc., 1993, Chapter 10.
73. (a) E. J. Corey and G. T. Kwiatkowski, *J. Am. Chem. Soc.* **88**, 5652, (1966), (b) *ibid.*, 5653.
74. E. J. Corey and G. T. Kwiatkowski, *J. Am. Chem. Soc.* **88**, 5654 (1966).
75. E. J. Corey and G. T. Kwiatkowski, *J. Am. Chem. Soc.* **98**, 1291 (1976).

76. P. Savignac and F. Mathey, *Tetrahedron Lett.*, 2829 (1976).
77. D. J. Burton, T. Ishihara, and M. Maruta, *Chem. Lett.*, 755 (1982).
78. G. M. Blackburn, D. Brown, S. J. Martin, and M. J. Parratt, *J. Chem. Soc., Perkin Trans. 1*, 181 (1987).
79. S. D. Lindell and R. M. Turner, *Tetrahedron Lett.* **31**, 5381 (1990).
80. G. M. Blackburn, D. E. Kent and F. Kolkmann, *J. Chem. Soc., Perkin Trans. 1*, 1119 (1984).
81. M.-P. Telaude and P. Savignac, *J. Organomet. Chem.* **338**, 295 (1988).
82. S. F. Martin and T. Chou, *J. Org. Chem.* **43**, 1027 (1978).
83. M.-P. Teulade, P. Savignac, E. E. Aboujaoude, S. Lietge and N. Collignon, *J. Organomet. Chem.* **312**, 283 (1986).
84. C. R. Hauser and T. M. Harris, *J. Am. Chem. Soc.* **80**, 6360 (1958).
85. M. H. B. Stowell, J. F. Witte and R. W. McClard, *Tetrahedron Lett.* **30**, 411 (1989).
86. A. R. P. M. Valentijn, G. A. van der Marel, L. H. Cohen and J. H. van Boom, *Synlett* 663 (1991).
87. S. Hanessian, Y. L. Bennani and D. Delorme, *Tetrahedron Lett.* **31**, 6461 (1990).
88. S. E. Denmark and R. L. Dorow, *J. Org. Chem.* **55**, 5926 (1990).
89. S. E. Denmark, K. A. Swiss, P. C. Miller and S. R. Wilson, *Heteroatom Chem.* **9**, 209 (1998).

APPENDIX 5.1

SYNTHESES OF PERTINENT COMPOUNDS IN *ORGANIC SYNTHESES* (OS) AND *INORGANIC SYNTHESES* (IS)

[1] (EtO)$_2$P(=O)—H
IS **4**, 58

[2] (C$_8$H$_{17}$O)$_2$P(=O)—H
IS **4**, 61

[3] Ph$_2$P(=O)—OH
IS **8**, 71

[4] PhP(=O)(OEt)—H
IS **31**, 144

[5] (EtO)$_2$P(=O)—NH$_2$
IS **4**, 77

[6] Me$_3$CC(=O)—C(=O)—O—P(=O)(OEt)$_2$
OS **VI**, 207

[7] Ph$_2$P(=O)—O—NH$_2$
OS **VII**, 8

[8] 5-methylcoprost-3-en-3-yl-O—P(=O)(OEt)$_2$
OS **VI**, 762

[9] Ph$_2$P(=O)—N$_3$
OS **VII**, 206

[10] (EtO)$_2$P(=O)—O—(cyclohexenyl-COOMe)
OS **VII**, 351

[11] cyclohexyl-C(=N-pyrrolidinyl)(N—P(=O)(OPh)$_2$)
OS **VII**, 135

[12] furan-2-yl-O—P(=O)Cl$_2$ · (NH$_2$)$_2$
OS **VIII**, 396

166

SYNTHESES OF PERTINENT COMPOUNDS 167

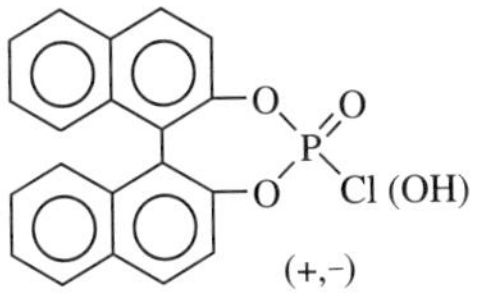

[13] binaphthyl-O₂P(=O)-Cl(OH) (+,-)
OS **VIII**, 58

[14] (Me)₂C=CHCH₂CH₂C(Me)=CHCH₂-O-P(=O)(O⁻)-O-P(=O)(O⁻)-O⁻ (NH₄⁺)₃
OS **VIII**, 616

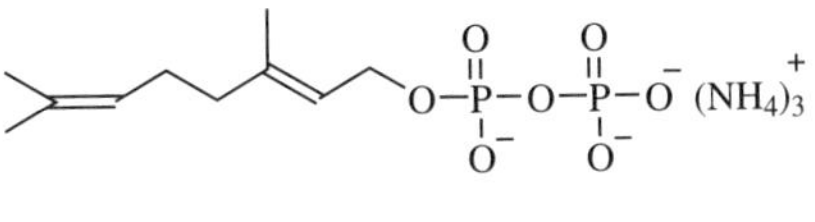

[15] RNH—P(=O)(OEt)₂
OS **71**, 246

[16] RO—P(=O)(OEt)₂
OS **71**, 241

[17] (Me₂N)₂P(=O)—Cl
IS **7**, 71

[18] Me—P(=O)(OR)₂
R= Et or *i*-Pr
OS **IV**, 325

[19] Et—P(=O)(OR)₂
R= Et or *i*-Pr
OS **IV**, 325

[20] (EtO)₂P(=O)—Cl
IS **4**, 78

[21] (EtO)₂P(=O)—CH₂CH(OEt)₂
OS **VI**, 448

[22] (EtO)₂P(=O)—CH₂CHO
OS **VI**, 448

[23] (EtO)₂P(=O)—CH=CH-NH-cyclohexyl
OS **VI**, 448

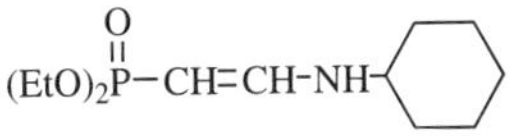

[24] phthalimide-N—CH₂—P(=O)(OEt)₂
OS **VIII**, 451

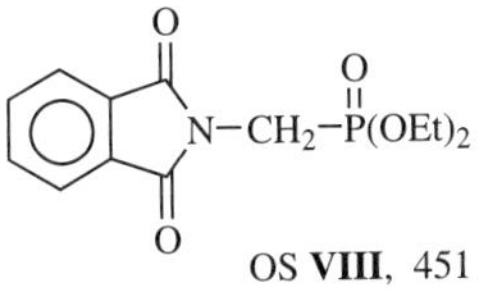

[25] Ph—P(=O)(OEt)₂
OS **VI**, 451

[26] (EtO)₂P(=O)—CH₂-(2,2-dimethyl-4-oxo-4H-1,3-dioxine)
OS **VIII**, 192

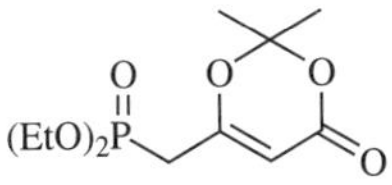

[27] (EtO)₂P(=O)—CH₂OH
OS **VII**, 160

[28] (EtO)₂P(=O)—CH₂—O—(tetrahydropyranyl)
OS **VII**, 160

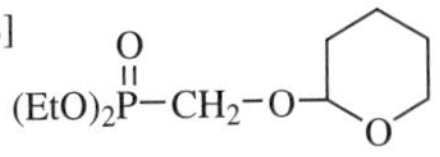

[29] PhCH=CHPOCl₂
OS **V**, 1005

[30] PhCH=CHPCl₄
OS **V**, 1005

[31] CCl₃POCl₂
OS **IV**, 950

[32] ClCH₂P(=S)Cl₂
OS **V**, 218

[33] (4-MeO-C₆H₄)P(=S)(S)(S)P(=S)(S)(4-MeO-C₆H₄) (Lawesson's reagent)
OS **7**, 372

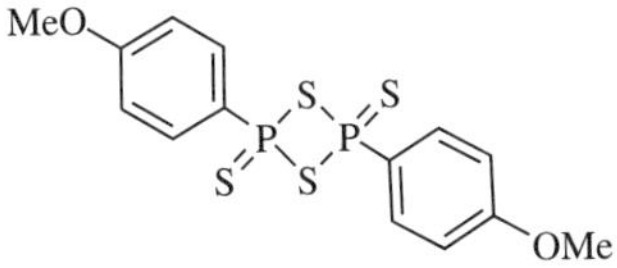

[34] EtO—P(=S)Cl₂
IS **4**, 75

[35] Me—P(=O)(OCH₂CF₃)₂

OS **73**, 152

[36] (CF₃CH₂O)₂P(=O)CH₂C(=O)OEt

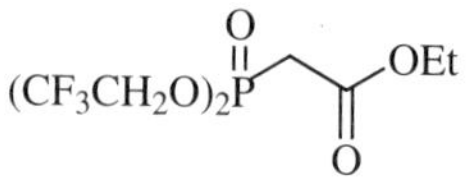

OS **73**, 152

[37] (EtO)₂P(=O)—CCl₃

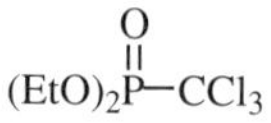

OS **74**, 108

[38] (EtO)₂P(=O)—CHCl₂

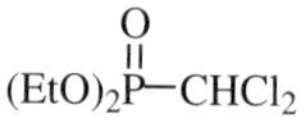

OS **74**, 108

[39] (EtO)₂P(=O)—C̄Cl₂

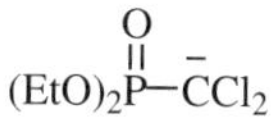

OS **74**, 108

[40] (EtO)₂P(=O)—CH(CH₂CH₂CH₃)—C(=O)H

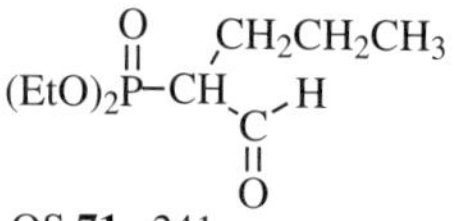

OS **71**, 241

CHAPTER 6

PHOSPHORUS-31 NMR SPECTROSCOPY

6.A. SOME GENERAL CONSIDERATIONS

Chemists conducting work with phosphorus compounds are fortunate to be dealing with a nucleus that has outstanding properties for the measurement of magnetic resonance phenomena. Phosphorus NMR spectra are easily prepared and are highly informative about the molecular environment; as will be seen, most common phosphorus compounds have chemical shifts that are spread over a range of about 500 ppm (the total range is about 2000 ppm)[1] in reasonably well defined characteristic regions for the numerous functional groups. The technique can also be used to determine the complexity of reaction mixtures or the purity of products, because separate signals will almost always be seen for each phosphorus compound when the spectrum is obtained under conditions of high resolution. Indeed, when sharp signals are obtained, it is quite common to see separate signals for members of a mixture of phosphorus compounds in which the chemical shifts differ by less than 0.1 ppm. An excellent illustration is provided by the spectra of the phosphoric acid derivatives in biological extracts as in Figure 6.1, which shows the spectrum for the phospholipid extract of broccoli, prepared under very carefully controlled conditions.[2a] Some 15 peaks can be observed over a span of only 2 ppm. Another valuable feature of phosphorus NMR spectroscopy is that, when run properly (vide infra), peak size can be used to obtain a quantitative ratio for members of a mixture, and by preparing a calibration curve with known samples, the concentration of particular compounds can be determined. Many other applications of phosphorus NMR have been made, such as performing conformational analysis with the aid of variable temperature measurements, characterizing phosphorus compounds in the solid state, and eluci-

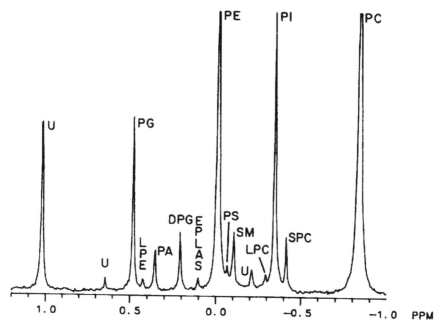

Figure 6.1. ^{31}P{^{1}H} NMR spectrum of extracted broccoli phospholipid preparation to which synthetic sphingosylphosphorylcholine (SPC) was added. *U*, unidentified phospholipids at 1.01, 0.64, and −0.17; *PG*, phosphatidylglycerol; *LPE*, lysophosphatidylethanolamine; *PA*, phosphatidic acid; *DPG*, diphosphatidylglycerol (cardiolipin); *EPLAS*, ethanolamine plasmalogen; *PE*, phosphatidylethanolamine; *PS*, phosphatidylserine; *SM*, sphingomyelin; *LPC*, lysophosphatidylcholine; *PI*, phosphatidylinositol; *PC*, phosphatidylcholine. T. Glonek in L. D. Quin and J. G. Verkade, eds., *Phosphorus-31 NMR Spectral Properties in Compound Characterization and Structural Analysis*, VCH Publishers, Inc., New York, 1994, p. 285. Reprinted by permission of John Wiley & Sons, Inc.

dating reaction mechanisms by following the appearance and disappearance of signals for intermediates.

Two properties of the ^{31}P nucleus have made the taking of phosphorus NMR spectra quite simple. First, there is only one naturally occurring isotope of the element, that with mass 31, and the problem of low nuclear concentration giving weak signals (as in the case of ^{13}C) is absent. Second, the spin quantum number for ^{31}P is $\frac{1}{2}$, just as for the proton, ^{13}C, ^{19}F, and other important nuclei, which insures that phosphorus has only two spin states. The magnetogyric constant for ^{31}P is 10.841×10^7 rad T^{-1} s^{-1}, compared to 26.7519×10^7 for ^{1}H. This means that an NMR instrument rated, for example, at 100 MHz for ^{1}H needs to be operated at 40.5 MHz for ^{31}P. There is, however, one problem with ^{31}P NMR: the nucleus is of rather low sensitivity, only 6.63% of that of the proton. With pure compounds, it is quite feasible to obtain ^{31}P spectra by a single scan run by the continuous wave (CW) technique, and this was done for hundreds of compounds starting very early in the NMR era (the first ^{31}P NMR spectra were reported in 1951,[3,4] just a few years after the original Bloch and Purcell reports of 1946). Multiple pulsing with data processing by the Fourier transform (FT) technique became common in NMR around 1970, and this revolutionized the field of ^{31}P NMR. Much sharper signals can be obtained on quite low concentrations with the FT method, especially when performed with broad-band decoupling of all the protons. The WALTZ[5] technique is in common current use for the decoupling experiment. Not only does the decoupling lead to single-line peaks, as opposed to the multiplets of proton-coupled spectra (which, however, will be seen to have their own importance), but a valuable signal enhancement by the nuclear Overhauser effect then takes place. Modern ^{31}P NMR is routinely performed by the FT method with ^{1}H decoupling, and this is to be assumed in the current literature even when not specified. A number of specialized multipulse and relaxation techniques have been applied to ^{31}P NMR.[1,5]

Modern spectrometers require the presence of deuterium nuclei for locking purposes; this is generally provided in ^{31}P NMR by using the solvents $CDCl_3$ or D_2O, but other deuterated solvents can also be used. The solvent must always be specified, because solvent effects can be important, especially when H-bonding can occur. This is especially true for phosphine oxides, which are strongly H-bonded in water but less so in chloroform; cases are known in which the chemical shifts will differ by several ppm, with water giving the downfield signals. Minor shift effects can occur with concentration or temperature changes, and unless conditions are closely duplicated, it is sometimes difficult to reproduce a published value to better than a few tenths of a ppm.

In ^{31}P NMR, the reference now in general use is 85% phosphoric acid, but in earlier years, other references were occasionally employed.[1] It is now the standard practice to assign shifts downfield of phosphoric acid as positive and upfield as negative. However, the opposite convention was in use for many years, and care must be taken when using data published before about 1980, by which time the new convention was in general use. Current tabulations of chemical shifts, as in the invaluable tables of Tebby,[6] contain values recalculated when necessary to the new convention.

As mentioned, the ^{31}P nucleus couples with ^{1}H, and both types of spectra show the effect. Coupling constants can be as small as a few Hertz or as many as several hundred Hertz for the direct P–H bond. Because the coupling effect is commonly seen on ^{1}H spectra, but usually avoided by decoupling in ^{31}P NMR, coupling constants are usually determined from the former spectra. Some representative values for $^1J_{PH}$ will be given as the different types of phosphorus functional groups are discussed in the remainder of this chapter. Steric structure can figure in establishing J_{PH}, and Karplus-type relations of dihedral angle and J_{PH} are known; these relations are discussed in connection with proton NMR spectra in chapter 7.B. A different situation prevails for the coupling of ^{31}P with ^{13}C; the effect is seen only on the ^{13}C NMR spectra, because the low natural abundance of ^{13}C (1.1%) is insufficient to lead to an observable number of coupled ^{31}P nuclei. We will, therefore, discuss J_{PC} values in connection with ^{13}C NMR in chapter 7.B. The same situation prevails for other low-abundance nuclei such as ^{17}O, ^{15}N, and ^{29}Si, etc. However, some other common elements do have a high natural abundance of

NMR-active nuclei, such as fluorine with ^{19}F (spin $\frac{1}{2}$) of abundance 100% and ^{77}Se with 7.58%; when these elements are present in organophosphorus compounds, the ^{31}P NMR spectra will show the coupling. Of course, isotopic enrichments can lead to ^{31}P signal splitting, the most common case being that of replacing H by D. Deuterated compounds have very nearly the same ^{31}P chemical shifts, but D has a spin of 1 and splits the ^{31}P signal to three lines (1 : 1 : 1), with coupling constants quite unlike those for the corresponding protonated compounds. They are in fact 0.163 (the ratio of the magnetogyric constants of D and P) times the size of the corresponding J_{PH} value. A valuable compilation of ^{31}P coupling constants to various nuclei has been published.[1]

The size of phosphorus signals depends heavily on their spin-lattice relaxation times (T_1) and nuclear Overhauser effect (NOE) differences. The T_1 values are determined by the nature of the phosphorus functional group and by the proximity of nonbonded protons and can vary over a range of about 1 to 30 s.[7a] Only when a spectrum of a mixture of phosphorus compounds is run under conditions that take relaxation time and NOE effects into consideration does relative peak area have any value for quantitative analysis. There are several techniques for improving the reliability of the peak area relations. The most common is to use a long delay between pulses to insure complete relaxation. This can, however, extend greatly the length of experimental time for data accumulation for dilute samples. All T_1 values can be made very small, thus eliminating the problem, by including relaxation agents such as the well-known complex Cr(acac) or the stable free radical 4-amino-2,2,6,6-tetramethylpiperidinooxy in the sample solution. Finally, with samples of high concentration, a single pulse can give a good spectrum, and here the relaxation time differences are no longer important.[7a] To eliminate the NOE differences, the spectrum can be run in the gated mode.[7a] For more detailed information on quantitative ^{31}P NMR, references 2b and 8 are helpful. Further discussion of T_1 and NOE effects are provided by McFarlane[7b] and Gorenstein.[5] Discussions by Tebby[7a] provide some additional information on various experimental aspects of ^{31}P NMR.

6.B. THE LITERATURE OF ^{31}P NMR SPECTROSCOPY

The first extensive review of ^{31}P NMR was provided as an entire volume in *Topics in Phosphorus Chemistry*[9] which appeared in 1967. In addition to in-depth theoretical treatment[9a] of the subject in the terms of those times, this issue contains a still-valuable compilation of the available data for individual compounds.[9b] The shifts were all obtained by the original continuous wave method and were reported with the reverse (downfield shifts being negative) of the current sign convention (downfield positive). Listings of compounds in both the Kosolapoff–Maier series[10a] of 1972, and in Edmundson's more recent (1988) *Dictionary of Organophosphorus Compounds*[10b] include ^{31}P NMR shift data or references, respectively, among the physical properties listed, and these are valuable sources of information. In 1981, a discussion of the interpretation of shifts in regard to heterocyclic compounds was published.[11] Included is a list of the available data for phosphorus heterocycles. Two books appeared in the 1980s that provided extensive discussions of theory, interpretive techniques, and applications of the method as well as compact tables of illustrative shift data, mostly obtained by the FT method. The first of these, edited by Gorenstein,[12] is particularly valuable in its emphasis on the importance of ^{31}P in biological chemistry. The other, edited by Verkade and Quin,[7] was published in 1987, with more emphasis on synthetic organophosphorus compounds, including metal complexes of various types. A more recent book[2] by the latter editors provides coverage of significant advances since the 1987 book. The latter two books include discussions of the latest attempts to develop the theory of ^{31}P shifts and also contain coverage of the emerging use of solid-state ^{31}P NMR spectroscopy. Another book exclusively devoted to biological applications of ^{31}P NMR also appeared in 1987.[13] The most recent review is that of Karaghiosoff,[1] which was published in 1997.

In 1991, the extremely valuable *Handbook of Phosphorus-31 NMR Nuclear Magnetic Resonance Data* with J. C. Tebby as editor appeared.[6] This book contains thousands of entries, with references, for compounds within all of the various phosphorus functional group families, and it now is the first place to go to obtain data on published compounds. Its contribution to the NMR field cannot be overstated. In addition, each chapter on a functional group contains a discussion on the molecular features that influence the chemical shifts. Metal complexes are not included in these tables, but data are available in reviews[14,15] in the Kosolapoff–Maier series and as special topics in the monographs of references 2 and 7.

A major advance in the cataloging of ^{31}P NMR data has been accomplished at Advanced Chemistry Development, Inc.,[16a] by the creation of computer programs available on CD-ROM, as well as by on-line connection to the database. At this writing, the shifts of about 18,500 compounds are listed, in many cases along with coupling constants to ^{1}H, ^{13}C, and other heteroatoms, with plans to expand the collection continually. Individual compounds may be located by drawing structural formulas as well as by IUPAC names and use of molecular formulas. The database can also be searched by substructure, chemical shift value or range, coupling constant value or range, and molecular weight. There is also the possibility of calculating approximate chemical shifts, making use of a structure-fragment algorithm obtained from the database (these calculations are not to be confused with the ab initio calculations of the Chesnut–Kutzelnigg–Schleyer, etc., type, vide infra). Computerized ^{31}P NMR data are also obtainable online from the spectroscopic database of STN International, where at present some 2000 ^{31}P NMR chemical shifts are included. This service is provided by Chemical Concepts;[16b] programs for spectra prediction by structure-fragment algorithms are also available.

6.C. GENERAL TRENDS AND STRUCTURAL INFLUENCES ON ^{31}P NMR SHIFTS

^{31}P NMR shifts span the enormous range of about 2000 ppm, although the vast majority fall in the region of about δ −200 to +300. Each type of functional group

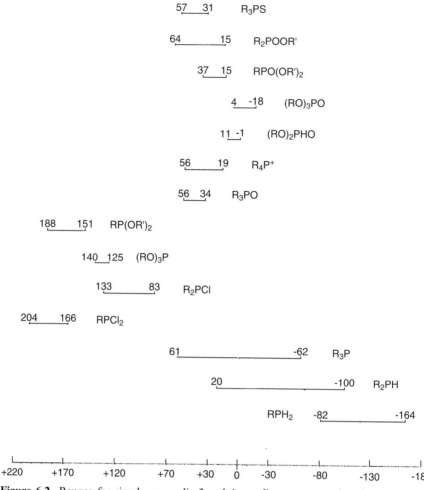

Figure 6.2. Ranges for simple noncyclic 3 and 4-coordinate compounds, where R may be alkyl or phenyl and R′ may be H, alkyl, or phenyl.

has its own range of shifts within this region. This is seen clearly in Figure 6.2 in which data for unsubstituted alkyl and phenyl derivatives of the common 3- and 4-coordinate functional groups are presented (ranges are much broader when substituted and cyclic structures are included, as seen in Tebby's range-inclusive figures[6b]). It will be noted that there is severe overlap of these functional group subregions, and it is rarely possible to use only the ^{31}P shift, with no other characterizing information, for a firm identification of the phosphorus group present in some unknown. It will also be noted that the ranges for phosphines are much broader than those for other 3-coordinate compounds and for 4-coordinate compounds, implying that there is much greater sensitivity to structural effects.

6.C. GENERAL TRENDS AND STRUCTURAL INFLUENCES ON ^{31}P NMR SHIFTS

Almost from the beginnings of ^{31}P NMR, attempts have been made to account for the change in shifts when one group is converted to another or when structural changes around a functional group type are made. Important theoretical explanations were published in 1966 by Letcher and Van Wazer[17] with emphasis on understanding why particular functional groups had their particular shifts, but this early theory did not address the needs of organic chemists trying to cope with the sometimes strange effects of structural changes on the shifts of compounds having a particular function. Theoretical work continued over the years, and a summary has been provided in a review by Gorenstein,[5] which also addresses work on the more practical aspects of structure change on shifts. Recently, however, with the advent of supercomputing capability, it has become possible to attempt the ab initio calculation of chemical shifts. This approach is now enjoying considerable success, as some representative reports to the work of Chesnut,[2c,7c,18] Kutzelnigg[19] and Schleyer[20] will show. The computational approach has already led to an understanding of certain of the factors influencing ^{31}P shielding as a function of molecular framework. In general, it is possible to perform a calculation that will give a value within ±30 ppm of the experimental value, although in many cases better matches have been obtained. Illustrative are calculations by Chesnut and co-workers[18] for the family of methyl phosphines: MePH$_2$, δ −160.3; Me$_2$PH, δ −126.7; Me$_3$P, δ −96.2. These shifts would correspond to experimental data taken on the gas phase, but few data of this type are available. Data in the Tebby tables for these compounds in solution are δ −163.5, δ −99.5, and δ −62, respectively. Although still in need of refinement, the calculations have clearly reproduced the effect of replacing H by CH$_3$ and have given reasonable values when one remembers that the range of shifts covered by the alkyl phosphines is some 225 ppm.

Some examples of the use of shift calculation to account for particular structural influences on observed shifts of heterocycles will be given in later sections. A more pragmatic and empirical approach to interpreting ^{31}P NMR spectra as used by organic chemists is simply to start with the functional group chemical shift ranges learned from experience, with little concern about why these ranges differ, and then attempt to account for variations within this range as the structure is modified. It is this approach that is taken in the present discussion, and we will make no strong attempt to explain *why* the group has its particular range relative to other groups.

Many factors have been considered to be important in establishing the shift for a particular structure. Most of these factors will be seen in the discussion to follow on the shifts for each of several common functional groups. In some cases, similarity will be found to interpretive ^{13}C NMR. In brief, practice if not always theory has given us the following factors to consider:

1. Electron withdrawal by electronegative groups, generally considered to act by contracting the p-orbitals at P and causing deshielding, as is familiar from the Karplus–Pople equation.[21]
2. Resonance interactions at phosphorus with unsaturated groups that change electron density on phosphorus in either direction.

3. Chain lengthening and branching effects, which can cause deshielding as the number of β-carbons to P increases or shielding as the number of γ-carbons increases.
4. Changes in bond angles at phosphorus, increases in which are said to cause deshielding of 3-coordinate phosphorus (but with exceptions), and shielding in phosphates.
5. Steric interactions in alicyclic systems, manifested by shielding at an axial, nonpolar phosphorus function, or in aromatic systems when a group is ortho to phosphorus, or in P heterocycles when a substituent is axial on P.
6. Nonbonded interactions of the lone pair on 3-coordinate phosphorus with properly oriented unsaturated groups.
7. Effects that cause lowering of the HOMO-LUMO energy gap, associated with deshielding, a concept only recently introduced and reviewed.[18]
8. The Gorenstein stereoelectronic effect,[5,22] arising in phosphates from interactions of lone pairs on oxygen.
9. Neighboring-atom and long-range anisotropic effects from electron circulation at nearby groups; being independent of the identity of the atom feeling the effect, these are limited to the same small size of a few ppm as found in ^{1}H and ^{13}C NMR spectroscopy.[23]

One explanation in phosphoryl structures that was favored for many years now seems to be in doubt: increased donation of electrons into d-orbitals causes shielding. With the current trend away from invoking d-orbitals in understanding the electronic structure of the phosphoryl and other groups, which was applied to ^{31}P shifts many years ago[9] and became widespread in the literature, it seems prudent to put aside the d-orbital shielding arguments until more progress is made in the continuing saga of electronic explanations for the phosphoryl group (chapter 2.D). There is little doubt, however, that a form of electron acceptance by the phosphoryl group can take place and that this is associated with shielding at phosphorus.

In the following sections, the ^{31}P NMR spectra for the 3- and 4-coordinate phosphorus functional groups discussed in chapters 3 to 5 will be analyzed, using the factors presented above that have been suggested to contribute to the observed shifts. Spectra for compounds with coordination states 1, 2, 5, and 6 are described in chapter 10, where the chemistry of these much less common compounds is treated. Heterocyclic compounds often show dramatic ^{31}P NMR shift effects; additional examples can be found in chapter 8. Similarly, ^{31}P NMR is useful in stereochemical analysis, and examples of this use are found in this chapter as well as in chapter 9.

With few exceptions, the ^{31}P NMR shifts referred to in the following discussion of the functional groups were taken from the extensive tables of Tebby.[6] The expedient has been taken of not giving the original references for the more routine structures, because these may be obtained from Tebby's book. In some cases, more than one shift value is found in the tables; generally the most recently reported value is used here, or when different solvents are the cause of shift differences, preference is given to CDCl$_3$ or to D$_2$O, as appropriate for the structural type. However, the

6.C. GENERAL TRENDS AND STRUCTURAL INFLUENCES ON ^{31}P NMR SHIFTS

Tebby tables and the original literature should be consulted if research is being contemplated in which ^{31}P NMR spectra for cited compounds are to be reproduced. To assist in developing an understanding of structural effects on ^{31}P NMR shifts, Tables 6.1 to 6.4, which contain data for representative compounds in a few of the major functional groups, are presented.

TABLE 6.1. Representative ^{31}P NMR Spectra of Primary Phosphines[a]

Compound	$\delta^{31}P$	$^1J_{PH}$, Hz
CH_3PH_2	−163.5	201
$CH_3CH_2PH_2$	−128	185
$(CH_3)_2CHPH_2$	−106	
$CH_3CH_2CH_2CH_2PH_2$	−140	
$(CH_3)_2CHCH_2PH_2$	−151	
$CH_3CH_2CH(CH_3)PH_2$	−120	
$(CH_3)_3CCH_2PH_2$	−162.4	
$(CH_3CH_2)_3CPH_2$	−111.6	190
$H_2NCH_2CH_2PH_2$	−150.4	194
$NCCH_2CH_2PH_2$	−135	195
$HOOCCH_2CH_2PH_2$	−136.9	194
$PhCH_2PH_2$	−120.9	184.5
$PH_2CH_2PH_2$	−121.8	190
$CH_3PHCH_2PH_2$	−138.9	201
$(CH_3)_2PCH_2PH_2$	−152.3	189.6
trans-4-t-Bu-cyclohexyl-PH$_2$ (equatorial)	−111.6	
cis-4-t-Bu-cyclohexyl-PH$_2$ (axial)	−131.3	
C$_6$H$_5$PH$_2$	−122.3	198
2-CH$_3$-C$_6$H$_4$-PH$_2$	−130.9	200
4-CH$_3$-C$_6$H$_4$-PH$_2$	−124.5	200

(*continued*)

TABLE 6.1. (*continued*)

Compound	$\delta^{31}P$	$^1J_{PH}$, Hz
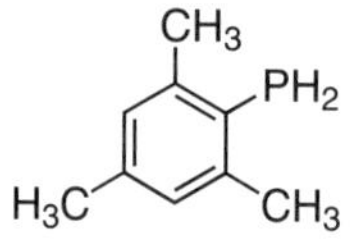 2,4,6-trimethylphenyl-PH$_2$	−160.2	203
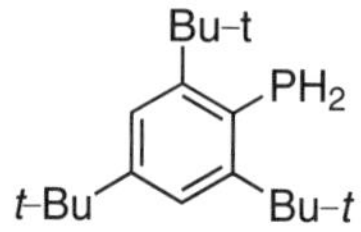 2,4,6-tri-*t*-butylphenyl-PH$_2$	−132.4	206
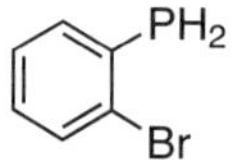 2-bromophenyl-PH$_2$	−130.6	204
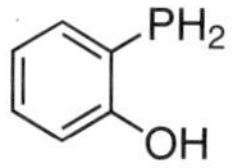 2-hydroxyphenyl-PH$_2$	−147.4	204
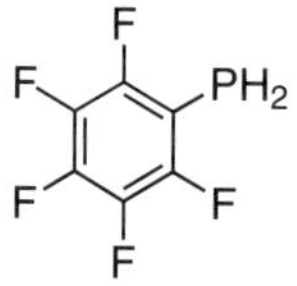 pentafluorophenyl-PH$_2$	−183.1	216

*Taken from Ref. 6c

TABLE 6.2. Representative ^{31}P NMR Spectra of Secondary Phosphines*

Compound	δ^{31}P NMR	$^1J_{PH}$
Me$_2$PH	−99.5	188
MeEtPH	−77.0	191
MePrPH	−87	196
Et$_2$PH	−55.5	190
Pr$_2$PH	−72	192
(Me$_3$C)$_2$PH	+20.1	197
(*n*-C$_8$H$_{17}$)$_2$PH	−71.5	196
i-Bu$_2$PH	−82.5	194
Ph$_2$PH	−43.8	239
o-Tolyl$_2$PH	−59.14	219
m-Tolyl$_2$PH	−40.2	214
p-Tolyl$_2$PH	−42.9	212
MePHCH$_2$PH$_2$	−69.1	201
PhHPCH$_2$CH$_2$PHPh	−50.8	
(Et$_2$NCH$_2$)$_2$PH	−101.9	194
(NCCH$_2$CH$_2$)$_2$PH	−75	195

*Taken from Ref. 6c.

TABLE 6.3. Representative ^{31}P NMR Spectra of Hetero-Substituted 3-Coordinate Phosphorus[a]

Compound	$\delta^{31}P$
Effect of Heteroatoms	
(MeO)$_3$P	+139.6–141
(MeS)$_3$P	+124.5
(Me$_2$N)$_3$P	+121.5
EtP(OMe)$_2$	+188.3
EtP(Set)$_2$	+83.5
MeP(Set)$_2$	+71.0
EtP(NMe)$_2$	+99.4
Tertiary Phosphites	
(EtO)$_3$P	+137.1–140
(PhO)$_3$P	+125–129
(MeO)(TMS–O)$_2$P	+117
(*t*-BuO)$_3$P	+138.2
Phosphonites	
MeP(OMe)$_2$	+182.5
EtP(OMe)$_2$	+188.3
n-BuP(OMe)$_2$	+187
Me$_2$CHP(OMe)$_2$	+157.9
PhP(OEt)$_2$	+153.5
MeP(OCHMe$_2$)$_2$	+173.0
MeP(OPh)$_2$	+178.5
MeP(O–TMS)$_2$	+171.9
(endo-Norbornyl)P(OMe)$_2$	+192.5
(exo-Norbornyl)P(OMe)$_2$	+185.3
Mixed Hetero-Substituents	
MeOPCl$_2$	+180.5
(EtO)$_2$PCl	+165
(Me$_2$N)$_2$PCl	+160
(*i*-Pr$_2$N)$_2$PCl	+140.8
(EtO)(EtS)$_2$P	+153.5
(MeO)$_2$(Me$_2$N)P	+148
(Me$_2$N)$_2$PBr	+184.2
(Me$_2$CH)$_2$NPCl	+140.8

[a] Taken from Ref. 6g,h.

TABLE 6.4. Representative ^{31}P NMR Spectra of Phosphine Oxides and Sulfides[a]

Compound	X=O, ^{31}P, δ	X=S, ^{31}P, δ
Me$_3$P=X	+36.2	+30.9
Et$_3$P=X	+48.3	+51.9
n-Bu$_3$P=X	+43.2, +45.8	+48.1
Ph$_3$P=X	+29.3	+42.6
(ClCH$_2$)$_2$EtP=X	+45.7	+55.2
Me$_2$PhP=X	+34.3	+32.5
n-Bu$_2$EtP=X	+45.7	
(ClCH$_2$)$_3$P=X	+38.1	
(Me$_2$NCH$_2$)$_3$P=X	+48.5	
(HOCH$_2$)$_3$P=X	+45.4	
(PhCH$_2$)$_3$P=X	+29.3	
(CNCH$_2$CH$_2$)$_3$P=X	+37.0	

[a]Taken from Ref. 6.

6.D. ^{31}P NMR SHIFTS OF PHOSPHINES

6.D.1. Primary Phosphines

The family of primary phosphines offers an excellent starting point for recognizing the operation of some of the factors that influence the chemical shift. The wide range of shifts is evident from examining the data in Table 6.1. We can start with the shift of phosphine itself (δ −230) and see that replacement of H by methyl causes a strong downfield shift to δ −163.5 in MePH$_2$. This is the equivalent, although much larger, of the α effect used in the interpretation of ^{13}C NMR shifts (and of ^{15}N), in which, for example, replacing H on methane by methyl gives a downfield shift of 9.1 ppm. Further replacement of H on P gives additional deshielding (Me$_2$PH, δ −99.5; Me$_3$P, δ −61.5). The effect of phenyl substitution is even more pronounced: PhPH$_2$, δ −118.7; Ph$_2$PH, δ −41; Ph$_3$P, δ −4.7.

The analogy to ^{13}C shift effects can be useful also in understanding other chain-lengthening and -branching effects of phosphines. Thus substitution of H on ethane by methyl gives a further downfied shift at the terminal carbon (9.4 ppm, the β effect), but substitution of H on propane by methyl gives an upfield shift (2.6 ppm, the γ effect).[24] The α, β and γ (and small δ) effects are additive and give quite good reproduction of ^{13}C NMR shifts. A form of the familiar Grant–Paul equation (Eq. 6.1) is used for this purpose, where α, β, and γ values are shown in parentheses.

$$\delta = -2.5 + n_\alpha\,(9.1) + n_\beta\,(9.4) + n_\gamma\,(-2.5) + n_\delta\,(0.3) \qquad (6.1)$$

Now considering primary phosphines, we can readily recognize the operation of β-deshielding and γ-shielding. Thus adding a β-carbon by methyl substitution on carbon in MePH$_2$ gives a downfield shift of about 35 ppm, to δ −128 in EtPH$_2$, whereas addition of a γ-carbon leads to an upfield shift of 12 ppm, to δ −140 in n-butylphosphine (data for n-propylphosphine are prefered but are not available; however, the δ-carbon of the butyl group causes only a very small (0 to 1 ppm) deshielding effect and does not obscure the γ-shielding). Trends in the ^{31}P NMR shifts of primary phosphines can then be readily explained, in some cases even calculated, from knowledge of these carbon substituent effects. Some examples include:

1. $(CH_3)_2CHCH_2PH_2$, δ −151; here we can say that a second carbon is γ to P, causing upfield shifting relative to $CH_3CH_2CH_2CH_2PH_2$. δ −140. The upfield shifting of 11 ppm is about the same as that (12 ppm) noted on moving from $CH_3CH_2PH_2$ to $CH_3CH_2CH_2CH_2PH_2$.
2. $(CH_3)_3CCH_2PH_2$, δ −162; another γ carbon is added, and the shift moves another 11 ppm upfield.
3. $(CH_3)_3CPH_2$, δ −81.2; only β-carbons are present, causing a net 82.2 ppm downfield shift from CH_3PH_2.
4. $(CH_3CH_2)_3CPH_2$, δ −111; the three terminal methyls constitute γ carbons, causing a net upfield shift of 30.4 ppm, close to the expected upfield γ shift of 11 to 12 ppm per carbon.

In ^{13}C NMR, the γ shielding is assumed to arise from steric compression between the observed carbon (C_1) and the γ-carbon (C_4), which in the preferred conformation are *anti* to each other as in *n*-butane. Because rotation occurs about the C_2–C_3 bond, eclipsing of C_1 and C_4 occurs to increase the steric compression. Considerations such as these led to the identification of steric compression as an important factor in explaining ^{31}P shift effects.[24] Many more examples, even with groups other than phosphines, are known and will be seen in further discussions. Just why steric compression causes upfield ^{31}P NMR shifts is not yet understood, and here perhaps is a case in which the new computational approach can be helpful. Shielding can be expected from increased electron density at the observed nucleus, and possibly the compression acts to contract the lone pair orbital toward the ^{31}P nucleus. Bond angle changes have frequently been used to explain shift trends in phosphines,[5,6c] but this is obviously without merit in explaining the pronounced fluctuations of primary (and other) phosphines. This subject has been mentioned as being amenable to study by the computational approach.[18]

An earlier empirical approach useful in phosphine NMR is to calculate the contribution to the shift made by a particular group (the group contributions, GC, of Grim, et al.[25]) from data for symmetrical tertiary phosphines and apply these with scaling to primary and secondary phosphines. The GC value for a particular substituent is derived from the ^{31}P shift of the corresponding symmetrical tertiary phosphine, simply by dividing the observed shift by 3. The chemical shift for any tertiary phosphine can then be calculated by adding the GC values for each substituent on P. Another approach is to add the GC value to +21 to give a term designated σ^P, which can be subtracted from the value of the parent of the series (for tertiary phosphines, trimethylphosphine with δ −62), to give the chemical shift. It is the σ^P value that is used in the calculation of the shifts for primary and secondary phosphines; it can be scaled for each phosphine type, and the value is then subtracted from the parent (methyl) compound of each series. To illustrate, the ^{31}P shift of tribenzylphosphine is δ −12, giving a GC of −4 and σ^P of 17. The latter is then scaled for the primary phosphine by a factor of 2.5, to give σ^P 42.5 (reflecting the much greater sensitivity of primary phosphines to structural changes). This is subtracted from the parent value of δ −163.5 (for CH_3PH_2) to give δ −121 for $PhCH_2PH_2$, the experimental shift of which is δ −120.9. It is remarkable, and extremely valuable, that a given substituent will have this predictable effect on the shifts of any type of phosphine. The original set[25] of tertiary phosphine GC values for alkyl and phenyl substituents has been greatly expanded[6c] (as shown in Table 6.5) to include some unsaturated groups and additional aryl groups, increasing the range of calculable shifts.

TABLE 6.5. Group Contribution Values for Tertiary Phosphines[a]

Group	GC	σ^P
Me	−21	0
Et	−7	14
n-Pr	−11	10
i-Pr	+6	27
n-Bu	−11	10
i-Bu	−14	7
sec-Bu	+2	23
t-Bu	+20	41
neopentyl	−18	3
t-amyl	−21	42
n-C_nH_{2n+1}, n > 4	−11	10
cyclo-C_5H_9	0	21
cyclo-C_6H_{11}	+2	23
$PhCH_2$	−4	17
Et_2NCH_2	−22	−1
$NCCH_2CH_2$	−8	13
CH_2=CH	−7	14
HC≡C	−31	−10
CH_3C≡C	−32	−11
Phenyl	−3	18
C_6F_5	−26	−5
2,6-$F_2C_6H_5$	−26.5	−5.5
2,6-$(RO)_2C_6H_5$	−22	−1
3-Furyl	−28	−7
CH_3CO	21	42
N≡C	−45.8	−24.8

[a]Taken from Ref. 6c, p. 23. The original signs[25] have been changed to conform with the present shift convention, using the equation $\sigma^P = 21 + GC$.

Experimental values for the replacement of a saturated alkyl on P by an unsaturated group are quite limited. Only a small upfield shift is seen in going from $CH_3CH_2PH_2$ (δ −128) to $CH_2=CHPH_2$ (δ −135.7), and is not easily accounted for. Were resonance involved (which is not thought to be of significance with primary phosphines), one might expect diminished electron density on P, as in enamines, from interaction of the lone pair with the double bond, and thus a downfield shift. Perhaps of more importance is the change to sp^2 hybridization at the attached C, which will cause some contraction of the C−P bond and other geometrical changes.

The sensitivity of the shift of PH_2 groups to the steric environment is also manifested in alicyclic phosphines. Thus PH_2 locked in the axial position of the cyclohexane ring by a 4-t-butyl group gives a shift of δ −131.3, whereas that for equatorial PH_2 is δ −111.6. This upfield shifting is also seen for an axial methyl on cyclohexane. It is an effect that is useful in the conformational analysis of some cyclohexyl P-substituted compounds.[26] When polar substituents are placed on the β-carbon of ethylphosphine, significant upfield shifts again occur, for example β-amino, δ −150.4; β-cyano, δ −135; and β-COOH, δ −138. It is tempting to attribute this shielding also to γ-steric compression, but polar interactions of other types may be involved. Thus the nonpolar PH_2 as a β-substituent leads only to a small γ upfield shift ($H_2PCH_2CH_2PH_2$, δ −130.8). Few data are available for α substitution, and no clear pattern is obvious. Thus a PH_2 group placed on CH_3PH_2 causes a strong downfield shift, possibly a β effect, in $H_2PCH_2PH_2$, δ −121.8, relative to CH_3PH_2 (δ −163.5), and phenyl gives also a downfield shift ($PhCH_2PH_2$, δ −120.9). But placing COOH on CH_3PH_2 gives a pronounced upfield shift ($HOOCCH_2PH_2$, δ −142.6). A clear γ-shielding effect can be seen on methyl substitution (introducing an atom γ to P) in $H_2PCH_2PH_2$; $CH_3PHCH_2PH_2$ has δ −138.9, and $(CH_3)_2PCH_2PH_2$ has δ −152.3.

Shifts due to steric compression can also be recognized in benzene derivatives. Thus placing a methyl group ortho to PH_2 as in o-tolylphosphine (δ −130.9) causes an upfield shift of 8.6 ppm relative to the shift for phenylphosphine (δ −122.3), whereas a para methyl causes a much smaller effect (p-tolylphosphine, δ −124.5). The upfield shifting is even more pronounced with two ortho methyls, as in 2,4,6-trimethylphenylphosphine (mesitylphosphine), with δ −160.2. As seen in Table 6.1, some polar substituents in the ortho position also cause upfield shifts; the effect may be a general one and of use in isomer assignment, but in these cases steric compression may not be the only cause of the shifting.

In practical work with primary phosphines, it can be helpful to record the spectrum under conditions of proton coupling. The presence of two H atoms causes the P signal to split into a 1:2:1 triplet, with very characteristic coupling constants. Typically these are about 180–220 Hz (Table 6.1), and constitute another identifying characteristic. Diphosphines where the P atoms have different substituents, as in i-PrPHCH$_2$PH$_2$, will give two different ^{31}P signals (here δ −29.41 and −138.0 for secondary and primary P, respectively), each showing the effect of ^{1}H coupling, but of ^{31}P-^{31}P coupling as well. For this particular compound, $^2J_{PP}$ is 14.8 Hz.[27]

6.D.2. Secondary Phosphines

As already noted, substitution of a second hydrogen on phosphine causes a further downfield shift, and the ^{31}P chemical shift range for secondary phosphines with simple alkyl and phenyl substituents (Fig. 6.2; δ −100 to +20) is decidedly downfield from that for primary phosphines (δ −164 to −82). Some typical data are given in Table 6.2, in which several cases can be seen of the operation of the β-deshielding effect (e.g., Me$_2$PH, δ −99.5; (CH$_3$CH$_2$)PHMe, δ −77) and the γ-shielding effect ((CH$_3$CH$_2$CH$_2$)PHMe, δ −87), and no further elaboration is needed. Little is known about the effect of polar substituents on carbon, but the situation may be similar to that in primary phosphines; CN groups on the β carbons of diethylphosphine cause the same upfield shift (from δ −55.5 to −75), and comparing CH$_3$CH$_2$PHPh (δ −44.9) to H$_2$NCH$_2$CH$_2$PHPh (δ −52) shows an upfield shift owing to NH$_2$. The GC concept is useful in calculating shifts of secondary phosphines. The σ^P values derived from the tertiary phosphines are increased by 1.5 because of the greater sensitivity of the secondaries; the sum of the scaled σ^P values for the two substituents on phosphorus is then subtracted from the shift for the parent Me$_2$PH, δ −99, to give a predicted shift.

Among aromatic secondary phosphines, we can see again the upfield shifting of an ortho substituent. Thus diphenylphosphine has δ −41 ($^1J_{PH}$ = 220 Hz), whereas di-ortho-tolylphosphine has δ −59.1.

We already noted that a di-phosphine with nonequivalent phosphorus substituents can give a doublet of doublets from the ^{31}P-^{31}P coupling. Because of the stability of the pyramidal structure at 3-coordinate P, each P is a chiral center, and the rules that apply to carbon compounds with two chiral centers are just as valid here. For the secondary diphosphines with the P atoms differently substituted, there will be four enantiomers, constituting two dl mixtures (structures **1** to **4**).

```
  ··              ··              ··              ··
H—P—R         R—P—H         H—P—R         R—P—H
  |             |             |             |
(CH₂)n        (CH₂)n        (CH₂)n        (CH₂)n
  |             |             |             |
H—P—R'        R'—P—H        R'—P—H        H—P—R'
  ··              ··              ··              ··
   1              2              3              4
```

Each dl pair will have its own ^{31}P NMR spectrum. If the substituent is the same at both P atoms, there will be one dl pair (**5** and **6**) and an optically inactive meso form (**7**).

```
  ··              ··              ··
H—P—R         R—P—H         H—P—R
  |             |             |
(CH₂)n        (CH₂)n        (CH₂)n
  |             |             |
R—P—H         H—P—R         H—P—R
  ··              ··              ··
   5              6              7
```

The latter situation is much more common because of the ease of synthesis of symmetrical diphosphines, and several examples are known in which the signal for both the dl and the meso form can be seen. Thus, MePHCH$_2$PMeH gives signals at δ −85.5 and δ −88.9, assigned to dl and meso, respectively.[27]

6.D.3. Tertiary Phosphines

Enormous sensitivity to structural effects is exhibited by the tertiary phosphines; the shift range runs from $\delta -135.7$ (for $(CN)_3P$) to $\delta +152.5$ in bridged bicyclic compound **8**.

8

Even the range just for saturated alkyl and phenyl substitution is large, extending from $\delta -62$ for Me_3P to $\delta +61.1$ for t-Bu_3P. The shift data are much more extensive for this compound class than for the other phosphine types. It was recognized some years ago that substitution effects in noncyclic compounds were additive and as already noted, good predictions of shifts from Group Contributions[25] were possible. Table 6.5 shows the GC values (as recently expanded[6c], a possibility for any tertiary phosphine derivative with three like groups) to be added for the groups present to give a predicted shift. Alternatively, the δ^P values may be summed and subtracted from the parent value, $\delta -62$ for Me_3P. To illustrate, the value for i-$Pr(t$-$Bu)_2P$ is calculated to be $\delta +48$, by adding the GC values for i-Pr of $+6$ and for t-Bu of $+21$ (taken twice). The experimental value is $\delta +45.9$. Matches are better in less-complicated phosphines. GC values have also been derived for tertiary phosphines in which P is in a simple ring system.[28] The chemical shift is calculated by using the ring GC along with the regular GC values for the P-substituent. With complex rings, some very large deshielding can occur.

The concept of β and γ carbons causing deshielding and shielding, respectively, is especially applicable to the tertiary phosphines, and constants can be developed for each effect that can be used to calculate the shift of a phosphine. The β constant used in the original reference[24] is $+13.5$, and the γ constant is -4. Using these constants to calculate the shift for $(CH_3CH_2CH_2CH_2)_3P$, we have 3 β carbons ($+40.5$) and 3 γ carbons (-12), which when added to the parent value -62 gives a shift of $\delta -33.5$. The experimental value is $\delta -32.3$. A feature of the β,γ concept is that it points out the importance of the upfield shifting due to steric compression, which is less visible in the GC values. The effect is much more dramatic in alicyclic systems, as seen in the values for the conformationally rigid cyclohexane derivatives with cis (**9**, $\delta -54.8$) and trans (**10**, $\delta -42.5$) PMe_2 groups.

9   **10**

In the former, the $-PMe_2$ group is axial and undergoes the 1,3-nonbonded interaction common for all axial substituents on cyclohexane. This steric compression is the apparent cause of the large ^{31}P shift difference, as was seen also for primary phosphines, and has been used to determine the A value ($-\Delta G°$, 1.6 kcal/mol[26]) for the Me_2P group.

Steric compression is also operative in ortho-substituted aromatic compounds, as may be seen on examining the cumulative upfield shifting effects of methylation on triphenylphosphine (**11** to **15**).

$(C_6H_5)_3P$	$(2-MeC_6H_4)PPh_2$	$(2-MeC_6H_4)_2PPh$
11	**12**	**13**
δ -8	δ -13	δ -21.4

$(2-MeC_6H_4)_3P$	$(2,4,6-triMeC_6H_2)_3P$
14	**15**
δ -29.9	δ -39.5

Methyl substitution at the para position causes no such shielding, supporting the proposal that the effect of ortho-substitution is not electronic. Many other examples exist of pronounced upfield shifting in ortho isomers, but of little shifting relative to triphenylphosphine in meta and para isomers (e.g., $2-MeOC_6H_4PPh_2$, δ -13.5; $3-MeOC_6H_4PPh_2$, δ -4.0; $4-MeOC_6H_4PPh_2$, δ -6.0).

Vinyl substitution seems to cause little change in shielding at phosphorus. Thus trivinylphosphine has δ -20.7, whereas triethylphosphine has δ -19.7. E and Z isomers of olefinic phosphines also show the steric compression effect, with the Z isomer more upfield (e.g., reference 29). However, a triple bond causes pronounced upfield shifting (δ $n-Bu_3P$, δ -33; $n-Bu_2$ethynylP, δ -56.0). It is tempting to attribute this to the anisotropic effect of the triple bond, but the size of the shielding is far too large for this to be the major explanation. The opposite effect, strong deshielding, is seen when acyl groups are placed on phosphorus, as in the series: Me_3P, δ -61.5; $(MeCO)PMe_2$, δ -19.7); $(MeCO)_2PMe$, δ $+31$; $(MeCO)_3P$, δ $+63.5$. Similar strong effects are seen on benzoyl substitution. Perhaps in part the effect may be attributed to inductive removal of electron density at phosphorus by the electronegative C=O, rather than by a resonance effect, which is generally weak for 3-coordinate phosphorus. Indeed, it has been determined that tribenzoylphosphine shows no significant flattening of the phosphorus pyramid, and NMR studies show no restriction of rotation around the P—C bond[30], both of which should be measurable consequences of any appreciable resonance effect. This interesting structural effect on ^{31}P shielding certainly calls for study from the theoretical standpoint.

Installing phosphorus in ring systems can lead to some strong and sometimes puzzling shift effects, as is made evident in reviews on the ^{31}P NMR shifts of heterocyclic compounds.[7d,11] Some of these effects will be described in later discussions on heterocyclic compounds (chapter 8), but we can at this time consider two unusual cases that illustrate how sensitive the shift in tertiary phosphines can be to structural effects. The first case is found in the shift relations of the simplest type of phosphorus heterocycle, the saturated monocyclic parents. Data are available for the series of P-phenyl derivatives of rings with 3 to 6 members (**16** to **19**), as follows:

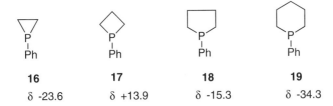

16	17	18	19
δ -23.6	δ +13.9	δ -15.3	δ -34.3

The shift of the 3-membered phosphirane is the most upfield known for any tertiary phosphine; the shielding has been associated with high s-character in the lone pair. This arises from the highly strained internal C—P bonds not tolerating much s-character, which is diverted to the lone pair orbital on phosphorus.[7e] To some extent, this must be true in the still-strained four-membered phosphetane ring, but the dramatic deshielding, to values beyond those for the five- and six-membered rings, defied explanation until recently.

There is also an anomaly in the shifts for the five- and six-membered rings; in phosphines, as in many other heterocyclic phosphorus compounds, deshielding is stronger in the smaller ring, quite inconsistent with an explanation based on the bond angle argument, where the larger angle in the six-membered ring should lead to greater deshielding. This has been tentatively attributed[24] in part to the fact that in the six-membered ring there is a built-in γ-shielding effect from the carbon in the 4-position, an effect that is absent in the five-membered ring. In fact, the value for the five-membered phospholane is practically the same as that for the open-chain counterpart, Et_2PPh, δ −16. From this analysis of the data, the major difficulty is explaining the dramatic deshielding in the four-membered ring, and this has been approached with the newly developed ab initio quantum-mechanical method for calculating ^{31}P NMR shifts.[31] This method is based on the Hartree–Fock calculation with gauge-including-atomic-orbitals (GIAO), which indeed reproduced the trend in the shifts for cyclic phosphines (within the limits of the calculation). Thus calculated values for the easier treated P—H derivatives were $HP(CH_2)_n$ ($n = 2$, δ −355.4; $n = 3$, δ −54; $n = 4$, δ −102.8; and $n = 5$, δ −92.2). The major accomplishment is the exact placement of the four-membered ring in the deshielded position relative to the other rings, just as in the experimental results for the P-phenyl series. The calculation did not reproduce the difference between the five- and six-membered rings, but the experimental difference is not great and is comparable to the error of the calculation. Further calculation of molecular orbitals with the STO-3G basis revealed that the phosphorus lone pair has nearly sp hybridization in the three-membered ring but nearly sp^2 in the four-membered ring, and gave values for the HOMO-LUMO gap for the entire series, which led to an explanation for the trend in the shifts. According to this concept, which we will see to be useful in explaining other shift phenomena, a high HOMO-LUMO gap in related compounds is associated with greater shielding, and a lower gap with diminished shielding, and it is a combination of the gap differences and the hybridization differences that leads to the ordering of the shifts. Relative to the five- and six-membered rings, the three-membered ring phosphorus had a larger HOMO-LUMO gap, and the four-membered ring a smaller gap, consistent with the shielding differences. The original paper should be consulted for a greater appreciation of the theory.

The other case from cyclic phosphines to consider is that of the bicyclic 7-phosphanorbornene system, which when the P-substituent is syn to the double bond, leads to the most downfield ^{31}P NMR shifts ever observed. This was noted in compound **8**, and thus the ^{31}P shifts for cyclic phosphines can fall anywhere over the enormous range of about δ −250 to δ +150! No other phosphorus function has anything like this dramatic dependence of shift on molecular structure. The effect in the 7-phosphanorbornene derivatives is critically dependent on the orientation of the phosphorus lone pair; the effect is much greater in the syn isomer (**20**, δ +100.8) than in the anti isomer (**21**, δ +30.2).

Expansion of the bridged ring causes diminution of the strong deshielding in the syn isomer, and saturation of the double bond eliminates the effect. Again, the computational approach has proved useful in developing an explanation for the strong deshielding in structures **20** and **21**.[32] Calculations of the same type that were used for the simple cyclic phosphines reproduced the strong deshielding in the syn isomer and further revealed that a major cause of the deshielding was the diminished value of the HOMO-LUMO gap for the syn P-substituent. As already noted, a reduced gap is associated with deshielding.

6.E. ^{31}P NMR SHIFTS OF HALOPHOSPHINES

The parent of the halophosphines may be considered to be PCl_3, whose ^{31}P shift is δ +219.5. Replacement of one chlorine by methyl gives $MePCl_2$ with δ +192.0, and a second methyl gives Me_2PCl with δ +96. These upfield shifts are consistent with the loss of the strong electron attraction of chlorine, which acts to contract the p-orbitals at phosphorus and causes a strong paramagnetic effect. The trend is seen also in the P-phenyl series ($PhPCl_2$, δ +166; Ph_2PCl, δ +81.5) and in many other series. It is curious and unexplained that replacement of Me by Ph causes a strong upfield shift in the halophosphines, but a downfield shift in the phosphines. A more recognizable electronegativity trend is seen when the shift effects of the different halogens are considered; for the series MePX$_2$, we have X = F, δ +244.2; X = Cl, δ +192.0; X = Br, δ +184.0; and X = I, δ +130.6. In general this trend is seen with other P-substituents, but exceptions do exist (e.g., CCl$_3$PCl$_2$, δ +147; CCl$_3$PF$_2$, δ +131).

Chain lengthening and branching can influence the shifts in the way seen for phosphines. The γ effect seems to be absent in simple phosphonous dichlorides but is noticeable in the iodides (MePI$_2$, δ +130.6; EtPI$_2$, δ +147.0; BuPI$_2$, δ +140.3) and in phosphinous chlorides (Me$_2$PCl, δ +96; MeEtPCl, δ +105.2; Me$_2$CHCH$_2$PMeCl, δ +95.5). When the PCl$_2$ group is placed on the cyclohexane ring, we find that the axial group (as in **22**) is downfield of the equatorial (**23**), the opposite of the relation for the Me$_2$P group, again suggesting other effects to be important along with those of steric compression.

22
δ +208.9

23
δ +194.6

Insufficient data are available to evaluate the shifts in aromatic compounds, but shielding effects of both the double (as in **24**) and the triple bond (as in **25** and **26**) can be detected.

PhHC=CPCl$_2$
24
δ +163.5

PhC≡CPClBu-*t*
25
δ +82

PCl$_2$C≡CPCl$_2$
26
δ +108.6

6.F. ^{31}P NMR SHIFTS OF 3-COORDINATE OXY, THIOXY, AND AMINO COMPOUNDS

Numerous combinations are possible when the chlorine of various phosphorus chlorides is replaced by oxy, thioxy, or amino groups, as seen in chapter 3. The trends in ^{31}P shifts can, however, be understood by first examining some structures in which the only variation is in the heteroatom substituent, and then some derivatives of the most commonly encountered forms, those with RO— substituents as in the tertiary phosphites and phosphonates. Representative data on which this discussion is based are given in Table 6.3. We note first that all of these heteroatoms exert very strong deshielding effects, and lead to derivatives with shifts far downfield of those for the tertiary phosphines. In a family in which the heteroatom (Y) is the only variant, as in the series (Me-Y)$_3$P, the oxy derivatives are generally seen to have the most strongly deshielded phosphorus (exceptions are MeOPCl$_2$, δ +180.3; MeSPCl$_2$, δ +206.0, and related structures) and are always downfield of the amino derivatives. On examining a family of oxy derivatives in which variations occur in either the O-substituent or the P-substituent, we note that most of the trends in shifts already discussed for other 3-coordinate compounds can be detected but are generally not as pronounced, and some are missing. At the P atom, β-deshielding can almost always be detected as substituents are added to the alkyl group, but γ shielding is weak or not observed. In the alkyl group attached to the heteroatom (which stands as the α-substituent), the γ effect can also be detected (MeP(OMe)$_2$, δ +182.5; MeP(OCHMe$_2$)$_2$, δ +173), but this is not always the case. Among phosphonites, replacement of alkyl on P by aryl is generally associated with a pronounced upfield shift (e.g., MeP(OEt)$_2$, δ +182.5; PhP(OEt)$_2$, δ +153.5), and this is true also for olefinic groups on P (e.g., Me$_2$C=CHP(OEt)$_2$, δ +158). Insufficient data are available to assess the effects of ring substitution in the P-aryl series. In the cyclic norbornane system, the effects of γ steric compression can be seen, causing for example the exo isomer (**27**, δ +185.3) to resonate upfield of the endo isomer (**28**, δ +192.4).

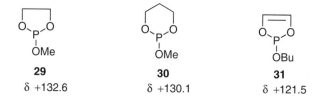

In phosphites, O-aryl substitution causes a pronounced upfield shift ((PhO)$_3$P, δ +125 or +129), but the effect is weak in phosphonites. Trimethylsilyl groups also lead to strong shielding in both structural types (e.g., (MeO)(TMS−O)$_2$, δ +117), which could be the result of the three methyl groups acting as γ substituents. Curiously, the γ effect is absent in the phosphite (*t*-BuO)$_3$P, δ +138.2. Placing the phosphorus of a phosphite in a saturated ring is accompanied by some shielding (8 to 10 ppm in **29** and **30**) relative to an acyclic model (e.g., (EtO)$_3$P, δ +137 to +140), but the size of the ring has little influence on the shifts even though bond angles and steric interactions would be quite different for these rings. As will be seen shortly, ring size does have a very pronounced effect on the shifts of the corresponding phosphates. Placement of a double bond in the 5-membered ring, as in **31**, causes a further shielding effect of about 10 ppm.

29	**30**	**31**
δ +132.6	δ +130.1	δ +121.5

Only two compounds are known in which OH is present on 3-coordinate phosphorus, because such structures rearrange to the 4-coordinate phosphoryl form (chapter 2.G). These compounds (**32** and **33**) have shifts of δ +142.5[33] and

δ +78[34] respectively, quite in keeping with the assigned 3-coordinate structures. This is, in fact, an excellent example of the use of ^{31}P NMR to define an otherwise unexpected structure.

[Structure 32: phthalimide-N–P̈(OH)$_2$]   [Structure 33: (CF$_3$)$_2$P̈–OH]

6.G. ^{31}P NMR SHIFTS OF TERTIARY PHOSPHINE OXIDES AND SULFIDES

Simple alkyl and phenyl phosphine oxides and sulfides have shifts that fall in the same rather small range of about δ +30 to +60 (Table 6.4), and with only the exception of some unsaturated bicyclic derivatives such as the oxide (with δ +91.4) of the 7-phosphanorbornene derivative **20**, they are always well downfield of the corresponding phosphines. The range for all known phosphine oxides is considerably broader, from δ −56 to δ +132, but few structures populate the regions near the extreme ends of the range. Generally, but not always, sulfides give shifts slightly more downfield than oxides. The data in Table 6.4 reveal the operation of the effects already seen in 3-coordinate compounds that occur on extending or branching the chain, effects that will also be seen in some other 4-coordinate structures. The steric compression that must be present in axial substituents on the cyclohexane ring does not lead to upfield shifting and must be over-ridden by (presumably) polar properties of the substituent, as was noted for the −PCl$_2$ group. Thus for the −PSMe$_2$ group on the 4-t-Bu-cyclohexyl ring, the axial group is downfield of the equatorial by 0.8 ppm.

We can also see that substitution of alkyl on P by aryl causes substantial upfield shifting, the opposite of the effect in phosphines. Stronger upfield shifting occurs in vinyl phosphine oxides, but not as much as is seen in acetylenic derivatives. The shift for triethynylphosphine oxide is a phenomenal δ −56, the most upfield of any yet reported for a tertiary phosphine oxide. It is remarkable and unexplained that this effect of the triple bond is so large and occurs in both the oxide and the corresponding triethynylphosphine (δ −91). As already noted, it is probably not appropriate to attribute an upfield shift of this magnitude entirely to magnetic anisotropy of the triple bond.[7e] Of course, a resonance interaction may also be involved in the unsaturated derivatives; electron release from these substituents to the electronegative P=O group could be associated with the upfield shifting.

Many cyclic phosphine oxides have been prepared in recent years, and these often exhibit pronounced shielding differences from simple acyclic models. Just as in ^{13}C NMR, phosphorus in a three-membered ring, with its unusual hybridization and highly contracted interatomic angles, is strongly shielded; at the extreme upfield end of the shielding range for saturated tertiary phosphine oxides is found the three-membered phosphirane oxide **34** with δ −1.3. At the other extreme are found the derivatives of 7-phosphanorbornenes, e.g., **35** with δ +93.1.

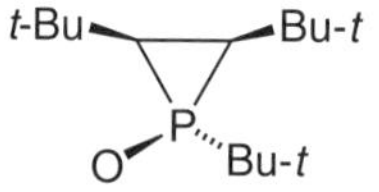

In the monocyclic six-membered phosphorinane ring, exemplified by **36**, the shift is in the range of that of an acyclic phosphine oxide (e.g., δ +48.3 for Et$_3$PO), but in the five-membered phospholane ring (as in **37**) substantial deshielding occurs.

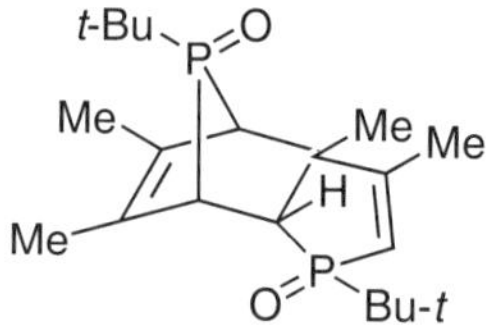

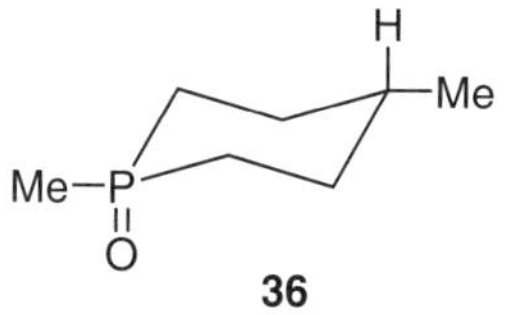

36
δ +40.9

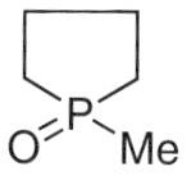

37
δ +65.8

Many other structural effects abound in cyclic phosphine oxides and sulfides, and predictions in the absence of models is precarious. Tebby's[6] tables should be consulted to appreciate more fully the wide variety of known cyclic structures and their influence on ^{31}P NMR shifts.

6.H. ^{31}P NMR SHIFTS OF QUATERNARY PHOSPHONIUM SALTS

When phosphines are protonated or quaternized, strong downfield shifts occur, with few exceptions, as the shielding effect of the lone pair is removed. As for phosphine oxides, there is a narrow range in which simple saturated alkyl and aryl derivatives resonate, but the range becomes extremely broad when certain cyclic and acetylenic structures are included. Thus, for the quaternary salts as an example, the "simple" range is δ +19 to +36, but far downfield will be found 7-phosphanorbornene derivatives (e.g., **38**, δ +98.8) and upfield will be found acetylenic derivatives (e.g., **39**, δ −22.3), just as in the case of the oxides.

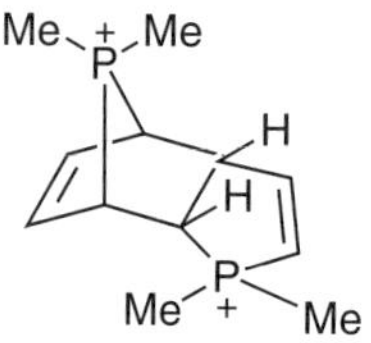

38

MePhP$^+$(C≡CH)$_2$ I$^-$

39

Other familiar trends are present; replacement of alkyl by aryl or vinyl causes upfield shifts, and β, γ effects are present. One must be especially careful in comparing data for salts in the literature, however; preferably the solvent and the counterion should be the same, because these can influence the shift. Frequently adequate data are not available. Eq. 6.2 has been presented[7e] that allows good approximation of the ^{31}P shifts using the same GC values as developed by Grim[25] for phosphines. Shift predictions are also possible from the β, γ constants.[24]

$$\delta = +21.5 + 0.26 \sum_{n=1}^{4} \sigma_n^P - 3.2\, m - 5.51 \qquad (6.2)$$

The use of these constants reveals an important consideration in the interpretation of shifts for acyclic phosphonium ions; the bond angles must all have the tetrahedral value in this series, yet clearly the shifts respond to the steric compression provided by γ-substituents. Thus in interpreting shift effects in cyclic compounds one must consider both bond angle variations and steric compression, perhaps in addition to other factors as well.

6.I. ^{31}P NMR SHIFTS OF HETEROPHOSPHONIUM IONS AND REACTION INTERMEDIATES

As the alkyls of a quaternary phosphonium ion are replaced by heteroatoms, quite pronounced chemical shift changes can be experienced, but surprisingly they can occur in either the upfield or (more commonly) the downfield direction. Downfield shifts would be predicted from electronegativity considerations, but acting in the opposite direction would be charge dispersal from phosphorus to the heteroatom. Other influences might also be present, such as bond angle differences and orbital interactions, and little study has been directed to the interpretation of the observed shift changes. The ambiguity is revealed by considering the effect of replacement of methyl by methoxy. Starting from $Me_4P^+Cl^-$ with δ +24.4, replacement of one methyl by methoxy leads to a strong downfield shift to δ +95.8 in $Me_3(MeO)P^+\ SbCl_4^-$, and a second methoxy causes an additional downfield shift to δ +98.2 in $Me_2(MeO)_2P^+\ SbCl_6^-$. However, a strong upfield shift occurs with a third methoxy $(Me(MeO)_3P^+\ BF_4^-, \delta$ +54), and a further shift occurs on going to $(MeO)_4P^+BF_4^-, \delta$ +1.9. (Note that the anion is not always the same in the salts being compared; there is occasionally a small effect, a few ppm at most, as the anion is changed. Solvent differences also can affect the shifts). Also at relatively high field is $(PhO)_4P^+, \delta$ −21; $(NH_2)_4P^+$, on the other hand, has δ +28.4. Some important chlorophosphonium ions include $MePCl_3^+Cl^-$ at δ +119 and $PhPCl_3^+AlCl_4^-$ at δ +101, both downfield of PCl_4^+ at δ +86.5. Just from these few data we can see that the shifts of heterophosphonium ions cover a very broad range and are very sensitive to changes in the atoms attached to phosphorus.

Heterophosphonium ions frequently appear as short-lived reaction intermediates, and ^{31}P NMR has played a most valuable role in detecting and characterizing such species for the elucidation of reaction mechanisms, as well as discriminating between them and the corresponding 5-coordinate forms from combination of the ions. Five-coordinate forms have shifts that are almost always strongly negative and are far upfield of those of the 4-coordinate heterophosphonium ions. Four examples of the use of ^{31}P NMR in defining reaction intermediates follow; others are to be found in reviews by Michalski, et al.,[7f] and by Hudson.[35]

6.I.1. The Michaelis–Arbusov Reaction

As was noted in chapter 5.C, the alkylation of 3-coordinate phosphorus esters proceeds by way of a phosphonium ion intermediate, which then undergoes dealkylation to give a phosphoryl derivative. The intermediate generally has a very short lifetime, but in a few cases it has been detected by ^{31}P NMR, which conclusively proved its ionic structure. The first structures established were **40**, δ +96.0, and **41**, δ +94.0, derived from methylation of phosphonites. The shifts correspond well to that of the related structure $Me_2(OMe)_2P^+$ $SbCl_4^-$ with δ +98.2.

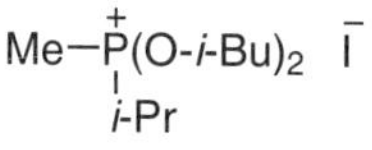

$$Me-\overset{+}{\underset{i\text{-Pr}}{P}}(O\text{-}i\text{-Bu})_2 \quad I^-$$

40

$$Me-\overset{+}{\underset{Et}{P}}(O\text{-}i\text{-Bu})_2 \quad I^-$$

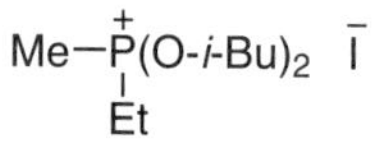

41

There are special variations of the Michaelis–Arbusov reaction in which both the usual 4-coordinate ion and the 5-coordinate phosphorane from combination of the ions have been detected. Thus in the reaction of ethyl *o*-phenylene phosphite with *N*-chlorodiethylamine at −50°, the ^{31}P NMR spectrum consists of four signals, δ +134.0 for the starting phosphite **42**, δ +32.6 for the final product **43**, the intermediate phosphonium ion **44** at δ +18.4, and an upfield signal at δ −38.5 which can be attributed to the phosphorane **45**. As the temperature is raised to −15°, the upfield signal disappears as the dissociation of the phosphorane occurs. Other cases[36] are known in which ^{31}P NMR spectroscopy has been used to detect the phosphonium ion–phosphorane equilibrium, made easily possible by the wide difference in their chemical shifts.

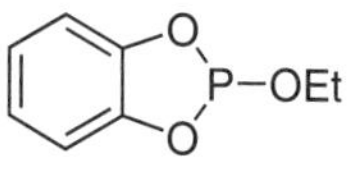

42 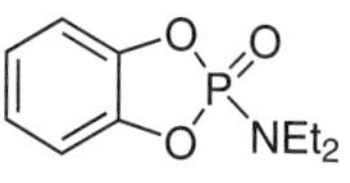   **43** 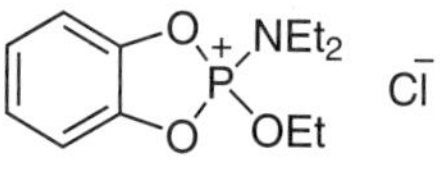   **44**

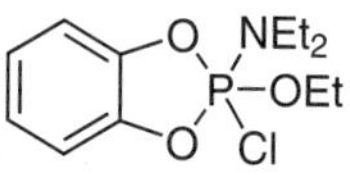

45

6.I.2. The Mitsunobu Reaction

The Mitsunobu reaction is a process that accomplishes the elimination of water from an alcohol and an acid to form an ester via phosphorus intermediates (chapter 3.C). The reactants involve triphenylphosphine and diethyl azodicarboxylate, along with the alcohol and the acid. Its mechanism has received considerable attention. That proposed by Jenkins and co-workers,[36] based on ^{31}P NMR studies, is presented here Eqn. 63. The first step is the addition of the phosphine to the N=N group, forming the heterophosphonium ion **46**, clearly identified by the ^{31}P NMR signal at δ +44. In the absence of the acid, a 5-coordinate species **47** is formed from the reaction with the alcohol, so indicated by the appearance of a ^{31}P NMR signal at δ −57.5. This species is in equilibrium with the ionic form **48**, δ +60. If, however, the acid is added along with the alcohol, the common procedure, only the ion **48** is formed, and the 5-coordinate species is not observed. Attack of the carboxylate ion on **48** at the site of the O−C bond then leads to the ester product.

$$\text{EtOOC-N=N-COOEt} \xrightarrow{Ph_3P} \underset{\underset{\text{COOEt}}{|}}{Ph_3\overset{+}{P}-N-\overset{-}{N}-COOEt} \xrightarrow[\text{or ROH}]{\text{R'COOH}}$$
$$\mathbf{46}$$

$$\underset{\underset{\text{COOEt}}{|}}{\overset{\overset{H}{|}}{Ph_3\overset{+}{P}-N-NCOOEt}} \xrightarrow{ROH} \left[\underset{\mathbf{47}}{Ph_3P(OR)_2} \rightleftharpoons \underset{\mathbf{48}}{Ph_3\overset{+}{P}OR} \right] \xrightarrow[\text{R'COOH}]{\text{R'COOH}} Ph_3P=O + R'COOR$$

$$\xrightarrow{\text{R'COOH, ROH}} Ph_3\overset{+}{P}OR$$

(6.3)

6.I.3. Study of the Friedel–Crafts Synthesis of Phenylphosphonous Dichloride

This important process, conducted at 90°, was followed by ^{31}P NMR.[37a] In accord with an earlier report of Symmes and Quin,[37b] no complex could be detected on mixing PCl$_3$ with AlCl$_3$ (although complexes of phosphonous and phosphinous halides have frequently been reported). On heating, three new species were observed. That at δ +69 was attributed to the heterophosphonium salt PhPCl$_2$H$^+$ AlCl$_4^-$, as indicated by a ^{1}H-^{31}P coupling constant of 751 Hz. The other species were simply the known complexes of the product (PhPCl$_2$−AlCl$_3$ with δ +83, (PhPCl$_2$)$_2$−AlCl$_3$ with δ +83), both of which reacted with dry HCl to produce PhPCl$_2$H$^+$ AlCl$_4^+$. The sequence is expressed by Eq. 6.4. No intermediates were detected in the initial reaction before the formation of the PhPCl$_2$H$^+$ species, and those that are formed must be of very short lifetime.

$$PCl_3 + AlCl_3 \rightleftharpoons \underset{\text{not observed}}{Cl_2\overset{\delta+}{P}--Cl--\overset{\delta-}{AlCl_3}} \xrightarrow{C_6H_6} Ph\overset{+}{P}Cl_2H\, \overline{AlCl_4} \underset{+HCl}{\overset{-HCl}{\rightleftharpoons}}$$

$$(PhPCl_2)AlCl_3 + (PhPCl_2)_2AlCl_3 \qquad (6.4)$$

6.I.4. Intermediates in Phosphorylations

Two recent studies have used ^{31}P NMR to examine the mechanism of the reaction of phosphoryl halides with alcohols in the presence of tertiary amines. The combination in THF of either $(PhO)_2P(O)Cl$ (δ -7.0) or $(PhO)_2P(O)Br$ (δ -16.8) with pyridine gave the same intermediate with δ -25.5, attributed to structure **49**,[38] which then reacted with the alcohol to give the ester. A similar type of intermediate **50**, with δ $+44.7$, was detected[39] when the phosphonochloridate Cbz-NHCH$_2$P(O)(OEt)Cl, with δ $+35.6$, was mixed with triethylamine; addition of an alcohol then gave the expected diester.

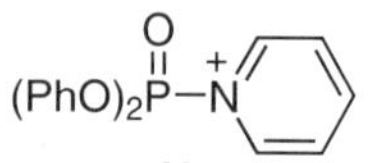

$$(PhO)_2\overset{O}{\overset{\|}{P}}-\overset{+}{N}\diagdown\!\!\diagup\!\!\!\diagdown$$
49

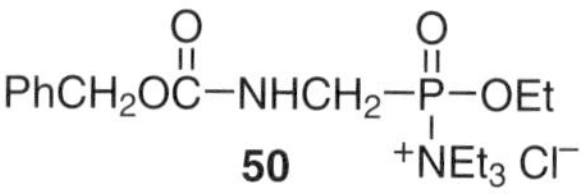

$$PhCH_2O\overset{O}{\overset{\|}{C}}-NHCH_2-\underset{\underset{^+NEt_3\ Cl^-}{|}}{\overset{O}{\overset{\|}{P}}}-OEt$$
50

6.J. ^{31}P NMR SPECTRA OF PHOSPHORUS ACIDS WITH P–C BONDS AND SOME DERIVATIVES

A great amount of NMR information is available on the P–C acid families, but only some broad comments can be offered here. Succinct summaries of important shift effects in phosphonic acids[6d] and phosphinic acids[6e] can be found in the Tebby handbook. We can see some of the main effects by examining structural changes starting with methylphosphonic acid, with δ $+30.5$. These trends are generally reproduced when starting with higher alkylphosphonic acids.

- Replacement of one OH by a second methyl to give dimethylphosphinic acid (Me$_2$P(O)OH) causes a downfield shift to δ $+48.6$.
- Replacement of one OH by H gives methylphosphonous acid (MeHP(O)OH) with a smaller downfield shift to δ $+35$ (with $^1J_{PH} = 557$ Hz).
- Esterification of methylphosphonic acid causes only small shifts (MePO(OMe)$_2$, δ $+31.5$; MePO(OEt)$_2$, δ $+27.7$). O-Aryl derivatives are substantially upfield of O-alkyl derivatives (MePO(OPh)$_2$, δ $+24$ (as are O-silyl derivatives, better seen with P-phenyl derivatives, vide infra).
- Replacement of both OH by chlorine leads to a large downfield shift (MePOCl$_2$, δ $+42.6$).
- Replacement of phosphoryl oxygen by sulfur causes a large downfield shift (MePSCl$_2$, δ $+77.5$).
- Phosphonamides are generally downfield of the acids and esters, although data are few (BuPO(NEt$_2$)$_2$, δ $+43.5$).
- Replacement of methyl by aryl causes a substantial upfield shift (PhPO(OH)$_2$, δ $+17$). The O-silylation effect may be seen here, PhPO(O-TMS)$_2$ δ 0.0. This may be due to a large cumulative γ effect from the methyls on silicon.

Phosphinic acids generally respond to these structural changes in the same way as do phosphonic acids (Me$_2$PO(OMe), δ +52.2; Me$_2$PO(O-TMS), δ +39.0; Me$_2$POCl, δ +58.7; and Ph$_2$PO(OH), δ +25.5). Some of these effects are known also for phosphonous acids, although few data are available (PhPH(O)(OH), δ +20; PhPH(O)(OMe), δ +27.3; and PhPH(O)(O-TMS), δ +10.1).

Much current research is concerned with structural changes in the carbon substituent of phosphonic acids and esters, and in the remainder of the discussion, some of the effects of typical structural modification will be described. In the first place, lengthening the carbon chain of methylphosphonic acid causes the expected downfield shift (the β effect) as seen in every class of phosphorus compounds. The shift is not large, however, and insufficient data are available to search for γ effects, which are probably very weak. It can safely be assumed that the parent alkyl chain will have a shift in the range δ +30 to +38 (the latter for Me$_3$CPO(OMe)$_2$), and this applies also to cycloalkyl derivatives (cyclohexylphosphonic acid, δ +33). It is the placing of polar substituents and unsaturation on the parent chain that most research is concerned with, however, and these can cause considerable shifting from the saturated parent provided the structural change occurs on the first three carbons of the chain. Polar substituents on methylphosphonic acid routinely cause upfield shifts, the only explanation for which must rest on the inductive effect of these substituents and an influence on the electronic structure at phosphorus. Data are not always available for the free acids, but ester derivatives can be used in witnessing the effect, which is significantly larger than that of the acid-ester exchange. Some examples are Cl–CH$_2$PO(OH)$_2$, δ +17.8; H$_2$N–CH$_2$PO(OH)$_2$, δ +10.1 (pH 10.5); O$_2$N–CH$_2$PO(OMe)$_2$, δ +10.2; NC–CH$_2$PO(OEt)$_2$, δ +15.0; MeC(O)–CH$_2$P(OEt)$_2$, δ +17.1; (HO)$_2$OP–CH$_2$PO(OH)$_2$, δ +16.7 (tetraethyl ester, δ + 19.0); and HO–CH$_2$PO(OH)$_2$, δ +20.3 (pH 1.9).

The same upfield shifting is seen when polar substituents are placed on the α carbon of ethylphosphonic acid and indeed on the α position of *any* P-substituent. Using only the amino group to illustrate, we find MeCH(NH$_2$)PO(OEt)$_2$ at δ +20.1, and EtCH(NH$_2$)PO(OH)$_2$, δ +14.5. When a second polar substituent is placed on the α carbon, there can be additional shielding: Cl$_2$CHPO(OEt)$_2$, δ +10.5; (Cl$_3$CPO(OEt)$_2$, δ +5.0. This may also be seen in the methylenediphosphonate series; the dichloro derivative **51** has a shift of δ +7.9. Alkyl substitution, however, gives the expected downfield shift (**52**, δ +23.0) relative to the methylene compound (δ +16.7). The addition of an NH$_2$ group to give **53** might then be expected to cause an upfield shift, but the effect is quite small (δ + 22.1), suggesting that caution is needed in making predictions in this series.

```
        PO(OH)2            Me   PO(OH)2           Me    PO(OK)2
   Cl2C                       C                      C
        PO(OH)2            H    PO(OH)2           H2N   PO(OK)2

         51                      52                     53
```

When the polar substituents are moved to the β carbon, the shifts are somewhat more downfield but remain upfield of the parent acid (ClCH$_2$CH$_2$PO(OEt)$_2$,

δ +24.4; (EtO)$_2$P(O)CH$_2$CH$_2$PO(OEt)$_2$, δ +28.6; H$_2$NCH$_2$CH$_2$PO(OH)$_2$ (in water[40]), δ +19.6). Placing sp^2 or sp carbons on phosphorus causes pronounced upfield shifts, the latter leading as might be expected to the most shielded of all phosphonic acids: CH$_2$=CHPO(OEt)$_2$, δ +17.3; HC≡CPO(OH)$_2$, δ −9.2; C$_6$H$_5$PO(OH)$_2$, δ +17; (MeO)$_2$P(O)COMe, δ +1.5.

Many arylphosphonic acids and derivatives are known, and several attempts have been made to develop correlations of ^{31}P shifts with electronic characteristics of the ring substituents (for references, see Tebby[6d,6e]). Gorenstein[12] provides an excellent summary of the (largely unsuccessful) attempts to find correlations with inductive and resonance effects. Much of the interpretation was based on changes in the d-orbital occupancy (with increases causing shielding), but we now know that this is not necessarily a valid concept. We do know that, in general, the shift effects are small at the meta and para positions, with inductive effects outweighing resonance effects. Electron-withdrawing groups can cause either deshielding or shielding. Upfield shifting by steric compression in the ortho isomers is a more reliable effect, and hydrogen bonding from ortho OH and NH groups can contribute a downfield component to the observed shifts. To explore the true effect of substituents on the ring, a series of related compounds that gives shifts not influenced by concentration or ionization effects would be the best models. Thus Grabiak et al.,[41] obtained useful data in the family of arylphosphonic dichlorides, which reveals that electron-withdrawing groups at the para position caused upfield shifts (e.g., in p-X–C$_6$H$_4$POCl$_2$, δ +36.0 for X=H but δ +32.65 for X=CF$_3$, and δ +37.6 for X=Me$_2$N; in m-X–C$_6$H$_4$POCl$_2$, δ +35.9 for CH$_3$ and +30.7 for CF$_3$). The data were correlated with Hammett σ^n or Taft σ^* constants. Further discussion of the effects of substituents on aromatic phosphonic and phosphinic acid derivatives is presented by Edmundson.[6d,6e]

6.K. ^{31}P NMR SHIFTS OF PHOSPHORIC ACID DERIVATIVES

Phosphoric acid derivatives are of great importance in biological processes and industrial chemistry, and ^{31}P NMR has played a major role in structural studies and analysis. One of the most exciting applications of the ^{31}P NMR technique is in the direct examination of living systems, which can give information on metabolic processes and diseased states. This intriguing work is the subject of reviews[12d,13] but is beyond the scope of this chapter.

With minor exception, the range of ^{31}P shifts for all alkyl derivatives (except tert-butyl) of phosphoric acid is defined by the values for the family of trialkyl phosphates, with trimethyl phosphate being the most deshielded at about δ +3, and tri-isopropyl phosphate at δ −3.3. Within these extremes will be found all derivatives with mixed alkyl groups, and the range includes monoalkyl and dialkyl derivatives as well. The very large number of phosphates isolated from natural sources are of the latter type; their alkyl groups are of considerable complexity but require that we extend the range of shifts to only about δ 0 ± 5. Extensive tables of data for individual biophosphates and structure-shift relationships are available.[2a,6f,12] Some clarification of the shift of the important parent compound, trimethyl phosphate, is required, however. A more recently reported value is $\delta + 1.97$ in a special lipid-extraction solvent consisting of chloroform-methanol and aqueous K-EDTA to complex interfering Mg^{+2} and Ca^{+2} ions;[2a] this shift was confirmed in the author's laboratory. Gorenstein shows a similar value of $\delta +2.1$ in a summary of phosphotriester shifts;[12b] however, the values in the Tebby tables are $\delta -3.0$ in chloroform and $\delta -2.4$ in dichloromethane,[6f] which must have the sign mistakenly reversed; this is born out by referring to the original article,[42] which used the old sign convention and indeed reported these values. However, a 1987 book[13b] also used negative values for trimethyl phosphate (e.g., $\delta -2.1$ in chloroform) and went on to make the observation that phosphotriesters in general are upfield of phosphodiesters (given a range of $\delta\ 0 \pm 1$) and mono esters (range $\delta +4$ to 0). Although some triesters do indeed have negative values, the generalization that all triesters are upfield of about $\delta -1$ is now seen to be too broad. An even more recent recent review[1] reports the incorrect value of $\delta -2.4$ for trimethyl phosphate, to compound the confusion.

Next, the shift for tri-t-butyl phosphate needs to be considered. The shift is $\delta -13.3$, and thus in a position well beyond the range for all other trialkyl esters. This compound has played a role in the development of an explanation of ^{31}P shift variations: as steric bulk increases, the alkoxy to P bond angles expand, and this is said to be the cause of the upfield shifting.[12a] Theoretical calculations are in support of this concept. We note, however, that the series of simple trialkyl phosphates have shifts that can also be interpreted by the β, γ substituent concept. Using the shift data summarized by Gorenstein,[12b] replacing the three H of phosphoric acid by methyls constitutes a β substitution (O being α to P), and indeed there is the expected downfield β effect. Converting methyl to ethyl adds a γ carbon, and the shift moves upfield to $\delta -1.0$. Going to isopropyl adds another γ carbon, and the shift goes farther upfield to $\delta -3.3$. By this approach, it could be argued that, at least in part, the upfield shifting in tri-t-butyl phosphate is derived from the large steric compression effect. The shift for trineopentyl phosphate seems to speak against the bond angle argument; surely with three such large groups the bond angles would be expected to undergo some expansion, but the observed shift ($\delta -0.6$) is essentially the same as that of triethyl phosphate. This observation fits better with the β, γ substituent argument; there is no difference between ethyl and neopentyl in regard to the number of these carbons.

Another sterically related effect is an important connection that has been made with the torsion angles in different conformations from rotation about the P—OR bonds; in the more extended conformation with the two R groups remote from each other (referred to as the trans,trans conformation), signals are downfield from those of more crowded conformations. These conformational effects, recognized by theoretical considerations and called stereoelectronic effects,[12b,22] may be seen experimentally when dealing with rigid cyclic systems; thus, the shifts of the bicyclic compounds **54** and **55** differ by several ppm, with the equatorial conformer **55** being the more downfield, consistent with the stretched-out relation of the OAr group relative to the ring C—O bonds. Other researchers, however, have noted that the steric compression effect would lead to the same prediction.[12c]

54
δ −13.0

55
δ −10.6

Structures **54** and **55** reveal another effect of cyclic structure: six-membered cyclic phosphates (2-oxo-1,3,2-dioxaphosphorinanes) have shifts that are noticeably upfield of acyclic models. In **54** and **55**, part of the upfield shifting comes from the phenoxy group, as will be noted below, but in general a six-membered ring is associated with an upfield shift of 5 to 6 ppm. Thus compound **56** has a shift of δ −6.7; an appropriate model to reveal the shielding might be methyl diethyl phosphate, δ −0.8. The shielding in a six-membered ring might in part come from the γ carbon being held in a gauche position relative to P, as was noted for six-membered cyclic phosphines. On the other hand, five-membered cyclic phosphates (2-oxo-1,3,2-dioxaphospholanes) are considerably downfield relative to acyclic phosphates; compound **57**, for example, has δ +18.8.

56

57

The structural and stereoelectronic changes resulting from the compressed internal O—P—O bond angle in the five-membered ring have been associated with this effect.[12]

Other important effects among the derivatives of phosphoric acid are as follows.

- ArO— groups lead to pronounced shielding: $(EtO)_2P(O)OPh$, δ −6.8; $(PhO)_3PO$, δ −18. There is uncertainty about the cause of this shift.
- Electron-withdrawing groups on the ArO substituent cause small upfield shifting, an effect recently analyzed with the aid of semiempirical MO calculations and attributed to an increase in the positive charge on phosphorus in turn causing an increase in the back-bonding from phosphoryl oxygen.[43]
- Me_3SiO— groups also lead to strong shielding: $(TMS-SiO)_3PO$, δ −26.5, predictable either by the bond angle expansion argument or the γ shielding effect of the 9 methyls.
- Amino and alkylamino groups on phosphorus lead to downfield shifts, below those of esters: $(EtO)_2P(O)NEt_2$, δ +9.8; $(Me_2N)_3PO$, δ +24.8. With aryl on nitrogen, the shielding effect as seen in O-aryl compounds counters the nitrogen deshielding, and the shifts by chance fall in the simple phosphate region: $(EtO)_2P(O)NHPh$, δ +2.0. Compound **58** exhibits an interesting

collection of effects in which phosphorus is deshielded by two N-substituents and a five-membered ring, but shielded by a six-membered ring. The shift is δ +24.8.[44]

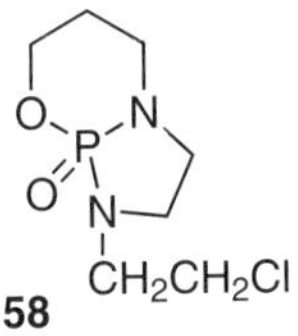

58

- Replacement of phosphoryl oxygen by sulfur causes very strong downfield shifts: $(EtO)_3PS$, δ +68.1. Replacement of one RO− by RS− also causes downfield shifts but of smaller size: $(EtO)_2P(O)SEt$, δ +26; the effect is additive, $[(EtS)_3PO, \delta +61]$. Tetrathiophosphates are as expected very far downfield: $(EtS)_3PS$, δ +92.
- Converting P−OH groups to anionic form has but a small effect on the ^{31}P shift: $(EtO)_2P(O)OH$, δ 0.0 and $(EtO)_2PO_2^-$, δ −1.8.
- Phosphorus in the pyrophosphate (diphosphate) structure is significantly shielded: $(EtO)_2(O)P-O-P(O)(OEt)_2$, δ −13; $(PhO)_2(O)P-O-P(O)(OPh)_2$, δ −24.8. In a triphosphate, the central phosphorus is additionally deshielded, whereas the terminal groups remain about the same. The cause of the upfield shifting has not been established, but it is a very useful effect in the study of naturally occurring phosphates. There are numerous biophosphates with the diphosphate and triphosphate structures; when the phosphorus atoms are differently substituted, two-bond coupling of the P nuclei is present. The best known example is adenosine triphosphate (ATP, **59**), which gives three ^{31}P signals;[45] the central phosphorus gives the most upfield of these signals at δ −21. The full spectrum taken at pH 8 consists of a doublet ($^2J_{PP}$ = 19 Hz) at δ −10.9 for P_α, a triplet ($^2J_{PP}$ = 19 Hz) at δ −21.3 for P_β, and a doublet ($^2J_{PP}$ = 19 Hz) at δ −6.0 for P_γ.

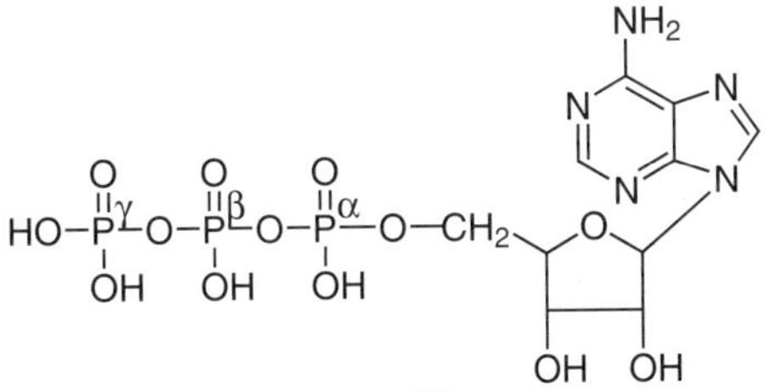

59

REFERENCES

1. K. Karaghiosoff in D. M. Grant and R. K. Harris, eds., *Encyclopedia of Nuclear Magnetic Resonance*, Vol. 12, John Wiley & Sons, Inc., New York, 1997, pp. 3612–3618.
2. L. D. Quin and J. G. Verkade, eds., *Phosphorus-31 NMR Spectral Properties in Compound Characterization and Structural Analysis*, VCH Publishers, Inc., New York, 1994,

(a) T. Glonek, Chapter 22, (b) J. K.Gard, Chapter 27, (c) D. B. Chesnut and B. E. Rusiloski, Chapter 1.
3. W. C. Dickenson, *Phys. Rev.* **81**, 717 (1951).
4. H. S. Gutowski and D. W. McCall, *Phys. Rev.* **82**, 748 (1951).
5. D. G. Gorenstein in R. Engel, ed., *Handbook of Organophosphorus Chemistry*, Marcel Dekker, New York Inc., 1992, pp 436–437.
6. J. C. Tebby, ed., *Handbook of Phosphorus-31 Nuclear Magnetic Resonance Data*, CRC Press, Boca Raton, FL, 1991, (a) J. C. Tebby, p. 8, (b) J. C. Tebby, Chapter 1, (c) L. Maier, P. J. Diehl and J. C. Tebby, p. 122, (d) R. S. Edmundson, Chapter 11, (e) R. S. Edmundson, Chapter 12, (f) R. S. Edmundson, Chapter 9, (g) A. W. G. Platt and S. G. Kleemann, Chapter 4, (h) Y. Leroux, R. Burgada, S. G. Kleemann and E. Fluck, Chapter 5.
7. J. G. Verkade and L. D. Quin, eds., *Phosphorus-31 NMR Spectroscopy in Stereochemical Analysis*, VCH Publishers, Inc., Deerfield Beach, FL, 1987, (a) J. C. Tebby, Chapter 1, (b) W. McFarlane, Chapter 3, (c) D. B. Chesnut, Chapter 5, (d) M. J. Gallagher, Chapter 9, (e) E. Fluck and G. Heckmann, Chapter 2, (f) J. Michalski, A. Skowronska and R. Bodalski, Chapter 8.
8. J. K. Gard, D. R. Gard, and C. F. Callis in E. N. Walsh, E. J. Griffith, R. W. Parry, and L. D. Quin, eds., *Phosphorus Chemistry*, American Chemical Society, Washington, DC, 1991, Chapter 3.
9. M. M. Crutchfield, C. H. Dungan, J. H. Letcher, V. Mark, and J. R. Van Wazer, *Top. Phosphorus Chem.*, **5**, 1967, (a) pp. 75–168, (b) pp. 227–457.
10. (a) G. M. Kosolapoff and L. Maier, eds., *Organic Phosphorus Compounds*, 7 vols. John Wiley & Sons, Inc., New York, 1972, (b) R. S. Edmundson, *Dictionary of Organophosphorus Compounds*, Chapman & Hall, London, 1988.
11. L. D. Quin, *The Heterocyclic Chemistry of Phosphorus*, John Wiley & Sons, Inc., New York, 1981, Chapter 5.
12. D. G. Gorenstein, ed., *Phosphorus-31 NMR: Principles and Applications*, Academic Press, Orlando, FL, 1984, (a) D. G. Gorenstein, Chapter 1, (b) D. G. Gorenstein, p. 13, (c) D. G. Gorenstein, p. 22 (d) M. Bárányi and T. Glonek, Chapter 17.
13. C. T. Burt, *Phosphorus NMR in Biology*, CRC Press, Inc., Boca Raton, FL, 1987, (a) pp. 1–227, (b) p. 10.
14. P. E. Garrou, *Chem. Rev.* **81**, 229 (1981).
15. P. S. Pregosin and R. W. Kung, *NMR Basic Principles Progr.* **60**, 1 (1979).
16. (a) Advanced Chemistry Development, Inc., "ACD/XNMR Predictor", 133 Richmond St. West, Suite 605, Toronto, Ontario M5H 2L3, Canada, (b) STN International, "SPECINFO FILE", Chemical Abstracts Service, 2540 Olentangy River Rd., Columbus, OH 43210.
17. J. H. Letcher and J. R. Van Wazer, *J. Chem. Phys.* **44**, 815 (1966); *ibid.*, **45**, 2916 (1966); *ibid.* **45**, 2926 (1966).
18. D. B. Chesnut and L. D. Quin in M. Hargittai and I. Hargittai, eds., *Advances in Molecular Structure Research*, Vol. 5, JAI Press, Greenwich, CT, 1999; Chapter 6, D. B. Chesnut, *Chem. Phys. Lett.* **246**, 235 (1995); D. B. Chesnut and E. F. C. Byrd, *Heteroatom Chem.* **7**, 307 (1996).
19. W. Kutzelnigg, U. Fleischer, and M. Schinder, *NMR Basic Principles Progr.* **23**, 165 (1990); K. Kruger, G. Grossmann, U. Fleischer, R. Franke, and W. Kutzelnigg, *Magn. Reson. Chem.* **32**, 596 (1994); U. Fleischer, C. Vanwullen, and W. Kutzelnigg, *Phosphorus Sulfur Silicon* **93**, 365 (1994); U. Fleischer and W. Kutzelnigg, *Phosphorus Sulfur Silicon* **77**, 657 (1993); U. Fleischer, C. Vanwullen, and W. Kutzelnigg, *Phosphorus Sulfur Silicon* **93**, 365 (1994).

20. A. Dransfeld and P. v. R. Schleyer, *Magn. Reson. Chem.* **36**, S29 (1998); A. Dransfeld, A. A. Korkin, U. Salzner, and P. v. R. Schleyer, *Phosphorus Sulfur Silicon* **77**, 260 (1993); A. E. Reed and P. v. R. Schleyer, *J. Am. Chem. Soc.* **109**, 7362 (1987).
21. M. Karplus and J. A. Pople, *J. Chem. Phys.* **38**, 2803 (1963).
22. D. G. Gorenstein and D. Kar, *Biochem. Biophys. Res. Commun.* **65**, 1073 (1975); *J. Am. Chem. Soc.* **99**, 672 (1977).
23. F. W. Wehrli and T. Wirthlin, *Interpretation of Carbon-13 NMR Spectra*, Heyden & Sons, Ltd., London, 1978, pp. 23–24.
24. L. D. Quin and J. J. Breen, *Org. Magn. Reson.* **5**, 17 (1973).
25. S. O. Grim, W. McFarlane, and E. F. Davidoff, *J. Org. Chem.* **32**, 781 (1966).
26. M. D. Gordon and L. D. Quin, *J. Am. Chem. Soc.* **98**, 15 1976).
27. S. Hietkamp, H. Sommer, and O. Stelzer, *Chem. Ber.* **117**, 3400 (1984).
28. J. J. Breen, J. F. Engel, D. K. Myers, and L. D. Quin, *Phosphorus* **2**, 55 (1972).
29. M. Duncan and M. J. Gallagher, *Org. Magn. Reson.* **15**, 37 (1981).
30. A. Cogne, L. Wiesenfeld, J. B. Robert and R. Tyka, *Org. Magn. Reson.* **13**, 72 (1980).
31. D. B. Chesnut, L. D. Quin and S. B. Wild, *Heteroatom Chem.* **8**, 451 (1997).
32. D. B. Chesnut, L. D. Quin and K. D. Moore, *J. Am. Chem. Soc.* **115**, 11984 (1993).
33. S. A. Mamedov, A. B. Kuliev, B. R. Gasanov and A. M. Mirmovsumova, *Zhur. Obshch. Khim.* **61**, 2669 (1991); *Chem. Abstr.* **117**, 69928 (1992).
34. P. Dagnac, R. Turpin, J.-L. Virlichis and D. Voigt, *Rev. Chim. Miner.* **14**, 370 (1977); Chem. Abstr. **88**, 43478 (1978).
35. H. R. Hudson, *Top. Phosphorus Chem.* **11**, 339–435 (1983).
36. D. Camp and I. D. Jenkins, *J. Org. Chem.* **54**, 3045 (189); M. von Itzstein and I. D. Jenkins, *Aust. J. Chem.* **36**, 557 (1983); D. A. Campbell and J. C. Bermak, *J. Org. Chem.* **59**, 658 (1994).
37. (a) T. R. Wu, W. Y. Chen, Y. S. Chiu, and Tg. C. Chang, *Phosphorus Sulfur Silicon* **122**, 197 (1997). (b) C. Symmes Jr, and L. D. Quin, *J. Org. Chem.* **43**, 1250 (1978).
38. J. W. Perich, P. F. Alewood, and R. B. Johns, *Phosphorus Sulfur Silicon* **105**, 1 (1995).
39. R. Hirschmann, K. M. Yager, C. M. Taylor, J. Witherington, P. A. Sprengeler, B. W. Phillips, W. Moore, and A. B. Smith III, *J. Am. Chem. Soc.* **119**, 8177 (1997).
40. R. Deslauriers, R. A. Byrd, H. C. Jarrell, and I. C. P. Smith, *Eur. J. Biochem.* **111**, 369 (1980).
41. R. C. Grabiak, J. A. Miles, and G. M. Schwenner, *Phosphorus Sulfur* **9**, 197 (1980).
42. A. C. Satterthwait and F. H. Westheimer, *J. Am. Chem. Soc.* **102**, 4464 (1980).
43. J. F. Cajaiba Da Silva, M. S. Pedrosa, H. T. Nakayama, and C. Costa Neto, *Phosphorus Sulfur Silicon* **131**, 97 (1997).
44. U. Niemeyer, B. Kutscher, J. Engel, I. Neda, A. Fischer, R. Schmutzler, P. G. Jones, M.-C. Malet-Martino, V. Gilard, and R. Martino, *Phosphorus Sulfur Silicon*, **107**, 4731 (1996).
45. E. K. Jaffe and M. Cohn, *Biochemistry* **17**, 652 (1978).

CHAPTER 7

OTHER SPECTROSCOPIC TECHNIQUES IN ORGANOPHOSPHORUS CHEMISTRY

7.A. ^{1}H NMR SPECTROSCOPY

7.A.1. General

Proton NMR spectroscopy became a standard technique for characterizing organophosphorus compounds soon after commercial instruments were introduced in the late 1950s, and many thousands have been studied over the years. Before about 1970, spectra were generally obtained at 60 to 100 MHz by the single-spectrum continuous wave method, but the introduction of the multipulse Fourier transform method and the movement to spectrometers operating at 300 to 500 MHz have greatly improved the sensitivity of the method and the resolution of the spectrum. This is an important consideration in organophosphorus chemistry, because the normal spin-spin splitting patterns of organic chemistry are made more complicated by additional coupling to the ^{31}P nucleus, with its spin quantum number of $\frac{1}{2}$. With the new high-field instruments, many spectra originally designated as second-order of the ABX type, where X = ^{31}P, and more complicated types as well, became easily interpreted. This point should be remembered when examining spectral interpretations published in the early days of NMR spectroscopy. Experimental conditions for obtaining ^{1}H NMR spectra have become rather standardized over the years; the reference for use with organic solvents is almost always internal tetramethylsilane, and the solvents must be deuterated to serve as locks for the spectrometers. Deuterochloroform is the solvent of choice; D_2O is used for water-soluble compounds. Other common deuterated solvents are deuterobenzene, deuteroethanol, deuteroacetone, etc. It is important to specify the solvent, as chemical shifts, especially those of phosphine oxides, can be modified by H-bonding effects, and aromatic solvents can induce shift changes, as is true for many organic compounds.

The review literature of ^{1}H NMR spectroscopy of organophosphorus compounds is surprisingly sparse, especially in recent years. There are two compilations of references for particular compounds that were published by Mavel in 1966[1] and then in 1973.[2] Similarly, literature references can be found in the extensive tables of compounds in the Kosolapoff–Maier series,[3] as well as in the Edmundson dictionary of compounds.[4] Many organophosphorus compounds are included in the spectral collections of Sadtler[5] and Aldrich.[6] A study of structural effects on the proton spectra of heterocyclic phosphorus compounds appeared in 1981,[7a] which provides some insight into the interpretation of spectra of noncyclic compounds as well.

In the sections to follow, the influence of the common phosphorus functional groups on carbon-bound protons will be considered. The most commonly encountered chemical shift effect is that derived from electronic induction; with an electronegativity just below that of carbon, phosphorus in the phosphine state actually causes weak upfield shifts of attached CH groups, but in the 4-coordinate compounds, all of which are moderate to strong electron attractors, substantial deshielding of attached CH occurs. With olefinic or aromatic groups, some form of resonance delocalization is present, easily seen by the greater downfield shifting of β olefinic protons than of α protons, just as is seen in α,β-conjugated carbonyl compounds. Again we will avoid specifying orbitals of phosphorus into which π electrons may be delocalized, although the literature has made exclusive use of the d-orbitals for this purpose for many years. A few examples of long-range, possibly anisotropic, effects are known; these are for the most part associated with the phosphoryl group and are of some importance for heterocyclic compounds in which geometric restrictions can cause phosphoryl groups to exhibit different deshielding on protons or alkyls on the two faces of the ring.

Phosphorus-proton coupling constants are as valuable in interpreting ^{1}H NMR spectra as are chemical shifts and should always be specified in reporting spectral data. In the following discussions, the coupling constants characteristic of the common phosphorus functional groups will be recorded. We will make no use of the sign of the coupling; some generalizations are, however, available.[7b] The coupling constants are enormous for protons directly bonded to phosphorus, ranging from about 180 to 220 Hz for 3-coordinate functions to about 350 to 1000 Hz in 4-coordinate functions. These and other P–H coupling constants are more easily obtained from the ^{31}P NMR spectra and can be of major diagnostic value in establishing the nature of CH groups bonded to P. An excellent discussion of this approach has been published by Gallagher.[8] The Tebby tables of ^{31}P NMR data include P–H coupling constants for phosphines[9a] and 4-coordinate PH derivatives.[9b] Both two-bond and three-bond P–H couplings are under stereochemical control, as will be discussed. Two-bond coupling is frequently smaller than three-bond coupling, which is the opposite of what one might expect. The magnitude of the observed coupling should not be the only criterion used in making structural assignments.

7.A.2. ^{1}H NMR Signals of P-Methyl Groups

In simple phosphines, the P-methyl group will give a signal at δ 0.7 to 1.0, almost always quite clearly distinguished on the spectrum from the slighly more downfield CH signals. The signal is a doublet with a coupling constant of 2 to 4 Hz. If phosphorus bears electronegative groups such as halogen or alkoxy, the doublet is shifted downfield by the inductive withdrawal of electron density from P and the coupling constant ($^2J_{PH}$) increases. Compounds **1** to **4** illustrate these effects.

(CH$_3$)$_3$P

1, $\delta = 0.89$, $J_{PH} = 2.7$)[10]

2, ($\delta = 0.82$, $J_{PH} = 3.5$)[11]

CH$_3$PCl$_2$

3, ($\delta = 2.18$, $J_{PH} = 17.6$)[12]

CH$_3$P(OMe)$_2$

4, ($\delta = 1.14$, $^2J_{PH} = 11.1$)[13]

Substitution of a methyl hydrogen by an electronegative heteroatom or by phenyl causes the expected downfield shifts; the signal for a P-benzyl group in a phosphine appears at about δ 3 (e.g., PhCH$_2$PMePh, δ 3.02).[14]

Conversion of a phosphine to a 4-coordinate derivative is always accompanied by diminution of electron density on P, and the shifts also move downfield with an increase in the magnitude of $^2J_{PH}$ (compounds **5** to **8**).

(CH$_3$)$_3$P=O   (CH$_3$)$_3$P=S

5, ($\delta = 1.93$, $J_{PH} = 13.4$)[10]   **6**, ($\delta = 1.74$, $J_{PH} = 13.0$)[10]

CH$_3$P(O)(OH)$_2$ (CH$_3$)$_4$P$^+$ I$^-$

7, ($\delta = 3.6$, $J_{PH} = 17.3$)[15]   **8**, ($\delta = 2.47$, $J_{PH} = 14.4$)[10]

7.A.3. ^{1}H NMR Signals of P-Alkyl Groups

The considerations made for the P-methyl signal in phosphines apply to larger alkyl groups, except that the α carbon is shifted farther downfield by the action of attached carbon atoms (e.g., to δ 1.2 in Et$_3$P)[10] Coupling from protons on the attached carbon will also be present; in an ethyl group, for example, the α–CH$_2$ signal will be split to a quartet by the methyl group ($^3J_{HH}$ 7 to 8 Hz), and all lines will be split by the ^{31}P nucleus, thus in principle giving an eight-line multiplet. Typical values for $^2J_{PH}$ and $^3J_{PH}$, seen in Et$_3$P, are 0.5 and 13.7 Hz. Alkyl groups on 4-coordinate phosphorus give more distinct signals for the protons on the α-carbon, because they are deshielded and have large coupling constants (e.g., in (CH$_3$CH$_2$)$_3$P=O, α-CH has δ 1.65 (J = 11.9 Hz) and β-CH has δ 1.10 (J = 16.3 Hz); in the P=S derivative, α-CH has δ 1.81 (J = 11.3 Hz) and β-CH has δ 1.17 (J = 18.1)).[10]

7.A.4. ¹H NMR Signals of Unsaturated Groups on Phosphorus

Proton signals for vinyl and alkenyl phosphines appear in the typical double-bond region (δ 5 to 6). Useful shift effects take place, however, when P has 4-coordination, especially in the phosphoryl derivatives. Both the α and β carbons are shifted downfield, with the β carbon the more so. This effect resembles closely that seen in α,β-unsaturated carbonyl compounds, and resonance comes to mind as the explanation, because this would reduce the electron density on the β carbon. This is represented by the canonical forms **9** and **10**, in which, as before, the nature of the interaction of the π electrons with the phosphoryl group, formerly attributed to d-orbital occupation, remains unspecified at this time.

$$RCH=CH-P(O)R_2 \longleftrightarrow R\overset{+}{C}H-CH=(POR_2)^-$$
$$\quad\quad\quad\quad 9 \quad\quad\quad\quad\quad\quad\quad 10$$

Three-bond PH coupling is stronger than two-bond coupling in conjugated systems. Compounds **11** to **13** are typical of those showing these effects.

11[16]

$$\text{CH}_3\text{-C=C-PO(OEt)}_2\text{, H}_B\text{, H}_A$$

(δ H$_A$ = 5.64, $^2J_{PH}$ = 19.5)
(δ H$_B$ = 6.64, $^3J_{PH}$ = 51.5)

12[11] H (δ = 6.29, J_{PH} = 25.7)

13[11] H (δ = 6.77, $^3J_{PH}$ = 40)

7.A.5. ¹H NMR Signals of P-Aryl Derivatives

Phosphine substituents have only a small effect on aromatic ring protons, because both inductive and resonance effects are weak. As might be expected, however, the phosphoryl group has a pronounced influence on the ¹H NMR shifts of aromatic ring protons. Deshielding is moderately strong at the para position and weaker at the meta position, again suggesting the importance of a resonance interaction. The deshielding is strongest at the ortho position, where steric effects also come into play. In fact, the ortho protons resonate significantly downfield of other ring protons and are generally apparent on a 300-MHz spectrum as a 2H multiplet (sometimes as four lines, as in PhPOCl$_2$) separated from the poorly resolved 3H multiplet for the other signals. Using dimethyl phenylphosphonate as an example,[17] we have ortho, δ 7.78; meta, δ 7.42; and para, δ 7.50. Similar shifts are observed for many phosphoryl compounds. The nature of the ortho deshielding requires a comment; it may arise from a long-range, perhaps anisotropic, effect of the P=O group, because this effect can be documented with certain nonaromatic compounds in which interacting groups are too remote for physical interaction (vide infra).

7.A.6. ¹H NMR Spectra of Alkoxy and Amino Groups on Phosphorus; Diastereotopic Protons

It is, of course, well known that electronegative oxygen and nitrogen cause strong deshielding of hydrogen on attached carbons. This effect is enhanced when the phosphoryl group, as found in any esters of the 4-coordinate types of phosphorus acids, replaces a proton on oxygen of ROH or on nitrogen of RNH_2 or R_2NH. In addition, coupling ($^3J_{PH}$) is of substantial size; for example, trimethyl phosphate[18] has δ 3.6, $^3J_{PH} = 11.0$ Hz, dimethyl methylphosphonate[19] has δ 3.72 and $(Me_2N)_3P=O$ has δ 2.65.[20] The P—O—CH groups are usually easily detected on a spectrum. Methyl groups give the expected doublet, but higher alkyls in certain structures have spectra that can be more complicated than a first-order prediction would suggest from the coupling with adjacent CH and ^{31}P. The problem is that of diastereotopic character (discussed in reference 21) that arises when an alkoxy group is attached to phosphorus that has three different substituents, a simple example being diethyl methylphosphonate. Here the two ethyl groups are chemically equivalent, but the prochiral phosphorus group (viewed from an ethoxy group as having three different substituents) renders the two α protons chemically nonequivalent (diastereotopic). Therefore, replacement of H_A or H_B (Eq. 7.1) by some other group would give diastereomeric products. These protons then in principle can give independent NMR signals and undergo coupling with each other to give more complicated spectra than expected.

$$(7.1)$$

In fact, simple esters do not always exhibit this effect on routine 300 MHz spectra; the H_A and H_B shifts are too close to each other to be observed, and also $^3J_{PH}$ can be close for H_A and for H_B. Frequently, the OCH_2 group appears as 5- or 6-line multiplet, an example[22] being $HP(O)(OEt)_2$ (6 lines at 300 MHz), but visible in many esters, as easily seen by perusing the very convenient collection of spectra by Aldrich.[6] Thiono derivatives, of course, have the same properties as their oxo equivalents, and alkylamino substituents on phosphorus acids can also exhibit the same effect.

The phenomenon of signal complexity arising from diastereotopic protons is not uncommon among other types of organophosphorus compounds, and interpreters need to keep this possibility in mind. An illustrative example[23] is provided by a family of dibenzylic phosphine oxides, $MeP(O)(CH_2-OAr)_2$. The CH_2 signal appears not as as the simple doublet one might expect, but as an 8-line ABX pattern. Here the benzylic protons are diastereotopic and thus nonequivalent, and their chemical shifts differ adequately to give separate subspectra. They also have different $^3J_{PH}$ values (6.0 and 7.0 Hz).

7.A.7. ^{1}H NMR Signals of Phosphorus-Bound Hydrogen

The chemical shifts of hydrogen directly bonded to phosphorus depend on the nature of the phosphorus functionality. In primary and secondary phosphines, the shifts are around δ 2 to 5 with $^1J_{PH}$ around 180 to 220 Hz. If protons are present on the α carbon, the signal will show additional small splitting (which can be useful in structural analysis). This doublet can sometimes be difficult to locate because one or both signals may occur in crowded spectral regions. If the coupling constant is desired, it is better to measure this on the ^{31}P NMR spectrum with the decoupler deactivated. Conversion of these phosphines to the oxides causes a strong change in the P—H signal. It is shifted downfield, and the coupling constant is greatly increased; frequently, one of the signals is outside the usual δ 0 to 10 scale. Again it is better to record the coupling on the ^{31}P NMR spectrum. The protons in H-phosphonates and secondary phosphine oxide have shifts between δ 6.7 and 7.6 $^1J_{PH}$ 670–715 Hz) and δ 6.5 and 8.0 ($^1J_{PH}$ 300 to 500 Hz), respectively. Proton-coupled ^{31}P NMR spectra can be valuable also in observing the much smaller three-bond coupling; knowing the multiplicity and coupling constants can be a great help in structural analysis. Compilations of various P—H shifts and couplings are available in reference 24. A detailed discussion of proton-coupled spectra has been prepared by Gallagher.[8]

7.A.8. Long-Range Effects on ^{1}H NMR Chemical Shifts

In cyclic phosphorus compounds, where groups are held in rigid or strongly conformationally biased positions, examples of shielding effects acting through space, rather than through bonds, are known. Such long-range shift effects more commonly arise from aromatic substituents, in which a proton might experience upfield shifting if it is located near the shielding cone emanating from the faces of the aromatic ring, or from the phosphoryl group, where deshielding may be felt. Many cases of these effects are known,[7c] and they have been especially useful in elucidating the structures of cis,trans isomers. To illustrate the effect, the spectrum[25] of compound **14** can be considered; here the two C-methyls give distinctly different signals, one at δ 1.32, which is the expected position, the other at δ 0.71, which shows a strong upfield shifting by the phenyl ring, thus identifying that methyl as being on the same face as the phenyl substituent.

CH$_3$ (δ = 1.32, $^3J_{PH}$ = 18.7)
CH$_3$ (δ = 0.71, $^3J_{PH}$ = 6.8)

14

An example of phosphoryl deshielding is provided by comparing the shifts for the olefinic protons in the syn, anti isomers **15** and **16**, respectively.[26] The placement of phosphoryl over the double bond in the six-membered ring component of **15** in the anti-isomer causes deshielding of the olefinic protons (δ 6.28) relative to their position in syn-isomer **16**, which has δ 6.08.

The substituents on P can also experience long-range effects from other functional groups. Thus deshielding by the OH group is believed to be responsible for the significant difference in the shifts of the P-methyl protons for the isomers **17** (δ 1.94) and **18** (δ 1.49).[27]

Although the origin of long-range shielding is not always known, whether it be from secondary magnetic fields or electric field effects of the acting group, the phenomenon is very real and can figure in the establishment of the spectrum for heterocyclic compounds. One should be alert to the possibilities for structure determination that such effects can supply.

7.A.9. Stereochemical Control of $^3J_{PH}$ and $^2J_{PH}$

Early in the studies of ^{1}H NMR spectra, it was recognized that the magnitude of three-bond ^{1}H-^{1}H coupling was in part controlled by the location of the interacting protons in space. In rigid systems, the extremes would be represented by a trans disposition of the hydrogens in a plane including the two carbons to which they are attached. In this plane, the hydrogens are defined by a dihedral angle of 180°, and this is associated with maximum values for the three-bond coupling. This would be the case, for example in trans hydrogens on a double bond, or diaxial hydrogens on the cyclohexane ring. The other extreme would have the cis orientation of the coupled hydrogens (dihedral angle 0°) and this leads to a sizeable but somewhat smaller coupling constant. When the dihedral angle is 90°, however, the coupling is virtually zero, and is rather small for the gauche angle of 60° through that of 120°. A curve can be established when $^3J_{PH}$ is plotted against the dihedral angle, using data from a collection of rigid compounds in which dihedral angles are known from measurements of structural parameters, frequently from X-ray analysis. Most fortunately for organophosphorus chemistry, the same type of Karplus relationship exists for three-bond ^{31}P-^{1}H coupling. Thus in phosphonates, $^{3}J_{PH}$ is about 35 to 40 Hz when the dihedral angle is 180°, then approaches zero at 90°, and rises back up to about 17 to 19 Hz at 0°. The nuclei through which the coupling takes place need not all be carbon; the effect holds true for such connections as H–C–O–P, H–C–N–P, and H–C–S–P, and this means that esters and amides, especially cyclic ones, give couplings under dihedral angle control. In phosphates, for example, $^{3}J_{POCH}$ is about 20 to 23 Hz at 180°, 0 Hz at 90°, and 15 to 17 Hz at 0°. This subject is well covered in a review by Bentrude and Setzer,[28] and many examples are provided of its use in stereochemical determinations. Six-membered cyclic phosphates (1,3,2-dioxaphosphorinanes) are especially well known and provide many examples of this coupling effect. This ring adopts a chair conformation, with C-substituents preferably in the equatorial positions. Thus one isomer of 1,3-dimethyl-1,3,2-dioxaphosphorinane is shown as structure **19**. The axial H (H$_A$) at C-4 has a dihedral angle of about 60° to P, while the equatorial H (H$_B$) has a value of about 180°. This leads to the observed coupling constants of 2.9 and 20 Hz, respectively.[29] Similar values are obtained for the isomer where the configuration at phosphorus is inverted.

19

A Karplus-type curve can be developed for 3-coordinate phosphorus functions as well, but here the configuration at phosphorus can play a role. For example, in compound **20**, with an axial lone pair, the coupling to the equatorial H$_B$ is much larger than to axial H$_A$[30a]. In compound **21**, the lone pair is equatorial, and again the couplings differ at H$_A$ and H$_B$[30b]. Note that the coupling to axial H$_A$ is quite small in both isomers.

20: Me$_2$N–P, H$_B$ (J_{PH} = 19.6 Hz), H$_A$ (J_{PH} = 2.5 Hz), Bu–t

21: OMe, H$_B$ (J_{PH} = 11.0 Hz), H$_A$ (J_{PH} = 2.9 Hz), Bu–t

Two-bond P–H coupling also depends on the orientation of the lone pair, which can be seen especially among phosphines. In essence, coupling is at a maximum when the lone pair is close to the coupled proton as was seen for $^{3}J_{PH}$ (dihedral angle for the lone pair orbital and C–H of 0°) and is quite small, even negligible, when the dihedral angle is between 60° and 180°. In fact, the sign of coupling changes from positive to negative as the dihedral angle increases to about 70 to 80°. The effect is well illustrated qualitatively by the cyclic phosphine **22**.[31]

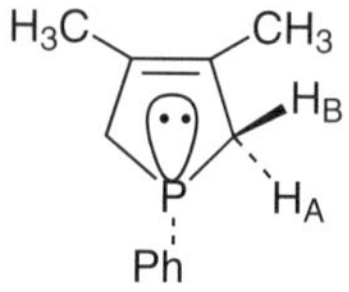

22

H_B is close to the lone pair, thus with a small dihedral angle, and has a very large $^2J_{PH}$ of +25 Hz. H_A, however, is remote from the lone pair (large dihedral angle) and has $^2J_{PH}$ of −6 Hz. Such effects have played a major role in stereochemical assignments of 3-coordinate phosphorus compounds. It is also known that in 4-coordinate compounds there is stereochemical control of $^2J_{PH}$. Thus in phosphine oxides there is an empirical relation with the dihedral angle formed between O of the P=O group and H on the attached carbon: $^2J_{PH}$ has a larger absolute size (actually more negative) at small dihedral angles than at large angles (less negative). This can be seen in isomers **23** ($^2J_{PH}$ = 6.5 Hz) and **24** ($^2J_{PH}$ = 13.5 Hz).[32] Stereochemical relationships also influence the magnitude of four-bond and one-bond coupling and are discussed in reference 28.

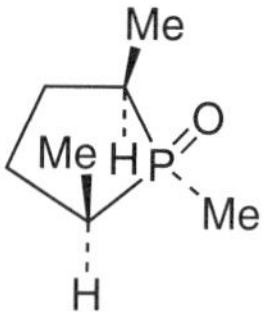

23 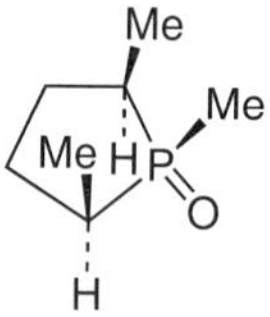   **24**

7.B. ^{13}C NMR SPECTROSCOPY IN ORGANOPHOSPHORUS CHEMISTRY

7.B.1. General

Spectrometers for efficient determination of ^{13}C NMR spectra became available around 1970 when multi-pulsing with data treated by the Fourier transform method was introduced. Since that time, the use of ^{13}C spectroscopy for compound characterization has become practically a standard procedure, in many cases more useful than ^{1}H NMR spectroscopy. This is especially so in organophosphorus chemistry; proton-decoupled ^{13}C NMR spectra are much simpler than ^{1}H NMR spectra, giving a single line for each different carbon that may then be split to a doublet by the ^{31}P nucleus. An enormous amount of structural information comes from the combined use of the chemical shift and the ^{13}C-^{31}P coupling constant. Another recent advance is the use of 2-dimensional (2-D) NMR techniques, especially that which correlates signals on ^{1}H spectra with those on ^{13}C spectra.

The first review of the ^{13}C NMR spectra of phosphorus compounds appeared in a book by Stothers,[33] although the field was small at that time (1972). A more detailed introduction to the interpretation of ^{13}C NMR spectra of phosphorus compounds, with emphasis on heterocyclic compounds (including tables of data) is found in reference 7d. A collection of fully interpreted spectra of heterocyclic phosphorus compounds is also available.[34] The Sadtler[5] and Aldrich collections[6] of interpreted ^{13}C NMR spectra also include many phosphorus compounds.

Carbon NMR spectra are presently being recorded in deuterochloroform wherever possible and are referenced against tetramethylsilane as 0 ppm. Unlike ^{31}P NMR spectra, the vast majority of the signals are downfield of the reference. Solvent effects are small (although not negligible), and other deuterated solvents are sometimes used.

It is the purpose of this chapter to outline only some of the major structural effects that operate to define a ^{13}C chemical shift and its coupling constant to ^{31}P and to make clear the possibilities for structure determination by this technique. The cited review literature should be consulted to develop a fuller understanding of the intricacies of ^{13}C NMR of phosphorus compounds.

7.B.2. ^{13}C NMR Spectra of Aliphatic Organophosphorus Compounds

In saturated compounds, carbon shifts are influenced by inductive effects of substituents, which can cause deshielding if they are electron attracting and cause contraction of the carbon p-orbitals, or shielding if they are electron releasing and expand the p-orbitals. As in ^{31}P NMR, these shift effects are very strong, much more so than those produced by the modification of the secondary magnetic field, through induction, that acts to shield a nucleus. This is the case for the proton. Steric effects also are in operation, and the generalization has developed that an atom that is in the γ-relation to the carbon of interest can cause steric compression as rotation about the central C—C bond brings these atoms into close proximity (as in the gauche and eclipsed conformations of *n*-butane). Steric compression may be considered to distort electron density to the mutually crowded atoms, and this causes shielding to occur. We have already seen that this is a useful concept also in interpreting ^{31}P NMR shifts. Although there may be other effects that contribute to the observed ^{13}C chemical shift, electronegativity and steric compression are adequate to interpret most spectral effects among saturated compounds, and this is so in organophosphorus chemistry. Phosphorus is less electronegative than carbon, and in methylphosphine the carbon shift is negative (δ -4.4). The shift becomes more positive (δ $+7.1$) in dimethylphosphine and trimethylphosphine (δ $+17.3$). Here the added carbons can be viewed as being in the β position (with P as α) to the methyl being observed, and we shall see that β substituents are generally associated with deshielding. Converting a phosphine to the oxide effects a change to an electronegative group, and thus in trimethylphosphine oxide the methyl shift moves downfield to δ 18.58.[35] In trimethylphosphine sulfide, a similar downfield shift to δ 22.8 is observed;[33] however, conversion of a tertiary phosphine to a quaternary phosphonium ion causes an upfield shift ($Me_4P^+I^-$, δ 11.3[33]). This is a well-known effect for quaternary ammonium ions as well. The methyl group when attached to the phosphonic or phosphinic acid groups is deshielded and appears at δ 9 to 10 (e.g., $MeP(O)(OMe)_2$, δ 9.8[36]). As the carbon chain is extended, a P-methyl group feels the steric compression provided by the added γ carbon, and the shift moves slightly upfield (e.g., $(CH_3)_2PC_4H_9$, δ 14.4; $CH_3P(C_4H_9)_2$, δ 11.9[37]).

The interpretation of ^{13}C NMR spectra of organic compounds in general is greatly aided by the Grant-Paul approach of using additive shift increments for carbons and functional groups attached to a carbon of interest in order to calculate its chemical shift. This approach works with well organophosphorus compounds, and constants have been determined for a variety of the common phosphorus functional groups that can be used in conjunction with the general Grant-Paul constants.[37] We find that the deshielding α constants of P-functions can be quite large (Me_2P, +19.4; PCl_2, +29.7; $(MeO)_2P$, +20.4; $Ph_2P(O)$, +16.8; Me_3P^+, +13.4; $Me_2P(S)$, +21.4), but the β constants are much smaller (and mostly negative) than is generally the case. Thus the β constant for Me_2P is +3.3 (compared to +9.4 for CH_3), and for all the groups whose α constants are listed above the β-constants range from −0.06 to −1.0. The γ constants are also relatively small; they are negative as expected but range only from −0.3 to −1.6 (for CH_3, −2.6). Delta-effects are small and all are positive (+0.4 to +0.8). Reference 7d should be consulted for the complete list of constants for the groups mentioned and for constants for other groups, as well as a rationalization of the sizes of the P substituent constants. To illustrate the use of these constants, consider the ^{13}C NMR spectrum of n-propylphosphonous dichloride with shifts at δ 14.9, 16.8, and 45.1. The shift of C-1 may be calculated from a version of the Grant-Paul equation (Eq. 7.2) where the summations are those for the various substituent constants acting on a carbon of interest.

$$^{13}C\,\delta = -2.5 + \Sigma_\alpha + \Sigma_\beta + \Sigma_\gamma + \Sigma_\delta \tag{7.2}$$

C-1 then has one α-C (+9.1) and PCl_2 (+29.7) attached to it, and it also has a β-carbon (+9.4). Putting these values in the equation allows a prediction of δ 45.7 for the position of C-1 on the spectrum. Similarly C-2 is calculated to be 15.8 (from 2 α-C and +0.1 as the β constant for PCl_2), and C-3 is calculated to be 14.6 (from 1 α-C, 1 β-C, and −1.9 as the γ constant for PCl_2). The signals on the spectrum are in reasonable agreement with the calculated values, and the correct structural assignment can be made. Coupling constants would have led to a faster assignment of the C-1 signal ($^1J_{PC}$ = 43 Hz), but the values for $^2J_{PC}$ (14 Hz) and $^3J_{PC}$ (12 Hz) are too close to use them as the basis for a signal assignment. Another approach that eliminates the need for the complete Grant–Paul calculation for each carbon is to work from the ^{13}C spectrum of a molecule before the introduction of the phosphorus function, and then apply the substituent constants[7d] to each signal. A set of substituent constants is also available for the effects of phosphorus groups attached to the cyclohexane ring,[7d,38] and qualitatively the effects on other alicyclic systems may be estimated from these constants. Steric compression effects on axial and equatorial substituents differ and can be used to assign the correct structure to a pair of isomers.

As mentioned, ^{13}C-^{31}P coupling constants are a great aid in the interpretation of spectra. The values for the various phosphorus groups on acyclic aliphatic compounds vary considerably, and many are collected in a published table.[7d] Taking one-bond coupling first, the largest values are found among phosphonic and phosphinic acids and their derivatives. The (MeO)$_2$(O)P– group for example has $^1J_{PC} = 143$ Hz, and this is typical of all phosphonic acids and their esters. Other 4-coordinate groups have much smaller values (50 to 75 Hz). With the exception of the Cl$_2$P– group, 3-coordinate groups have $^1J_{PC}$ values in the range of 7 to 25 Hz. The Cl$_2$P– group has a larger value (43 to 45 Hz). Two-bond coupling has some peculiarities; phosphines have $^2J_{PC}$ values (11 to 19 Hz, similarly for Cl$_2$P–) that are larger than $^1J_{PC}$ values (6 to 12 Hz), whereas just the opposite is true for 4-coordinate functions, all of which have very small two-bond coupling values of 0 to 5 Hz. Three-bond coupling for the 4-coordinate functions of 13 to 17 Hz are common. Among phosphines, however, three-bond coupling (8 to 12 Hz) is slightly smaller than is two-bond coupling. These confusing relations are discussed in more detail elsewhere,[7d] but when understood and used in connection with chemical shift values, they are a powerful aid in structural analysis.

Studies with saturated acyclic and heterocyclic compounds have revealed that there are significant stereochemical influences on the magnitude of two- and three-bond coupling. These are discussed in detail in reference 39. Just as in ^1H NMR spectroscopy, the influences on three-bond coupling can be analyzed with the aid of dihedral angles that relate P and ^{13}C through a two-atom fragment, as in P–C–C–^{13}C, P–O–C–^{13}C, P–C–O–^{13}C, etc. Although each type of phosphorus function will respond differently in its coupling to ^{13}C as the angular relation is changed, in general it is found that for 4-coordinate functions coupling is large and of similar magnitude at dihedral angles of 180° and 0°, and nearly zero at 90°. The plot of $^3J_{PC}$ vs. dihedral angle gives the familiar Karplus curve. (Note: in reference 7d, the typical plots shown for $^3J_{PC}$ and $^2J_{PC}$ are reversed; this led to the incorrect labelling of the ^2J plot for the Me$_2$P– group in reference 39 as that for the Me$_2$P(S) group). The dihedral angle relation has been widely used in stereochemical analysis. However, another effect enters in when 3-coordinate P functions are considered, the orientation of the lone pair. It has been found that the lone pair has a profound effect on the $^3J_{PC}$ values. In general, coupling in P–C–C–C is at a maximum when the lone pair orbital has a dihedral angle to the coupled carbon of 0° or 180° (with the former being larger: for Me$_2$P–, J at 0° is 2 to 3 times that for J at 180°), and reaches a minimum near 0 Hz at an angle of 100 to 110°. Again, once properly appreciated, the lone pair effect on ^3J can be quite useful in stereochemical analysis. As an example, the equatorial phenyl in **25** can be confirmed by the $^3J_{PC}$ value of 10.1 Hz to the equatorial C-methyl; in the isomer **26**, which differs only in the configuration at P, the lone pair is remote from the C-methyl, and this carbon exhibits no coupling to ^{31}P.[40]

The stereochemical control of $^2J_{PC}$ is also derived from the lone pair orientation. When the lone pair on P is close to the coupled carbon, $^2J_{PC}$ is large; when remote it can be small or even zero. An example is provided by structure **27**, which has two different two-bond connections at the bridging phosphorus. For the sp^2 carbon, the lone pair is remote, and $^2J_{PC}$ is only 4.9 Hz, whereas the lone pair is closer to the sp^3 carbon and $^2J_{PC}$ is 28.3 Hz. If the configuration is inverted at P (**28**), the $^2J_{PC}$ values are 20.5 and 3.9 Hz, respectively. With this information alone, the identity of these syn and anti isomers could be established.[41] Many other examples of the use of the lone pair effect are known.[39]

Stereospecificity in one- and four-bond coupling is also known, and both are discussed in reference 39.

7.B.3. ^{13}C NMR Spectra of Aromatic and Unsaturated Phosphorus Compounds

Considerable data (summarized in reference 7d) are available for the shifts and coupling constants of aromatic compounds with phosphorus substituents. With 4-coordinate groups including Cl$_2$P, the ortho carbons are routinely downfield-shifted relative to benzene; the meta carbons are but weakly affected, and the para carbon shows a downfield shift. At the ipso carbon, deshielding can be substantial. Ortho effects are, as always, associated with a combination of electronic and steric interactions, but meta and para shiftings arise from inductive and/or resonance effects. Modro[42] used the "dual substituent parameter" approach to separate inductive from resonance effects. With several 4-coordinate groups, it was shown that electron withdrawal by resonance did account for some of the deshielding at the para position. In dimethylphenylphosphine oxide, for example, the para carbon signal was shifted downfield from the benzene signal (δ 128.5) by 2.7 ppm, suggesting the involvement of resonance form **29** in Eq. 7.3. Some of the data from reference 7d are reproduced in Table 7.1. It can be expected that similar effects will prevail in other aromatic and heteroaromatic systems. The lone pair on phosphines does not give any indication of significant shielding at the para position, as is common with electron releasing groups such as H$_2$N (-9.5 ppm) and HO–

TABLE 7.1. Effects of Phosphorus Groups on Aromatic Ring Carbons[a]

Group	C-1	C-2	C-3	C-4
Me_2P-		+2.5 (18.4)	+0.07 (5.8)	−0.33 (∼0)
Cl_2P-	+12.6 (51.8)	+2.0 (31.2)	+0.57 (8.0)	+4.2 (∼0)
$(EtO)_2P-$	+13.9 (23.2)	+1.6 (18.4)	−0.39 (5.0)	+1.0 (∼0)
$(Et_2N)_2P-$		+2.9 (16.6)	−0.31 (2.8)	−1.2 (∼0)
$Me_2P(O)-$		+1.7 (8.8)	+0.10 (11.8)	+2.7 (2.2)
$Me_2P(S)-$	+6.7 (82.0)	+2.0 (10.6)	+0.19 (11.8)	+2.9 (3.0)
$Ph_2P(O)-$	+5.8 (102.8)	+3.9 (9.6)	−0.10 (11.6)	+3.0 (1.8)
$(EtO)_2P(O)-$	+1.6 (187.4)	+3.6 (10.0)	−0.17 (14.8)	+3.4 (3.0)

[a]Data given in Reference 7d. Positive increments are ppm downfield of benzene, δ 128.5. Values in parentheses are J_{PC} in Hz.

(−7.3). Thus for the Me_2P- group, the upfield shifting at the para carbon is only −0.33 ppm. This is consistent with the general view that phosphines are but weakly involved in resonance interactions with unsaturated centers.

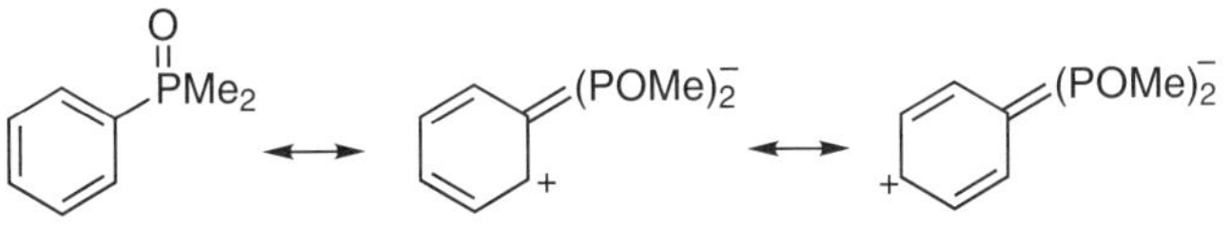

$$\text{Ph-P(O)Me}_2 \leftrightarrow \text{(POMe)}_2^- \leftrightarrow \text{(POMe)}_2^- \qquad (7.3)$$

29

Resonance interactions can also be identified in alkenyl phosphorus derivatives. An example is seen in diphenylvinylphosphine oxide;[43] the β carbon is shifted to δ 134.4 from 127.8 for ethylene. This effect is well known in α,β-unsaturated carbonyl compounds. A stronger downfield shift occurs in triphenylvinylphosphonium ion,[44] in which the β carbon is shifted to δ 145.2. Unsaturated heterocyclic compounds provide many examples of NMR effects from polarization of double bonds, as in the spectrum for 2-phospholene oxide **30**[45]. Note here the significantly larger value of $^1J_{PC}$ when P is attached to an sp^2 carbon, a general effect.

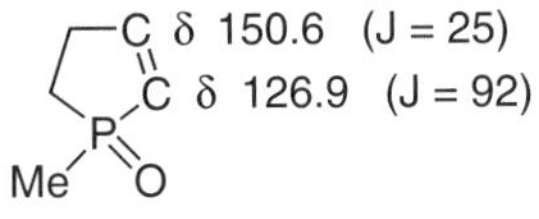

C δ 150.6 (J = 25)
C δ 126.9 (J = 92)

30

7.C. INFRARED SPECTROSCOPY

Until the advent of the various NMR techniques, infrared spectroscopy was the major tool for the identification or confirmation of phosphorus functional groups, and for studying physical chemical matters. Research papers still appear in which the infrared spectra are reported as characteristics of compounds, especially for novel functional groups such as those with phosphorus in unusual coordination states. However, its use as a primary structural investigative tool has diminished. It remains as a very useful tool for performing quantitative analysis and physical studies, a capability exploited from the start of practical IR spectroscopy in the 1940s. The modern Fourier transform IR spectrometers have brought renewed interest to the technique for such studies. It is frequently possible to locate a meaningful peak on the IR spectrum and to measure its intensity as a function of the concentration of a solution. Linear plots are usually obtained. We describe in this section only the major absorption signals directly associated with the common phosphorus functional groups, but for quantitative analysis, any of the numerous signals, even if not fully identified, can in principle be used.

The review literature for interpreting IR spectra is extensive. Comprehensive surveys and discussions by Daasch and Smith,[46] Bellamy[47] and Corbridge[48] are especially valuable, and the spectra of many phosphorus compounds appear in the extensive Sadtler and Aldrich IR spectral collections.[49] A simple summary of important absorption bands for phosphorus compounds is included in the tables of data for structure determination by Pretsch et al.[50]

The stretching of the P—H bond in phosphines, as well as for H attached to phosphoryl groups, gives rise to an easily detected signal in an uncluttered spectral region, of immediate value in characterizing these groups. The signals are in the region of 2440 to 2350 cm^{-1}; they are sharp and of weak to medium intensity. A bending vibration is found at 1090 to 910 cm^{-1}, and is especially recognizable for H-phosphonates.

The phosphoryl group, because of its polarity, gives a very strong stretching signal, widely studied and used in structural analysis. It may be found in a rather wide spectral region (1365 to 1090 cm^{-1}), but is easily recognized from its dominance in this region. Fortunately, the various structural forms in which the P=O group appears have their own positions within this range, and the signal has been much used to characterize the phosphorus functional group. The position of the signal may vary, however, in media where hydrogen bonding can occur; a low-frequency shift with signal intensification can take place. Some of the important correlations for the various bonding modes of P=O, taken from reference 50, are given below.

Phosphine oxides	1190–1150 cm^{-1}
Phosphonic acids	1220–1150 cm^{-1}
Dialkyl phosphonates	1280–1240 cm^{-1}
Phosphonic dichlorides	1365–1260 cm^{-1}
Phosphinic acids	1205–1090 cm^{-1}
Alkyl phosphinates	1265–1200 cm^{-1}
Trialkyl phosphates	1300–1260 cm^{-1}
Dialkyl phosphates (phosphodiesters)	1250–1210 cm^{-1}
Monoalkyl phosphates	~1250 cm^{-1}
Pyrophosphates	1310–1250 cm^{-1}

Studies of phosphoryl absorptions have been of importance in clarifying some important structural features of phosphorus compounds. An early example concerned the keto-enol tautomeric shift of H−P(P)(OR)$_2$ to (HO)P(OR)$_2$; IR spectroscopy confirmed the presence of the P−H and P=O bonds and the absence of P−OH signals. The possibility of resonance interaction of P=O with conjugated double bonds was also studied, and although the C=C stretching band shifted to lower frequency as in conjugated carbonyl compounds, thus suggesting some form of resonance interaction, this was not accompanied by any significant shifting of the P=O stretching signal. Much attention has been given to the shift of the P=O signal when changes in the electronegativity of attached groups are made; the more electronegative groups cause strong shifts to higher frequency, an effect early noted and attributed to an increase in the bond order arising from increased occupation of the (now discredited) d-orbitals. The effect continues to attract attention as the saga of the electronic structure of the phosphoryl group moves on (chapter 2.D). In recent years, the true phosphorus to oxygen double bond using p-orbitals on both elements has been synthesized, albeit in highly unstable species (chapter 10.C). This σ^2, λ^3 bonding is known in structures Y−P=O, where Y is halogen, H, OH, or aryloxy. The P=O stretching occurs in the same region (1185 to 1292 cm^{-1}) as for the common phosphoryl group.

Another signal of value in practical IR spectroscopy arises from stretching of the single P−O bond in various esters, again strong and recognizable. Using Bellamy's data[47] for alkyl esters, the P−O signal appears at 1190 to 1170 cm^{-1} for methyl esters, 1170 to 1150 cm^{-1} for ethyl esters, and 1050 to 990 cm^{-1} for higher esters. Aryl esters give signals at higher frequency, typically 1240 to 1190 cm^{-1}. Used in conjunction with the P=O stretching signal, the P−O signal is quite valuable for confirming the presence of an ester function in a phosphorus compound.

Various other bonds to phosphorus have characteristic signals, but are harder to recognize through weakness or appearance in crowded spectral regions. Of occasional use are the signals for P-Me and P-phenyl at 1320 to 1280 and 1450 to 1435 cm^{-1}, respectively, and of P=S at 840 to 600 cm^{-1}. The P−OH functions give broad, shallow O−H stretching signals at 2700 to 2250 cm^{-1}. This group is sensitive to hydrogen-bonding effects. Another signal may appear at around 1740 to 1600 cm^{-1}. IR signals also can be associated with P-halogen and P-nitrogen bonds.[47,48]

7.D. MASS SPECTROMETRY OF ORGANOPHOSPHORUS COMPOUNDS

Mass spectrometry (MS) has aided many studies of organophosphorus compounds, especially in providing molecular weight information. In making this determination by the customary electron impact method, however, it is important to remember that some phosphorus compounds, notably esters (vide infra), undergo fragmentations with great ease, making the molecular ion of very low intensity or even absent on the spectrum. This problem can be overcome by special techniques such as chemical ionization (CI) or fast atom bombardment (FAB) for solids, and others. Some esters are thermally unstable, and chemical reactions in the spectrometer can precede ionization and fragmentation. Again this problem can be avoided by the use of special techniques.

Most of the information on fragmentation patterns of phosphorus compounds has been developed from the study of simple monofunctional compounds; with the more complex molecules that are usually the subject of research, the patterns derived from fragmentations elsewhere in the molecule need to be considered. These fragmentations can lead to many ions and to quite complex spectra, and seldom do research papers provide an extensive analysis of the fragmentation patterns. Reviews by Granoth,[51] Gillis and Occolowitz,[52] and most recently by O'Hair[53] are very helpful in providing basic information on the fragmentations that occur at the fundamental functional groups, only a few of which will be outlined here. Heterocyclic compounds and multifunctional compounds will not be treated, but many examples can be found in the literature.

As in the MS of carbon compounds, one can frequently observe peaks for phosphorus-containing ions that would be highly reactive or have no existence outside of the environment of the spectrometer. Of particular interest are ions that correspond to molecules classifiable as low-coordination species, which are the object of much current research. Thus the phosphinidene ion (RP^+) and the oxophosphine ion ($RP=O^+$), among others, can be cited. The phosphacylium ion ($R_2P=O^+$), another sought-after species, is common in the spectra of esters of dialkylphosphinic acids.

7.D.1. Phosphines

In general, phosphines give strong molecular ion peaks, which in some cases are the base peak. Cleavage of a C—P bond then occurs by elimination of an alkene, as shown in Eq. 7.4. Several other peaks are formed from further degradations.

$$(RCH_2CH_2)_2PCH_2CH_2R \xrightarrow{-e} (RCH_2CH_2)_2\overset{+\bullet}{P} \overset{H}{\underset{CH_2-CHR}{\diagup}} \longrightarrow (RCH_2CH_2)_2\overset{+\bullet}{PH} + CH_2=CHR$$

(7.4)

The spectrum of trimethylphosphine has been well worked out and is informative in revealing other common modes of fragmentation in addition to the cleavage of the C—P bond. The molecular ion may lose H· to form $Me_2P^+=CH_2$, which may lose a molecule of hydrogen, or the joining of carbon fragments may occur as a result of the elimination of PH_3 or C_2H_4 molecules. The fragmentation pattern[54] is shown as Scheme 7.1. Triaryl phosphines also give a detectable molecular ion, but this undergoes loss of H· with the formation of a cyclic ion (**31**), a derivative of the dibenzophosphole system. Cleavage of the P-aryl group gives ion **32** (Eq. 7.5) and this is followed by further fragmentation.

$$(C_6H_5)_3\overset{+\bullet}{P} \xrightarrow{-H\bullet} \underset{\underset{\mathbf{31}}{C_6H_5\ \ H}}{[\text{structure}]} \xrightarrow{-C_6H_6} \underset{\mathbf{32}}{[\text{structure}]}$$

(7.5)

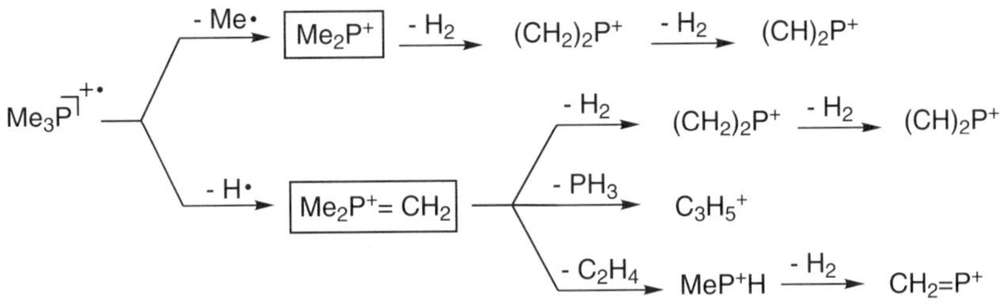

Scheme 7.1

7.D.2. Phosphine Oxides

Trialkyl phosphine oxides give strong molecular ions, which fragment by cleavage of a C−P bond. The resulting fragment is frequently the base peak. This fragmentation is depicted in Eq. 7.6, and, of course, is followed by many other events.

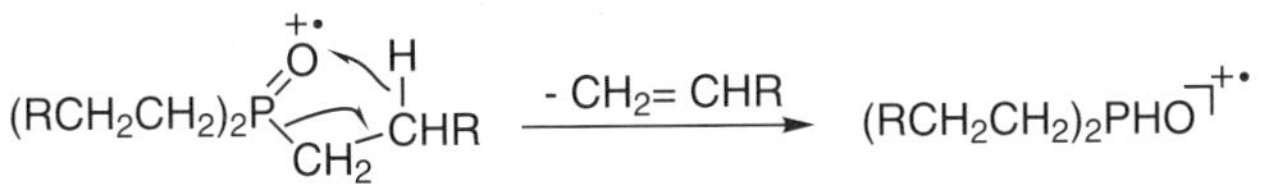

$$(RCH_2CH_2)_2P(O)(CH_2)(CHR) \xrightarrow{-CH_2=CHR} (RCH_2CH_2)_2PHO^{+\bullet} \quad (7.6)$$

In triaryl phosphine oxides, the base peak is frequently $M^+ - 1$ and is assigned the benzophosphole framework (**33**). An interesting fragment that is lost is ArO·, which implies a transfer of aryl from phosphorus to oxygen before further fragmentation. This can be represented as in Eq. 7.7, and can be formally described as a reverse Michaelis–Arbusov reaction.

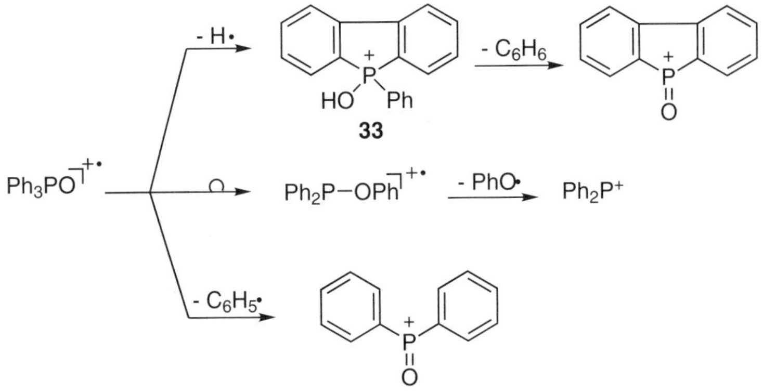

$$Ph_3PO^{+\bullet} \longrightarrow Ph_2P-OPh^{+\bullet} \xrightarrow{-PhO\cdot} Ph_2P^+ \quad (7.7)$$

7.D.3. Phosphorus Acids and Esters

Phosphonic and phosphinic acids give reasonably strong molecular ions, and follow several paths of degradation. A P-methyl group is easily lost as Me·, and hydroxy is lost as HO·. P-Ethyl groups are eliminated as ethylene to form a radical cation, which undergoes further cleavage, illustrated in Eq. 7.8 for diethylphosphinic acid.

$$\text{Et}\underset{\text{HO}}{\overset{\text{O}}{\underset{\|}{\text{P}}}}\overset{\text{H}}{\underset{\text{CH}_2}{\overset{+\bullet}{\rceil}}}\overset{}{\underset{\text{CH}_2}{}} \xrightarrow{-\text{CH}_2=\text{CH}_2} \text{Et}-\underset{\text{OH}}{\overset{\text{O}}{\underset{\|}{\text{P}}}}-\text{H}\; \rceil^{+\bullet} \xrightarrow{-\text{Et}\bullet} \text{HO}-\underset{\text{H}}{\overset{\text{O}}{\underset{\|}{\text{P}}}}^+ \quad (7.8)$$

With longer-chain P-alkyl groups, P–C cleavage by a version of the well-known McLafferty rearrangement can be detected, as shown for a butyl group in Eq. 7.9.

$$\text{R}-\underset{\text{HO}}{\overset{\text{O}}{\underset{\|}{\text{P}}}}\overset{\text{H}}{\underset{\text{CH}_2}{\overset{+\bullet \rceil}{}}}\overset{\text{CH-CH}_3}{\underset{\text{CH}_2}{}} \longrightarrow \text{R}-\underset{\text{HO}}{\overset{\text{OH}}{\underset{\|}{\text{P}}}}\overset{\rceil^{+\bullet}}{\underset{\text{CH}_2}{}} + \text{C}_3\text{H}_6 \quad (7.9)$$

Ester derivatives rarely give a detectable molecular ion; with methylphosphonic and ethylphosphonic acid esters, cleavages of the alkoxy groups are facile, and the base peak frequently has the form RP(OH)_3^+ (Eq. 7.10).

$$\text{R}'\text{CHCH}_2\text{O}\underset{}{\overset{+\bullet}{\underset{\|}{\text{P}}}}\cdots \xrightarrow{-\text{R}'\text{C}_2\text{H}_2} \text{R}-\overset{\text{OH}\;\rceil^{+\bullet}}{\underset{\text{O}-\text{CH}_2}{\text{P}=\text{OH}}} \xrightarrow{-\text{R}'\text{C}_2\text{H}_3\bullet} \text{R}-\overset{\text{OH}}{\underset{\text{OH}}{\text{P}=\text{OH}}}^+ \longleftrightarrow \text{R}-\overset{\text{OH}}{\underset{\text{OH}}{\text{P}-\text{OH}}}^+ \quad (7.10)$$

The loss of an aldehyde is another observable fragmentation mode of the alkoxy groups (Eq. 7.11).

$$\overset{\text{O}}{\underset{\|}{\text{P}}}\overset{\text{O}}{\underset{\text{H}}{\text{CHR}}}\;\rceil^{+\bullet} \longrightarrow \overset{=\text{O}}{\underset{\text{H}}{\text{P}}}\;\rceil^{+\bullet} + \text{R}\overset{\text{O}}{\underset{\|}{\text{CH}}} \quad (7.11)$$

The RP(OH)_3^+ peak also occurs in the spectra of esters with longer chain P-alkyls, but here fragmentation of the P-alkyl chain by a series of McLafferty rearrangements is prominent, and a major fragment has the structure $\text{MeP(O)(OH)}_2^{+\bullet}$. Using diethyl butylphosphonate as an example, the formation of this product can be accounted for as in Eq. 7.12.

$$(\text{EtO})_2\text{P}\cdots \xrightarrow{-\text{C}_3\text{H}_6} \text{EtO}-\text{P}\cdots \xrightarrow{-\text{C}_2\text{H}_4} \text{HO}-\text{P}\cdots \xrightarrow{-\text{C}_2\text{H}_4} \text{HO}-\underset{\text{CH}_3}{\overset{\text{O}}{\underset{\|}{\text{P}}}}-\text{OH} \quad (7.12)$$

Alkyl esters of phosphoric acid rarely give a discernible peak for M$^+$; a series of rearrangements and fragmentations takes place and the most prominent peaks are of the form **34**, **35**, and **36**, shown in Eq. 7.13.

$$(RO)_2\overset{\overset{\overset{+\cdot}{O}}{\|}}{P}\underset{H}{\overset{H}{\underset{O}{\diagdown}}}\underset{CHR'}{CH} \xrightarrow{-R'C_2H_2\cdot} RO-\overset{\overset{OH}{|}}{\underset{O-CH_2}{\overset{+}{P}=OH}} \xrightarrow{-R'C_2H_3} RO-\overset{\overset{OH}{|}}{\underset{OH}{\overset{+}{P}=OH}} \xrightarrow{-R'C_2H_3}$$

$$\underset{H-CHR'}{}$$
$$\mathbf{34}\mathbf{35}$$
$$(R = R'CH_2CH_2)$$

$$HO-\overset{\overset{OH}{|}}{\underset{OH}{\overset{+}{P}=OH}} \longleftrightarrow HO-\overset{\overset{OH}{|}}{\underset{OH}{\overset{|+}{P}-OH}} \tag{7.13}$$

$$\mathbf{36}$$

Another frequently observed fragment is that from loss of an aldehyde group from the alkoxy substituent, as noted in the spectra of phosphonates and phosphinates (Eqn 7.11) to give the ion $(RO)_2PHO^+$.

Triaryl phosphates do give an observable molecular ion, but fragments from loss of aryl and of aryl-O are prominent, as are fragments from rearrangements involving carbon–carbon and carbon–oxygen bond formation. Thus from triphenyl phosphate the fragments $Ph^+\cdot$, $C_{12}H_8^+\cdot$, $Ph_2O^+\cdot$, $C_{12}H_8O_2P^+\cdot$, and $C_{18}H_{12}^+\cdot$ are formed.

7.E. SPECIALIZED SPECTROSCOPIC METHODS USEFUL IN ORGANOPHOSPHORUS CHEMISTRY

7.E.1. ^{17}O NMR Spectroscopy

One of the isotopes of oxygen, ^{17}O, is NMR active (spin of $\frac{5}{2}$); and although it has a natural abundance of only 0.037%, the high-field Fourier transform NMR instruments developed in the last three decades have made it possible to obtain oxygen spectra without isotopic enrichment of a sample. Of course, when enrichment is feasible, one obtains spectra with much fewer scans and that have higher quality, but the price of ^{17}O is quite expensive and the availability of ^{17}O-labelled starting materials is limited to elemental oxygen, CO_2 and H_2O. With highly concentrated solutions, it is sometimes possible to obtain useful natural abundance spectra in 30 min or less on a 400 or 500-MHz instrument, which has made it possible to apply the technique to the many phosphorus compounds that have P=O and P—O bonds. Some valuable structural information can come from such studies.

The reference for ^{17}O NMR is water, and for phosphorus compounds all signals appear in the downfield direction. As a quadrupolar nucleus, ^{17}O signals are broader than those one is used to from ^{13}C or ^{31}P, but this is generally not a problem, unless one is dealing with a compound that contains more than one phosphorus-oxygen or that gives close-lying signals from other types of oxygen-containing organic functional groups. Line broadness can be controlled to some extent by the proper choice of solvent. Preferred solvents are CD_3CN and toluene-d_6, and running the spectrometer at moderately elevated temperatures (e.g., 70 to 80° and 95 to 100°, respectively) helps reduce the viscosity and hence the rotational correlation time, leading to sharper signals. Solvent effects can be strong on ^{17}O; thus triphenylphosphine oxide has δ 51.90 in toluene and δ 44.8 in CDCl$_3$,[55] and this needs to be remembered when comparing data obtained in different solvents. In general, hydrogen-bonding solvents have signals shifted to higher field. The P=O bond gives especially good spectra, found over a very large range (δ 18 to 259), which suggests the high sensitivity of the nucleus to structural effects. The signal is a doublet from coupling to ^{31}P, and thus we have another parameter to consider for characterizing a phosphorus-oxygen compound. This can be quite useful, because the ^{31}P coupling to ^{17}O in the P=O group is generally larger than 150 Hz and easily measured, whereas for P—O, the coupling is less than 90 Hz and is frequently seen only as a broad, poorly resolved absorption band. Indeed, most of the ^{17}O spectra that have been reported for phosphorus compounds are of the P=O group. An excellent review of ^{17}O NMR of phosphorus compounds has been provided by Evans within a monograph by Boykin[56] that covers all aspects of ^{17}O NMR spectroscopy. Tables of data and applications in structural studies may be found there. In the following discussion of some selected spectral data, the ^{17}O-^{31}P coupling constants are given in parentheses.

Trimethyl phosphine oxide in CDCl$_3$ has δ 66.3 (150.9 Hz), whereas lengthening the chain to n-Bu$_3$P=O reduces the shift (CDCl$_3$) to δ 43 (152 Hz),[57] suggesting the operation of the familiar γ-shielding effect on the oxygen. More recently ^{17}O NMR data were reported for a series of nine para-substituted triarylphosphine oxides with 10% isotopic enrichment[55] in both toluene and CDCl$_3$. The former data will be used in the following discussion. The chemical shifts were sensitive to the nature of the para substituent and were spread over the range δ 51.8 to 55.7 (with the range of 159.6 to 168.6 Hz for ^{31}P-^{17}O coupling). Although not a large shift range, the data do indicate a sensitivity to both resonance and inductive effects. Triphenylphosphine oxide had δ 51.90 (166.6 Hz); with the exception of the COOMe group (δ 51.84), all other substituents caused downfield shifts. It was found that the data could be correlated with the Taft Dual Substituent Parameter Equation, which takes into account both the (sometimes competing) inductive and resonance effects of the substituents. The most important point is that there is indeed sensitivity at the oxygen to electronic changes at the remote para-position, and that they are transmitted through the P atom. The authors point out that the results are consistent with the Gilheany view (chapter 2.D) that there can be the equivalent of triple bond character in the phosphoryl group and propose that the inductive and resonance effects can operate separately through two pπ-σ* orbitals.

Much more data are available for cyclic phosphine oxides, and other examples of γ shielding can be seen there, e.g., in the isomeric 1,2-dimethyl-3-phospholene oxides **37** and **38**.[57] In both, the 2-methyl is γ to ^{17}O and causes an upfield shift relative to a 3-methyl-3-phospholene oxide **39**, but in **37**, the 2-methyl is close to ^{17}O, enhancing the strength of the γ interaction. This effect is seen in some other heterocycles[56] and can be useful in making structural assignments. (Note: compound **38** appears in Table 12 of ref. 56 as "Entry 8," but the CH$_3$ and H at C-2 in this structure should be interchanged. The original article should be consulted in connection with this table.)

37, $\delta^{17}O = 53.1$ **38**, $\delta^{17}O = 59.2$ **39**, $\delta^{17}O = 66.0$

A striking effect is seen in the ^{17}O NMR spectra of Diels–Alder dimers of phosphole oxides. Using compound **40** as a typical example, it is first seen that the two different $^{17}O=P$ groups give quite distinct signals, δ 42.1 at the bridging P (P_A) and δ 71.5 at the 2-phospholene P (P_B). In the diastereomer **41** in which the configuration at P_A is reversed, $^{17}O=P_A$ is far downfield (δ 100.8) of $^{17}O=P_B$, which is in the same position (δ 71) as for **40**. This enormous difference, here 60 ppm, between diastereoisomers having the 7-phosphanorbornene moiety is reproduced in other compounds,[57] and is a useful effect for assigning the structure at the bridging P, even though the origin of the effect remains unknown.

40   **41**

A study of a number of six-membered cyclic phosphates has revealed that when the $^{17}O=P$ group is axial, its NMR signal is consistently downfield of that of the diastereomer with equatorial $^{17}O=P$ (e.g., as in isomers **42** and **43**, for which ring flipping is prevented by the two equatorial C-methyls[58]).

42, $^{17}O=P$, δ 77.7   **43**, $^{17}O=P$, δ 85.6

This effect prevails in the highly important bicyclic phosphate structure of cyclic nucleotides, as in isomers **44** and **45** (T = thymidine) and again is of value in assigning stereostructure at P in the esters, as well as in other P-derivatives.[59]

44, $^{17}O=P$, δ 87.7   **45**, $^{17}O=P$, δ 78.2

Exactly the same shift relation has been observed in the anion of cyclic phosphodiesters such as that from cyclic 2′-deoxyadenosine monophosphate. In the anion, the two oxygens on P have the same bond order but can be differentiated by synthetically labelling one with ^{17}O and the other with ^{18}O, thus giving the diastereomers **46** and **47** (A = adenine). NMR then shows that the axial ^{17}O resonates slightly downfield (δ 92.8) of the equatorial ^{17}O (δ 91.2).[60]

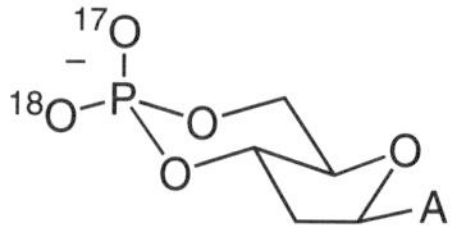

46 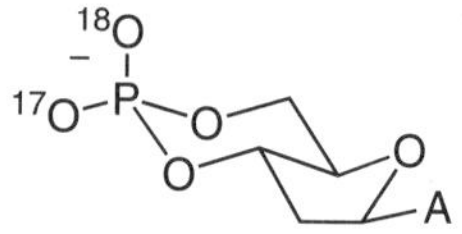   **47**

The examples of ^{17}O NMR applications shown here were selected to emphasize the sensitivity of the method to small structural differences and hence its value as a rising technique for structure determination. Other NMR data and applications do exist in the literature, especially for biophosphates (e.g., reference 61) and the technique is a valuable addition, if specialized, to the much more widely practiced ^{31}P, ^{13}C and ^{1}H spectroscopies.

7.E.2. X-Ray Diffraction Analysis of Organophosphorus Compounds

Just as in all areas of organic and inorganic chemistry, single-crystal X-ray diffraction analysis plays an essential role in organophosphorus chemistry by proving molecular structures, determining bond lengths and angles, and defining shapes and dihedral (torsional) angles within molecules. Much useful structural data can be found in several reviews.[62–64] Table 7.2 shows some data from the literature[65] on the length of common bonds to phosphorus. A special review of X-ray structural analysis of heterocyclic phosphorus compounds was published in 1981.[7e] These cited references provide a strong background of structural data, although in the course of research on new molecules, exceptional parameters do appear.

7.E.3. Electron Spin Resonance Spectroscopy

Free radicals based on phosphorus can be generated by several methods and studied directly, or they may appear as transient intermediates in reactions of organophosphorus compounds. As with organic radicals, phosphorus radicals can be identified and mechanisms elucidated very effectively with the aid of electron spin resonance (ESR), also known as electron paramagnetic resonance (EPR). Many examples are known, and useful orienting reviews for this aspect of phosphorus chemistry are those of Schipper, Jansen and Buck[66] and Tordo.[67] The ESR signals for phosphorus-centered radicals show splitting from coupling of the electron with the ^{31}P nucleus. Some radical types that have been studied, with the coupling values,[66] include: $Ph_2P\cdot$, 12 G; $(p\text{-}Me_2NC_6H_4)_3P^{+}\cdot$, 12 G; $Ph_2PK-\cdot$, 8.4 G; $Ph_3(OK)P\cdot$, 5.25 G; $Me_2(O)P\cdot$, 375 G; $(RO)_2(O)P\cdot$, 375 G; and $MeCl_3P\cdot$, 1077 G.

TABLE 7.2. Lengths of Some Common Bonds to Phosphorus[a]

Compound	d^b	m^c	$q_1{}^d$	$q_u{}^e$
$R_3P^+-C_{sp3}$	1.800	1.802	1.790	1.812
$R_2(O)P-CH_3$	1.791	1.790	1.786	1.795
$R_2(O)P-CH_2R$	1.806	1.806	1.801	1.813
$R_2(O)P-CMe_3$	1.841	1.842	1.835	1.847
R_2P-C_{sp3}	1.855	1.857	1.840	1.870
$R_3P^+-C_{arom}$	1.793	1.792	1.786	1.800
$R_2(O)P-C_{arom}$	1.801	1.802	1.796	1.807
R_2P-C_{arom}	1.836	1.837	1.830	1.844
$X(RO)_3P-OR_{apical}$	1.689	1.685	1.675	1.712
$X(OR_3P-OR_{equat}$	1.619	1.622	1.604	1.628
$R_3P=O$	1.489	1.486	1.481	1.496
$(R_2N)_3=O$	1.461	1.462	1.449	1.470
$(RO)_3P=O$	1.449	1.448	1.446	1.452
$(RO)P-O_3{}^{--}$	1.513	1.513	1.508	1.518
$(RO)_2P-O_2{}^-$	1.483	1.485	1.474	1.490
$R_3P=S$	1.954	1.952	1.950	1.957
$R(RO)_2P=S$	1.922	1.924	1.913	1.927

[a]Data from reference 65.
[b]The mean of several values.
[c]The median of several values (50% are above and 50% are below this value).
[d]The bottom quartile; 25% are of lower value.
[e]The top quartile; 25% are of higher value.

7.E.4. Ultraviolet and Visible Spectroscopy

Ultraviolet-visble spectroscopy was of some importance in early studies on electronic interactions of phosphorus groups with unsaturated centers and remains so today, but its use has somewhat declined with the coming of other techniques that bear on this subject. None of the common 3- and 4-coordinate phosphorus functional groups alone function as useful chromophores when irradiated with light in the UV and Vis regions of the electromagnetic spectrum (although phosphines do exhibit a weak absorption (log $\varepsilon \sim 2$) around 220 to 230 nm). The P=O group has no absorption even into the far UV region. As will be noted in chapter 10, some 2-coordinate compounds (e.g., $R_2C=PR$, $ArO-P=O$, $RP=PR$) containing double bonds to phosphorus do exhibit UV–Vis absorption. The use of UV-Vis spectroscopy is, therefore, limited to studies based on absorptions by the organic framework surrounding the phosphorus functional group and the possible conjugative interaction of the phosphorus group with multiple bonds.

Absorptions of aryl groups with phosphorus substituents have been extensively studied; a review of the fundamental work appears in the monograph of Hudson,[68] and the more recent work on the important family of phosphine chalcogenides is treated by Davidson in the Hartley series,[69] and work on phosphines is reviewed there by Gilheany.[70]

In some early work,[71] it was shown that little or no change in the UV spectrum of an alkene accompanied the attachment of a phosphonate group, even though, as we have noted, there are indications of a conjugative interaction from other techniques (NMR, IR, and chemical reactivity). More recent studies on styrene derivatives (**48**) do show a bathochromic shift in bands for the styrene fragment due to the interaction with the $Ph_2P(O)$ group.[72]

48

Absorptions of the benzene ring with phosphorus substituents have been much more extensively studied,[72] and definite spectral perturbations arise from phosphonic and phosphinic acid substituents and phosphine oxides; a list of compounds studied is provided in reference 69. A noticeable effect is seen on the B-band of benzene (centered at 255 nm with vibrational fine structure), which is shifted a few nm to longer wave lengths and somewhat intensified. The primary band of benzene at 203 nm is similarly shifted. Substitution of the para position of phenyl-substituted phosphine oxides with electron-releasing amino and alkoxy groups causes more substantial shifts, especially of the primary band, suggesting excited state stabilization by conjugative interaction (Eq. 7.14), originally attributed to occupation of phosphorus d-orbitals.[74] Phosphonium ions and phosphine sulfides also interact with the benzene ring, especially when electron-releasing substituents are present. The spectra of some heterocyclic systems with phosphorus substituents have also been recorded (e.g., tri-2-pyrrylphosphine oxide).[75]

(7.14)

Conjugative interaction of a different type must be present when 3-coordinate phosphorus groups are attached to the benzene ring, and these compounds have been of more interest in recent years.[76] Significant changes in the benzene spectrum are found in aryl phosphines; the primary band is shifted to about 220 nm, and the B-band is also shifted (e.g., triphenylphosphine, 261 nm), both undergoing intensification. Stabilization of the excited state by release of the phosphorus lone pair to the aromatic system may be involved (Eq. 7.15). A weak, sometimes hidden absorption around 290 to 333 nm has been attributed to electron transfer from the lone pair orbital to an antibonding π orbital of the benzene ring (an n to π^* transition).

(7.15)

In the five-membered phosphole ring, substantial changes in the spectrum relative to cyclopentadiene (π to π* at 239 nm) take place; 1-methylphosphole and other simple phospholes absorb around 285 nm,[77] and polyphenyl phospholes are yellow.[68a] Aryl phosphide anions also have very significant absorption, so that they are colored, and this seems attributable to resonance stabilization of the form seen in Eq. 7.16.

$$\text{(7.16)}$$

The anions derived from phospholes also have conjugation by electron release from phosphorus and when extended by the presence of C-aryl substitution or benzo-fusion, they absorb at such long wave length as to be highly colored. Potassium 2,5-diphenylphospholide (**49**), for example, is red-violet, and lithium dibenzophospholide (**50**) is red-orange.

49   **50**

In contemporary organophosphorus chemistry, UV spectroscopy also plays an important analytical role, by which absorptions from the organic fragment can be monitored. This is especially so with some biological phosphates. For example, quantitative analysis of nucleosides and nucleotides may be accomplished by monitoring the absorption of the pyrimidine or purine base component, typically around 260 to 270 nm. Thus adenosine triphosphate has a strong (ε_{max} 15,400) band at 259 nm that is used for this purpose.[78]

REFERENCES

1. G. Mavel in J. W. Emsley, J. Feeney, and L. H. Sutcliffe, *Progress in Nuclear Magnetic Resonance Spectroscopy*, Pergamon Press, Oxford, 1966, Chapter 4.
2. G. Mavel in E. F. Mooney, *Annual Reports on NMR Spectroscopy*, Vol. 1. 5B, Academic Press, New York, 1973.
3. G. M. Kosolapoff and L. Maier, eds., *Organic Phosphorus Compounds*, John Wiley & Sons, Inc., New York, 1972.
4. R. S. Edmundson, *Dictionary of Organophosphorus Compounds*, Chapman & Hall, London, 1988.
5. *Sadtler Nuclear Magnetic Resonance Spectra*, Sadtler Research Laboratories, Philadelphia PA.
6. C. J. Pouchert and J. Behnke, *Aldrich Library of ^{13}C and FT-NMR Spectra*, Aldrich Chemical Co., Milwaukee, WI, 1993.

7. L. D. Quin, *The Heterocyclic Chemistry of Phosphorus*, John Wiley & Sons, Inc., New York, 1981, (a) Chapter 7, (b) p. 320, (c) pp. 349–356, (d) Chapter 6, (e) Chapter 8.
8. M. J. Gallagher in L. D. Quin and J. G. Verkade, eds., *Phosphorus-31 NMR Spectral Properties in Compound Characterization and Structural Analysis*, VCH Publishers, Inc., New York, 1974, Chapter 29.
9. J. C. Tebby, ed., *Handbook of Phosphorus Nuclear Magnetic Resonance Data*, CRC Press, Boca Raton, FL. 1991, (a) Chapter 6, (b) Chapter 10.
10. J. B. Hendrickson, M. L. Maddox, J. J. Sims, and H. D. Kaesz, *Tetrahedron* **20**, 449 (1964).
11. L. D. Quin, J. P. Gratz, and T. P. Barket, *J. Org. Chem.* **33**, 1034 (1968).
12. J. F. Nixon and R. Schmutzler, *Spectrochim. Acta* **22**, 565 (1966).
13. H. Vahrenkamp and H. Nöth, *J. Organomet. Chem.* **12**, 281 (1968).
14. K. Naumann, G. Zon, and K. Mislow, *J. Am. Chem. Soc.* **91**, 7012 (1969).
15. D. Fiat, M. Halmann, L. Kugel, and J. Reuben, *J. Chem. Soc.*, 3837 (1962).
16. T. N. Timofeeva, B. I. Ionin, and A. A. Petrov, *Zhur. Obshch. Khim.* **39**, 354 (1969); *Chem. Abstr.* **71**, 26438 (1969).
17. E. Pretsch, J. Seibl, W. Simon, and T. Clerc, *Tables of Spectral Data for Structure Determination of Organic Compounds*, Springer-Verlag, Berlin, 1981, p. H260.
18. A. A. Bothner-By and W.-P. Trautwein, *J. Am. Chem. Soc.* **93**, 2189 (1971).
19. Reference 5a, Proton NMR Spectrum 34, 254.
20. Reference 5b, Proton NMR Spectrum 1501A.
21. J. B. Lambert, *Organic Structural Analysis*, Macmillan Publishing Co., Inc., New York, 1976, pp. 54–56, 86–87.
22. Reference 5a, Proton NMR Spectrum 29632.
23. S. Varbanov, V. Vasileva, V. Zlateva, T. Radeva, and G. Borisov, *Phosphorus Sulfur* **21**, 17 (1984).
24. J. F. Brazier, D. Houalla, M. Koenig, and R. Wolf, *Top. Phosphorus Chem.* **8**, 99 (1976); D. Houalla, R. Marty, and R. Wolf, *Z. Naturforsch. B*, **25**, 451 (1970).
25. G. A. Gray, S. E. Cremer, and K. L. Marsi, *J. Am. Chem. Soc.*, **98**, 2109 (1976).
26. E. W. Turnblom and T. J. Katz, *J. Am. Chem. Soc.* **95**, 4292 (1973).
27. I. Granoth, Y. Segall, and H. Leader, *J. Chem. Soc., Perkin Trans.* 1, 465 (1978).
28. W. G. Bentrude and W. N. Setzer in J. G. Verkade and L. D. Quin, eds., *Phosphorus-31 NMR Spectroscopy in Stereochemical Analysis*, VCH Publishers, Inc., Deerfield Beach, FL, 1987, Chapter 11.
29. J. A. Mosbo, *Org. Magn. Reson.* **11**, 281 (1978).
30. (a) A. Cogne, A. G. Guimaraes, J. Martin, R. Nardin, J.-B. Robert, and W. J. Stec; *Org. Magn. Reson.* **6**, 629, (1974); (b) W. G. Bentrude and J. H. Hargis *J. Am. Chem. Soc.* **92**, 7136 (1970).
31. J.-P. Albrand, D. Gagnaire, J. Martin, and J.-B. Robert, *Bull. Soc. Chim. Fr.*, 40 (1969).
32. A. Bond, M. Green, and S. C. Pearson, *J. Chem. Soc. B*, 929 (1968).
33. J. B. Stothers, *Carbon-13 NMR Spectroscopy*, Academic Press, New York, 1972, pp. 158–161.
34. L. D. Quin, ^{13}C *Nuclear Magnetic Resonance Spectral Data of Heterocyclic Phosphorus Compounds*, Thermodynamics Research Center, College Station, Texas, 1983.
35. G. A. Gray and S. E. Cremer, *J. Org. Chem.* **37**, 3458 (1972).
36. Reference 5a, Carbon NMR Spectrum 13792C.
37. L. D. Quin, M. D. Gordon, and S.O. Lee, *Org. Magn. Reson.* **6**, 503 (1974).
38. M. D. Gordon and L. D. Quin, *J. Org. Chem.* **41**, 1690 (1976).
39. L. D. Quin in J. G. Verkade and L. D. Quin, eds., *Phosphorus-31 NMR Spectroscopy in Stereochemical Analysis*, VCH Publishers, Inc., Deerfield Beach, FL, 1987, Chapter 12.

40. K. M. Pietrusiewicz, *Org. Magn. Reson.* **21**, 345 (1983).
41. L. D. Quin, K. C. Caster, J. C. Kisalus, and K. A. Mesch, *J. Am. Chem. Soc.* **106**, 7021 (1984).
42. T. A. Modro, *Can. J. Chem.* **55**, 3681 (1977).
43. T. A. Albright, W. J. Freeman, and E. E. Schweizer, *J. Org. Chem.* **40**, 3437 (1975).
44. T. A. Albright, W. J. Freeman, and E. E. Schweizer, *J. Am. Chem. Soc.* **97**, 2946 (1975).
45. C. Symmes Jr., and L. D. Quin, *J. Org. Chem.* **41**, 1548 (1978).
46. L. W. Daasch and D. C. Smith, *Anal. Chem.* **23**, 853 (1951).
47. L. J. Bellamy, *The Infra-Red Spectra of Complex Molecules*, 2nd. ed., Methuen & Co., Ltd., London, 1958, Chapter 18.
48. D. E. C. Corbridge, *Top. Phosphorus Chem.* **6**, 235 (1969).
49. (a) *Sadtler Standard Infrared Grating Spectra*, Sadtler Research Labs., Philadelphia, PA, (b) *Aldrich Library of FT-IR Spectra*, Aldrich Chemical Co., Milwaukee, WI, 1985.
50. Reference 17, pp. I235–I245.
51. I. Granoth, *Top. Phosphorus Chem.* **8**, 41 (1976).
52. R. G. Gillis and J. L. Occolowitz in M. Halmann, ed., *Analytical Chemistry of Phosphorus*, Wiley-Interscience, New York, 1972.
53. R. A. J. O'Hair in F. R. Hartley, ed., *The Chemistry of Organophosphorus Compounds*, John Wiley & Sons, Inc., New York, 1996, Chapter 4.
54. R. G. Gillis and G. J. Long, *Org. Mass. Spectr.* **2**, 1315 (1969).
55. M. Pomerantz, M. S. Terrazas, Y. Cheng, and X. Gu, *Phosphorus Sulfur Silicon* **109–110**, 505 (1996).
56. S. A. Evans Jr. in D. W. Boykin, *^{17}O NMR Spectroscopy in Organic Chemistry*, CRC Press, Boca Raton, FL, 1990, Chapter 10.
57. L. D. Quin, J. Szewczyk, K. Linehan, and D. L. Harris, *Magn. Reson. Chem.* **25**, 271 (1987).
58. P. L. Bock, J. A. Mosbo, and J. L. Redmon, *Org. Magn. Reson.* **21**, 491 (1983).
59. A. E. Sopchik, S. M. Cairns, and W. G. Bentrude, *Tetrahedron Lett.*, **30**, 1221 (1989).
60. J. A. Coderre, S. Mehdi, P. C. Demou, R. Weber, D. D. Traficante, and J. A. Gerlt, *J. Am. Chem. Soc.* **103**, 1870 (1981).
61. M.-D. Tsai and K. Bruzik in L. J. Berliner and J. Reuben, eds., *Biological Magnetic Resonance*, Vol. 5, Plenum Press, New York, 1983, p. 129.
62. J. J. Daly in J. D. Dunitz and J. A. Ibers, eds., *Perspectives in Structural Chemistry*, Vol. 3, John Wiley & Sons, Inc., New York, 1970, pp. 165–228.
63. D. E. C. Corbridge, *Top. Phosphorus Chem.* **3**, 57 (1966).
64. D. E. C. Corbridge, *The Structural Chemistry of Phosphorus*, Elsevier, Amsterdam, 1974.
65. F. H. Allen, O. Kennard, D. G. Watson, L. Brammer, A. G. Orpen, and R. Taylor, *J. Chem. Soc. Perkin Trans.* 2, S1 (1987).
66. P. Schipper, E. H. J. M. Jansen, and H. M. Buck, *Top. Phosphorus Chem.* **9**, 407 (1977).
67. P. Tordo in F. R. Hartley, ed., *The Chemistry of Organic Phosphorus Compounds*, Vol. 1, John Wiley & Sons, Inc., New York, 1996, Chapter 6.
68. R. F. Hudson, *Structure and Mechanism in Organo-Phosphorus Chemistry*, Academic Press, New York, 1965, (a) pp. 24–28, (b) pp. 76–80.
69. G. Davidson in F. R. Hartley, ed., *The Chemistry of Organic Phosphorus Compounds*, Vol. 2, John Wiley & Sons, Inc., 1992, pp. 169–171.
70. D. Gilheany in F. R. Hartley, ed., *The Chemistry of Organic Phosphorus Compounds*, Vol. 1, John Wiley & Sons, Inc., 1990, pp. 40–43.
71. M. I. Kabachnik, *Tetrahedron* **20**, 655 (1964).

72. K. G. Berndt and D. Gloyna, *Acta Chim. Acad. Sci. Hung.* **110**, 145 (1982); *Chem. Abstr.* **98**, 53063 (1983); quoted in Reference 69.
73. H. H. Jaffé and L. D. Freedman, *J. Am. Chem. Soc.* **74**, 1069, 2930 (1952).
74. H. Goetz, F. Nerdel, and K. H. Wiechel, *Liebigs Ann. Chem.* **665**, 1 (1963).
75. C. E. Griffin and R. A. Polsky, *J. Org. Chem.* **26**, 4772 (1961).
76. D. J. Fife, K. W. Morse, and W. M. Moore, *J. Photochem.* **24**, 249 (1984).
77. L. D. Quin, J. G. Bryson, and C. G. Moreland, *J. Am. Chem. Soc.* **91**, 3308 (1969).
78. J. Feder, *Top. Phosphorus Chem.* **11**, 1 (1983).

CHAPTER 8

HETEROCYCLIC PHOSPHORUS COMPOUNDS

8.A. GENERAL

There are many ways in which phosphorus enters into the construction of rings; a count made in 1981[1] included 243 ring systems, and more have been created since this time. In these systems, phosphorus may be present in any of the familiar 3- and 4-coordinate functional groups, and in more recent years, a few structures with 5-coordinate phosphorus have been synthesized. Novel heterocycles based on 2-coordinate phosphorus, in which true double bonding to P is present, are discoveries of only the last 2–3 decades, but they now play a prominent role in contemporary research. The number of compounds based on phosphorus in rings, with the several possibilities for P-functionality and with various forms of ring substitution, is very large indeed, and still rapidly growing. This has not always been the case, however, and detailed study of phosphorus heterocycles was slow to develop. Phosphorus did not share in the great research surge started in the late 1800s on heterocyclic systems, during which many fundamentals of rings containing oxygen, nitrogen and sulfur were laid down. In fact, the first heterocycle containing phosphorus and carbon was not made until 1915,[2] and research was very slow in this area until about 1950. It is regrettable that still little or no mention of phosphorus heterocycles appears in the textbooks of heterocyclic chemistry, because the field, with its great number of structural varieties and potential applications, remains relatively unknown to many organic chemists. However, the second edition (1996) of *Comprehensive Heterocyclic Chemistry*[3] (CHEC II), has recognized the contemporary importance of phosphorus heterocycles and has devoted several chapters to them. This series, therefore, is the best source of information on the fundamental ring systems as well as on fused, bicyclic, and spiro systems. The older review literature on phosphorus ring systems remains of great value in supplying more complete information. The status of the entire field up to about 1970 was summarized in an excellent review by a pioneer in the field, F. G. Mann.[4] A book devoted primarily to phosphorus heterocycles with carbon-phosphorus bonds appeared in 1981.[1] Three general review articles, by Berlin and co-workers in 1969[5] and in 1977[6] and Dimroth in 1984,[7] and several specialized reviews have appeared over the years. Some topics of the reviews have included cyclic phosphines,[8] phospholes,[9] phosphinines,[10] phosphiranes[11] and 7-phosphanorbornenes.[12] Reviews on some other ring systems will be cited later in this chapter.

It is the goal of this chapter to illustrate the structural variety possible for phosphorus heterocycles by presenting short summaries of the major synthetic methods and of the properties of some of the more important ring systems. Many novel ring systems were created in the 1980s and 1990s as a result of the intensive research on compounds with low coordination states; some of these systems will be seen in chapter 10. The information to be presented in the present chapter on a particular cyclic system is selective, and the review or primary literature should be consulted for greater detail. In the discussion of a particular ring system, a synthesis might be seen to produce phosphorus in a particular functionality. Thus it is quite common for phosphine oxides to be the product of a ring synthesis. It is generally possible with little difficulty, however, to convert this function to the phosphine, typically with trichlorosilane, which then gives access to the cyclic phosphine sulfide, phosphonium salts, etc. Conversions of other functional groups are also possible.

In section 8.B the fundamental monocyclic systems are presented. The correct IUPAC name is provided for the parent cycle containing only carbon and phosphorus; the names for other cycles with additional hetero atoms (O, N, S, P, etc.) in the ring and the numbering system are derived by standard heterocyclic naming techniques,[13] in which the parent name is prefaced by the substituting atoms and their location. Thus the parent five-membered ring is phospholane; if the two carbons attached to P in the ring are replaced by oxygen, the system is named 1,3,2-dioxaphospholane. The numbering priority is that of the Hantzsch-Widman system, which places the common heteroatoms above phosphorus (O > S > N > P).

8.B. THE FUNDAMENTAL RING SYSTEMS

8.B.1. Three-Membered Rings

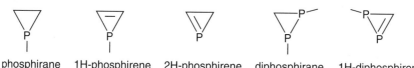

Figure 8.1. Three-membered rings.

The 3-membered ring remained unknown in phosphorus chemistry until the synthesis of phosphiranes was mentioned in a U.S. patent in 1963[14] and confirmed in 1967.[15] The method of synthesis is surprisingly simple and direct, involving the reaction of a metallic phosphide with a 1,2-dichloride, as generalized in Eq. 8.1.

$$\underset{\text{ClCHCH}_2\text{Cl}}{\overset{R}{|}} + \text{R'PHNa} \xrightarrow{\text{liq. NH}_3} \underset{\mathbf{1}}{\overset{R}{\underset{R'}{\bigtriangledown_{P}}}} \qquad (8.1)$$

This approach has been used for the preparation of a number of substituted phosphiranes and remains of interest; an example is the recent modification of the synthesis of 1-phenylphosphirane (**1**, R = H, R′ = Ph) by reacting PhPLi$_2$ with 1,2-dichloroethane.[16] Compound **1** is a distillable liquid that is stable at 0° for over 1 month, but as is typical of unstabilized phosphiranes, it undergoes decomposition at higher temperatures. The phosphiranes remained chemical oddities for many years after the initial reports[14,15] until two important techniques for stabilizing them were devised. As for several other types of unstable phosphorus species, attaching a large bulky group such as *t*-butyl to phosphorus was found to be an effective way to stabilize the ring. Thus *trans*-1-*t*-butyl-2-methylphosphirane (Eq. 8.2) is a distillable liquid.[17]

$$\underset{\text{ClCHCH}_2\text{Cl}}{\overset{\text{Me}}{|}} + t\text{-BuPLi}_2 \xrightarrow[\text{hexane}]{\text{liq. NH}_3} \underset{t\text{-Bu}}{\overset{\text{Me}}{\bigtriangledown_{P}}} \qquad (8.2)$$

Stabilization has also been effected by coordination of the phosphirane to metal centers, notably as a ligand to Mo(CO)$_5$, W(CO)$_5$, and Cr(CO$_5$); in these complexes chemical reactions can be performed on the phosphirane ring while in the coordination sphere of the metal without its decomposition, and the new phosphirane released by decomplexation procedures. Following these discoveries, new syntheses of phosphiranes were devised, a highly significant one[18] being the creation of the ring by addition to an alkene of a terminal phosphinidene complex (a term indicating that an RP unit is bonded to a single metal center, as in PhP=W(CO)$_5$). The phosphinidene complex is easily generated by decomposition of a Mo(CO)$_5$ complex (**3**) of a 7-phosphanorbornadiene and is immediately trapped by the alkene that is included in the reaction mixture. The entire process, starting with a well-known complexed phosphole (**2**, vide infra) and concluding with release of the free phosphirane by oxidative decomplexation with a halogen, is illustrated in Eq. 8.3. The synthesis is quite versatile, and several phosphiranes have been prepared with its use. Several other syntheses for phosphiranes have been devised, and are summarized in the 1990 review by Mathey.[11a] With stabilized derivatives on hand, it has become possible to extend the knowledge of physical and chemical properties of

236 HETEROCYCLIC PHOSPHORUS COMPOUNDS

$$[\text{Reaction scheme: Compound } \mathbf{2} + \text{MeOOCC}\equiv\text{CCOOMe} \longrightarrow \mathbf{3} \xrightarrow{\Delta \text{ or CuCl}}$$

$$[\text{RP=Mo(CO)}_5] \xrightarrow{\text{C=C}} \text{(phosphirane-Mo(CO)}_5) \xrightarrow{1.\ X_2,\ 2.\ \text{base}} \text{(phosphirane)} \quad (8.3)$$

the phosphirane ring. The C–P–C internal angle is about 49°, and phosphorus has highly pyramidal character. There does not appear to be any experimental determination of the inversion barrier, but a recent theoretical calculation for the parent 1H-phosphirane gives a value of 69.35 kcal/mol.[19] The inversion barrier is certainly high enough that stable cis,trans derivatives are allowed, even when P has only hydrogen attached (e.g., compounds **4** and **5**[20]).

4, $\delta\ ^{31}P = -271$ **5**, $\delta\ ^{31}P = -288$

The ^{31}P NMR shift for simple phosphiranes is far upfield; some values include $\delta -341$ for 1H-phosphirane, $\delta -251$ for 1-methylphosphirane, and $\delta -234$ for 1-phenylphosphirane. Cis,trans isomers give well-separated signals, with trans upfield of cis; many examples are known (e.g., references 17 and 21). The ^{13}C NMR shift also appears at quite high field, which is typical of other types of 3-membered rings (e.g., δ 0.7 for 1H-phosphirane). As in many other phosphorus heterocycles, the stereospecificity of ^{31}P-^{13}C coupling to C-substituents is a useful aid for assigning cis,trans structure. Thus proximity of the lone pair to a C-substituent (the trans isomer) leads to a larger $^2J_{PC}$ value. Compounds **6** and **7**[21] illustrate this effect as well as that on $^3J_{PC}$ and the effect of crowding in the cis isomer on the ^{13}C shift.

6
$\delta\ ^{13}C^a = 140.4$
$^2J_{PCa} = 11.5,\ ^3J_{PCb} = 7.3\ \text{Hz}$

7
$\delta\ ^{13}C^a = 137.9$
$^2J_{PCa} = 2.3,\ ^3J_{PCb} = 0\ \text{Hz}$

A summary of these and other properties of the phosphirane system are provided by Mathey.[11a] Theoretical calculations on the ring are also covered there.

Various chemical properties of the phosphirane ring have been studied. Here we will consider only those properties that pertain to their behavior as phosphines; numerous other reactions, especially those of their metal complexes, are described by Mathey[15]. The lone pair in the phosphirane system has high s-character, and in a qualitative sense it is less available than in acyclic counterparts for forming a new bond so as to provide 4-coordinate derivatives. In addition, these products would have phosphorus in sp^3 condition, further increasing ring strain. The result is that nucleophilicity in phosphiranes is quite low; neither protonation nor quaternization with alkyl halides takes place under the normal conditions. The first phosphiranium ion was not prepared until 1995; 1-phenylphosphirane formed a crystalline salt with methyl triflate that was characterized by X-ray diffraction analysis.[22] Phosphine oxides are not readily formed by oxidative methods. They became known only as a result of ring formation where phosphorus is already present in that state, as in Eq. 8.4, for the synthesis of the highly hindered oxide **8**.[23] Exceptions are known for certain bicyclic phosphiranes, where examples of oxide formation[24] with *t*-butyl hydrogen peroxide and sulfide formation[25] with S_8 have been reported.

$$t\text{-Bu-CHCl}\diagdown_{\substack{P\\\diagup\diagdown\\O\quad Bu\text{-}t}}\diagup \text{CH}_2\text{-Bu-}t \quad \xrightarrow[-78°]{\text{LiNEt}_2} \quad \underset{\mathbf{8}}{\overset{t\text{-Bu}\diagdown\diagup\text{Bu-}t}{\underset{O}{\diagdown}\overset{}{P}\diagup\text{Bu-}t}} \tag{8.4}$$

On the other hand, complexation with metal carbonyls is possible, although such compounds are more commonly prepared by indirect methods (e.g., as seen in Eq. 8.3) and the well-known conversion of 3- to 5-coordinate phosphorus by addition of a peroxide has been demonstrated in one case[26] (Eq. 8.5). The adduct **9**, however, while stable at −80°, eliminated ethylene on warming.

$$\underset{\text{Ph}}{\triangledown\!\!\!-\!P} + \underset{\substack{|\quad|\\O\text{-}O}}{\text{Me}_2\text{C-CHMe}} \xrightarrow{-80°} \underset{\mathbf{9}}{\overset{\triangledown}{\underset{\text{Ph}}{\overset{|}{P}}\!\!-\!\!O\diagdown\underset{\text{Me}_2}{\overset{}{\diagup}\text{Me}}}} \xrightarrow{\Delta}$$

$$\text{H}_2\text{C=CH}_2 \quad + \quad \underset{\text{Me}_2}{\overset{\text{Ph-P-O}}{\underset{O\diagdown\diagup\text{Me}}{|}}} \tag{8.5}$$

It may follow that if nucleophilicity at P is low, electrophilicity should be high, and that seems to be the case. 1-Chlorophosphiranes have recently become available with the aid of steric stabilization at carbon, as in **10**[27] and **11**,[28] and also as metal complexes; they were found to undergo the normal substitution reactions of a phosphinous chloride with organometallics as well as with O, N, and P nucleophiles.

238 HETEROCYCLIC PHOSPHORUS COMPOUNDS

[Structures 10 and 11: compound 10 shows a three-membered ring with Ph, Ph, TMS, Ph substituents and P-Cl; compound 11 shows TMS, TMS, TMS, TMS substituents with P-Cl]

Derivatives of the 1H-phosphirene system were first prepared in 1982 by the trapping with an alkyne of the terminal phosphinidene complex released by decomposition of the 7-phosphanorbornadiene complexes seen in Eq. 8.3. This highly significant work,[29] outlined in Eq. 8.6 for the synthesis of 1,2,3-triphenylphosphirene, opened a new wave of research on the three-membered ring, particularly because it was discovered that the phosphirene ring, after release from its complexed form,[29a,30] was surprisingly stable and did not require the presence of sterically protecting substituents. Thus, 1,2,3-triphenylphosphirene[29b] has m.p. 73° and ^{31}P δ −190.3. Its stability, even with an internal C−P−C angle of only 41.8°, is demonstrated by its mass spectrum, where the molecular ion was the base peak.[30] The phosphirene resisted oxidation and quaternization with an alkyl halide, but it did add sulfur to form the phosphine sulfide. Calculations showed that the HOMO has low localized electron density at P, consistent with this behavior.

$$\text{(CO)}_5\text{W-P(Ph)(Me)(Me)-norbornadiene-COOMe} \xrightarrow{\Delta} [\text{PhP=W(CO)}_5] \xrightarrow{\text{PhC≡CPh}} \text{Ph-C=C(Ph)-P(Ph)-W(CO)}_5 \quad (8.6)$$

Another technique for generating terminal phosphinidene complexes for addition to a triple bond involves the reactions outlined in Eq. 8.7.[30]

$$(\text{Et}_2\text{N})_2\text{P-H} \cdot \text{W(CO)}_5 \xrightarrow[25°]{\text{Br}_2} [\text{Et}_2\text{NP=W(CO)}_5] \xrightarrow{\text{PhC≡CPh}} \text{Ph-C=C(Ph)-P(NEt}_2\text{)-W(CO)}_5 \quad (8.7)$$

A quite different approach to the phosphirene ring is provided by the reaction of RPCl$_2$−AlCl$_3$ complexes with alkynes; this produces, in a single step, phosphirenes with phosphonium phosphorus, having ^{31}P NMR shifts at very high field for an ion, typically δ −57 to −70[31] (Eq. 8.8).

$$\text{RC≡CR} \xrightarrow[\text{AlCl}_3]{\text{RPCl}_2} \text{R-C=C(R)-P}^+(\text{R})(\text{Cl}) \cdot \text{AlCl}_4^- \quad (8.8)$$

Variations on this procedure include the use of phosphinous halides or PCl$_3$ with AlCl$_3$, and of a true phosphenium ion.[32]

The chlorophosphirenium ions of Eq. 8.8, although of limited stability, were later found to be reducible in situ in the way seen before for heterophosphonium ions to form the free phosphines (chapter 3.C) by action of tributylphosphine, and this appears to be the most straightforward way to generate simple P-alkyl or aryl phosphirenes[33] (Eq. 8.9). Thus 1,2,3-triphenylphosphirene was obtained in 43% yield after column chromatography and recrystallization.

$$\underset{R\overset{+}{P}Cl\ AlCl_4^-}{R\diagdown\!\!=\!\!\diagup R} \xrightarrow{Bu_3P} \underset{\underset{R}{P}}{R\diagdown\!\!=\!\!\diagup R} \qquad (8.9)$$

The same method was used[34] in a synthesis of a 3-coordinate phosphirene with the bulky mesityl group on phosphorus; the stabilizing effect of the mesityl substituent allowed a reaction with butyllithium that converted a methyl group to a carbanion, which then could be condensed with carbonyl compounds to provide C-functionalized phosphirenes (Eq. 8.10). With unprotected phosphirenes, ring opening occurs with butyllithium, forming allenylphosphine anions.

$$MeC\equiv CMe \xrightarrow[AlCl_3]{MesPCl_2} \underset{Mes\overset{+}{P}Cl\ AlCl_4^-}{Me\diagdown\!\!=\!\!\diagup Me} \xrightarrow{Bu_3P} \underset{\underset{Mes}{P}}{Me\diagdown\!\!=\!\!\diagup Me} \xrightarrow[2.\ R_2CO]{1.\ BuLi}$$

$$\underset{\underset{Mes}{P}}{Me\diagdown\!\!=\!\!\diagup CH_2\text{-}\underset{OH}{\overset{|}{C}R_2}} \qquad (8.10)$$

Phosphirenes with imino groups on phosphorus are formed in a cyclization reaction occurring between acetylenic compounds and iminophosphines[35] (12, Eq. 8.11).

$$\underset{R}{\overset{Ar}{P=N}} + PhC\equiv CPh \longrightarrow \underset{R\overset{P}{\diagdown}NAr}{Ph\diagdown\!\!=\!\!\diagup Ph} \qquad (8.11)$$

12
Ar = 2,4,6-tri-*t*-butylphenyl

As seen for the phosphiranes, nucleophilicity is low in the phosphirenes, but the 1-chloro derivatives have normal electrophilic character. One addition reaction to form a stable 5-coordinate product, proven by X-ray analysis, is known[36] (Eq. 8.12).

$$\text{(8.12)}$$

Much of the intricate chemistry of 1H-phosphirenes has been developed through the use of metal complexes and is well summarized in the review by Mathey.[11a] This continues to be an active area of research. The 1H-phosphirene system attracts attention from another standpoint, that of its possible involvement in aromaticity phenomena. Theoretically, the ring with 3-coordinate phosphorus constitutes a 4 π-electron cyclic system and could be antiaromatic. In the phosphirenylium ion, with 2-coordinate phosphorus, the system resembles the cyclopropylium ion with two π-electrons and could show aromatic character. It is not clear, according to a recent discussion[9d] of the available physical data, if these characteristics are present in phosphirene derivatives, and continued research into this matter is required.

The 2H-phosphirene ring system, containing the carbon-phosphorus double bond, was first prepared in a spirocyclic structure (**15**), which provided stability against rearrangement to the 1H-phosphirene.[37] The reaction (Eq. 8.13) involves a 1,3-dipolar cycloaddition of a phosphaalkyne (**13**) (chapter 10.B) with an azo compound; the cycloadduct (**14**) loses N_2 on photolysis to form the phosphirene in low yield (11%). The structure was established by X-ray crystallography.

$$\text{(8.13)}$$

The reaction of a chlorocarbene precursor (**16**) with phosphaalkynes gives as a final product a 1-chlorophosphirene, and it appears that the initial product may be the 2-phosphirene (**17**) that undergoes a 1,3-shift of chlorine (Eq. 8.14).[38] Theoretical calculations have shown, however, that unsubstituted 2H-phosphirene is more stable than its isomer 1H-phosphirene by 12.3 kcal/mol.[39]

$$\text{(8.14)}$$

8.B. THE FUNDAMENTAL RING SYSTEMS

Another part of the rapidly developing and expanding three-membered phosphorus heterocycle field is based on systems with the P—P bond. Many diphosphiranes have been prepared since their initial (1977) observation by Baudler and Carlsohn,[40] who employed the chemistry of Eq. 8.15.

$$\text{PhP}-\bar{\text{P}}\text{Ph} \; 2\text{K}^+ \;+\; \text{Cl}_2\text{CMe}_2 \xrightarrow{-30°} \text{diphosphirane} \tag{8.15}$$

Several other methods, some using precursors with P=P or P=C bonds, were later developed.[11a] 1H-Diphosphirenes became known in 1989,[41] from the addition of t-butylphosphaacetylene to the carbene-like phosphorus atom of iminophosphines (Eq. 8.16) and remain under development (e.g., reference 42).

$$\text{R-P=N-R'} \;+\; t\text{-BuC}\equiv\text{P} \xrightarrow{-30°} \text{product} \tag{8.16}$$

The field of three-membered phosphorus heterocycles has expanded so much in the last two decades that it is clearly one of the fastest moving of the various aspects of heterocyclic phosphorus chemistry. It has not been possible to include in this discussion many structures in which the three-membered ring is embodied in bicyclic or multicyclic form, either as a product or a reactant, but this fascinating and frequently complex chemistry can be appreciated from the Mathey review.[11a] A number of novel heterocyclic structures have emerged from this work. Another valuable aspect of phosphirane chemistry is the use of ring fragmentation methods to generate low-coordination phosphorus species. The first example is that of Quast and Heuschmann[23] in which the species R—P=O was generated and trapped with an ortho-quinone (Eq. 8.17), but other examples now exist[24] and will receive attention in chapter 10.C.

$$\xrightarrow{> 60°} \; + \; t\text{-Bu}-\text{P}=\text{O} \longrightarrow \tag{8.17}$$

8.B.2. Four-Membered rings

No member of the family of four-membered rings was known until 1962, when a method for phosphetane synthesis based on the reaction of olefins with PCl_3 or phosphonous dichlorides in the presence of $AlCl_3$ was announced.[43] Because of the cheapness and ready availability of the starting materials, this process immediately

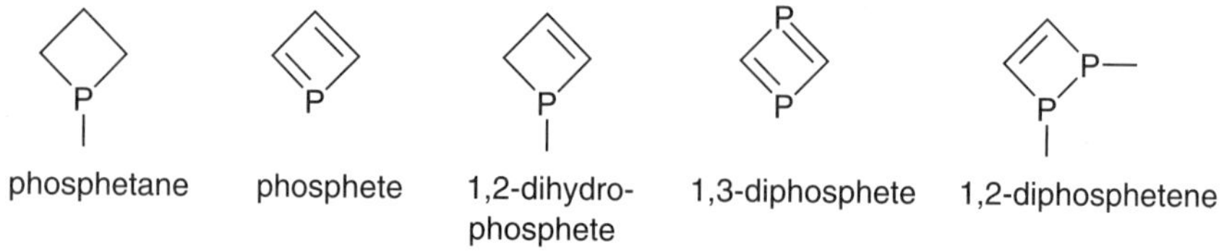

242 HETEROCYCLIC PHOSPHORUS COMPOUNDS

phosphetane   phosphete   1,2-dihydro-   1,3-diphosphete   1,2-diphosphetene
                            phosphete

Figure 8.2. Four-membered rings.

attracted attention and encouraged a great deal of work on the basic chemistry of the phosphetane ring. Until recently, this method remained the source of most of the phosphetane derivatives used in basic studies. The research in this area has been reviewed in detail,[44] and will be described here only in abbreviated form. First we notice that the combination PCl_3 or $RPCl_2$ with $AlCl_3$ is the same as that used to add to alkynes to form the three-membered ring (Eq. 8.8), and one might think that addition to an alkene to form a three-membered ring would be possible. Simple alkenes do not react to give cyclic products, however, and only with alkenes bearing multiple substituents on one or both of the carbons attached to the double bond does phosphetane formation take place. Typical examples are shown in Eqs. 8.18 and 8.19.

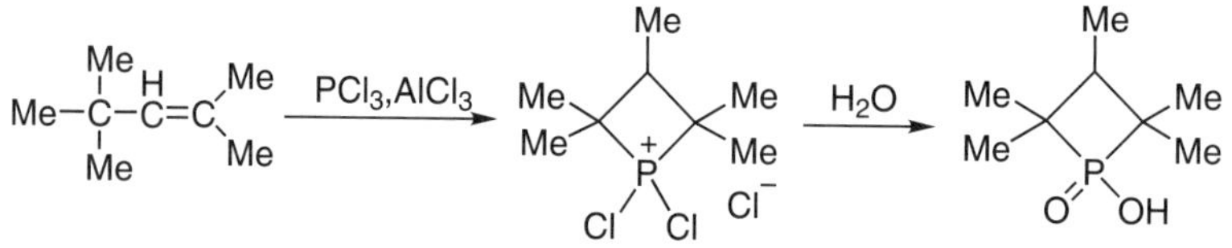

$$ (8.18) $$

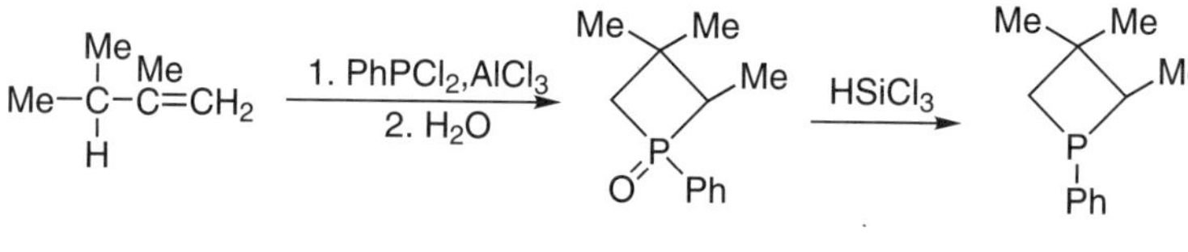

$$ (8.19) $$

It is obvious that a methyl group has migrated in Eq. 8.18 (and hydrogen in Eq. 19), and this gives a clue that a likely mechanism involves carbocation-like intermediates from initial attack of $RPCl_2/AlCl_3$ on the double bond, followed by a 1,2-shift (not always giving the most stable carbocation) and then ring closure (Eq. 8.20).

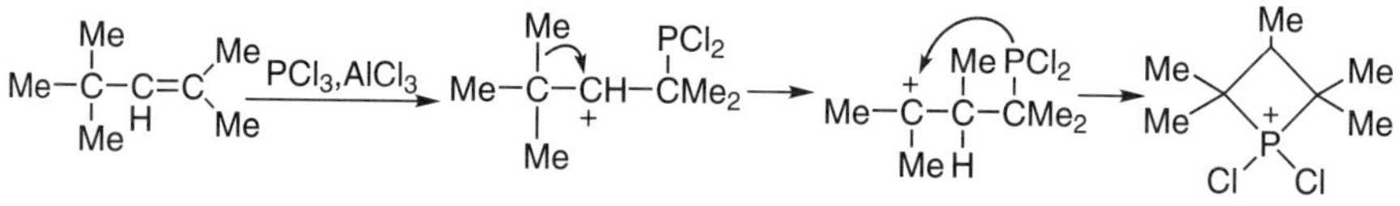

$$ (8.20) $$

As a consequence of the need for the 1,2-shift, all phosphetanes made by this method are multiply substituted by methyl groups. This, in fact, may have been an aid in these early studies in that the substituents add to the stabilization of the ring. Years later, as will be seen, phosphetanes with only a P-substituent became available and were relatively unstable. A considerable number of phosphetanes have been prepared from branched alkenes (including a bromoalkene, giving a bromophosphetane oxide[45]) and obtained with phosphorus in all of the common 3-coordinate and 4-coordinate functionalities and certain 5-coordinate functions as well. The process remains of value, as exemplified by its recent use in the synthesis of P-chiral phosphetanes in which several phosphonous dichlorides with optically active substituents derived from terpene derivatives were employed. Examples are the use of *endo*-bornylphosphonous dichloride[46] (Eq. 8.21) or L-menthylphosphonous dichloride.[47] Phosphines can be obtained from phosphetane oxides by the conventional reduction techniques; in these cases, optically active phosphetanes were obtained that were used as ligands in the preparation of homogeneous catalyst systems.

$$\text{Me-}\underset{\underset{H}{|}}{\overset{\overset{MeMe}{|}}{C}}\text{-C=CH}_2 + \text{[bornyl-PCl}_2\text{]} \xrightarrow{\text{then H}_2\text{O}} \text{[product]} \quad (8.21)$$

exo, endo
40 : 60

With the phosphetanes in hand from the reduction of the various oxides, all of the usual transformation methods of phosphines can be applied with success, quite unlike the case of phosphiranes. Similarly, the phosphinic acids resulting from hydrolysis of the alkene-PCl_3 adducts give the usual transformations possible for this functional group; no elaboration is needed. Ring-opening reactions are, however, known for phosphetane derivatives, being especially likely on exposing phosphine oxides and phosphonium salts to strong base. Also the insertion of oxygen into a C−P bond by peroxy acids,[48] discussed in chapter 4.A, occurs in the phosphetane oxides, because ring strain is high due to the contracted bond angles. Phosphetane derivatives were used extensively to probe stereochemical aspects of the mechanism of nucleophilic substitution at phosphorus. A major observation was the occurrence of nucleophilic substitution on phosphinic acid derivatives of the series with retention of configuration, rather than the normal inversion with larger rings or acyclic compounds. Retention implies that the 5-coordinate intermediate undergoes a pseudorotation before the departure of the leaving group. This is associated with the fact that the four-membered ring has a very strong preference for apical and equatorial positions, and thus the group to leave is initially in the equatorial position. Pseudorotation places it in the desired apical position, resulting in retention of configuration.

The polymethyl phosphetanes with different P-functions were used in fundamental X-ray analysis studies that established the shape and parameters of the ring. As for other four-membered rings, the ring is folded with a "puckering" angle[44] of

17 to 30°. The internal bond C—P—C angle is 80 to 83°. However, a more recent X-ray analysis of phosphetanium salts with no C-substituents (**18**, vide infra) revealed that the ring was planar, suggesting that the puckering in the substituted rings may arise from partial relief of nonbonded interactions.[49]

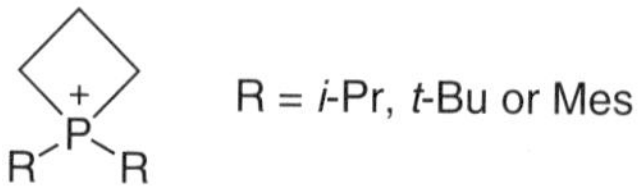

R = i-Pr, t-Bu or Mes

18

A more obvious method for preparing phosphetanes would be the reaction of 1,3-dihalides with a metallic phosphide, the approach used for phosphirane synthesis from 1,2-dihalides, and there are scattered reports in the literature on the use of this reaction for special compounds. Some reported cases are summarized in Eqs. 8.22,[50] 8.23,[51] 8.24[52] and 8.25.[16]

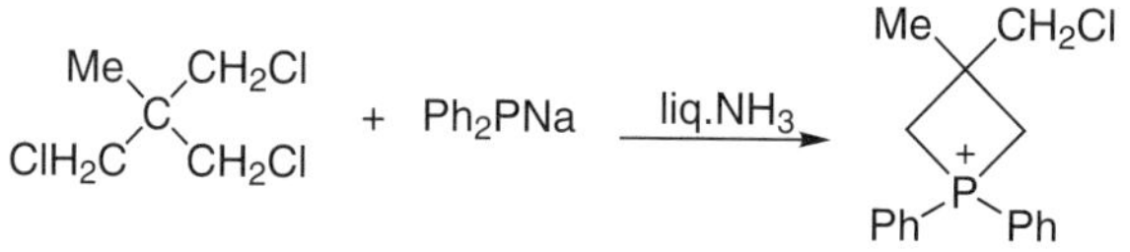

(8.22)

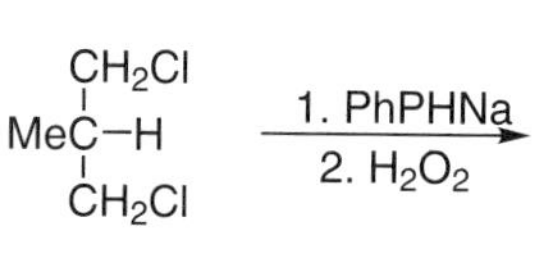

(8.23)

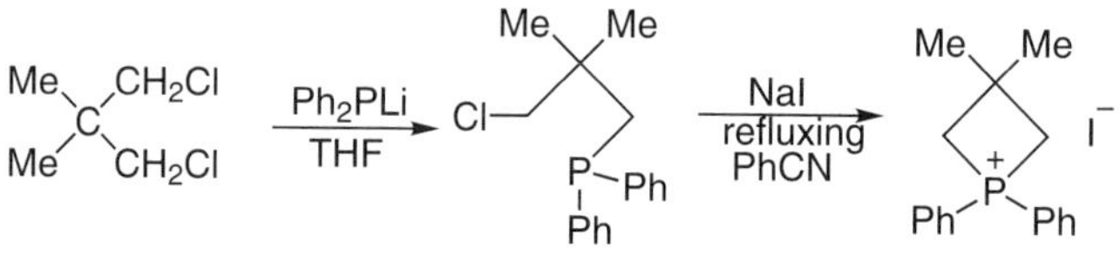

(8.24)

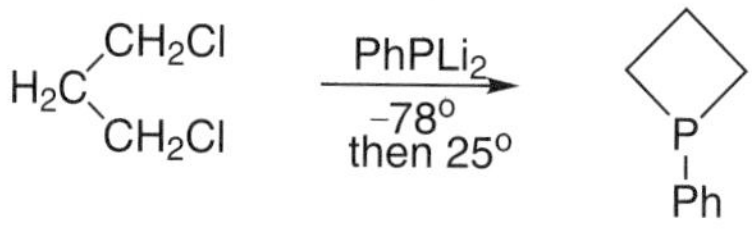

(8.25)

The last of these is important as the first reaction that has ever given a 3- coordinate phosphetane lacking substituents on the ring carbons. Thus 1-phenylphosphetane was obtained in 13% yield as a distillable oil (b.p 63° at 0.05 mm Hg), with a remarkably downfield ^{31}P NMR shift of δ +13.9, already noted in chapter 6 with a theoretical explanation. Unless stabilized by dilution with benzene, however, the compound rapidly polymerizes to form poly(phenylphosphino)propane, giving a ^{31}P

NMR singlet at δ −27.2. The phosphine formed complexes with metals, e.g., of form fac-[Mo(CO)$_3$(R$_3$P)], but no other phosphine reactions were reported. 1-Phenylphosphetane was also constructed within a metal complex in much higher yield (85%) by reacting a complex (19) of phenylphosphine with 1,3-dibromopropane[53] (Eq. 8.26), and thus the concept of stabilizing phosphorus heterocycles by metal coordination, so effective with phosphiranes, is seen to be useful with phosphetanes as well.

$$\text{(19)} \xrightarrow{(Br_2CH_2)_2CH_2} \text{product} \tag{8.26}$$

The dihalide reaction has also been applied to silylphosphines; the silyl group is displaced, and the 3-halopropyl group reacts at the phosphine center to close the ring[49] (Eq. 8.27). Secondary phosphines were also reported to give phosphetanium salts with 1,3-dihalides.

$$R_2PSiMe_3 + BrCH_2CH_2CH_2Br \longrightarrow R_2P(CH_2CH_2CH_2Br) \longrightarrow [\text{phosphetanium}]^+ Br^- \tag{8.27}$$

R = i-Pr, t-Bu, Mes

A quite different approach is outlined in Eq. 8.28, which gave the first phosphetanes with functional groups.[54] No further use of this method seems to have been made.

$$\tag{8.28}$$

In the 1990s, there was increased interest in phosphetanes, and several new synthetic methods were announced. In the first of these,[55] 2-coordinate phosphenium ions (20) insert into substituted cyclopropanes to give adducts that can be hydrolyzed to phosphinamides (21, Eq. 8.29). The combination PhPCl$_2$/AlCl$_3$, which is known to accomplish the equivalence of the transfer of PhPCl$^+$, also led to the formation of the phosphetane ring (Eq. 8.29). Both reactions proceed in high yield and seem to have some generality.

$$i\text{-Pr}_2\text{N-PCl}_2 \xrightarrow{\text{AlCl}_3} (i\text{-Pr}_2\text{N})\text{ClP}^+ \xrightarrow{\triangle\text{Me}_2} \underset{\mathbf{20}}{\overset{\text{Me}}{\underset{\text{Cl}}{\square}}\overset{\text{Me}}{\underset{\text{N}(i\text{-Pr})_2}{}}} \xrightarrow{\text{H}_2\text{O}} \underset{\mathbf{21}}{\overset{\text{Me}}{\underset{\text{O}}{\square}}\overset{\text{Me}}{\underset{\text{N}(i\text{-Pr})_2}{}}} \quad (8.29)$$

$$\triangle\text{Me}_2 \xrightarrow[\text{AlCl}_3]{\text{PhPCl}_2} \underset{\text{Ph}}{\overset{+}{\square}}\overset{\text{Me}_2}{\underset{\text{Cl}}{}} \xrightarrow{\text{H}_2\text{O}} \underset{\text{Ph}}{\overset{}{\square}}\overset{\text{Me}_2}{\underset{\text{O}}{}} \quad (8.30)$$

Metal complex chemistry is involved in the new method presented by Marinetti and Mathey.[56] Certain complexes (**22**) of phosphaalkenes were found to add to alkenes and alkynes, the former process giving phosphetane derivatives (Eq. 8.31). The free phosphetane could presumably be released from the complex using the halogen-base treatment effective in phosphirane chemistry, which the authors demonstrated for a dihydrophosphete (vide infra).

$$\underset{\mathbf{22}}{\overset{\text{PhP=CMeCOOEt}}{\underset{\text{W(CO)}_5}{\downarrow}}} + \underset{\text{Et}}{\overset{\text{H}}{\text{C=C}}}\overset{\text{N}\langle\text{pip}\rangle}{\underset{\text{H}}{}} \longrightarrow \underset{\text{Ph}}{\overset{\text{Et}}{\square}}\overset{\text{N}\langle\text{pip}\rangle}{\underset{\text{P}\rightarrow\text{W(CO)}_5}{}}\text{Me,COOEt} \quad (8.31)$$

3-*tert*-Butyl-1-phenylphosphetane was prepared in another new approach[57] by the reaction of a titanacyclobutane (**23**) and PhPCl$_2$ (Eq. 8.32). This method seems extendable to other P-substituted phosphetanes. The product was obtained as a mixture of cis,trans isomers, characterized by their ^{31}P NMR spectra (δ −38.3 and δ −6.8, respectively); it had been established years ago[58] that in phosphetanes cis resonates at higher field than does trans.

$$\underset{\mathbf{23}}{\overset{t\text{-Bu}}{\underset{\text{Cl}}{\square}\overset{}{\underset{\text{Ti}}{}}\text{Cp}}} \xrightarrow{\text{PhPCl}_2} \overset{t\text{-Bu}}{\underset{\underset{\text{Ph}}{|}}{\square}\overset{}{\underset{\text{P}}{}}} \quad (8.32)$$

Finally, a synthesis of the parent acid in the phosphetane series was recently reported.[59] This was accomplished by reacting a di-Grignard reagent with ethyl phosphorodichloridate and then acid hydrolysis of the ester (Eq. 8.33). The yield in the first step was only 7.5%, however.

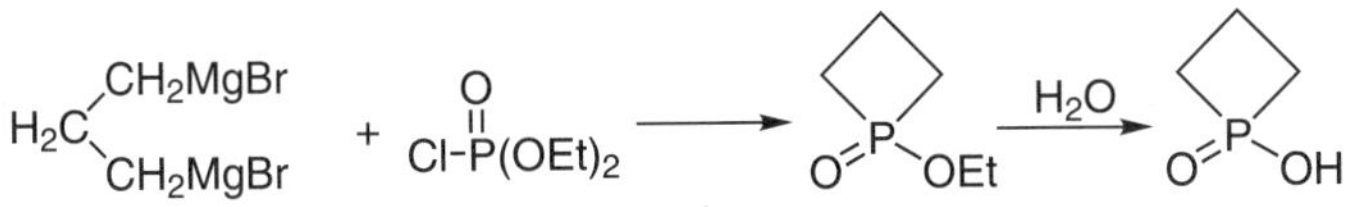

$$\underset{H_2C}{\overset{CH_2MgBr}{\diagdown}}\underset{CH_2MgBr}{\diagup} + \underset{Cl-\overset{O}{\overset{\|}{P}}(OEt)_2}{} \longrightarrow \underset{O}{\diagup}\!\!\!\overset{\square}{\underset{P}{\diagdown}}\!\!\!_{OEt} \xrightarrow{H_2O} \underset{O}{\diagup}\!\!\!\overset{\square}{\underset{P}{\diagdown}}\!\!\!_{OH} \quad (8.33)$$

Unsaturated derivatives of the four-membered ring were completely unknown until convincing evidence was presented by Marinetti, et al., in 1985[60] for the existence of the keto derivative **25** of the 1,2-dihydrophosphete system. The synthesis was accomplished by insertion of CO into a P–C bond of a metal-coordinated phosphirene derivative (e.g., **24**, Eq. 8.34). Decomplexation was performed in the usual way to release the heterocycle in the form of its P-oxide, a stable solid with m.p. 130° and ^{31}P NMR δ +63. The infrared stretching of the C=O group occurred at 1722 cm^{-1} and the ^{13}C NMR of the carbonyl carbon at δ 202.13 ($^{1}J_{PC}$ = 65.7 Hz).

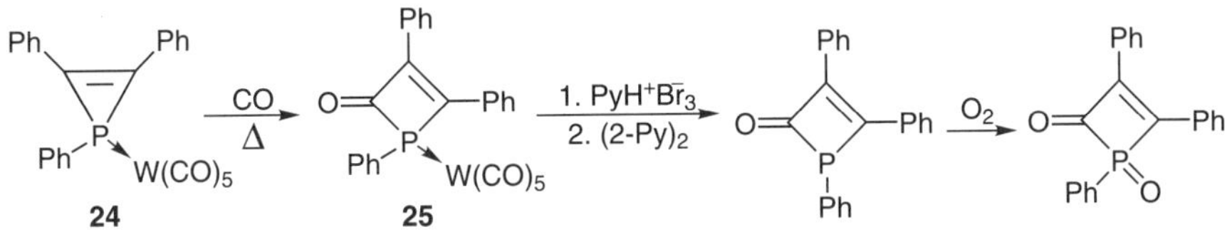

[Scheme showing compounds 24 and 25, and subsequent transformations with 1. PyH⁺Br₃⁻ 2. (2-Py)₂, then O₂]

24 → (CO, Δ) → **25** → (1. PyH⁺Br₃⁻; 2. (2-Py)₂) → phosphete → (O₂) → P-oxide

(8.34)

The 1,2-dihydrophosphete ring is available in the phosphine form from the reaction of 1-titanacyclobut-2-ene derivatives (**26**) with phosphonous dihalides.[61,62] The synthesis in 66% yield of 1,3,4-triphenyl-1,2-dihydrophosphete (Eq. 8.35) is an example.[63] The compound is a crystalline solid, ^{31}P NMR δ −16.04.

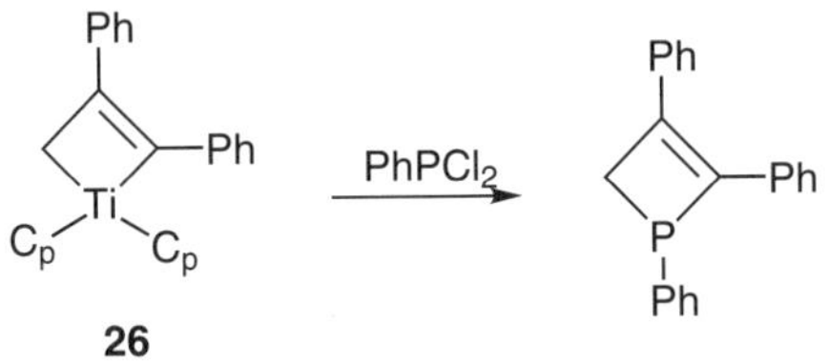

[Scheme: compound 26 (titanacycle with Cp₂Ti) + PhPCl₂ → 1,3,4-triphenyl-1,2-dihydrophosphete]

(8.35)

Phosphorus in these compounds is surprisingly unreactive to oxygen, although the P-oxide can be formed after prolonged reaction times or at high temperatures.[61] Metal complexation occurs more readily, however.[63] Nucleophilic behavior is seen in the ring-opening addition reaction (nonconcerted) with Michael acceptors,[64] an example of which is seen in Eq. 8.36.

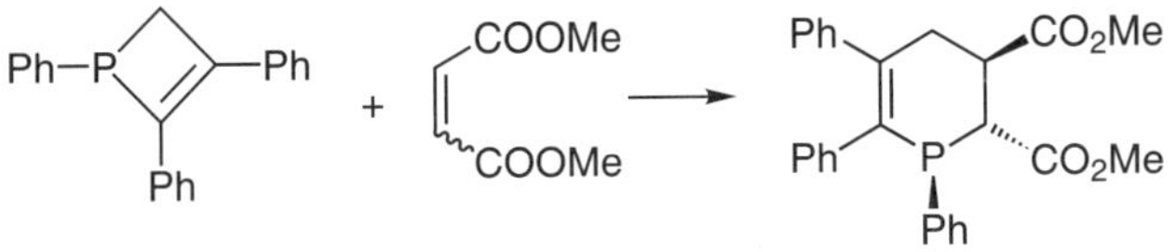

[Scheme: 1,2,3-triphenyl-1,2-dihydrophosphete + dimethyl maleate → six-membered ring product with Ph, CO₂Me, P-Ph substituents]

(8.36)

Several structural studies of the 1,2-dihydrophosphete ring have been made (e.g., reference 64). The ring is nearly planar, with a C—P—C internal angle of about 74°. The phosphorus atom retains its pyramidal character, and theoretical studies[65] on 1H-1,2-dihydrophosphete have indicated a large inversion barrier of 46.62 kcal/mol. Calculations also provide structural parameters and reveal that the ring should undergo conrotatory opening with near thermal neutrality.[65b] There has, however, been no experimental verification of this prediction.

1,2-Dihydrophosphetes have also been synthesized in the form of metallic complexes,[56] from which the phosphine ligand can be liberated and isolated as the P-oxide (Eq. 8.37, path a). A formal [4+2] concerted cycloaddition reaction takes place between the phosphete complexes and activated alkynes,[66] in which the elements of the alkyne are inserted into a P—C bond. The phosphete complex may be in equilibrium with the ring-opened form (a phosphadiene). The products are 1,4-dihydrophosphinine derivatives (Eq. 8.37, path b). Other related reactions are known.

(8.37)

Other forms of complexed 1,2-dihydrophosphetes have been reported (see, for example references 64b, 67 and 68) and their chemistry has been studied.

The fully unsaturated phosphete ring system has been detected only as a transient species or as a ligand in transition metal complexes. As noted, this ring has a four π-electron system and could be antiaromatic and of low stability. Its chemistry, as well as that of the phosphetanes and phosphetenes, has recently been reviewed.[69]

Little is yet known about the 1,3-diphosphete system; it is antiaromatic and has so far been created only within a metal complex[70] (**27**) and in the form **28**, in which phosphorus is in the ylidic σ^4, λ^4 condition[71] without true double bonds. The isomeric 1,2-diphosphete system is also antiaromatic and unstable.

The 1,2-diphosphetene ring was first prepared as early as 1964[72] by heating a mixture of $(CF_3P)_{4,5}$ with hexafluorobut-2-yne (Eq. 8.38), a type of reaction also used in later studies[73,74]. These compounds are relatively stable.

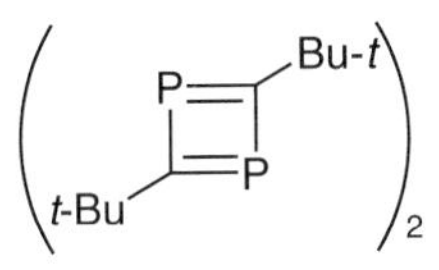

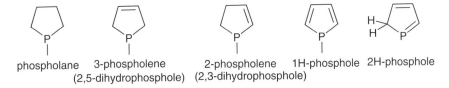

27 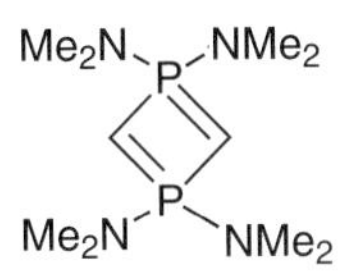 **28**

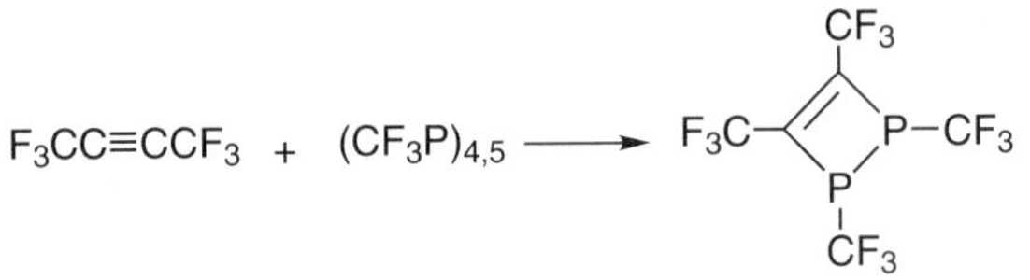

$$F_3CC\equiv CCF_3 + (CF_3P)_{4,5} \longrightarrow \text{[structure]} \quad (8.38)$$

The properties of the diphosphetenes and diphosphetes have been discussed in a CHEC II chapter,[75] and a review of ylidic four-membered phosphorus heterocycles as antiaromatic substances is also available.[76]

8.B.3. Five-Membered Rings

phospholane   3-phospholene   2-phospholene   1H-phosphole   2H-phosphole
             (2,5-dihydrophosphole) (2,3-dihydrophosphole)

Figure 8.3. Five-membered rings.

Knowledge of the five-membered ring, in its various levels of unsaturation, far exceeds that of any of the other fundamental ring systems, and the development of its chemistry has been an important part of phosphorus research for many years. Furthermore, with this ring size we find reasonable stability for systems in which heteroatoms, especially oxygen or nitrogen, replace carbon in the ring.

The literature on the five-membered ring is extensive, and only a brief outline of the subject can be presented here. Numerous reviews, however, are available that provide much detail on syntheses and properties of the several types of derivatives. A major review, covering all literature including that on oxa- and aza-substituted rings as well as the rings based only on carbon, is that published in 1970 by F. G. Mann.[4] The five-membered C—P rings were reviewed in 1981;[1] the phosphole system has been the subject of other recent reviews,[9] as already noted.

Tertiary phosphines incorporated in the saturated five-membered ring are quite stable and exhibit all of the usual properties of this function. This is also true of the two isomeric systems that contain one double bond (the phospholenes). The various forms of 4-coordinate structures are all easily synthesized and these too possess the typical stability and reactivity of noncyclic counterparts. It is only with the phosphole system that special comment, provided later in this section, is required.

250 HETEROCYCLIC PHOSPHORUS COMPOUNDS

Numerous techniques have been devised for the construction of the saturated phospholane ring. Generally, these are based on two different approaches: (1) the reaction of a 1,4-difunctional carbon compound with a phosphorus species, e.g. 1,4-dihalides with phosphine derivatives, or 1,4-di-metallic reagents with P halides, and (2) intramolecular cyclization reactions, several of which are known and discussed in reference 1 and exemplified by Eqs. 8.39 to 8.41.

$$Br\text{-}CH_2CH_2CH_2CH_2\text{-}Br \xrightarrow{PhP(OEt)_2 \; (Arbusov)} Br\text{-}CH_2CH_2CH_2CH_2\text{-}P(=O)(OEt)(Ph) \xrightarrow{NaAlH_2(OR)_2}$$

$$Br\text{-}CH_2CH_2CH_2CH_2\text{-}P(=O)(H)(Ph) \xrightarrow{NaAlH_2(OR)_2} \text{phospholane P(=O)Ph} \qquad (8.39)$$

$$(PhCH_2)_2\overset{O}{\overset{\|}{P}}H + PhHC=CHCOOEt \xrightarrow{\overline{O}Et} (PhCH_2)_2\overset{O}{\overset{\|}{P}}\text{-}CH\overline{C}HCOOEt \longrightarrow$$
$$ Ph$$

PhHC̄-P(O)-CH-CH₂ → Ph, Ph, PhH₂C, O substituted cyclopentanone phospholane oxide   (8.40)
PhH₂C   Ph

$$H_2C=CH\text{-}CH_2\text{-}\overset{O}{\overset{\|}{C}}\text{-}Cl + PhPH_2 \xrightarrow{K_2CO_3} H\text{-}P(Ph)\text{-}C(=O)\text{-}CH_2CH=CH_2 \xrightarrow{AIBN} \text{phospholanone} \qquad (8.41)$$

Eqs. 8.40 and 8.41 are shown to make the point that C-functional groups can be built into the phospholane system. In addition to these carbonyl derivatives (the 3-keto derivative being unique in having unusually high enolic character), many other functions are found among the numerous known phospholanes, including OH, OR, NO₂, halogen, COOR, and COR. In many of these, phosphorus can be in any of its common functionality forms, especially as phosphines and phosphine oxides, which are easily interchanged by oxidation and reduction. In both of these functions, and others as well, phosphorus can act as a chiral center, and many examples of cis,trans isomerism are known. The isomers of phosphines, of course, depend on the high inversion barrier of phosphorus; isomers such as **29** and **30** (Eq. 8.42), for example, are stable distillable liquids with different spectral properties.

$$\text{3-keto-1-methylphospholane} \xrightarrow{LiAlH_4} \text{3-OH, 1-Me phospholane (cis)} + \text{3-OH, 1-Me phospholane (trans)} \qquad (8.42)$$

29, cis   **30**, trans

The highly versatile McCormack reaction, the cycloaddition of dienes with 3-coordinate phosphorus halides introduced in 1953 (see chapter 3.A), has provided a very large number of phospholene derivatives and has been instrumental in the rapid development of five-membered ring chemistry. Indeed, the phospholenes made this way are simple precursors, on hydrogenation, of phospholanes, and this process should be included as one of the major routes to these compounds. The McCormack reaction first leads to a cycloadduct that can be classed as a halophosphonium halide and has all the expected properties of such a species, which were seen in chapter 4.D. In particular, the adducts are readily hydrolyzed to give phosphoryl derivatives, and it is to prepare this type of product for which the McCormack reaction is most commonly used. The process is illustrated by Eq. 8.43 with the use of methylphosphonous dichloride and butadiene, a combination that was noted in chapter 4.A to be used commercially for the production of 1-methyl-3-phospholene 1-oxide.

$$\text{MePCl}_2 \longrightarrow \left[\begin{array}{c} \diagup\!\!\!\diagdown \\ \text{Me}^{-\overset{|}{\text{P}}-\text{Cl}} \\ \text{Cl} \end{array} \right] \longrightarrow \underset{\text{Me}\;\;\;\text{Cl}}{\overset{+}{\text{P}}}\;\text{Cl}^- \xrightarrow{\text{H}_2\text{O}} \underset{\text{Me}\;\;\;\text{O}}{\text{P}{=}} \quad (8.43)$$

When phosphorus tribromide is used in this reaction, an adduct is formed that on hydrolysis gives the cyclic phosphinic acid **31** (Eq. 8.44).

$$\diagup\!\!\!\diagdown + \text{PBr}_3 \longrightarrow \underset{\text{Br}\;\;\;\text{Br}}{\overset{+}{\text{P}}}\;\text{Br}^- \xrightarrow{\text{H}_2\text{O}} \underset{\text{O}{=}\;\;\;\text{OH}}{\text{P}} \quad (8.44)$$
$$\mathbf{31}$$

The initial adduct in some cases undergoes a rearrangement of the double bond from the 3-position to the 2-position, thus into conjugation with the phosphorus center; this rearrangement can also be effected intentionally if desired, either at the adduct stage or in the phosphoryl hydrolysis products (thermally or with base). There are some generalities that are useful in predicting the major isomer structure when the hydrolysis is conducted at or below room temperature, and in the presence of a base to avoid high acidity; thus, from butadiene or mono-substituted butadienes, 3-phospholenes are obtained with alkylphosphonous dichlorides or dibromides and phosphorus tribromide, whereas 2-phospholenes are obtained with arylphosphonous dichlorides and phosphorus trichloride. 2,3-Disubstituted butadienes give only 3-phospholenes. The occurrence of the rearrangement was not recognized in the initial studies of the cycloaddition reaction, and this led to incorrect reports (e.g., in *Organic Syntheses*,[77a] later presented correctly[77b]) of the 3-phospholene structure for the products from phenylphosphonous dichloride (Eq. 8.45).

$$\underset{\text{Me}}{\diagup\!\!\!\diagdown} + \text{PhPCl}_2 \longrightarrow \underset{\text{Ph}\;\;\;\text{Cl}}{\overset{+}{\text{P}}}\;\text{Cl}^- \longrightarrow \underset{\text{Ph}\;\;\;\text{Cl}}{\overset{+}{\text{P}}}\;\text{Cl}^- \xrightarrow{\text{H}_2\text{O}} \underset{\text{Ph}\;\;\;\text{O}}{\text{P}{=}} \quad (8.45)$$

Several studies have been conducted on the conditions giving the best results in the cycloaddition.[78] The reactions generally proceed at room temperature but can be quite slow, in some cases taking weeks for a high yield. Polymerization of the diene is avoided with an inhibitor, frequently copper stearate. Faster reactions are achieved at higher temperatures or under pressure. The results of a multi-variant study[79] are particularly helpful in selecting the proper conditions for a cycloaddition.

Many dienes and phosphorus halides have been used successfully in the McCormack cycloaddition; the 1981 review[1] tabulated the use of 17 acyclic dienes and 19 phosphonous and phosphinous halides, and others have been added to the list since then. The diene system can also be incorporated in rings, and as noted, the phosphorus trihalides can be used as well, as can their mono- and di-alkoxy derivatives. But the great number of phospholene derivatives made directly by this process does not alone reveal its importance in heterocyclic phosphorus chemistry; many transformations based on reactions of the double bond, at CH groups α to phosphorus, and at phosphorus itself have been described, and covered by reviews,[1,9c] and it has been of great importance in the rapid development of the chemistry of the five-membered ring, especially of the fully unsaturated (phosphole) form. With proper manipulations, the McCormack cycloadducts and hydrolysis products are excellent starting materials for the construction of the phosphole ring. Although other routes do exist to this highly important system, a useful example being seen in Eq. 8.46, none approach the McCormack route in utility and versatility.

$$R-C\equiv C-C\equiv C-R + R'PH_2 \xrightarrow{\text{base}} \underset{\underset{R'}{|}}{\underset{P}{\overset{R\diagup\!\!\diagdown R}{\bigcirc}}} \quad (8.46)$$

The most direct path to a phosphole is shown in Eq. 8.47, in which the cycloadduct is simply subjected to dehydrohalogenation with a non-nucleophilic base (originally DBU, but methylpyridines are often superior[80]). In the case shown, the yield from the adduct is 85%.[80]

$$\text{Me}_2\text{C}=\text{CH-CH}=\text{CMe}_2 \xrightarrow{\text{PhPBr}_2} \underset{\underset{\text{Ph}}{\overset{+}{P}}\diagdown\text{Br}}{\overset{\text{Me}\diagup\!\!\diagdown\text{Me}}{\bigcirc}} \text{Br}^- \xrightarrow[(-2\text{HBr})]{\alpha\text{-picoline}} \underset{\underset{\text{Ph}}{|}}{\underset{P}{\overset{\text{Me}\diagup\!\!\diagdown\text{Me}}{\bigcirc}}} \quad (8.47)$$

Other routes make use of the cycloadduct hydrolysis products, the 3-phospholene oxides. It was from 1-methyl-3-phospholene 1-oxide that 1-methylphosphole was first synthesized (Eq. 8.48).[81]

The importance of the phosphole ring stems from its resemblance to the heteroaromatic substances pyrrole, furan, and thiophene, and years of research have gone into answering the question if phosphole should be included in this

$$\text{[Me-P=O cyclopentene]} \xrightarrow{Br_2} \text{[Br,Br-Me-P=O]} \xrightarrow{HSiCl_3} \text{[Br,Br-Me-P]} \xrightarrow{KOBu\text{-}t} \text{[Me-P pyrrole-like]} \quad (8.48)$$

aromatic series. Much has been written on this subject and the literature has been thoroughly reviewed.[9] The view has evolved, and is now supported by strong theoretical computations, that some electron delocalization of the type seen in the true heteroaromatics (as would be expressed by the resonance hybrid of Eq. 8.49) may be present, but not nearly to the extent seen in the other heterocycles and is less than that in furan.

$$\text{[resonance structures of phosphole]} \quad (8.49)$$

We cannot go into all of the arguments based on experimental data that have been presented both supporting and opposing delocalization, but a few salient points may be presented. The most important is that the P atom in simple phospholes retains its pyramidal character, as has been established in solution by ^{1}H NMR studies on appropriate derivatives,[82] and in the solid state by several X-ray crystallographic analyses (e.g., reference 83 on 1-benzylphosphole). The latter showed some modifications of the bond lengths in the direction of a delocalized system (shortening of P—C and CH—CH) but much less than seen in the true heteroaromatic systems. The barrier to inversion is reduced in phospholes, possibly due to resonance delocalization in the planar transition state and the resulting stabilization as the pyramid flattens and passes to the inverted form, but in the pyramidal form itself efficient orbital overlap is lacking. By comparison, the usually pyramidal nitrogen adopts planar sp^2 geometry in the pyrrole ring system as a result of the extensive electron delocalization. Nevertheless, phospholes are remarkably stable substances. Simple phospholes are usually distillable liquids and are easily preserved at room temperature. Higher derivatives can be crystallizable solids. Phospholes are far less easily oxidized, quaternized, and protonated than are normal phosphines; when formed, most of the oxides and some of the quaternary salts dimerize in the fashion of cyclopentadiene, a point to which we will return.

Phospholes have some quite characteristic spectral features. The ^{31}P NMR signals are found somewhat downfield compared to those of comparable saturated tertiary phosphines (e.g., δ −8.7 for 1-methylphosphole and δ +6.4 for 1-phenylphosphole). Pronounced shift effects are seen in their ^{1}H NMR spectra. Thus the 2,5-protons absorb at about δ 6.7 to 7.1 and have very large values for $^2J_{PH}$, about 34 to 42 Hz. The 3,4-protons are also strongly deshielded (about δ 7 to 7.3, with $^3J_{PH}$ about 12 to 17 Hz). Simple phospholes without aryl substituents show a pronounced maximum in their ultraviolet spectra, at about 280 to 290 nm. No electrophilic substitution at a ring carbon of a simple phosphole has ever been achieved (but see below for a special case), quite the opposite of the chemistry of the true heteroaromatic compounds, although substitution has been demonstrated by Mathey[9a] on the phospholide ion when coordinated to certain metals (Eq. 8.50).

$$\text{phospholide with Ph} \xrightarrow{Mn_2(CO)_{10}} \text{Mn(CO)}_3 \text{ complex} \xrightarrow[AlCl_3,\ 25°]{CH_3COCl} \text{acetylated Mn(CO)}_3 \text{ complex} \quad (8.50)$$

When the phosphorus atom bears very large substituents, some flattening of the pyramid can occur. Theory predicts that delocalization should then be more extensive, and indeed this has been demonstrated to be the case in phosphole **32**, which is unique in bearing the extremely space-demanding tri-*t*-butylphenyl group.[84] X-ray analysis proved that the pyramid was significantly flattened and that the highly indicative C_3-C_4 bond, as well as the P–C bond were considerably shorter than those of any other phosphole. This molecule has proved to be the first free phosphole that can undergo electrophilic substitution on a ring carbon (Eq. 8.51).

$$\text{P-Cl phosphole} + \text{LiAr} \longrightarrow \text{P-Ar} \xrightarrow{Br_2} \text{dibromide} \xrightarrow{\text{2-picoline}} \text{P-Ar phosphole } \mathbf{32} \xrightarrow[AlCl_3]{CH_3COCl} \text{acylated product} \quad (8.51)$$

$$\text{Ar} = \text{2,4,6-tri-}t\text{-butylphenyl}$$

Phospholes exhibit some unique chemical properties that are useful in forming specialized structures. Thus phospholes are not very useful dienes in the Diels–Alder reaction, but they become extremely active if they are first converted to σ-type complexes with certain metal carbonyls. This allows the construction of the complexed 7-phosphanorbornadiene system (as yet unknown in the free form), as in Eq. 8.52; the complex is of great value as an agent for the transfer of the moiety RP–M(CO)$_5$ to other groups, e.g., to alkenes to give phosphiranes (Eq. 8.3) and to alkynes to give phosphirenes. Other reactions of coordinated phospholes are reviewed elsewhere.[9]

$$\text{Ph-P-M(CO)}_5 \text{ phosphole} + \text{MeOOCC≡CCOOMe} \longrightarrow \text{7-phosphanorbornadiene complex} \quad (8.52)$$

Another unique property of phospholes is the occurrence at high temperature of a rearrangement of a P-substituent to an adjacent carbon.[85] This has been attributed to the operation of a [1,5]-sigmatropic shift, which results in the creation of the 2H-phosphole system. The 2H-phosphole is, however, but of transient existence under these conditions; it is not isolatable, but it can be trapped with various reagents included in the reaction mixture. The most important is an alkyne, which leads to the novel 1-phosphanorbornadiene system (**33**), itself a heterocycle of considerable importance. The entire sequence is shown in Eq. 8.53.

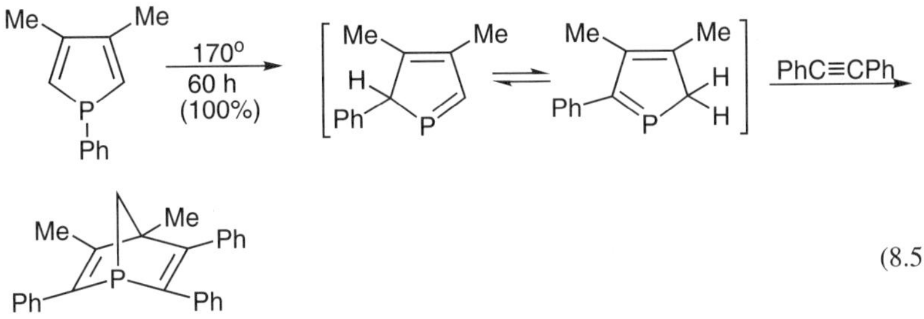

(8.53)

The [1,5]-sigmatropic shift occurs much more readily when hydrogen is the migrating group, and this has given great instability to phospholes with hydrogen on phosphorus, unless the ring is heavily substituted or the molecule is stabilized in a coordination complex. The sequence of Eq. 8.54 illustrates the high reactivity of 3,4-dimethyl-1H- phosphole.[86]

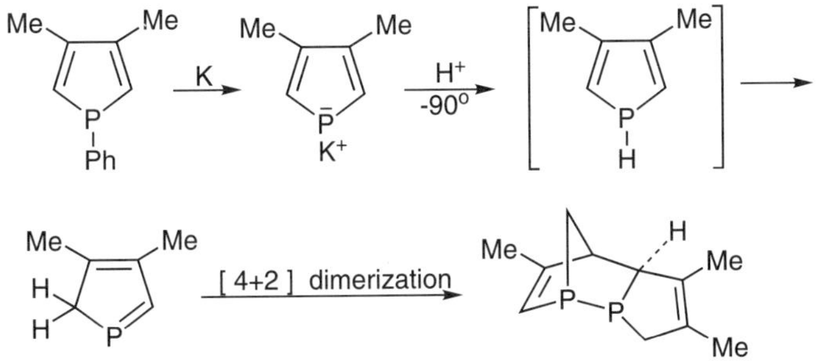

(8.54)

The 2H-phosphole moiety can be preserved in special structures. This can be accomplished by synthesizing a highly substituted derivative (**34**[87]) or a metal complex (**35**[88]). The P=C unit has also been stabilized in a bicyclic molecule (**36**[89]).

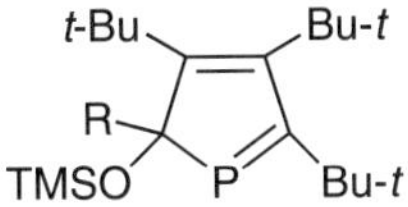

34
^{31}P δ + 159

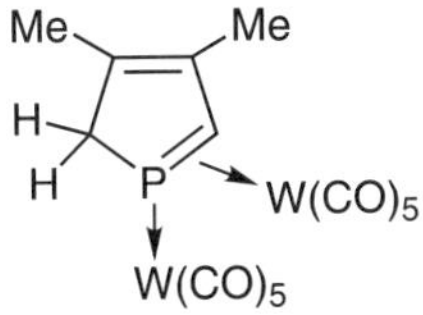

35

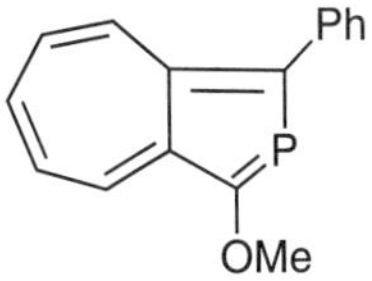

36
^{31}P δ + 177.9

The carbon-phosphorus double bond can be also found in molecules in which heteroatoms take the place of ring carbon. The first structures of this type[90] were

1,2,3-triazaphospholes (**37**), which proved to be stable distillable liquids or crystallizable solids (Eq. 8.55). Stabilization of the C=P unit is the consequence of the presence of the pyrrole-like resonance shown.

(8.55)

This discovery of 1967 heralded a new and fascinating aspect of heterocyclic phosphorus chemistry; many new hetero-substituted phospholes have since been synthesized, and the concept continues to be exploited. These compounds can be of good stability, and the number of known structures is considerable, as evidenced in recent reviews,[91a–91c,92] and continues to grow. We show here only some of the structures for the known mono-heteraphospholes (**37–41**; the diphosphole ring is also known) to draw attention to this new field. From these, one can visualize other structural possibilities that could result from further heteroatom substitution.

To complete this brief outline of five-membered phosphorus heterocycles, mention is required of the many known derivatives that can come from substitution of saturated ring carbon in phospholanes and phospholenes by heteroatoms, oxygen being especially important. Much of the foundation work was in place by 1970 and is covered in great detail in the monograph by Mann.[4] The structural possibilities are too numerous to recite here, and attention is directed only to the best known heterocyclic system of this type, 1,3,2-dioxaphospholane. This ring system is known with phosphorus in various functional forms: phosphite (**42**), phosphate (**43**), phosphonite (**44**), phosphonate (**45**), and H-phosphonate (**46**).

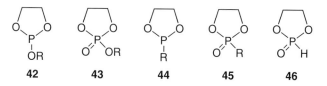

The 1,3,2-dioxaphospholane system has been widely used in mechanistic studies and has aided greatly in developing understanding of nucleophilic substitution phenomena. Derivatives of the phosphate (**43**) system are far more reactive to hydrolysis than are open-chain models; loss of the exocyclic alkoxy group is much faster than cleavage of a ring P—O bond. Consideration of such effects was highly instrumental in the development of the concept of 5-coordinate, trigonal pyramidal intermediates in nucleophilic substitution (see chapter 2.F). The rings of **42** to **45** are generally constructed from glycols and phosphorus dihalides as starting materials (e.g., Eq. 8.56); the ring of **46** comes from the use of phosphorus trichloride with glycols to form the chlorophosphite, which is then carefuly hydrolyzed.

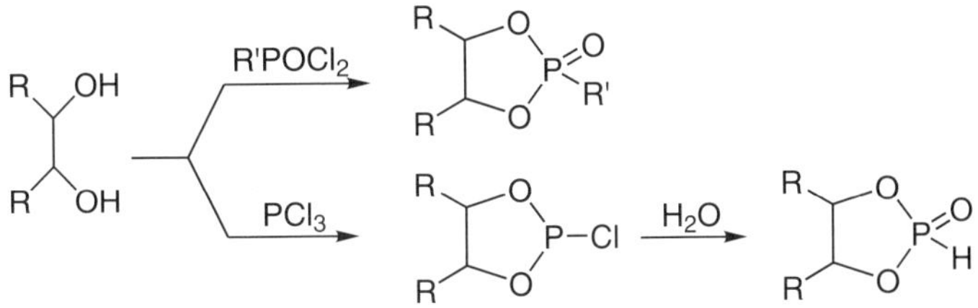

(8.56)

8.B.4. Six-Membered Rings

phosphorinane or phosphinane   1,2,5,6-tetrahydrophosphinine   1,2-dihydrophosphinine λ^3-phosphinine

Figure 8.4. Six-membered rings.

The saturated phosphorinane ring (phosphinane according to recent IUPAC changes,[93] but not in general use) can be synthesized in phosphine form from a variety of 1,5-pentane derivatives with appropriate phosphorus reactants. Probably the most widely used reaction type is that in Eq. 8.57; information on the other methods in this relatively inactive field is available in reviews by Mann,[4] Quin,[1] and Hewitt.[10a]

$$BrMg(CH_2)_5MgBr + RPCl_2 \longrightarrow$$ (8.57)

The chemistry of these cyclic phosphines is unexceptional, and they serve as the main source of 4-coordinate phosphorinane derivatives, which are rarely synthesized directly. Phosphorinanes with C-functions are also well known, a useful approach being intramolecular cyclizations of the Thorpe (Eq. 8.58) or Dieckmann (Eq. 8.59) types. The Thorpe reaction is especially useful, because the requisite cyanoethyl phosphine derivatives are available from the addition of phosphines to acrylonitrile. The 4-phosphorinanones exhibit normal carbonyl chemistry and have been used to

$CH_2=CHCN + PhPH_2 \xrightarrow{\text{conc. KOH}} PhP(CH_2CH_2CN)_2 \xrightarrow{\text{KOBu-}t}$

[structure with Ph, CN, =NH] $\xrightarrow[\Delta]{H_2O, HCl}$ [structure with Ph, COOH, =O] $\xrightarrow{-CO_2}$ [4-phosphorinanone with Ph] (8.58)

$CH_2=\overset{Me}{\underset{}{C}}COOMe \xrightarrow{PhPH_2} PhP(CH_2CHMeCOOMe)_2 \xrightarrow{NaOR}$ [phosphorinanone with Me, COOMe, Me, Ph] (8.59)

make a great variety of other 3-coordinate compounds, which, of course, can be easily converted to the usual 4-coordinate derivatives.

Another approach to C-functional phosphorinanes involves generation of a carbanion with strong base at a ring α position, followed by attack of an electrophile. Two examples are shown in Eqs. 8.60[94] and 8.61.[95] In addition to the use of CO_2 as the electrophile in Eq. 8.60, condensations have also been effected with the keto and ester groups, as well as with halogens.

[phosphorinane with dioxolane, P(=O)Ph] $\xrightarrow[-30°]{n\text{-BuLi}}$ [anion] $\xrightarrow{CO_2}$ [COOH product] (8.60)

[phosphorinene with Bu-t, P(=S)Ph] $\xrightarrow{t\text{-BuMgBr}}$ [anion] $\xrightarrow{Me_2CO}$ [product with OH, CMe$_2$] (8.61)

As noted, the configurational stability of the phosphorus atom leads to the existence of cis,trans isomers when a ring carbon bears an exocyclic substituent. The easily formed alcohols from reduction of the C=O group of 4-phosphorinanones or from addition of organometallics were the first phosphorinane compounds recognized to have cis,trans isomerism.[96] The isomers have different NMR spectral properties, as noted for structures **47** and **48** in Eq. 8.62. As expected, the conformation adopted by the phosphorinane ring is that of the chair; several X-ray analyses have confirmed that this is true of both 3- and 4-coordinate derivatives.[97] In 4-substituted 3-coordinate phosphorinanes, the substituent at carbon has

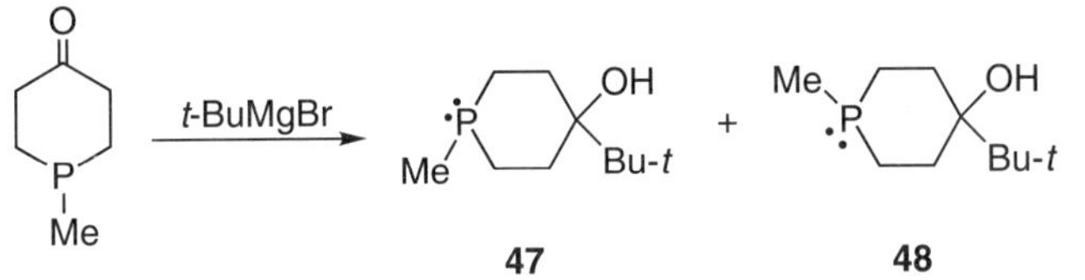

8.B. THE FUNDAMENTAL RING SYSTEMS

$$\text{(8.62)}$$

47
$^{31}P\ \delta\ -67.3$

48
$^{31}P\ \delta\ -57.7$

greater demand for the equatorial position than does the substituent at phosphorus. Even when the small OH group is at the 4-position, a P-substituent of the size of phenyl is found in the axial position in the cis isomer, as seen in structure **49**. In the trans isomer **50**, both substituents are in equatorial positions.

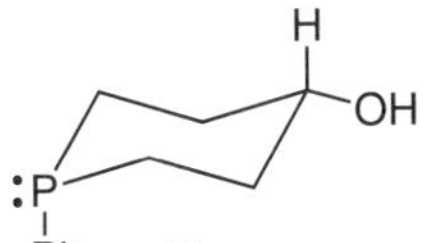

49 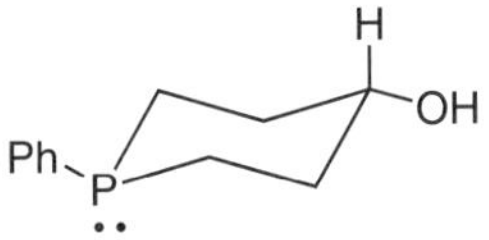   **50**

The longer C—P bond diminishes some of the strain from the 1,3-interactions in the P-axial conformer, as does some flattening of the phosphorus pyramid, which is quite noticeable in the X-ray analyses of some solid compounds.[97] In fact, even without a C-substituent, the P-methyl group prefers (2 : 1) the axial position **52** at room temperature in the conformational equilibrium of Eq. 8.63.[98] The ring inversion barrier (ΔG of activation) is a reasonable 8.7 kcal/mol. At $-140°$ the usual equatorial preference (2 : 1) is seen.[98] An entropy effect accounts for the small change in the positional preference. Both ^{31}P and ^1H NMR can be used to observe the frozen conformations **51** and **52**. The axial preference is even stronger in the P—H derivative, about 9 : 1 at room temperature.[99]

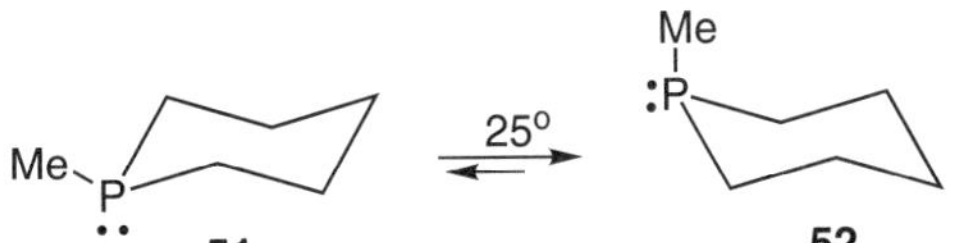

$$\text{(8.63)}$$

51   **52**

We can see in structures **47** and **48** that the axial P-substituent leads to a more upfield ^{31}P shift, and this is true of other alkyl derivatives as well. This shift is a result of the greater γ-interaction (with C-3) in the axial conformer; the effect is quite useful for making isomer assignments. The opposite relation, however, is known for P-aryl compounds (e.g., **53** and **54**), and caution is called for in using only the ^{31}P NMR shift for isomer assignment.

These unusual conformational properties need to be kept in mind in other phosphorinane derivatives, and in fused bicyclic compounds as well. The dominat-

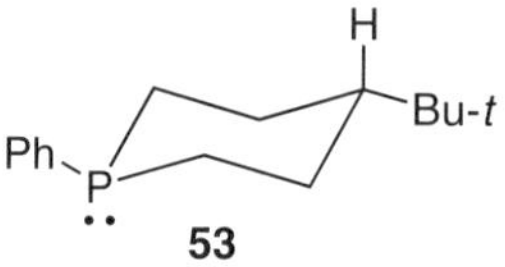

[Structures 53 and 54 shown, with Ph-P and :P-Ph substituents on cyclohexane rings bearing Bu-t and H]

53
^{31}P δ -38.6

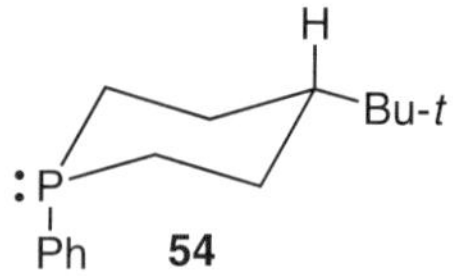

54
^{31}P δ -32.9

ing influence on conformer preference by the C-substituent also prevails in 4-coordinate derivatives, so that the configuration at the carbon is the same in both cis and trans isomers, as in **55** and **56**.[100]

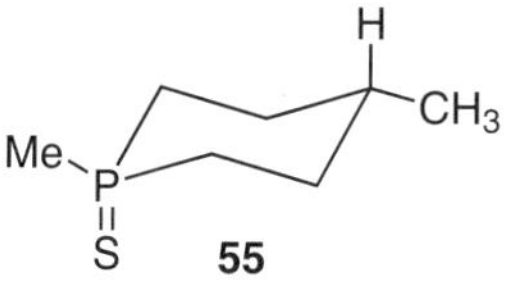

[Structures 55 and 56 shown]

55
56 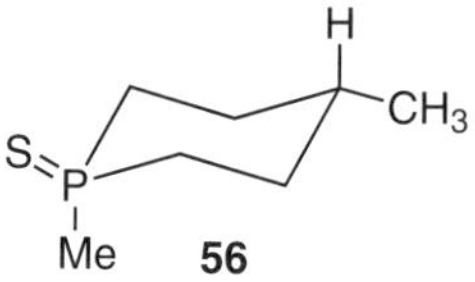

The fully unsaturated phosphinine (the present IUPAC name,[93] replacing the older phosphorin) system has attracted more attention than reduced forms of the six-membered ring, because of its structural resemblance to the pyridine ring system. The phosphinine ring was first constructed in 1966 by Märkl,[101] and indeed this represented a major step in the development of the chemistry of the p_π-p_π carbon-phosphorus double bond in later years (chapter 10.C). The first phosphinines were stabilized by substitution of aryl groups at the 2, 4, and 6 positions; the triphenyl derivative (**57**) is an air-stable, yellow crystalline compound with m.p. 172–174°, prepared as in Eq. 8.64 from a pyrylium fluoroborate.

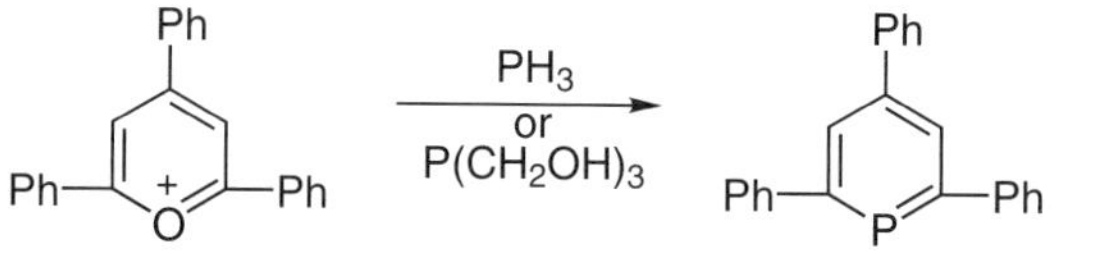

$$\text{pyrylium} \xrightarrow[\text{P(CH}_2\text{OH)}_3]{\text{PH}_3 \text{ or}} \textbf{57} \tag{8.64}$$

X-ray analysis, as well as other techniques, provided confirmation of cyclic electron delocalization in the system (Eq. 8.65); in 2,6-dimethyl-4-phenylphosphinine,[102] the ring is planar and the two P–C bond lengths are essentially equal (1.746 and 1.734 Å), as are those for the two C_2–C_3 bonds (1.396 and 1.410 Å) and the C_5–C_6 bonds (1.372 and 1.404 Å).

$$\text{[resonance structures of 2,4,6-triphenylphosphinine]} \tag{8.65}$$

8.B. THE FUNDAMENTAL RING SYSTEMS

The parent phosphinine was prepared several years later by Ashe,[103] using the reactions of Eq. 8.66.

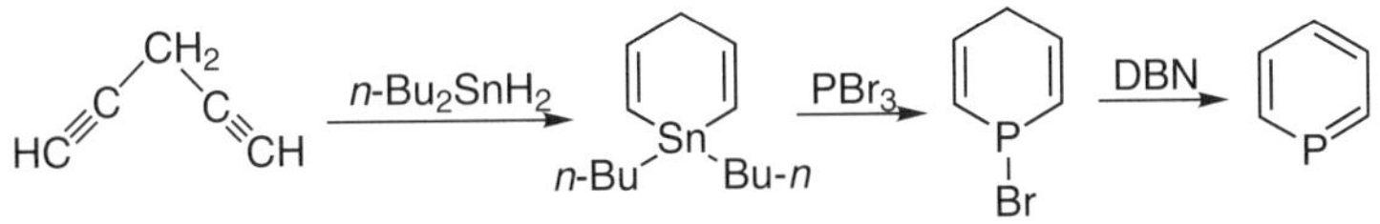

$$\text{HC}{\equiv}\text{C-CH}_2\text{-C}{\equiv}\text{CH} \xrightarrow{n\text{-Bu}_2\text{SnH}_2} \underset{n\text{-Bu}\ \ \text{Bu-}n}{\text{[stannacycle]}} \xrightarrow{\text{PBr}_3} \underset{\text{Br}}{\text{[P-Br cycle]}} \xrightarrow{\text{DBN}} \text{[phosphinine]} \quad (8.66)$$

This compound, a liquid, is much less stable than the substituted derivatives but nevertheless could be purified by gas chromatography and characterized spectroscopically, providing basic data on the ring system (reviewed in reference 104). The ^{31}P NMR signal was at δ +211; the three ^{13}C NMR signals (doublets) had δ 154.1 ($C_{2,6}$, J = 53 Hz), δ 133.6 ($C_{3,5}$, J = 14 Hz), and δ 128.8 (C_4, J = 22 Hz). The ^{1}H NMR signals at δ 7.38 (J_{PH} = 3.5 Hz) for C–H$_4$, δ 7.72 (J_{PH} = 8 Hz) for C–H$_{3,5}$, and δ 8.6 (J_{PH} = 38 Hz) for C–H$_{2,6}$ are in the downfield region expected for an aromatic system (cf. pyridine, δ 6.78, 7.10, and 8.56). The UV spectrum[10b] of phosphinine in cyclohexane consisted of π to π^* transitions at 213 nm (log ε = 4.28) and 261 nm (log ε = 3.93), and an n to π^* transition at 290 nm (sh.) with ε about 250.

Several other methods for the construction of the phosphinine ring have been described and are reviewed in references 8 and 10. Some of these methods are similar in involving thermal elimination of a small molecule from dihydrophosphinine to install the third double bond. Representative syntheses of this type are shown as Eqs. 8.67 to 8.69. One of the more recent approaches involves a cycloaddition of a phosphaalkyne with an α-pyrone[105] or a phosphole sulfide[106] (Eq. 8.70) to give a bridged six-membered ring that can be caused to lose the bridge thermally.

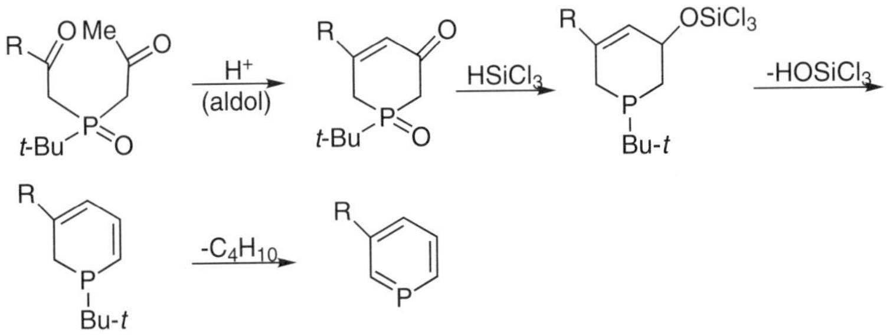

(8.67)

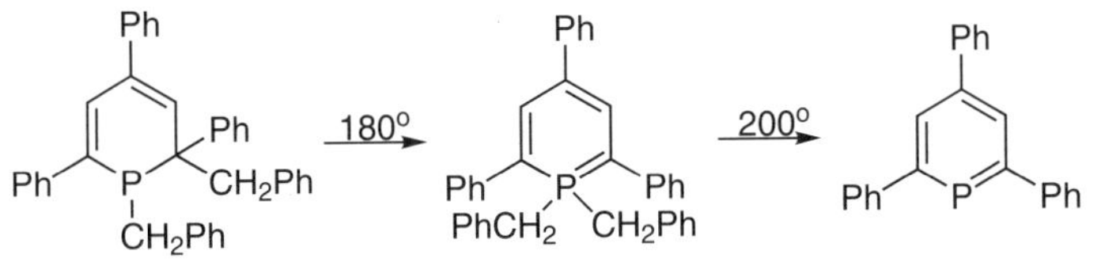

(8.68)

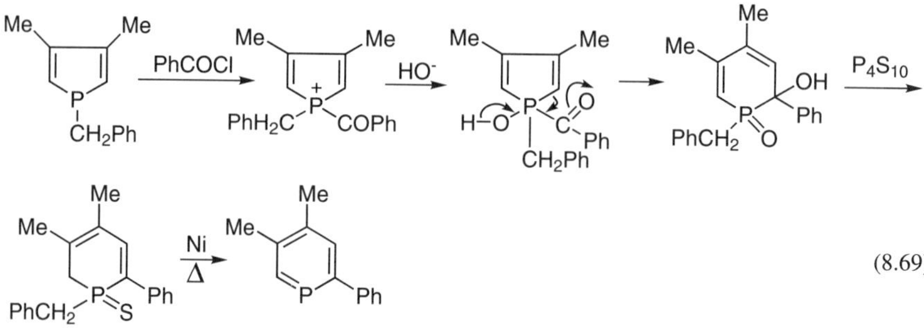

[Scheme for Eq. (8.69) and (8.70)]

(8.69)

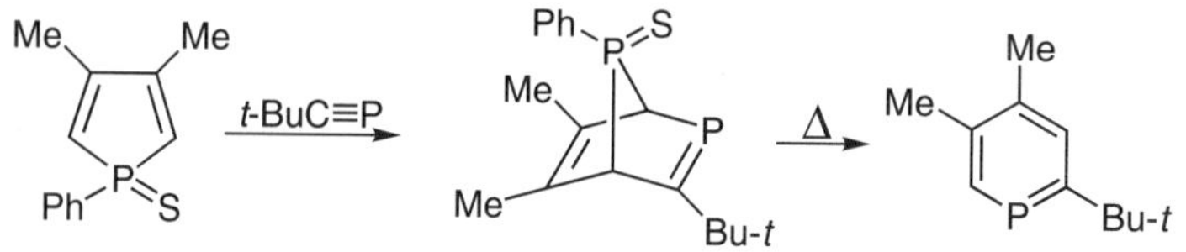

(8.70)

Important chemical properties known for the phosphinine ring system include the following.

1. The lone pair on phosphorus has little of the character of a normal phosphine. It cannot be protonated or alkylated; however, a few examples of direct formation of a P-sulfide are known.[107] This lack of reactivity of the lone pair has been attributed to inability of the internal C–P–C angle to undergo the expansion required in the formation of these 3-coordinate species and also to the unusual orbital energy sequence that places the energy of the nonbonding orbital below that of the HOMO.[108] Oxidizing agents (e.g., hydrogen peroxide) do attack the ring, possibly forming P-oxides as intermediates, but the products are dihydro derivatives with phosphorus in the phosphinic acid form (e.g., Eq. 8.71). No P-oxide seems yet to have been isolated.

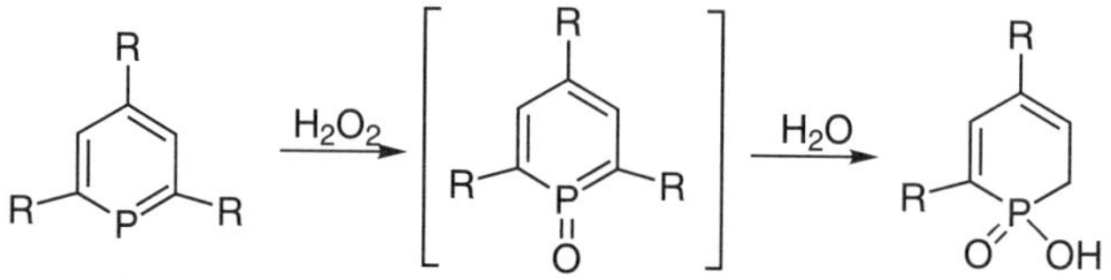

(8.71)

2. Halogens add to phosphorus to form covalent dihalides, e.g. **58**, but hydrolysis gives nonaromatic phosphinic acids (Eq. 8.72).

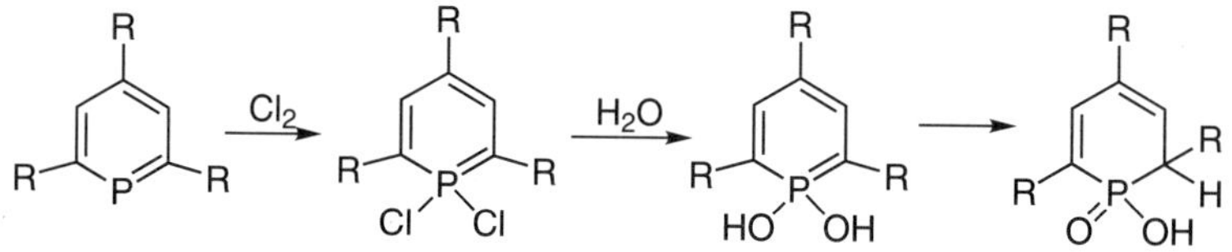

(8.72)

58

3. The phosphinine ring can act as a diene in Diels–Alder reactions, leading to some novel bridged systems, as illustrated by the reaction of Eq. 8.73. Several examples are known.

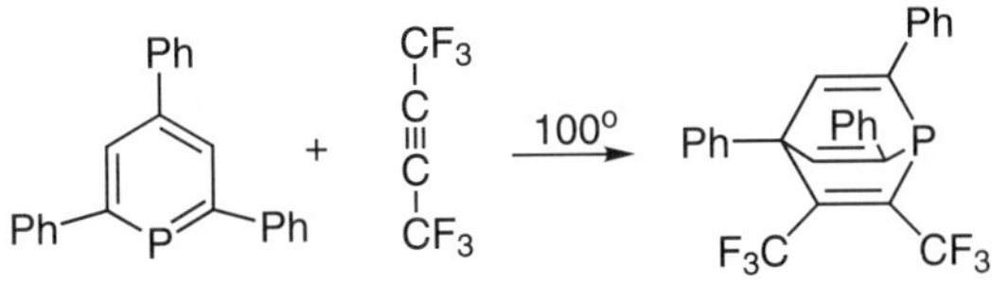

$$\text{(8.73)}$$

4. Anions and cations are formed with ease, the former from reaction with alkali metals (**59**, a radical anion) or organometallic reagents (**60**), the latter (**61**, a radical cation) from electron removal with organic free radicals and certain oxidizing metal compounds (e.g., mercuric acetate). These reactive intermediates participate in various valuable reactions.

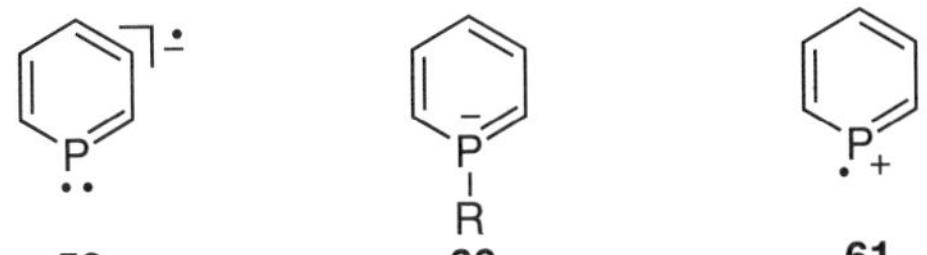

59   **60**   **61**

5. Metal complexes of various types are readily formed, by use either of the lone pair or of the π system. Useful reactions can be carried out on the ring while in complexed form, as in Eq. 8.74.[109]

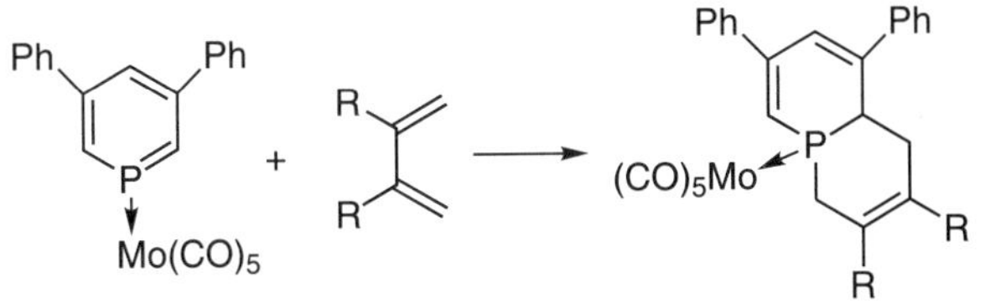

$$\text{(8.74)}$$

The phosphinine system discussed above is described as having σ^2, λ^3 bonding. There is another form of phosphinine that can be described as σ^4, λ^4, in which the P atom is 4-coordinate and formally in the state of an ylide (**63**), perhaps with a contribution from the ylene form (**62**). The first synthesis of this structure[110] actually preceded that of the λ^3-phosphinines and is illustrated in Eq. 8.75.

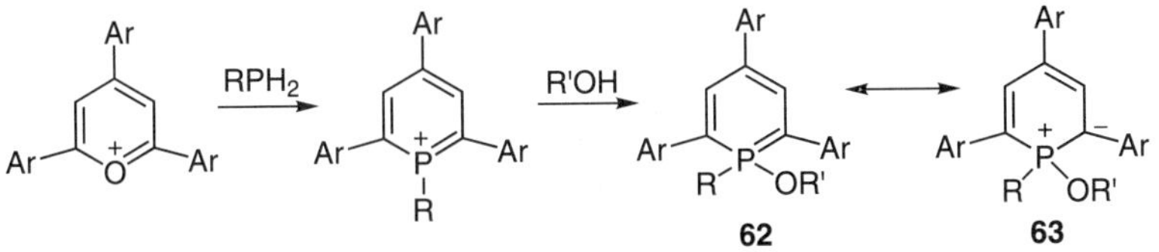

$$\text{(8.75)}$$

62   **63**

The ylide structure is extensively delocalized, as depicted in Eq. 8.76; by X-ray analysis, carbons 2 and 6 are seen to be identical, as are carbons 3 and 5. The ^{31}P NMR shift of these phosphinines is consistent with the ylide structure (e.g., $\delta -6.5$

for the 1-methyl-1,2,4,6-tetraphenyl derivative[111]). The bonding and the possibility of the use of phosphorus d-orbitals (now to be regarded with reservation) to permit a type of cyclic delocalization was discussed in detail in reviews.[10] Descriptions of the chemistry of these reactive substances will also be found in the reviews.

$$\left[\begin{array}{c} \text{structures} \end{array} \right] \equiv \begin{array}{c} \text{structure} \end{array} \quad (8.76)$$

Heteroatoms may take the place of carbon in both the phosphorinanes and phosphinines; many examples are already known but research continues in this area. Some of these structures are shown as **64** to **71**, which include the anticancer drug **65** (cyclophosphamide) and other types (**67** to **71**) of stable aromatic systems.

8.B.5. The Larger Ring Systems

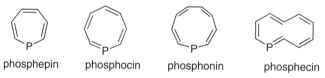

Figure 8.5. Large ring systems.

The literature up to about 1980 on the synthesis and properties of the large ring systems and their reduced forms, which is not extensive, has been reviewed,[1] and more recent work is discussed by Pabel and Wild.[91d] Conformational properties have also been reviewed.[112] These ring systems are of less importance than the systems of three to six members, and only a few comments on their synthetic methods will be given here. In fact, very little is known about the fully unsaturated ring systems. The phosphepin ring was first created in 1970[113] in the form of a P-oxide (1-phenyl), but not as a phosphine until 1984[114] (**72**, Eq. 8.77). This ring system requires stabilization by the bulky *t*-butyl groups, which prevent intermolecular interactions.

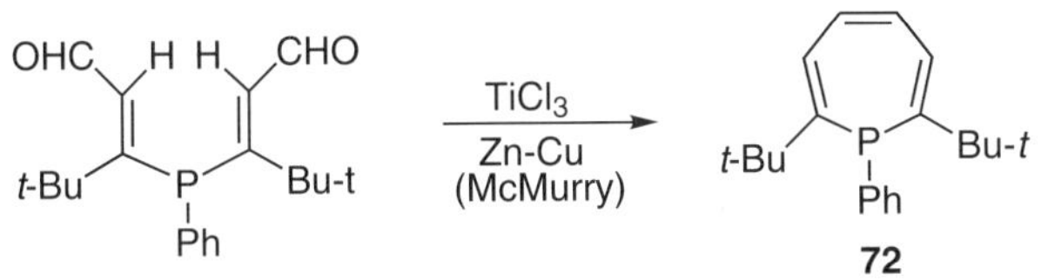

(8.77)

One phosphonin derivative (**73**) has been claimed as a product from a reaction series starting with 1,2,5-triphenylphosphole[115] (Eq. 8.78). The phosphonin ring is of some interest as a potentially aromatic system, making use of the lone pair as in phosphole to create the Hückel-allowed complement of 10 π electrons, but this property is obscured by the complex substitution in compound **73**.

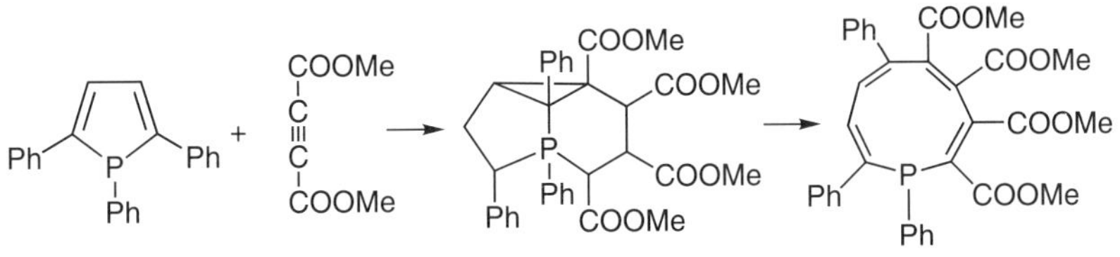

(8.78)

The fully reduced forms of the heterocycles of seven or more members are much more common. They exhibit normal phosphine character and are easily converted to the usual 4-coordinate derivatives. Several partially reduced forms are also known, and are of current interest (e.g., reference 116).

8.C. MULTICYCLIC PHOSPHORUS RING SYSTEMS

Here we can do no more than observe some of the more important systems; the review literature[1,3–6,12] should be consulted for information on properties and synthesis. Examining these complex systems, even briefly, is important in gaining appreciation for the very large and complex field that heterocyclic phosphorus chemistry has become in a relatively short time period. The systems are classified as: benzo derivatives, mostly from five- and six-membered rings (Fig. 8.6); saturated or partially saturated fused systems (Fig. 8.7); bridged bicyclic systems (Fig. 8.8); polycycle systems (Fig. 8.9). Bridged tricyclic and tetracyclic systems are not included, but do exist.

Some of the multicyclic systems have special properties worthy of note. Thus system **74** with P in 4-coordinate structures is a member of the family of atranes, in which strong intermolecular interaction involving the bridgehead nitrogen lone pair and the P function can be present.[117] The parent phosphines are very strong bases; the basicity is enhanced when an imino group is attached to P,[118] and such structures

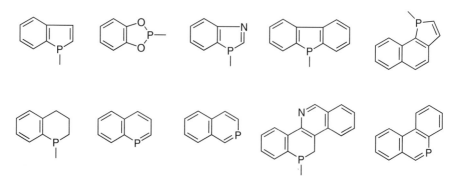

Figure 8.6. Benzo derivatives.

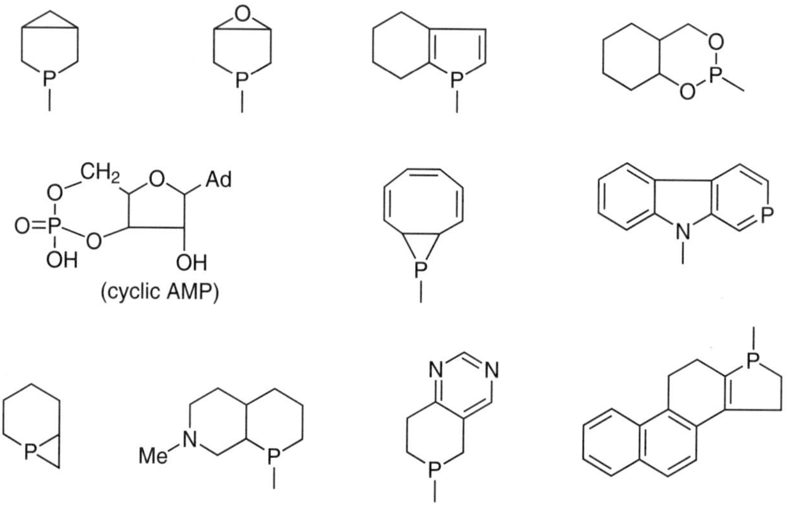

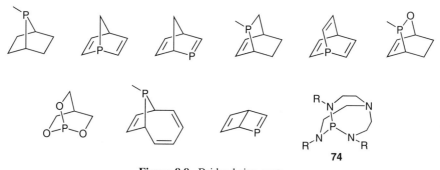

Figure 8.7. Fused systems.

74

Figure 8.8. Bridged ring systems.

8.C. MULTICYCLIC PHOSPHORUS RING SYSTEMS

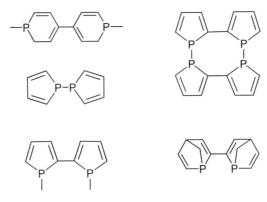

Figure 8.9. Poly-cycle systems.

are currently finding practical applications, as in the abstraction of H from phosphonium salts to create ylides for Wittig reactions.[119] Ring system **75** with biradical character is formed from dimerization of a diphosphirenyl radical (**76**) and is formally a valence isomer of tetraphosphabenzene.[120] Between the three-membered rings there is the longest P–P bond (2.634 Å by X-ray analysis) ever observed. To account for these and other properties, a novel one-electron P–P bond is postulated, which is proposed to result from a π^*-π^* interaction between the two diphosphirenyl radicals (**76**). The intriguing point is raised that one-electron bonds may play a wider role in organophosphorus compounds, and future studies on heterocyclic phosphorus compounds may well be directed to testing this new bonding concept. Ring system **77** is an example of a quite unusual type of macrocyclic ligand, bearing a resemblance to the calixarenes (as suggested by its name, silacalix-[4]-phosphaarene).[121] A system with three phosphinine rings is also known. These new molecules differ from calixarenes (σ-donors) in having strong π-accepting properties at the P=C units, and useful applications can be anticipated for them.

REFERENCES

1. L. D. Quin, *The Heterocyclic Chemistry of Phosphorus*, John Wiley & Sons, Inc., New York, 1981.
2. G. Grüttner and M. Wiernik, *Ber.* **48**, 1473 (1915).
3. A. R. Katritzky, C. W. Rees, and E. F. V. Scriven, eds., *Comprehensive Heterocyclic Chemistry II*, Pergamon Press, Oxford 1996.
4. F. G. Mann, *The Heterocyclic Derivatives of Phosphorus, Arsenic, Antimony and Bismuth*, John Wiley & Sons, Inc., New York, 1970, pp. 3–354.
5. K. D. Berlin and D. M. Hellwege, *Top. Phosphorus Chem.* **6**, 1 (1969).
6. S. D. Venkataramu, G. D. Macdonell, W. R. Purdum, M. El-Deek, and K. D. Berlin, *Chem. Rev.* **77**, 121 (1977).
7. K. Dimroth in A. R. Katritzky and C. W. Rees, eds., *Comprehensive Heterocyclic Chemistry*, Pergamon Press, Oxford, 1984, Chapter 17.
8. L. D. Quin and A. N. Hughes in F. R. Hartley, ed., *The Chemistry of Organophosphorus Compounds*, Vol. 1, John Wiley & Sons, Inc., New York, 1990, Chapter 10.
9. (a) F. Mathey, *Chem. Rev.* **88**, 429 (1988), (b) A. N. Hughes in J. Engel, ed., *Handbook of Organophosphorus Chemistry*, Marcel Dekker, Inc., New York, 1992, Chapter 10, (c) L. D. Quin in Ref. 3, Vol. 2, Chapter 15, (d) K. B. Dillon, F. Mathey, and J. R. Nixon, *Phosphorus: The Carbon Copy*, John Wiley & Sons, Inc., New York, 1998, Chapter 8.
10. (a) D. G. Hewitt in Ref. 3, Vol. 5, Chapter 12, (b) G. Märkl in M. Regitz and O. J. Scherer, eds., *Multiple Bonds and Low Coordination in Phosphorus Chemistry*, Georg Thieme Verlag, Stuttgart, Germany, 1990, pp. 220–257.
11. (a) F. Mathey, *Chem. Rev.* **90**, 997 (1990), (b) F. Mathey and M. Regitz, in Ref. 3, Vol. 1A, Chapter 8.
12. L. D. Quin, *Rev. Heteroatom Chem.* **3**, 39 (1990).
13. T. L. Gilchrist, *Heterocyclic Chemistry*, 3rd ed., Addison Wesley Longman, Harlow, UK, 1997, Chapter 11.
14. R. I. Wagner, (to American Potash and Chemical Corp.) U.S. Patent 3,086,053 (April 16, 1963); *Chem. Abstr.* **59**, 10124 (1963).
15. R. I. Wagner, L. V. D. Freeman, H. Goldwhite, and D. G. Rowsell, *J. Am. Chem. Soc.* **89**, 1102 (1967).
16. Y. B. Kang, M. Pabel, A. C. Willis, and S. B. Wild, *J. Chem. Soc., Chem. Commun.*, 475 (1994).
17. M. Baudler and J. Germeshausen, *Chem. Ber.* **118**, 4285 (1985).
18. A. Marinetti and F. Mathey, *Organometallics* **3**, 456 (1984).
19. S. M. Bachrach, *J. Phys. Chem.* **93**, 7780 (1989).
20. S. Chan, H. Goldwhite, H. Keyzer, D. G. Rowsell, and R. Tang, *Tetrahedron* **25**, 1097 (1969).
21. W. J. Richter, *Chem. Ber.* **116**, 3293 (1983).
22. D. C. R. Hockless, M. A. McDonald, M. Pabel, and S. B. Wild, *J. Chem. Soc., Chem. Commun.*, 257 (1995); *J. Organomet. Chem.* **529**, 189 (1997).
23. H. Quast and M. Heuschmann, *Angew. Chem., Int. Ed. Engl.* **17**, 867 (1978).
24. L. D. Quin, E.-Y. Yao, and J. Szewczyk, *Tetrahedron Lett.* **28**, 1077 (1987).
25. B. A. Arbusov, E. N. Dianova, and Y. Y. Samitov, *Dokl. Akad. Nauk SSSR* **244**, 117 (1979); *Chem. Abstr.* **90**, 187058 (1979).
26. B. C. Campbell, D. B. Denney, D. Z. Denney, and L. S. Shih, *J. Chem. Soc., Chem. Commun.*, 854 (1978).

27. G. Märkl, W. Hölzl, and I. Trötsch-Schaller, *Tetrahedron Lett.* **28**, 2693 (1987).
28. E. Niecke, M. Leuer, and M. Nieger, *Chem. Ber.* **122**, 453 (1989).
29. (a) A. Marinetti and F. Mathey, *J. Am. Chem. Soc.* **107**, 4700 (1985), (b) A. Marinetti, F. Mathey, J. Fischer, and A. Mitschler, *J. Chem. Soc., Chem. Commun.*, 45 (1984).
30. F. Mercier and F. Mathey, *Tetrahedron Lett.* **27**, 1323 (1986).
31. K. S. Fongers, H. Hogeveen, and R. F. Kingma, *Tetrahedron Lett.* **24**, 643 (1983).
32. S. A. Weissman and S. G. Baxter, *Tetrahedron Lett.* **31**, 819 (1990).
33. S. Lochschmidt, F. Mathey, and A. Schmidpeter, *Tetrahedron Lett.* **27**, 2635 (1986).
34. F. Nief and F. Mathey, *Tetrahedron* **47**, 6673 (1991).
35. E. Niecke and M. Lysek, *Tetrahedron Lett.* **29**, 605 (1988).
36. M. Ehle, O. Wagner, U. Bergsträsser, and M. Regitz, *Tetrahedron Lett.* **31**, 3429 (1990).
37. O. Wagner, G. Maas, and M. Regitz, *Angew. Chem., Int. Ed. Engl.* **26**, 1257 (1987).
38. O. Wagner, M. Ehle, and M. Regitz, *Angew. Chem., Int. Ed. Engl.* **28**, 225 (1989).
39. S. M. Bachrach, *J. Org. Chem.* **56**, 2205 (1991).
40. M. Baudler and B. Carlsohn, *Z. Naturforsch. B* **32**, 1490 (1977).
41. E. Niecke and D. Barion, *Tetrahedron Lett.* **30**, 459 (1989).
42. R. Streubel, E. Niecke, and M. Nieger, *Phosphorus Sulfur Silicon* **65**, 115 (1992).
43. E. Jungermann, J. J. McBride Jr., R. Clutter, and A. Mais, *J. Org. Chem.* **27**, 606 (1962).
44. Ref. 1, p. 364.
45. S. E. Cremer, P. W. Kramer, and P. K. Kafarski, *J. Chem. Soc., Perkin Trans.* 2, 1138 (1981).
46. A. Marinetti, F.-X. Buzin, and L. Ricard, *J. Org. Chem.* **62**, 297 (1997).
47. A. Marinetti and L. Ricard, *Tetrahedron* **49**, 10291 (1993).
48. J. Szewczyk, E.-Y. Yao, and L. D. Quin, *Phosphorus Sulfur Silicon* **54**, 1 (1990).
49. D. J. Brauer, G. Hessler, and O. Stelzer, *Chem. Ber.* **125**, 1987 (1992).
50. D. Berglund and D. W. Meek, *J. Am. Chem. Soc.* **90**, 518 (1968).
51. G. Zon and K. Mislow, *Fortschr. Chem. Forsch.* **19**, 61 (1971).
52. G. A. Gray, S. E. Cremer, and K. L. Marsi, *J. Am. Chem. Soc.* **98**, 2109 (1976).
53. A. Bader, D. D. Pathak, S. B. Wild, and A. C. Willis, *J. Chem. Soc., Dalton Trans.*, 1751 (1992).
54. T. A. Zyablikova, A. R. Panteleeva, and I. M. Shermergon, *Izv. Akad. Nauk SSSR Ser. Khim.*, 373 (1969); *Chem. Abstr.* **70**, 115228 (1969).
55. S. A. Weissman and S. G. Baxter, *Tetrahedron Lett.* **29**, 1219 (1988).
56. A. Marinetti and F. Mathey, *J. Chem. Soc., Chem. Commun.*, 153 (1990).
57. W. Tumas, J. C. Huang, P. E. Fanwick, and C. P. Kubiak, *Organometallics* **11**, 2944 (1992).
58. S. E. Cremer, *J. Chem. Soc., Chem. Commun.*, 616 (1970).
59. S.-C. Wong, N. I. Carruthers, and T.-M. Chan, *J. Chem. Res. (S)*, 268 (1993).
60. A. Marinetti, J. Fischer, and F. Mathey, *J. Am. Chem. Soc.* **107**, 5001 (1985).
61. K. M. Doxsee, G. S. Shen, and C. B. Knobler, *J. Am. Chem. Soc.* **111**, 9129 (1989).
62. W. Tumas, J. A. Suriano, and R. L. Harlow, *Angew. Chem., Int. Ed. Engl.* **29**, 75 (1990).
63. K. M. Doxsee, E. M. Hanawalt, G. S. Shen, T. J. R. Weakley, H. Hope, and C. B. Knobler, *Inorg. Chem.* **30**, 3381 (1991).
64. (a) K. M. Doxsee, G. S. Shen, and C. B. Knobler, *J. Chem. Soc., Chem. Commun.*, 1649 (1990); E. M. Hanawalt, K. M. Doxsee, G. S. Shen, T. J. R. Weakley, C. B. Knobler, and H. Hope, *Heteroatom Chem.* **9**, 9 (1998), (b) E. M. Hanawalt, K. M. Doxsee, G. S. Shen, and H. Hope, *Heteroatom Chem.* **9**, 21 (1998).
65. (a) S. M. Bachrach, *J. Org. Chem.* **56**, 2205 (1991), (b) S. M. Bachrach and M. Liu, *J. Org. Chem.* **57**, 209 (1992).

66. (a) N. H. Tran Huy and F. Mathey, *Tetrahedron Lett.* **29**, 3077 (1988), (b) H. Trauner, E. Delacuestra, A. Marinetti, and F. Mathey, *Bull. Soc. Chim. Fr.* **132**, 384 (1995).
67. R. H. Neilson, B. A. Boyd, D. A. Dubois, R. Hani, G. M. Scheide, J. T. Shore, and U. G. Wettermark, *Phosphorus Sulfur* **30**, 463 (1987).
68. A. C. Dema, X. Li, C. M. Lukehart, and M. D. Owen, *Organometallics* **10**, 1197 (1991).
69. T. Kawashima and R. Okazaki in Ref. 3, Vol. 1B, Chapter 27.
70. T. Wettling, G. Wolmershäuser, P. Binger, and M. Regitz, *J. Chem. Soc., Chem. Commun.*, 1541 (1990).
71. J. Svara, E. Fluck, and H. Riffel, *Z. Naturforsch. B* **40**, 1258 (1985).
72. W. Mahler, *J. Am. Chem. Soc.* **86**, 2306 (1964).
73. C. Charrier, N. Maigrot, F. Mathey, F. Robert, and Y. Jeannin, *Organometallics* **5**, 623 (1986).
74. I. G. Phillips, R. G. Ball, and R. G. Cavell, *Inorg. Chem.* **27**, 2269 (1988).
75. S. Kummer and U. Zennock in Ref. 3, Vol. 1B, Chapter 37.
76. G. Bertrand, *Angew. Chem., Int. Ed. Engl.* **37**, 270 (1998).
77. *Organic Syntheses*, (a) **43**, 73 (1963), (b) *Coll.* Vol. V, 787 (1973).
78. Ref. 9c, p. 828.
79. G. V. Coleman, D. Price, A. R. Horrocks, and J. E. Stephenson, *J. Chem. Soc., Perkin Trans. 2*, 629 (1993).
80. A. Breque, F. Mathey, and P. Savignac, *Synthesis* 983 (1981).
81. L. D. Quin, J. G. Bryson, and C. G. Moreland, *J. Am. Chem. Soc.* **91**, 3308 (1969).
82. W. Egan, R. Tang, G. Zon, and K. Mislow, *J. Am. Chem. Soc.* **92**, 1447 (1970).
83. P. Coggon, J. E. Engel, A. T. McPhail, and L. D. Quin, *J. Am. Chem. Soc.* **92**, 5779 (1970).
84. G. Keglevich, Z. Böcskei, G. M. Keserü, K. Ujszaszy, and L. D. Quin, *J. Am. Chem. Soc.* **119**, 5095 (1997).
85. F. Mathey, F. Mercier, C. Charrier, J. Fischer, and A. Marinetti, *J. Am. Chem. Soc.* **103**, 4595 (1981).
86. G. de Lauzon, C. Charrier, H. Bonnard, and F. Mathey, *Tetrahedron Lett.* **23**, 511 (1982).
87. F. Zurmühlen and M. Regitz, *J. Organometallic Chem.* **332**, C1 (1987).
88. S. Holand, C. Charrier, F. Mathey, J. Fischer, and A. Mitschler, *J. Am. Chem. Soc.* **106**, 826 (1984).
89. G. Märkl, E. Seidel, and I. Trötsch, *Angew. Chem., Int. Ed. Engl.* **22**, 571 (1983).
90. N. P. Ignatova, N. N. Melnikov, and N. I. Shvetsov-Shilovskii, *Khim. Geterotsikl. Soedin.*, 753 (1967); *Chem. Abstr.* **68**, 78367 (1968).
91. Ref. 3, (a) A. Schmidpeter, Vol. 3, Chapter 15, (b) A. Schmidpeter, Vol. 3, Chapter 16, (c) A. Schmidpeter, Vol. 4, Chapter 22, (d) M. Pabel and S. B. Wild, Vol. 9, Chapter 34.
92. A. Schmidpeter and K. Karaghiosoff in M. Regitz and O. J. Scherer, eds., *Multiple Bonds and Low Coordination in Phosphorus Chemistry*, Georg Thieme, Stuttgart, Germany, 1990, pp. 258–283.
93. W. H. Powell, *Pure Appl. Chem.* **55**, 410 (1983).
94. F. Mathey and G. Muller, *Bull. Soc. Chim. Fr.*, 4021 (1972).
95. F. Mathey and C. Santini, *Can. J. Chem.* **57**, 723 (1979).
96. L. D. Quin and J. H. Somers, *J. Org. Chem.* **37**, 1217 (1972).
97. Ref. 1, pp. 368–370.
98. S. I. Featherman and L. D. Quin, *J. Am. Chem. Soc.* **97**, 4349 (1975).
99. J. B. Lambert and W. L. Oliver, Jr., *Tetrahedron* **27**, 4245 (1971).
100. L. D. Quin and S. O. Lee, *J. Org. Chem.* **43**, 1424 (1978).
101. G. Märkl, *Angew. Chem., Int. Ed. Engl.* **5**, 846 (1966).
102. J. C. J. Bart and J. J. Daly, *Angew. Chem., Int. Ed. Engl.* **7**, 811 (1968).

103. A. J. Ashe III, *J. Am. Chem. Soc.* **93**, 3293 (1971).
104. A. J. Ashe III, *Acc. Chem. Res.* **11**, 153 (1978).
105. G. Märkl, G. Y. Jin, and E. Silbereisen, *Angew. Chem., Int. Ed. Engl.* **21**, 370 (1982).
106. W. Rösch and M. Regitz, *Z. Naturforsch. B* **41**, 931 (1986).
107. J.-M. Alcaraz and F. Mathey, *J. Chem. Soc., Chem. Commun.*, 508 (1984); D. G. Holah, A. N. Hughes, and K. L. Knudsen, *J. Chem. Soc., Chem. Commun.*, 493 (1988).
108. H. Oehling and A. Schweig, *Tetrahedron Lett.* 4941 (1970); H. Oehling and A. Schweig, *Phosphorus* **1**, 203 (1972).
109. G. Märkl and H.-J. Beckh, *Tetrahedron Lett.* **28**, 3475 (1987).
110. K. Dimroth and P. Hoffmann, *Chem. Ber.* **99**, 1325 (1966).
111. G. Märkl and K.-H. Heier, *Angew. Chem., Int. Ed. Engl.* **11**, 1017 (1972).
112. L. D. Quin in R. S. Glass, ed., *Conformational Analysis of Medium- Sized Heterocycles*, VCH Publishers, Inc., Deerfield Beach, FL, 1988, pp. 181–216.
113. G. Märkl and H. Schubert, *Tetrahedron Lett.*, 1273 (1970).
114. G. Märkl and W. Burger, *Angew. Chem., Int. Ed. Engl.* **23**, 894 (1984).
115. N. E. Waite and J. C. Tebby, *J. Chem. Soc. C*, 386 (1970).
116. G. Keglevich, *Synthesis*, 931 (1993).
117. J. G. Verkade, *Acc. Chem. Res.* **26**, 483 (1993).
118. J. Tang, J. Dopke, and J. G. Verkade, *J. Am. Chem. Soc.* **115**, 5015 (1993).
119. Z. Wang and J. G. Verkade, *Tetrahedron Lett.* **39**, 9331 (1998); Z. Wang and J. G. Verkade, *Heteroatom Chem.* **9**, 687 (1998).
120. Y. Canac, D. Bourissau, A. Baceiredo, H. Gornitzka, W. W. Schoeler, and G. Bertrand, *Science* **279**, 2080 (1998).
121. N. Avaran, N. Mézailles, L. Ricard, P. Le Floch, and F. Mathey, *Science* **280**, 1587 (1998).

CHAPTER 9

OPTICALLY ACTIVE ORGANOPHOSPHORUS COMPOUNDS

9.A. GENERAL

In principle, any 4-coordinate tetrahedral phosphorus group that has four different substituents is a chiral center that is capable of existence in two enantiomeric, optically active forms. This property was recognized many years ago, and a phosphine oxide was resolved into enantiomers by an early pioneer of stereochemistry, J. Meisenheimer,[1] in 1911. The enantiomeric phosphine oxides are quite resistant to racemization when in pure form, although a mechanism for racemization does exist that depends on addition of a nucleophile and passage through a 5-coordinate intermediate that isomerizes. The resolution of a phosphine sulfide was accomplished in 1944 by Davies and Mann[2] and of a simple quaternary phosphonium salt by Kumli, McEwen and Vander Werf[3] in 1959 (a heterocyclic phosphonium salt with molecular dissymmetry but achiral phosphorus had been resolved in 1947)[4]. It was not until 1961, however, that tertiary phosphines were obtained in optically active forms, a dramatic discovery made in the laboratory of L. Horner.[5] This property depends on the stability of the pyramidal structure (Eq. 9.1) of 3-coordinate phosphorus under laboratory conditions, for if the pyramid inverted readily, as is the case with tertiary amines, there would be no possibility to observe the enantiomers.

$$B-\underset{C}{\overset{A}{P}}: \rightleftharpoons :\underset{C}{\overset{A}{P}}-B \qquad (9.1)$$

In fact, the free energy of activation for the inversion of simple tertiary phosphines is 30 to 35 kcal/mol; the inversion is immeasurably slow at room temperature, and even at 130° the half-life of a simple phosphine such as n-PrMePhP is about 3 h 20 min in methylnaphthalene solution.[5] In later years, optically active derivatives of other 3-coordinate derivatives were prepared, and now this family plays a major role as ligands in the construction of metal coordination complexes.

Of major significance in phosphorus chemistry was the development of techniques for procuring optically active forms of derivatives of phosphoric, phosphonic, and phosphinic acids. The first optically active phosphorus acid was prepared in 1956, when the resolution of an O-alkyl alkylphosphonothioic acid was announced by Aaron and Miller.[6] Since that time, dozens of phosphorus acids have been obtained in optically active form.

From this historical introduction, it can be seen that phosphorus stereochemistry really had its beginnings in the 1950s, but it rapidly became one of the fastest-growing and most important components of contemporary phosphorus chemistry. We have, of course, already seen some of the consequences of configurational stability in many phosphorus functional groups in our considerations of heterocyclic phosphorus compounds (chapter 8); cis,trans or syn,anti isomers are common in this field, being especially well known for cyclic phosphines, phosphine oxides, and phosphates. But phosphorus stereochemistry is of great importance in other contemporary studies. Many reaction mechanisms have been elucidated with the aid of optically active phosphorus compounds. The metal coordination complexes prepared from optically active phosphines as ligands are employed as catalysts for asymmetric syntheses, especially those involving homogeneous hydrogenation of alkenes (chapter 11.E). In biological chemistry, enantiomers may have quite different reaction rates or show stereospecificity in binding or toxicity, and this needs to be taken into consideration in the design of new medicinal agents. The mechanisms of biological transformations involving phosphorus compounds are also effectively studied with the use of chiral 4-coordinate derivatives.

Several valuable reviews exist on the subject of chirality at phosphorus. The most recent (1994) is that by Pietrusiewicz and Zablocka.[7] Those by Valentine[8] and Imamoto[9] are also of great value. Earlier surveys by McEwen in *Topics in Phosphorus Chemistry*[10] and also in a book that collects various papers making advances in the field and provides commentary[11] remain of value. Optically active phosphines are reviewed by Kagan and Sisaki in the Hartley series,[12] and optically active phosphine chalcogenides by Gallagher also in this series.[13] The stereochemistry of phosphorus acid derivatives is included in a review by Hall and Inch.[14] The stereochemistry of heterocyclic phosphorus compounds is reviewed by Gallagher[15] and by Quin.[16]

The emphasis in this chapter is placed on techniques for obtaining optically active forms of phosphorus compounds. One method consists of direct racemate separation (resolution), including the conversion of racemates to separable diastereomeric derivatives. Another involves reaction chemistry where the phosphorus chirality is created with one enantiomer in excess (indicated by the notation "ee") in an asymmetric synthesis with an optically active reagent, which (unless ee is close to 100%) is followed by separation of the unequal mixture of diastereoisomers that is formed. A summary of the stereochemistry of some important reactions that take place with stereochemical retention or inversion is also provided.

In specifying absolute configurations at phosphorus, the usual Cahn-Ingold Prelog rules for group sequence determination are used, with the following additions to handle phosphorus functional groups: the lone pair is always the group of lowest priority, and the P=O group is of lower priority than P−O−C.

9.B. RACEMIC MIXTURE SEPARATION TECHNIQUES

9.B.1. Phosphine Oxides

Much work has been done on the preparation of optically active phosphine oxides, mostly because they are immediate precursors, on reduction, of optically active phosphines. There are important methods for obtaining the phosphine oxides by synthetic procedures, to be discussed separately, but direct resolution is an option that has been demonstrated on many occasions. These compounds generally respond well to the modern techniques of separation on a chiral chromatographic column, and this resolution method already appears to be the preferred one for phosphine oxide resolution. Several types of columns have been employed for this purpose. The well-known, commercially available Pirkle chiral packing (CSP-1), has been used effectively; this consists of the optically active group 3,5-$(NO_2)_2C_6H_3CONHC*HPhCONHCH_2CH_2CH_2-$ bonded to silica gel. Other types of optically active column packings that have been used include cyclodextrin[17] and poly(tritylmethacrylate).[18] Some specially designed Pirkle-type packings are also described in papers[19,20] that give full details on their preparation and the procedures for conducting the HPLC separations. For example, silica gel to which the 3-aminopropyl group was covalently bonded was treated to form a chiral salt structure (**1** and **2**) or to form a covalently bonded chiral amide group (**3**); chiral centers are marked with *, and DNP is the common Pirkle aryl substituent (3,5-dinitrophenyl).

The solvent consisted of various mixtures of 2-propanol in *n*-hexane. Several (R-naphthyl)methylphenylphosphine oxides were separated quite cleanly on these columns. The preparative scale separation of 1-naphthylmethylphenylphosphine oxide on the packing prepared from compound **1** gave the two isomers with $[\alpha]_D^{22}$ +21° and −17.3°. A previously reported[21] value was +18.6°. A more recent chromatographic advance[22] consists of the application of subcritical fluid chromatography on a preparative scale with a Pirkle-type chiral stationary phase; the heterocyclic phosphine oxides **4, 5,** and **6** were separated with an optical purity exceeding 95% at a production rate of 510 mg/h. The mobile phase was a mixture of carbon dioxide and ethanol.

The older technique of performing a resolution by converting a racemic mixture into a pair of diastereomers by reaction with an optically active reagent has been applied in several instances to phosphine oxides. The first reported resolution employed this technique, taking advantage of the weak basicity of phosphine oxides that allows them to form salts with an optically active acid. Meisenheimer[1] employed (+)-bromocamphorsulfonic acid (BCSH in Eq. 9.2) for this purpose, and others have made use of tartaric acid.[23]

$$(+),(-)\text{-MeEtPhP=O} + (+)\text{-BCSH} \longrightarrow$$

$$(+)\text{-MeEtPhPOH}^+(+)\text{-BCS}^- + (-)\text{-MeEtPhPOH}^+(+)\text{BCS}^- \quad (9.2)$$

These separations, which require fractional crystallization of the diastereomeric salts, can sometimes be laborious and are complicated by dissociation of the salt into the components due to the weakness of the phosphine oxide basicity. However, the method continues to be of interest, and has been used recently by two groups to effect the resolution of highly important diphosphine dioxides, **7** (norphos dioxide) with tartaric acid[24] and **8** (BINAP dioxide, which has molecular dissymmetry from restricted rotation) with camphoric acid;[25] both dioxides on reduction serve as precursors of diphosphines used as metal complex catalytic agents.

A related process[26] depends on the formation of inclusion-type complexes with the optically active (through molecular dissymetry) phenolic compound **9**. The 1:1 complexes are separated by fractional crystallization, and then the free oxide is released on passage through a silica gel column. This method was used to separate two arylethylmethylphosphine oxides and later was successfully applied to the resolution of the diphosphine dioxide **8**.[26]

When the phosphine oxide contains other reactive functional groups, diastereomers can be formed by appropriate reactions at this site, and after the separation, the products are converted to the original form. An excellent example, outlined in Eq. 9.3, of this approach is that of Bodalski, et al.,[27] who performed the resolution of the hydroxy phospholene oxide **10** by reacting the OH group with (−)-camphanyl chloride to form diasteromeric esters that were separated by crystallization. The optically active esters were hydrolyzed to recover the original hydroxy phospholene oxide; the isomer with $[\alpha]^{21}_D$ +68.8° was determined by X-ray crystallography to have the absolute configuration of S at phosphorus and R at carbon. Oxidation then gave the optically active ketone **11** ($[\alpha]^{20}_D$ +285.5°).

(9.3)

9.B.2. Phosphine Sulfides

The phosphine sulfide **12** was one of the earliest phosphorus compounds to be resolved.[2] This was accomplished by reaction with (+)-α-phenethylamine to form diasteromeric carboxylates; fractional crystallization and regeneration of the free carboxylic acid gave an enantiomer with $[\alpha]_D$ −9.7°.

HOOCCH₂—C₆H₄—P(S)PhBu

12

9.B.3. Phosphonium Salts

The common method for resolving phosphonium salts is to exchange the anion for one that is optically active, and then to separate by fractional crystallization the mixture of diastereomeric salts. The first resolution employed silver hydrogen (−)-dibenzoyltartrate in reaction with the iodide of benzylethylmethylphenylphosphonium ion to obtain the diastereomeric salt mixture (Eq. 9.4). The salts were converted back to the iodide form with ammonium iodide; they had $[\alpha]^{25}_D$ +24° and −23.8°. More recently, several other optically active anionic resolving agents for phosphonium salts have been introduced, including silver menthoxy acetate, potassium hydrogen dibenzoyl tartrate, and silver camphorsulfonate. References to the considerable literature on these resolutions may be found in review articles.[7,8]

$$(\pm)\text{-PBzEtMePh} \xrightarrow{(-)\text{-anion}} (+)\text{-PBzEtMePh} \cdot (-)\text{-anion} + (-)\text{-PBzEtMePh} \cdot (-)\text{-anion} \quad (9.4)$$

9.B.4. Phosphines

Although optically active tertiary phosphines are generally prepared by reduction of the corresponding phosphine oxides and by other synthetic methods (vide infra), techniques do exist for the resolution of racemic mixtures of phosphines. The resolution may be accomplished by forming a metal coordination complex with an optically active coordinating agent, followed by fractional crystallization of the mixture of diastereomeric complexes, and finally decomplexation to release the optically active phosphine. Many such separations have been perfomed, making use of optically active complexes based on Pd(II), Pt(II), and Rh(I). The first reported example[28] serves to illustrate the technique (Eq. 9.5). Here the *trans*-platinum (II) complex **13** of *t*-butylmethylphenylphosphine is converted to the binuclear complex **14**, which on reaction with (+)-deoxyephedrine gave a mixture of diastereomers of structure **15**. This mixture was separated into the diastereomers by crystallization, and the optically active phosphines recovered by decomplexation with KCN.

$$t\text{-BuMePhP} \xrightarrow{K_2PtCl_2} \underset{\mathbf{13}}{[R_3P\text{-}PtCl_2\text{-}PR_3]} \xrightarrow{PtCl_2} \underset{\mathbf{14}}{[R_3P\text{-}PtCl_2\text{-}PtCl_2\text{-}PR_3]} \xrightarrow{(+)\text{-deoxyephedrine}}$$
(R,S)

$$\underset{\mathbf{15}}{[R_3P\text{-}Pt(Cl)(CH_2Ph)\text{-}NH\text{-}CH(Me)(Me)]} \xrightarrow[\text{2. KCN}]{\text{1. separate}} \underset{(S)}{\text{Me}\cdots\overset{..}{P}(t\text{-Bu})(Ph)} \quad (9.5)$$

Diphosphines important as ligands in asymmetric hydrogenation catalysts have been so separated, as in the example of the (+, −) isomers of phosphine **16**, which were resolved by use of the chiral agent **17**.[29] Both isomers were obtained in optically pure form (Eq. 9.6).

The same resolving agent was later used to separate the enantiomers of the racemic form of the valuable chelating tetraphosphine **18**;[30a] a related resolving agent with chirality both at C and N, as in structure **17** where one *N*-methyl is replaced by H, was used to obtain the enantiomers of *t*-butylmethylphenylphosphine.[30b]

Selective complexation of one enantiomer in racemic benzylmethylphenylphosphine with a chiral ruthenium porphyrin complex has also been reported.[31] This process is suggested to be of importance as a form of molecular recognition exhibited by the porphyrin complex.

A major advance in phosphine resolution made recently is based on the phenomenon of crystallization-induced asymmetric transformation.[32] Here an (R,S)-phosphine with a covalently attached chiral auxiliary is heated to accelerate pyramidal equilibration at phosphorus; if the melting point of one diastereomer is higher than the temperature of rapid pyramidal inversion (typically well over 100° for tertiary phosphines but closer to room temperature for acylphosphines), then that diastereomer may crystallize first, with the supply of this form in the liquid continually being replenished by the inversion process. The reactions used in this study are shown in Eq. 9.7. The preferred chiral auxiliary is the alcohol (R)-pantolactone (Pant-OH), which is converted to the chloroformate **19** for reaction with *ortho*-anisylphenylphosphine; this gives the desired acylphosphine as a 1 : 1 mixture of the diastereoisomers **20** and **21** (*o*-An = *ortho*-anisyl).

9.B. RACEMIC MIXTURE SEPARATION TECHNIQUES

$$o\text{-AnPhPH} + \underset{\mathbf{19}}{\text{[ClC(=O)-O-CH(Me}_2\text{)-CH-C(=O)-O]}} \longrightarrow \underset{\mathbf{20}}{o\text{-An--P(Ph)(COO-Pant)}} + \underset{\mathbf{21}}{\text{Ph--P(}o\text{-An)(COO-Pant)}} \quad (9.7)$$

Crystallization of the neat mixture of **20** and **21** indeed afforded a single diastereomer, whereas crystallization from ethanol at room temperature gave a product consisting of the isomers in a ratio of 25–32 : 1. Such a ratio could be achieved only if pyramidal inversion of **21** to **20** had occurred. Even on storing a 25 : 1 mixture at room temperature, some asymmetric transformation occurred, improving the ratio to 91 : 1 after 6 weeks. The resolved product from these experiments was subjected to further transformations (Eq. 9.8), resulting in the final isolation of a tertiary phosphine in the stabilized form of its borane complex (**22**). Conversion to the free optically active phosphine can be accomplished by decomplexation with diethylamine. The particular phosphine synthesized, in multigram amounts, is known as PAMP and is of major importance as a ligand in asymmetric homogeneous hydrogenation catalysts.

$$\mathbf{20} \xrightarrow{CF_3SO_2OMe} o\text{-An--P}^+(\text{O-Pant})(\text{Me})(\text{Ph}) \xrightarrow{H_2O} \left[o\text{-An--P(Me)(Ph)} \right]_{R_P\text{-PAMP}} \xrightarrow{BH_3} \underset{\mathbf{22}}{o\text{-An--P}^+(\text{BH}_3^-)(\text{Me})(\text{Ph})} \quad (9.8)$$

The first optically active secondary phosphine was not isolated until 1994.[33] It is to be expected that such compounds would be subject to ready racemization in acid media (losing chirality by addition of a second proton) and perhaps in basic media as well (losing a proton to form the achiral phosphide anion), and precautions are required to prevent its occurrence. A solution to the racemization problem was presented by Wild and co-workers,[33a,b] who achieved the resolution of (−)-menthylmesitylphosphine (**23**). Although **23** was actually separated from an unequal mixture (43 : 57) of diastereomers formed by the asymmetric synthesis method (Eq. 9.9), the technique is included here to suggest a way that racemate resolution of secondary phosphines might be approached in the future. The diastereoisomers of **23** had remarkably well separated ^{31}P NMR signals (δ −62.62 and −84.24), which were used to monitor the separation. Attempted separation by crystallization from acetonitrile was initially unsuccessful, but it was found that including a small amount of sodium acetylacetonate (Na[acac]) in the medium controlled the acidity and prevented the racemization. Thus was obtained an enantiomer (**23A**) of the secondary phosphine having ^{31}P δ −85.16 in 94% purity, with $[\alpha]_D$ −186°. The resolution was also accomplished quite readily on the borane adduct of the diastereomeric phosphines; the borane group prevents the protonation that appears to cause the racemization. The adduct isomer isolated by crystallization gave the same enantiomer of **23A** when it was decomposed with diethylamine. X-ray crystallography established the conformation of the borane complex, which on deboronation (retention) gave the phosphine enantiomer **23A**, thus providing the proof of its absolute configuration as R. That racemization of a pure secondary phosphine is indeed a major problem was demonstrated by attempted recrystallization of an enantiomer of 94% purity from acetonitrile in the absence of Na[acac], whereupon the original 43 : 57 diastereoisomer mixture was obtained.

$$\text{MesPCl}_2 + \underset{\text{Pr-}i}{\overset{\text{Me}}{\text{cyclohexyl-MgCl}}} \longrightarrow \underset{\text{Pr-}i}{\overset{\text{Me}}{\text{cyclohexyl-P(Cl)Mes}}} \xrightarrow{\text{LiAlH}_4}$$

$$\underset{\underset{\textbf{23A (R}_P\textbf{),43\%}}{\text{Pr-}i}}{\overset{\text{Me}}{\text{cyclohexyl-P(H)Mes}}} + \underset{\underset{\textbf{23B (S}_P\textbf{),57\%}}{\text{Pr-}i}}{\overset{\text{Me}}{\text{cyclohexyl-P(H)Mes}}} \qquad (9.9)$$

In principle, any 3-coordinate derivative of phosphorus with three different substituents would be chiral, provided the derivative has sufficiently high pyramidal stability. This point was first demonstrated for a phosphite ester in 1985.[34] The compound P(OPh)(OC$_6$H$_4$–p-Cl)(OC$_6$H$_4$–p-Me) was resolved as a complex with an optically active platinum derivative; one isomer was reported to have $[\alpha]_{589}^{25}$ +21.4 in chloroform, but the optical activity was lost after a few weeks at room temperature.

9.B.5. Phosphorus Acids

Because proton exchange occurs rapidly in phosphorus acids with a P(O)OH group, which has the effect of equilibrating the two oxygen atoms, it is not possible to obtain such acids in optically form. Attempted resolution by the formation of diastereomeric salts with optically active bases will fail because the two oxygens of the phosphorus anion are equivalent through resonance (Eq. 9.10).

$$\left[\overset{O}{\underset{}{P}}-OH \rightleftharpoons \overset{OH}{\underset{}{P}}=O \right] \xrightarrow{\text{base}} \left[\overset{O}{\underset{}{P}}-O^- \longleftrightarrow \overset{O^-}{\underset{}{P}}=O \right] \qquad (9.10)$$

Esterification of such acids eliminates these problems, and suitable esters can be obtained in optically active form by asymmetric synthesis procedures, to be discussed in the next section. If one oxygen of the acid group is replaced by sulfur or selenium, however, stable optically forms are possible and are indeed obtained quite readily by racemate resolution. Aaron et al.,[6,35] described the resolution of *O*-ethyl ethylphosphonothioic acid by fractional crystallization of the diastereomers formed with quinine or brucine. A facile tautomeric shift of the proton between oxygen and sulfur (Eq. 9.11) does not allow the specification of the exact structure of the resolved compound, although the dominant form in such acids is that with the proton on oxygen.

$$\text{Et}-\overset{\overset{\displaystyle O}{\|}}{\underset{\underset{\displaystyle \text{OEt}}{|}}{\text{P}}}-\text{SH} \rightleftarrows \text{Et}-\overset{\overset{\displaystyle \text{OH}}{|}}{\underset{\underset{\displaystyle \text{OEt}}{|}}{\text{P}}}=\text{S} \qquad (9.11)$$

The same research group reported several other similar resolutions in 1960.[36] The method was adopted by other groups and by 1979, as noted in a review,[37] as many as 23 thio acids and three selenoacids had been resolved, with others in later years. The most effective base is the commercially available α-phenethylamine, but various alkaloids (quinine, brucine, cinchonidine, strychnine and ephedrine) have all been used. These resolutions are probably the easiest to perform among those of organophosphorus compounds, and as will be seen, the enantiomeric forms are valuable precursors to other optically active derivatives. The diastereomeric salt method has also been used to resolve some derivatives of thiophosphinic acids, e.g., RR′P(O)SH, using quinine as the base.[38] One member of this series deserves special mention: (−)-(S) or (+)-(R)-t-butylphenylphosphinothioic acid has found considerable use as a reagent that allows analysis of the composition of mixtures of enantiomeric phosphine oxides or phosphinates that originate from various partial resolutions or asymmetric synthetic procedures, as described in the next section. This reagent was introduced by Harger in 1980,[39] who found that, through hydrogen bonding, the phosphinothioic acid formed diastereomeric complexes in solution that gave distinctly different proton NMR signals for groups attached directly to phosphorus. For example, racemic MePhP(O)(OMe) with one equivalent of the phosphinothioic acid gave two 1:1 doublets for the methoxy protons; the doublets were separated from each other by 7 Hz. The literature contains many examples of this type of analysis.[40]

Another type of resolution was performed on racemic methyl methylphenylphosphinate.[26] This consisted of complex formation with optically active 2,2′-dihydroxy-1,1′-binaphthyl (9), followed by the usual fractional crystallization, a procedure previously noted to be useful in phosphine oxide resolution.

Numerous resolutions have been performed on appropriate derivatives of phosphoric acid. Some of these resolutions are summarized below, showing the structure of the diastereomer that was used for the separation by crystallization. We should note, however, that superior methods involving asymmetric synthesis procedures have become of more importance in recent years (vide infra).

(MeO)(p-nitrophenoxy)P(S)(O)⁻ (strychnine-Me)⁺ (ref. 41)
(MeO)(1-naphthyloxy)P(S)(O)⁻ (ephedrine-H)⁺ (ref. 42)
(MeO)(MeS)P(S)(O)⁻ (strychnine-Me)⁺ (ref. 43)
(ArO)(RNH)P(S)(O)⁻ (quinine-H)⁺ (ref. 44)
(MeO)(c-hexylNH)P(S)(O)⁻ (quinine-H)⁺ (ref. 45)
(MeO)(O-1-naphthyl)P(S)(O)⁻ (ephedrine-H)⁺ (ref. 46)

9.C. OPTICALLY ACTIVE PHOSPHORUS COMPOUNDS FROM TRANSFORMATIONS OF OTHER P-CHIRAL PRECURSORS AND FROM ASYMMETRIC SYNTHESIS METHODS

9.C.1. Phosphine Oxides

As noted in the preceding section, optically active phosphonium salts and phosphines are available, and both serve as stereospecific precursors of optically active phosphine oxides. McEwen, et al.,[47] converted their optically active quaternary phosphonium salts to phosphine oxides by the action of strong alkali and also by making the ylide and performing Wittig olefination (Scheme 9.1). The former proceeds with inversion, the latter with retention. These procedures are most efficient when a benzyl group is present on phosphorus that can be sacrificed in the reactions, as shown in the equations.

Optically active phosphines give oxides with simple oxidation procedures, most commonly with peroxy compounds. These oxidations proceed with retention of configuration. Optically active phosphine sulfides are readily obtained by addition of sulfur to the phosphines, which proceeds with retention of configuration.

We have already noted that O-alkyl alkylphosphonothioic acids are readily resolved into their enantiomeric forms, and these have proved useful as precursors of optically active phosphine oxides. Thus, (+, −)-O-isopropyl methylphosphonothioic acid was easily resolved, in two recrystallizations, as the diastereoisomeric salt from (+) or (−) α-phenethylamine.[48] The separated salts were subsequently[49] alkylated and the resulting thioates (**24**) subjected to displacement with benzyl Grignard reagent (retention is known for such reactions); the resulting phosphinate was reacted with a different Grignard reagent (inversion) to form tertiary phosphine oxides (Eq. 9.12).

$$\text{(9.12)}$$

Optically active phosphine oxides, of course, are of major importance as precursors of optically active phosphines, and a number of other methods are available for their generation that take advantage of reactions with chiral reagents, both natural and synthetic, to induce an enantiomeric excess at a phosphorus function, followed by diastereomer separation. Optically active phosphinates, as encountered in Eq. 9.12, are generally the initial targets in this method; they are then

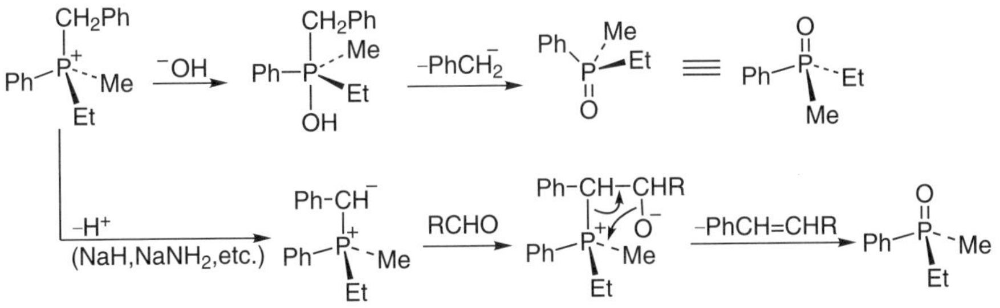

9.C. OPTICALLY ACTIVE PHOSPHORUS COMPOUNDS

Scheme 9.1

converted to the optically active phosphine oxides. An excellent early example of this approach is that of Mislow and co-workers,[50] who used (−)-menthol (MenOH) to prepare (−)-menthyl phosphinates with unequal amounts of the R- and S-enantiomeric forms at phosphorus (Eq. 9.13). These diastereomers were separated by fractional crystallization, and the individual forms converted to phosphine oxides by reaction with Grignard reagents (as well as alkyl lithiums). This displacement of the ester group is known to occur with inversion of configuration.

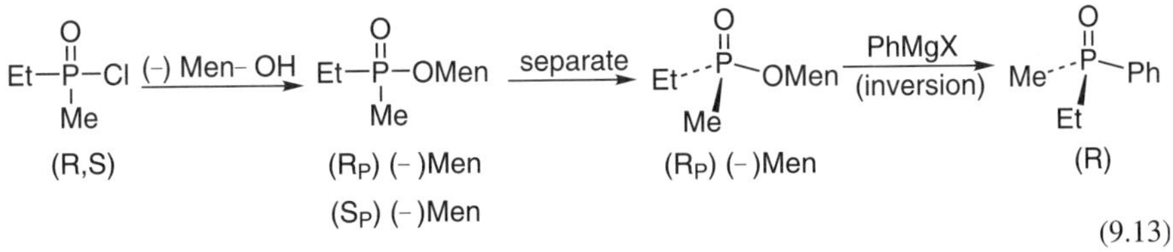

$$(9.13)$$

By this method, a number of simple optically active oxides, such as o-anisylmethylphenylphosphine oxide, have been obtained. This oxide is of particular significance, because it is a precursor of the ligand PAMP as well as the chelating diphosphine diPAMP, which also finds use in asymmetric homogeneous catalytic hydrogenation. The synthesis of diPAMP is shown in Eq. 9.14; it has been suggested[51] that this may constitute a commercial synthesis of this material.

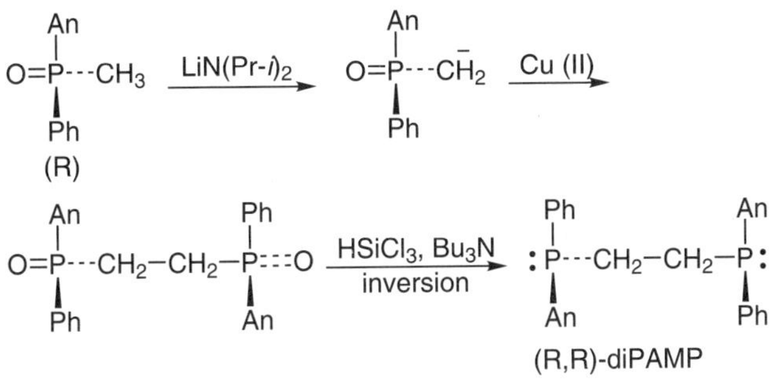

$$(9.14)$$

A related process[52] is to prepare a H-phosphinate with a chiral auxillary, and then replace hydrogen by forming the anion followed by alkylation (retention) to form the phosphinate. The ester group is then also replaced by alkyl to form the phosphine oxide with inversion. As seen in Eq. 9.15, the (−)-menthyl ester of an H-phosphinate can be used for this purpose; the mixture of diastereomeric H-phosphinates is partially separated by fractional crystallization, achieving a ratio of R to S isomers of 95 : 5. The authors used this mixture for conversion to the dialkyl phosphonate, and then conversion to the phosphine oxide with methyllithium.

$$\underset{H\ (R_P)}{\overset{OMen(-)}{O=P\text{-}Ph}} \xrightarrow[\text{(retention)}]{1.\ NaH, DMF}{2.\ i\text{-}PrI} \underset{i\text{-}Pr\ (R_P)}{\overset{OMen(-)}{O=P\text{-}Ph}} \xrightarrow[\text{(inversion)}]{MeLi} \underset{Me\ (R)}{\overset{i\text{-}Pr}{O=P\text{-}Ph}} \quad (9.15)$$

Another member of the chiral pool that has been used in an asymmetric synthesis of phosphinates for service as precursors to tertiary phosphine oxides is the ethyl ester of L-proline (**25**).[53] This is reacted with a phosphinic chloride (Eq. 9.16), and the resulting mixture of diastereomeric phosphinamides separated on a chromatographic column.

$$\text{25 (L)} + \text{(R,S) PhP(=O)(Cl)Me} \longrightarrow \text{(prolinyl-N-P(=O)(Ph)Me)} \xrightarrow[2.\ EtOH, H_2SO_4]{1.\ \text{separate}} \text{Ph-P(=O)(Me)OEt (R)} + \text{Ph-P(=O)(Me)OEt (S)} \quad (9.16)$$

The use of a carbohydrate derivative as a chiral auxillary is exemplified by Eq. 9.17.[36,54] Here a chiral cyclic phosphonate is prepared by reaction of a phosphonic dichloride with a blocked carbohydrate having only two free OH groups. Unequal amounts of the R_P and S_P configurations are formed, and these diastereomers are separated by crystallization (more easily than are the menthyl phosphinates[55]). Reaction with one mole of a Grignard reagent opens the cyclic phosphonate ring at C-6 of the glucose unit to give phosphinate **26**, and reaction with a second Grignard reagent displaces the remaining oxy function to give the phosphine oxide.

$$MePOF_2 + \text{(sugar with HO-CH}_2, \text{HO, MeO, MeO, MeO)} \xrightarrow{Et_3N} \text{(cyclic: Me-P(=O)-O-CH}_2\text{-sugar)} \ (R_P, S_P) \xrightarrow[2.\ RMgX]{1.\ \text{separate}}$$

$$\underset{\textbf{26}\ (R_P\ \text{or}\ S_P)}{\text{R-P(=O)(Me)-O-CH}_2\text{-sugar(HO, MeO, MeO, MeO)}} \xrightarrow{R'MgX} \underset{(R\ \text{or}\ S)}{Me\text{-P(=O)(R)(R')}} \quad (9.17)$$

A quite different approach to optically active phosphinates involves the synthesis of a cyclic phosphonite by reaction of an optically active diol with a phosphonous dichloride,[56–58] and then subjecting the phosphonite to a Michaelis–Arbusov reaction with an alkyl halide. In one method,[58] the product is transesterified to the methyl ester, which is then reacted with a Grignard reagent to displace the methoxy group and give the phosphine oxide (Eq. 9.18). The enantiomeric excess at phosphorus is quite good.

$$\text{PhPCl}_2 + \begin{array}{c}\text{HO--CH(Ph)}\\\text{HO--CH(Me)}\end{array} \xrightarrow{\text{then separation}} \text{Ph-P}\begin{array}{c}\text{O--CHPh}\\\text{O--CHMe}\end{array} \xrightarrow{\text{MeI}}$$

(R_P or S_P)

$$\underset{\underset{\text{Me Ph Me}}{|\ \ |\ \ |}}{\text{Ph--P(=O)--O--CH--CH}_2\text{I}} \xrightarrow[\text{H}^+]{\text{MeOH}} \underset{\underset{\text{Me}}{|}}{\text{Ph--P(=O)--OMe}} \xrightarrow{\text{PhCH}_2\text{MgBr}} \underset{\underset{\text{CH}_2\text{Ph}}{|}}{\text{Ph--P(=O)--Me}} \quad (9.18)$$

(R or S)

(−)-Ephedrine is another useful chiral building block, providing optically active 1,3,2-oxazaphospholidines on reaction with phosphonous dihalides. Methods exist for converting these to phosphinates and then to oxides[59–61] (as well as to phosphinite boranes, which will be seen later to be useful in chiral phosphine synthesis).

$$\text{PhP(NEt}_2)_2 + \begin{array}{c}\text{HO--CHPh}\\\text{HN--CHMe}\\|\\\text{Me}\end{array} \xrightarrow{\text{then separation}} \begin{array}{c}\text{Ph}\diagdown\ \ \text{O--CHPh}\\\text{P}\\\diagup\ \diagdown\\\text{N--CHMe}\\|\\\text{Me}\end{array} \xrightarrow{\text{RX (Arbusov)}}$$

$$\underset{\underset{\text{Ph Me}}{|\ \ \ |}}{\text{R--P(=O)--N--CHPhX--Me}} \xrightarrow[\text{(inversion)}]{\text{MeOH, H}^+} \underset{\underset{\text{Ph}}{|}}{\text{MeO--P(=O)--R}} \xrightarrow[\text{(inversion)}]{\text{R'MgX}} \underset{\underset{\text{Ph}}{|}}{\text{R--P(=O)--R'}} \quad (9.19)$$

In principle, a secondary phosphine oxide with two different carbon substituents can be converted to a chiral tertiary phosphine oxide, and when this step is accomplished by alkylation of the derived anion with an optically active halide a separable mixture of diastereomers would result. This sequence has in fact been successfully demonstrated[62] (Eq. 9.20). The optically active halide is the (−)-menthyl ester of chloroacetic acid, which gives a tertiary phosphine oxide with a

$$\underset{\underset{t\text{-Bu}}{|}}{\text{Ph--P(=O)--H}} + \text{ClCH}_2\text{COOMen-}(-) \xrightarrow{\text{NaH}} \underset{\underset{t\text{-Bu}}{|}}{\text{Ph--P(=O)--CH}_2\text{COOMen}} \xrightarrow[\text{2. KOH}]{\text{1. fract.cryst.}}$$

(R,S) \hspace{4cm} (R_P,S_P)

$$\underset{\underset{t\text{-Bu}}{|}}{\text{Ph--P(=O)--CH}_2\text{COOH}} \xrightarrow{160°} \underset{\underset{t\text{-Bu}}{|}}{\text{Ph--P(=O)--CH}_3} \quad (9.20)$$

(R and S)

carboxymethyl substituent. This can be simplified to the methyl group by hydrolysis to the carboxylic acid, followed by decarboxylation.

Optically active secondary phosphine oxides can also be obtained by reduction of resolved P=S or P=Se acids[63,64] (Eq. 9.21) and used to form active tertiary phosphine oxides. These secondary oxides can be de-protonated with retention of configuration, and then subjected to alkylation with a halide to give the optically active tertiary phosphine oxide with the same configuration as the starting secondary oxide.

$$\text{Ph}\overset{X}{\underset{t\text{-Bu}}{-P-}}\text{OH} \xrightarrow{\text{Raney Ni}} \text{Ph}\overset{O}{\underset{t\text{-Bu}}{-P-}}\text{H} \xrightarrow[\text{2. RX}]{\text{1. LiNEt}_2} \text{Ph}\overset{O}{\underset{t\text{-Bu}}{-P-}}\text{R} \quad (9.21)$$

(S) (X=S or Se)

Once obtained, optically active phosphine oxides with a reactive site on a carbon substituent can be used to generate new compounds. An excellent example is that provided by a phosphine oxide with a P-vinyl substituent.[65] Thus (−)-(S)-methylphenylvinylphosphine oxide (**27**) was prepared by the reactions of Eq. 9.22,[65] and then subjected to a variety of typical conjugated double-bond reactions, such as Michael additions (including secondary phosphine oxides, valuable in providing access to optically active bis(phosphinyl)ethanes), Diels–Alder, and 1,3-dipolar cycloadditions, as well as halogen addition and PdAc$_2$ catalyzed coupling with aryl halides (the Heck reaction). The diversity of products obtainable from these reactions is quite wide, and they are well summarized in a recent review.[66] We show here only the 1,3-cycloaddition of a nitrone,[67,68] because this type of reaction was later applied in admirable fashion to provide optically active cyclic phosphine oxides (vide infra). In this reaction, a 71 : 29 mixture of diastereomeric adducts was formed that was found to be readily separable into the isomers.

$$\text{Ph-P}\overset{\diagup\!=}{\underset{\text{OBu}}{}} \xrightarrow[\text{(Arbusov)}]{\text{ClCH}_2\text{COO}((\text{-})\text{-Menthyl})} \text{Ph-}\overset{O}{\underset{\text{CH}_2\text{COOMen}}{\overset{\|}{P}}}\!\!\diagup\!= \xrightarrow[\text{2. H}_2\text{O}]{\text{1. separate}} \text{Ph-}\overset{O}{\underset{\text{CH}_2\text{COOH}}{\overset{\|}{P}}}\!\!\diagup\!= \xrightarrow{-\text{CO}_2}$$

27 (S) (29%) (71%) (9.22)

Optically active nitrones (e.g., **28**) cycloadd in a doubly asymmetric, highly stereoselective process to give the erytho (**29**) and threo (**30**) adducts[69] (Eq. 9.23). The erythro adduct heavily dominates (for R=Me, 96 : 4); of its two configurations possible at C-5, the endo is favored over exo, e.g., by 93 : 3 for R=Me.

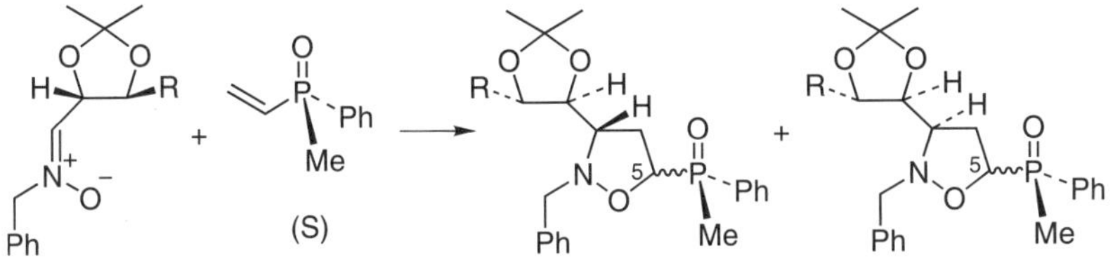

[Scheme showing reaction 9.23 with compounds 28, 29(erythro), 30(threo)]

The authors also applied this methodology to a racemic cyclic vinyl phosphine oxide, the well-known 1-phenyl-2-phospholene oxide.[70] When a 50% excess of the phospholene oxide was used in reaction with optically active nitrile oxide 31 (derived from L-tartaric acid), the unreacted amount was almost completely in the R form, and thus a kinetic resolution of the phospholene oxide had been accomplished. This is represented by Eq. 9.24, which also reveals the high selectivity in the formation of the cycloadduct.

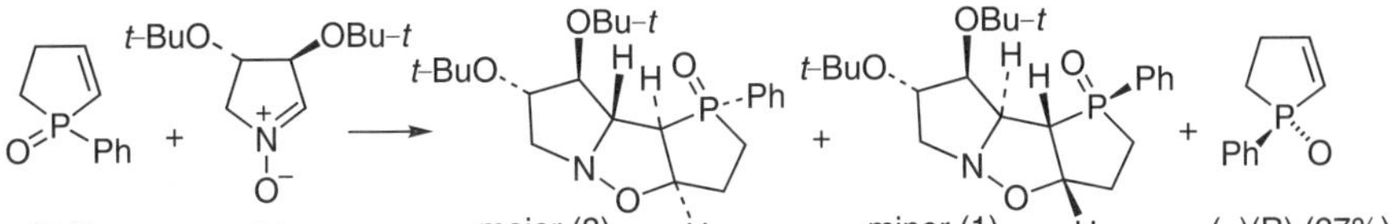

[Scheme for equation 9.24 showing (R,S) phospholene oxide + 31 → major (3) + minor (1) + (−)(R), (27%)]

(9.24)

Several other derivatives of the 1-phenyl-2-phospholene oxide system (32 to 34) were also resolved kinetically by the dipolar cycloaddition process.[71] Their absolute configurations were determined to be R. The P-ethoxy derivative 35 has been resolved (to give the S-form) with a different nitrone, and it is obvious that this method constitutes a major breakthrough in the stereochemistry of heterocyclic phosphorus compounds, since many compounds can be formed from phospholenes.

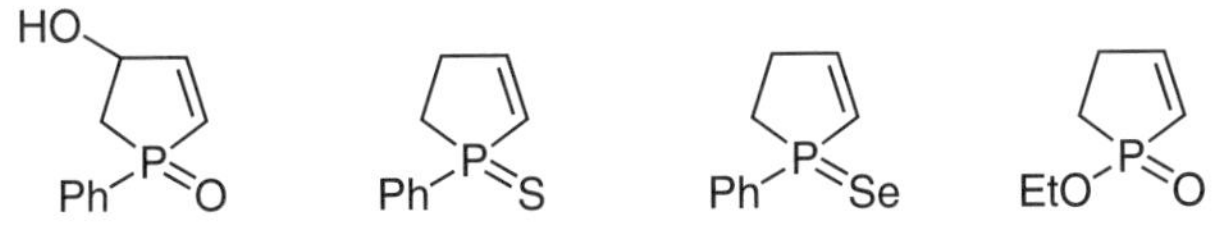

[Structures 32, 33, 34, 35]

Another advance in the stereochemistry of cyclic phosphine oxides was made by Brèque, et al.,[72] who carried out a Diels–Alder reaction (Eq. 9.25) on the product of thermal [1,5] sigmatropic rearrangement of a phosphole (as seen in chapter 8.B) with an acetylene derivative. Isomeric reaction products were separated and oxidized to the phosphine oxide, and the enantiomers resolved by HPLC on a chiral column (5-thioDNBTyr-E).

So much effort has gone into the development of practical methods for the synthesis of optically active phosphine oxides that the discussion in this section is somewhat limited and is designed more to be instructive of the principles involved

$$\text{(9.25)}$$

rather than to be a complete survey. The detailed recent review by Pietrusiewicz[7] provides additional references to advances in this important subject.

9.C.2. Phosphines

One of the most significant events in phosphorus stereochemistry was the discovery by Horner, et al., in 1961[5] that tertiary phosphines with three different substituents could exist in stable optically active forms. Horner's group found that optically active quaternary phosphonium salts containing a benzyl group, prepared by the resolution technique of Kumli, et al.,[3] using D-dibenzoyl tartrates, could be electrolytically reduced at a mercury cathode with loss of the benzyl group to give optically active phosphines (Eq. 9.26). Three optically active phosphines (n-propylMePhP; allylMePhP; MeEtPhP) were reported in this ground-breaking paper, and many other examples in later papers (cited in reference 8).

$$\text{(9.26)}$$

A later discovery was that cyanide ion caused cleavage of a benzyl or allyl group from a quaternary salt, giving the phosphine with predominant retention. Lithium aluminum hydride also could be used for the dealkylation, but racemization accompanied the retention pathway. These techniques were also applied successfully to the synthesis of optically active aminophosphines from resolved aminophosphonium salts[73] (Eq. 9.27).

$$\text{(9.27)}$$

Many of the methods already outlined for the synthesis of optically active phosphine oxides were developed with the goal in mind of using them as precursors of phosphines, because it is far easier experimentally to work with the phosphines as final products rather than as early reactants in a complex scheme. The task of deoxygenating phosphine oxides in a stereospecific manner can be accomplished by using silane derivatives as reducing agents. Many examples have been reported and reviewed.[7–9] Thus phenylsilane, trichlorosilane or the milder trichlorosilane-pyridine complex are frequently used for this purpose, giving the phosphines with high enantiomeric purity and with retention of the oxide configuration. On the other hand, reduction with inversion may be accomplished with hexachlorodisilane or a mixture of triethylamine (or amines of comparable basicity) and trichlorosilane. Mechanisms that account for these stereochemical results were presented in chapter 3.C.

Phosphine boranes, R_3P-BH_3, are another class of compounds finding use as a form of blocked phosphine derivative on which useful chemistry can be performed, following which de-blocking is accomplished by reaction with a large excess of diethylamine (or morpholine) at 50°.[74] The concept can be illustrated by the reactions in Eq. 9.28, in which a mixture of diastereomeric (−)-menthyl phosphinites (**36**) is prepared, converted to BH_3 adducts and separated by preparative TLC, giving isomer **37** in 100% ee and isomer **38** in 93% ee. Each isomer then undergoes replacement of the menthyloxy group by methyl with retention (95% ee) on reaction with methyllithium (Eq. 9.29). Removal of the borane group was accomplished with diethylamine to give the phosphine with 89% ee. Here the synthesis of the popular target phosphine PAMP is shown.

$$\text{PhPCl}_2 + o\text{-AnisylMgBr} \longrightarrow \text{Ph-P(Cl)-An} \xrightarrow[\text{pyridine}]{(-)\text{menthol}} \underset{\textbf{36 }(R_P, S_P)}{\text{Ph-P(O-Men)-An-}o} \xrightarrow{BH_3 \cdot THF}$$

$$\text{Ph-P(BH}_3\text{)(OMen)(An-}o\text{)} \xrightarrow[\text{(TLC)}]{\text{separate}} \underset{\textbf{37 }(S)}{o\text{-An-P(BH}_3\text{)(OMen)(Ph)}} + \underset{\textbf{38 }(R)}{\text{Ph-P(BH}_3\text{)(OMen)(An-}o\text{)}} \quad (9.28)$$

$$\textbf{37 or 38} \xrightarrow{\text{MeLi}} \underset{(R \text{ or } S)}{\text{Ph-P(BH}_3\text{)(Me)(}o\text{-An)}} \xrightarrow{Et_2NH} \underset{(R \text{ or } S)}{\text{Ph-P(Me)(}o\text{-An)}} \quad (9.29)$$

Borane complexes are also used in a chiral phosphine synthesis based on phosphonous acid derivatives in which ephedrine is used as the chiral auxiliary to obtain separable diastereomers (Eq. 9.30).[75] Thus the cyclic phosphonamidite **39** (a

$$\text{PhP(NEt}_2)_2 + \underset{\text{Me}}{\underset{|}{\text{HN}}}\text{-}\underset{\text{Me}}{\overset{\text{HO-}}{\text{-Ph}}} \longrightarrow \underset{\textbf{39}}{\overset{\text{Ph}}{\underset{\text{Me}}{\text{P-O-}}}\text{-Ph}} \xrightarrow{BH_3 \cdot H_2S} \underset{BH_3}{\overset{\text{Ph}}{\text{P-O-Ph}}}\underset{\text{Me}}{\text{N-Me}} \xrightarrow[\text{retention}]{RLi}$$

$$\underset{BH_3}{\overset{\text{Ph}}{\text{P}}}\underset{\underset{\text{Me}}{|}}{\overset{\text{HO}}{\text{N}}}\text{-Ph} \xrightarrow[\text{inversion}]{\text{MeOH, HCl}} \underset{BH_3}{\overset{\text{MeO}}{\text{P-Ph}}}\text{R} \xrightarrow[\text{inversion}]{R'Li} \underset{BH_3}{\overset{\text{Ph}}{\text{P}}}\text{R-R'} \xrightarrow{Et_2NH} \overset{\text{Ph}}{\text{P}}\text{R-R'}$$

$$(9.30)$$

1,3,2-oxazapholidine) is first formed from ephedrine and then complexed with borane. The BH_3 complex is obtained in diastereomerically pure form. Two displacements with different alkyllithiums (retention) give the optically active phosphineborane, from which the phosphine is released with diethylamine.

Yet another application of phosphine boranes is seen in the report by Muci, et al.,[76] where an achiral phosphine-borane with two P—Me groups (**40**) undergoes enantioselective deprotonation with a chiral base. The base employed is the (−)-sparteine complex (**41**) of *sec.*-butyllithium. The anion is condensed with benzophenone to form an optically active phosphine-borane, having the (S) configuration. Enantiomeric excesses were in the 79 to 87% range for several different P—Ar groups. The phosphine-boranes were readily stripped of borane by reaction with diethylamine at 55° (Eq. 9.31). A similar chiral base was also applied to achiral phosphine sulfides and the anion condensed with benzophenone, again to give a chiral product (60 to 81% ee). Sparteine also has been used to effect a dynamic resolution of a racemic secondary phosphine-BH_3 adduct (*t*-BuPhHP—BH_3); this was accomplished by preparing the anion of the adduct and subjecting it to alkylation with *n*-BuLi in the presence of sparteine. The tertiary phosphine-BH_3 adduct was obtained with 95% ee.[77]

$$\text{(9.31)}$$

A novel type of optically active cyclic phosphonous acid derivative useful for the synthesis of phosphines and phosphine-boranes has structure **42**; it is prepared (stereospecifically) as shown in Eq. 9.32 and subjected to two displacement reactions

$$\text{(9.32)}$$

9.C. OPTICALLY ACTIVE PHOSPHORUS COMPOUNDS

with lithium reagents, first to open the P—S bond (retention) and then to replace the alkoxy group (inversion) to give the phosphine sulfide. The phosphine was prepared by Si_2Cl_6 reduction (on sulfides with retention,[7] unlike oxides) and isolated as the borane complex.[78] PAMP was prepared by this method. Compound **42** can also be used in the double displacements to give the tertiary phosphine.

In the several papers[7] in which PAMP was prepared from the phosphine-boranes as well as one in which it was approached through reduction of the corresponding optically active phosphine sulfide (seen in Eq. 9.32) and then isolated as the borane complex, the further transformation to the important chelating ligand diPAMP was achieved. This makes use of a coupling of the carbanion formed at the methyl group, promoted by the action of $CuCl_2$ (Eq. 9.33). These various methods are of great value for the production of diPAMP in high optical purity as well as related diphosphines with different P-substituents.

$$\text{o-An} \overset{BH_3}{\underset{Ph}{-P-CH_3}} \xrightarrow{RLi} \text{o-An} \overset{BH_3}{\underset{Ph}{-P-\bar{C}H_2}} \xrightarrow{CuCl_2}$$

$$\text{o-An} \overset{BH_3}{\underset{Ph}{-P-CH_2-CH_2-}} \overset{BH_3}{\underset{An\text{-}o}{-P-Ph}} \xrightarrow{Et_2NH} \text{o-An} \overset{..}{\underset{Ph}{-P}} \overset{..}{\underset{An\text{-}o}{-P-Ph}} \quad (9.33)$$

(R,R)-di PAMP

Processes that lead directly to optically active phosphines are not unknown, and an early one is that of Chodkiewicz, et al.,[79] which started with optically active phosphinites (**43**) prepared from the alkaloidal alcohol cinchonine. The diastereoisomers, although greatly enriched in one isomer and useful in this form, could be separated by crystallization of their CuCN complexes. After decomplexation with KCN, the optically active phosphinites were alkylated with various Grignard reagents to provide the optically active phosphines in up to 80% ee (Eq. 9.34).

$$\underset{R}{Ph\overset{..}{P}-Cl} \xrightarrow{\text{cinchonine}} \underset{R}{Ph-\overset{..}{P}-\text{O-cinch}} \xrightarrow[\text{(inversion)}]{R'MgX} \underset{Ph}{R-\overset{..}{P}-R'} \quad (9.34)$$

43

Another group[80] later used the di-(−)-menthyl ester of phenylphosphonous acid as a prochiral reagent, which was found to give the phosphinite **44** in high diastereomeric purity on reaction with alkyllithiums (Eq. 9.35). Further reaction with an alkyllithium (inversion) gave the phosphine, but with some loss of optical activity.

$$PhP(OMen)_2 \xrightarrow{RLi} Ph-\overset{..}{\underset{OMen}{P}}-R \xrightarrow[\text{(inversion)}]{R'Li} Ph-\overset{..}{\underset{R}{P}}-R' \quad (9.35)$$

44

An excellent technique for converting a primary phosphine into an optically active cyclic tertiary phosphine was developed by Burk and co-workers.[81] The method was designed to produce bis-phosphino derivatives as ligands for coordination complexes of value in homogeneous catalysis. The chiral auxiliary is the cyclic sulfate **45**; this is reacted with the bis-phosphide ion produced from a bis-phosphine to give secondary phosphines of structure **46** from nucleophilic ring opening with inversion. The secondary phosphine groups are then deprotonated, and a second attack occurs on the sulfate groups to form the phospholane ring at each phosphorus atom. The yield and enantiomerc purity of the products are very high. The process is illustrated in Scheme 9.2; as described in reference 81, the method can be applied both to bis-phosphinoethane and 1,2-bis-phosphinobenzene (forming **47**), both of which are commercially available. The cyclic sulfate can be constructed with various ring substituents (Me, Et, Pr, *i*-Pr) that then appear on the phospholane ring that is formed. Compounds of structure **47** are very important ligands known as the DuPHOS family.

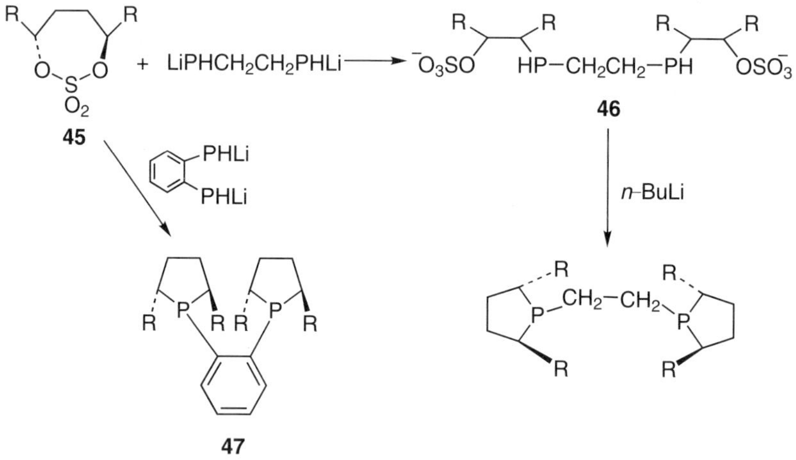

Scheme 9.2

A later extension of the method to a bis-phosphine (**48**) based on a ferrocene skeleton[82] provides an example of the versatility of the process, and other applications seem possible.

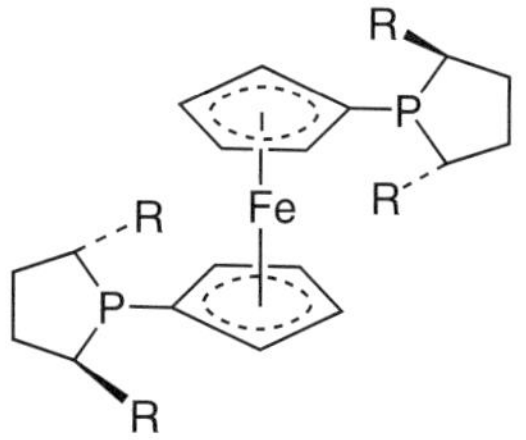

An optically active secondary phosphine figures in a synthesis of diphosphines and triphosphines developed by King, et al.[83] In this process (Eq. 9.36), optically active neomenthylmethylphosphine is prepared and reacted with diphenylvinylphosphine. The adduct **49** is a mixture of diastereomers that is separated by fractional crystallization. Divinylphenylphosphine was also used in this process, giving a triphosphine **50**, indeed the first triphosphine to have chirality.

$$\text{neo-Men}-\underset{\text{Ph}}{\overset{|}{\text{P}}}-\text{H} \quad \underset{\text{KOBu-}t}{\overset{\text{Ph}_2\text{PCH}=\text{CH}_2}{\diagup}} \quad \text{neo-Men}-\underset{\text{Ph}}{\overset{|}{\text{P}}}\text{CH}_2\text{CH}_2\text{PPh}_2 \quad \mathbf{49}$$

$$\underset{\text{KOBu-}t}{\overset{\text{PhP(CH=CH}_2)_2}{\diagdown}} \quad \text{PhP}\left[\text{CH}_2\text{CH}_2\text{PPh (neo-Men)}\right]_2 \quad \mathbf{50}$$

(9.36)

Examples of kinetic resolution of phosphines are known, but this approach has not yet received the attention that it deserves. The first attempt was reported by Wittig and co-workers,[84] who carried out an alkylation of a racemic phosphine (*p*-biphenylyl-1-naphthylphenyl; Eq. 9.37) with only half of the necessary amount of paraformaldehyde and, as an acid catalyst, (+)-camphor-10-sulfonic acid (CSA). It was then found that the unreacted phosphine had a modest excess (as confirmed by later workers[85]) of the (−)-enantiomer. The salt product was decomposed by triethylamine to regenerate the phosphine, with an excess of the other enantiomer.

(9.37)

An approach reported by Verfürth and Ugi[86] for the kinetic resolution of some phosphite derivatives, consisting of partial oxidation with optically active oxidizing agents (camphor and fenchone *N*-sulfonyloxaziridines) seems applicable to phosphines.

In principle, other types of 3-coordinate phosphorus compounds are potentially capable of existence in enantiomeric forms, provided the barrier to pyramidal inversion is sufficiently high to prevent rapid racemization. In fact, we have already noted many instances in which certain phosphonous and phosphinous acid derivatives have been synthesized in optically active form, and other derivatives are also known (summarized in reference 87). Phosphinous chlorides are also potentially capable of existence in enantiomeric form, and indeed the synthesis of optically active *t*-butylphenylphosphinous chloride as well as methyl *t*-butylphenylphosphinothioite has been described (Eq. 9.38).[87] The chloride was of low configurational stability, and the optical purity (O.P.) was not high. In a recent paper,[88] examples of stable diastereomeric forms of cyclic chlorodiaminophosphines (1,3,2-diazaphosphorinane derivatives) bearing a chiral auxiliary have been described, and their conversion to diastereomerically pure triaminophosphines and other derivatives accomplished.

$$\underset{\underset{(-)\,(S)\,(63\%\,o.p.)}{}}{\overset{Ph}{\underset{Cl}{t\text{-}Bu\text{---}\overset{|}{P}\text{=}S}}} \xrightarrow{CF_3SO_3Me} \underset{}{\overset{Ph}{\underset{Cl}{t\text{-}Bu\text{---}\overset{|+}{P}\text{--}SMe}}}\;CF_3S\bar{O}_3 \xrightarrow[-70°]{P(NMe_2)_3} \underset{(+)\,(S)}{\overset{Ph}{\underset{Cl}{t\text{-}Bu\text{---}\overset{|}{P}\text{--}:}}} \quad (9.38)$$

A Pd(II) complex of iso-propylphenylphosphinous chloride was later resolved by Wild and co-workers,[89] but the free phosphinous chloride could not be successfully released from the complex. However, reaction of the complex with methanol gave methyl iso-propylphenylphosphinite with complete stereoselectivity (inversion). Optically active phosphinous esters and amides, as made by methods described previously in this chapter and by other methods, in general appear to have lower pyramidal stability than do tertiary phosphines; some esters racemize fairly readily even at room temperature. A notable process[90] that gave methyl t-butylphenylphosphinite in high optical purity is shown as Eq. 9.39.

$$\underset{(-)\,(S)}{\overset{Se}{\underset{t\text{-}Bu}{Ph\text{---}\overset{\|}{\underset{|}{P}}\text{--}OH}}} \xrightarrow[2.\,MeI]{1.\,Et_3N} \underset{}{\overset{Se}{\underset{t\text{-}Bu}{Ph\text{---}\overset{\|}{\underset{|}{P}}\text{--}OMe}}} \xrightarrow{CF_3SO_3Me}$$

$$\underset{}{\overset{SeMe}{\underset{t\text{-}Bu}{Ph\text{---}\overset{|+}{\underset{|}{P}}\text{--}OMe}}}\;CF_3S\bar{O}_3 \xrightarrow{Et\bar{S}} \underset{(+)\,(R)}{\overset{\cdot\cdot}{\underset{t\text{-}Bu}{Ph\text{---}\overset{|}{P}\text{--}OMe}}} \quad (9.39)$$

Phosphinous chlorides in racemic form can figure as precursors of optically active phosphinites. Thus, when reacted with an alcohol in the presence of an optically active amine,[91] an optically active phosphinite is formed. This result suggests that the amine may have first complexed with the phosphinous chloride to give a stereochemical bias to the ensuing reaction with the alcohol (Eq. 9.40).

$$\underset{(R,S)}{\overset{\cdot\cdot}{\underset{Ph}{Et\text{--}\overset{|}{P}\text{--}Cl}}} \xrightarrow[\underset{Me}{(-)PhCHNMe_2}]{MeOH} \underset{(+)\,(R)}{\overset{\cdot\cdot}{\underset{Et}{Ph\text{---}\overset{|}{P}\text{--}OMe}}} \quad (9.40)$$

9.C.3. Phosphoric Acid Derivatives

Efficient and versatile techniques have been developed for the synthesis of optically active phosphoric acid derivatives through the use of chiral auxiliaries. The natural aminoalcohol (−)-ephedrine and the diol **51** synthesized from D-glucose are the most commonly used; these are reacted with $PSCl_3$ or $POCl_3$ to form cyclic diastereomeric

9.C. OPTICALLY ACTIVE PHOSPHORUS COMPOUNDS

chlorophosphate derivatives, which are useful precursors of other structures (Eq. 41 and 42 respectively).

$$PXCl_3 + \begin{array}{c}HO\text{-}\overset{\text{\textbackslash}\text{Ph}}{\underset{Me}{HN}}\text{-}Me\end{array} \longrightarrow \begin{array}{c}\text{Cl-P(X)(O)(N(Me))}\text{-}\text{Ph, Me}\end{array} \quad (9.41)$$

$$PXCl_3 + \mathbf{51} \longrightarrow \text{(cyclic chlorophosphate of sugar derivative)} \quad (9.42)$$

Ephedrine appears to be the reagent of choice; it is readily available and gives easily separated[92] diastereomeric derivatives. Norephedrine is also useful. These reactions have led to the synthesis of a great number of optically active phosphates, many of which are listed in a review by the originators of these techniques, Hall and Inch,[14] as well as that of Valentine.[8] To illustrate the chemistry behind this approach, the synthesis based on ephedrine of a thiophosphate derivative,[93] a common target of such work, is presented in Eq. 9.43.

$$PSCl_3 + \text{HN(Me)CH(Me)CH(OH)Ph} \longrightarrow \text{cyclic Cl-P(=S)} \xrightarrow{\text{crystallization (60\%)}} (R_P, S_P) \xrightarrow{\text{NaOEt (retention)}}$$

$$\text{EtO-P(=S) cyclic } \mathbf{52} \xrightarrow{\text{NaOH, H}_2\text{O (retention)}} \text{EtO-P(=S)(O}^-\text{)-N(Me)CH(Me)CH(OH)Ph} \xrightarrow{\text{1. MeI} \atop \text{2. Ac}_2\text{O}} \text{EtO-P(=O)(SMe)-N(Me)CH(Me)CH(OAc)Ph}$$

$$\xrightarrow{\text{CH}_3\text{OH, HCl (inversion)}} \text{MeO-P(=O)(SMe)(OEt)} \quad (S) \quad (9.43)$$

Optically active phosphoric acid esters, such as ethyl methyl iso-propyl phosphate, can also be obtained from oxazaphospholidine **52**, as shown in Eq. 9.44, and other syntheses are also possible.

$$\mathbf{52} \xrightarrow[\text{2. NaOH, 3.H}^+]{\text{1. MeOH,HCl}} \text{HS-P(=O)(OMe)(OEt)} \xrightarrow{PCl_5} \text{Cl-P(=S)(OMe)(OEt)} \xrightarrow{\text{NaOPr-}i \text{ (inversion)}}$$
$$(R) \qquad (R)$$

$$i\text{-PrO-P(=S)(OEt)(OMe)} \xrightarrow[\text{retention (80\%)}]{MCPBA} i\text{-PrO-P(=O)(OEt)(OMe)} \quad (9.44)$$
$$(R) \qquad\qquad (S)$$

It has even proved possible to synthesize an optically active phosphate monoester by the chiral template approach. To achieve the necessary chirality, the three singly bonded oxygen atoms must be differentiated by isotopic labeling with ^{16}O, ^{17}O, and ^{18}O, thus providing the species $RO-P(^{16}O)(^{17}O)(^{18}O)^{2-}$. This remarkable structure was required for the elucidation of certain biochemical phospho transfer processes, in which it has proved quite effective (reference 8 provides many references, as does a review by Knowles[94]). It was first synthesized in 1978[95] by an ingenious process involving (−)-ephedrine as the chiral template (Eq. 9.45) and as the source of ^{16}O. This was reacted with $^{17}OPCl_3$ to provide the 1,3,2-oxazaphospholidine (**53**). The desired alkoxy group is then attached by replacement of the chlorine, and the P−N bond selectively hydrolyzed with $^{18}OH_3^+$ to install the third required oxygen isotope. Catalytic hydrogenation then cleaves the ephedrine moiety at its benzylic carbon, releasing the labelled phospho monoester.

$$(9.45)$$

Cullis and Lowe[96] announced another process starting with ^{18}O-labelled optically active benzoin (**54**) as the chiral auxiliary. Reduction to the diol is followed by reaction with $^{17}OPCl_3$ to give a cyclic chlorophosphate, then converted to the cyclic triester (Eq. 9.46). Because the two carbons in the ring are benzylic in character, catalytic hydrogenation can be used to cleave the C−O bonds, which gives the desired labelled phosphomonoester.

$$(9.46)$$

The isotopically labeled phosphomonoesters have small but measurable optical rotations. ^{31}P NMR has played a major role in the stereochemical analyses of the isotopically labelled phosphates. Replacement of O^{16} by O^{18} leads to a small but detectable high-field shift (e.g., 1.27 Hz in (MeO)$_3$P=O at 36.43 MHz[97]). Another aid is the broadening effect of the quadrupolar ^{17}O on the ^{31}P signal.

A similar method (Eq. 9.47) was used by Lowe[98] to give an improved route to [^{16}O,^{17}O,^{18}O]thiophosphate, which had previously been reported by two other groups.[99]

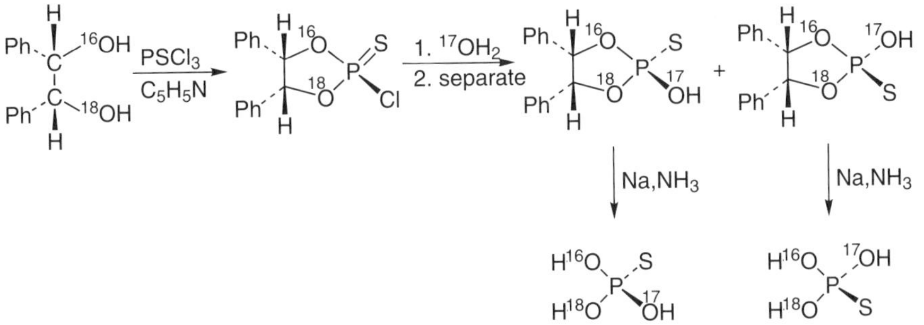

(9.47)

9.C.4. Phosphorane Derivatives

It has been noted, for example by Gallagher,[100] that an acyclic phosphorane with five different substituents will lack symmetry and would be capable of optical isomerism; however, there will be 20 possible isomers and 10 pairs of enantiomers! Complicated also by rapid isomerization among many of these trigonal bipyramidal compounds (chapter 2.F), the difficulties in obtaining an individual optically active structure of this type are indeed great. The problem is simplified, however, in spirocyclic structures in which pseudorotation is energetically more difficult, and indeed examples are known of such compounds that are optically active. Thus Hellwinkel[101] succeeded in preparing phosphorane **56** in optically active form by acidification of an optically active 6-coordinate salt **55** (Eq. 9.48). This salt had been obtained by a resolution procedure and had $[\alpha]^{24}_{578} = \pm 1870°$. Compound **56** was recovered from a mixture of products and found to have $[\alpha]^{24}_{578} = \pm 94.1°$. In this special case, geometrical constraints prevent the development of symmetry in the pseudorotation process, and thus optical activity is retained even though pseudorotation may still be occurring. This point is further developed by Gallagher.[100]

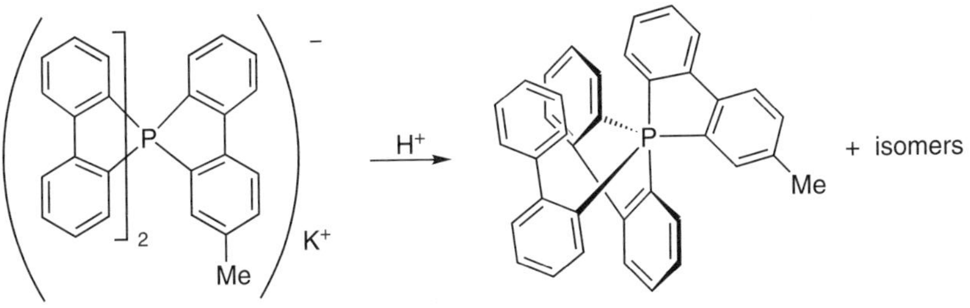

(9.48)

55   **56**

Other 5-coordinate compounds are known that have chirality in attached substituents, some of which exist in diastereomeric forms due to additional chirality at phosphorus. Because of the rarity of this phenomenon, only some structures (**57**,[102] **58**,[103] and **59**[104]) and references will be given here.

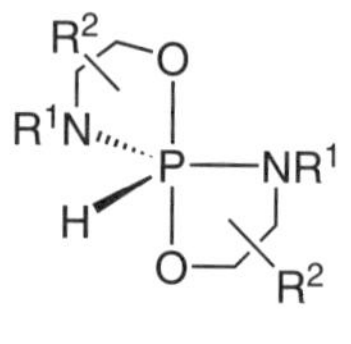

57

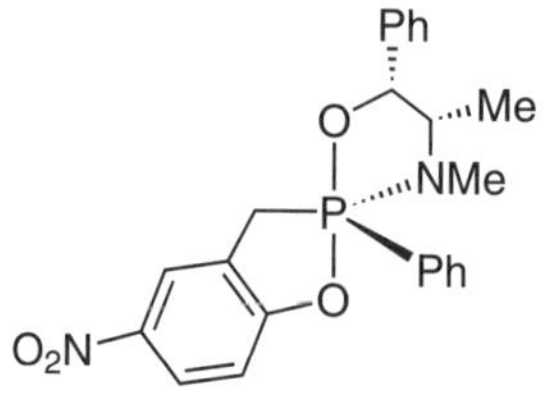

58

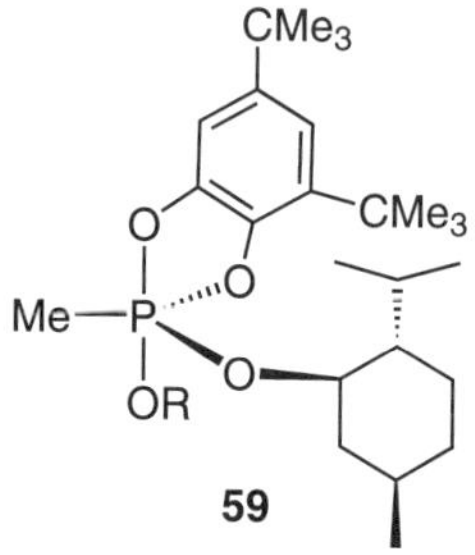

59

9.D. KINETIC RESOLUTION OF PHOSPHORUS COMPOUNDS BY ENZYMES

Enzymatic systems have been used to perform kinetic resolution of many carbon compounds, in which one enantiomer of a racemic mixture is chemically changed with little or no effect on the other. A recent review[40] points out some examples of enzymatic reactions on heteroatom functional groups, and stresses some of the valuable features to be found in kinetic resolutions of this type. There appears, however, to be only one case on record in which an enzyme has been used to selectively catalyze a reaction at a phosphorus center. This report is of interest for another reason; it constituted the first isolation of an optically active phosphotriester. Dudman and Zerner[105] found that when racemic n-butyl methyl p-nitrophenyl phosphate was hydrolyzed with either beef or horse serum hydrolase, the hydrolysis (Eq. 9.49) proceeded with greater than 99% enantioselectivity for the (−) enantiomer; the unchanged (+)-enantiomer was recovered and had $[\alpha]_D^{25}$ +8.48°.

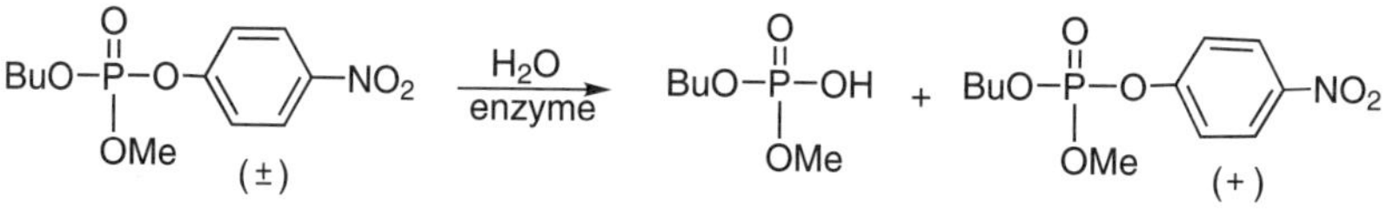

$$\text{BuO-P(=O)(OMe)-O-C}_6\text{H}_4\text{-NO}_2 \; (\pm) \xrightarrow[\text{enzyme}]{\text{H}_2\text{O}} \text{BuO-P(=O)(OMe)-OH} + \text{BuO-P(=O)(OMe)-O-C}_6\text{H}_4\text{-NO}_2 \; (+) \quad (9.49)$$

Despite this impressive accomplishment, it was some 20 years before another resolution by a stereoselective enzymatic reaction was reported.[106] In this case, racemic phosphine oxides (e.g., **60**) bearing a carboxylate group were selectively hydrolyzed with pig liver esterase (PLE); the unreacted substrates were optically active and were recovered in enantiomeric excesses of 80 to 100%. The acids from the hydrolysis were also recovered with about 80% ee (Eq. 9.50).

$$\underset{\underset{\underset{60}{(\pm)}}{\overset{\overset{O}{\|}}{\underset{Ph}{Me-P-CH_2COOMe}}}}{} \xrightarrow[\text{50\% conversion}]{H_2O, \text{PLE}} \underset{Ph}{\overset{\overset{O}{\|}}{Me^{\cdots}P-CH_2COOH}} + \underset{\underset{(+)(R)}{Me}}{\overset{\overset{O}{\|}}{Ph^{\cdots}P-CH_2COOMe}} \quad (9.50)$$

The process was extended to other types of phosphoryl derivatives, of structures PhP(O)(OR)CH₂COOMe, EtP(O)(OR)CH₂COOMe, PhOP(O)(OEt)CH₂COOMe, and Et₂NP(O)(OMe)CH₂COOMe. In these cases, the products were isolated with lower ee values, with the exception of PhP(O)(OMe)CH₂COOMe, in which the ester was recovered with 95% ee. The recovered acids could be decarboxylated, converting the —CH₂COOH group to CH₃. In some cases this gave compounds of already known absolute configurations.

In another 1994 study,[107] both phosphines and phosphine oxides of structures **61** (and some related structures) were hydrolyzed at the ester group with the lipase from *Candida rugosa* and cholesterol esterase. The enantioselectivity with both enzymes was quite high. On a 1-g scale using the former enzyme, compound **61**, X=O, gave the phenol product **62** in 88% ee (Eq. 9.51); the unreacted (S)-(+) form of **61** was recovered with 90% ee.

It is of interest that the enzyme shows selectivity on enantiomers in which the group being acted on is rather remote from the chiral center. These results, and those of the other two studies cited, do suggest that other useful applications of kinetic resolution by enzymes can be expected in the future.

9.E. SUMMARY OF SOME STEREOSELECTIVE REACTIONS OCCURRING AT PHOSPHORUS FUNCTIONS

From a combination of studies based on optically active compounds and on heterocyclic compounds with known stereo structure, much information has been developed on the stereochemical pathway of reactions occurring at phosphorus, some of which have been seen in this as well as in earlier chapters. As in carbon chemistry, stereochemical studies have played a major role in the elucidation of the mechanism of reactions at phosphorus centers. The mechanisms of most of the important reactions were outlined in chapters 3 to 5 as the various functional groups were presented and will not be repeated here; however, these mechanisms might be re-examined for a better appreciation of the stereochemical consequences of the reactions. Also, detailed summaries of the stereochemistry and mechanism of many reactions can be found in the review literature.[7–9] It is the purpose of this section, then, merely to state the stereochemical outcome of important reactions at phosphorus in summary form to serve as a rapid guide in research planning (Scheme 9.3). A statement that a reaction occurs with retention (or inversion) should not be construed to mean that the reaction has 100% stereospecificity, but that the reaction proceeds with very high stereoselectivity. Few reactions in fact proceed with complete stereospecificity.

Scheme 9.3

A. Reduction of Phosphine Oxide

Retention (except inversion with phosphetane oxides) with $PhSiH_3$; $HSiCl_3$; $HSiCl_3$–Pyridine

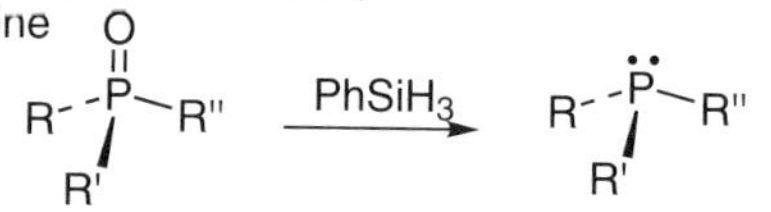

$$R(R')(R'')P=O \xrightarrow{PhSiH_3} R(R')(R'')P:$$

Inversion with Si_2Cl_6; $HSiCl_3$–Et_3N ($LiAlH_4$ causes racemization)

$$R(R')(R'')P=O \xrightarrow{Si_2Cl_6} R(R''')(R')P:$$

B. Reduction of Sulfides and Selenides

Retention with Si_2Cl_6; $LiAlH_4$

$$R(R')(R'')P=S \xrightarrow{Si_2Cl_6} R(R')(R'')P:$$

C. Oxidation of Phosphines

Retention with H_2O_2; t-BuOOH; $ArCO_3H$; O_2; O_3

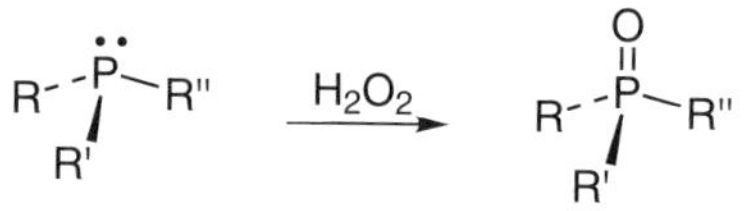

$$R(R')(R'')P: \xrightarrow{H_2O_2} R(R')(R'')P=O$$

Inversion with I_2 + H_2O; HNO_3; Me_2SeO

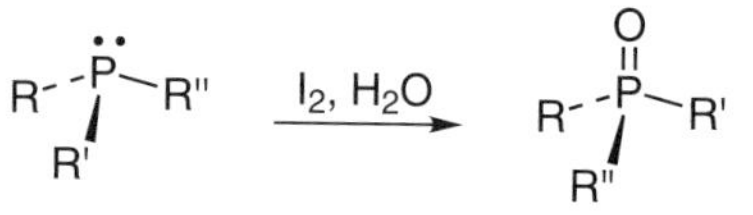

$$R(R')(R'')P: \xrightarrow{I_2, H_2O} R(R''')(R')P=O$$

D. Sulfuration of Phosphines

Retention with S_8

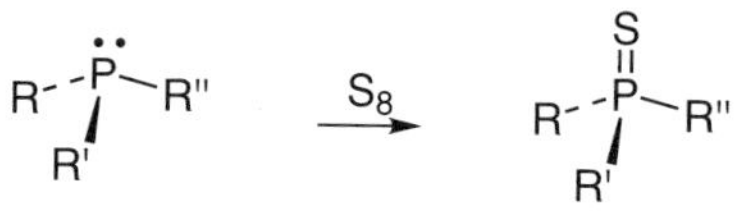

$$R(R')(R'')P: \xrightarrow{S_8} R(R')(R'')P=S$$

9.E. SUMMARY OF SOME STEREOSELECTIVE REACTIONS

Scheme 9.3 (*continued*)

E. Quaternization of Phosphines

Retention with alkyl halides

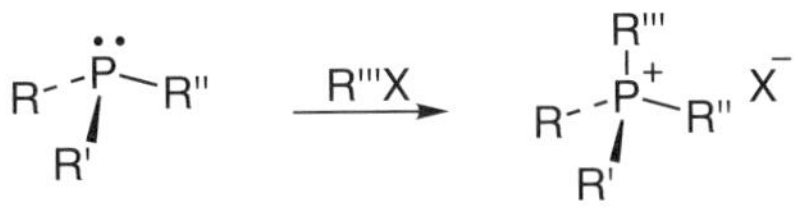

$$R\text{-}PR'R'' + R'''X \longrightarrow [R\text{-}PR'R''R''']^+ X^-$$

F. Decomposition of Quaternary Ions

Inversion with HO⁻

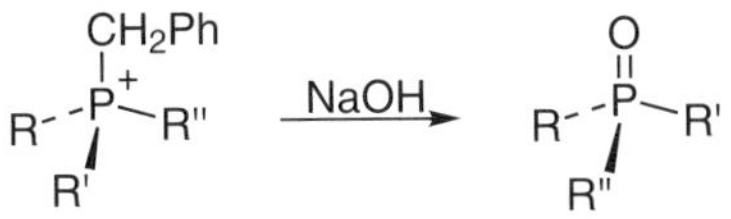

$$[R\text{-}P(CH_2Ph)R'R'']^+ \xrightarrow{NaOH} R\text{-}P(=O)R'R''$$

G. Wittig Reaction of Quaternary Ions

Retention

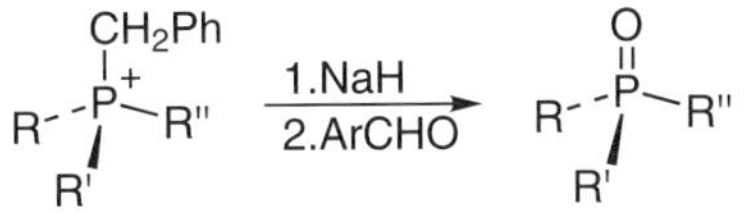

$$[R\text{-}P(CH_2Ph)R'R'']^+ \xrightarrow[\text{2. ArCHO}]{\text{1. NaH}} R\text{-}P(=O)R'R''$$

H. Conversion of Phosphine Oxides to Sulfides

Retention with B_2S_3

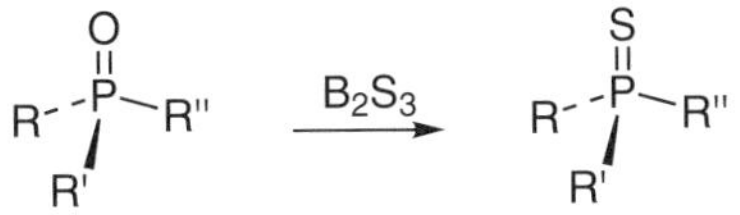

$$R\text{-}P(=O)R'R'' \xrightarrow{B_2S_3} R\text{-}P(=S)R'R''$$

I. Conversion of Phosphine Sulfides or Selenides to Oxides

Retention with $KMnO_4$

$$R\text{-}P(=S \text{ or } Se)R'R'' \xrightarrow{KMnO_4} R\text{-}P(=O)R'R''$$

Inversion with $(Me_3SiO)_2$; $Me_2SO + I_2$

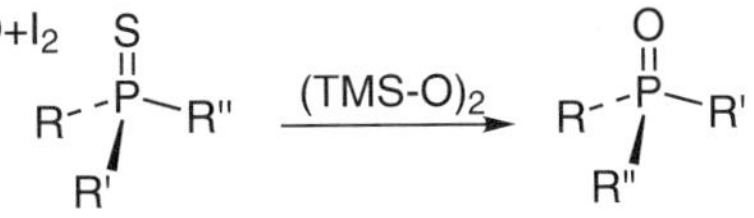

$$R\text{-}P(=S)R'R'' \xrightarrow{(TMS\text{-}O)_2} R\text{-}P(=O)R'R''$$

M. Conversion of Phosphine Imines to Oxides

Inversion on hydrolysis

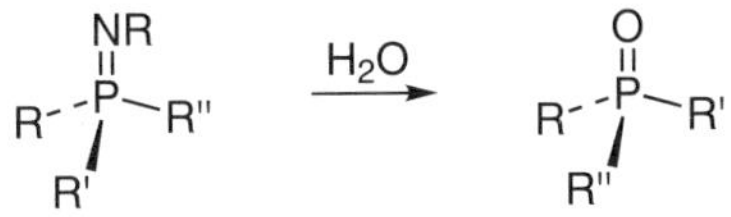

$$R\text{-}P(=NR)R'R'' \xrightarrow{H_2O} R\text{-}P(=O)R'R''$$

Scheme 9.3 (*continued*)

N. Decomplexation of Phosphine Boranes

Retention with Et₂NH; morpholine

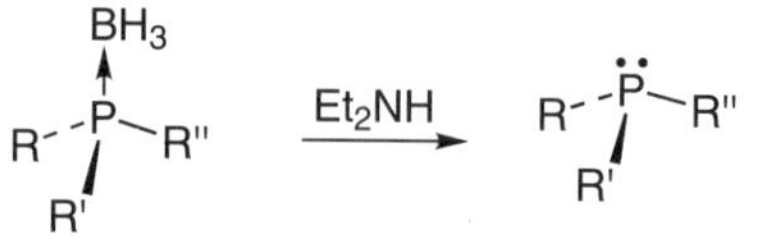

O. Alkylation of Phosphites, Phosphonites, and Phosphinites (Arbusov Reactions)

Retention with RX; MeOSO₂CF₃

P. Dealkylation of Oxyphosphonium Ions

Retention with $\bar{X}$

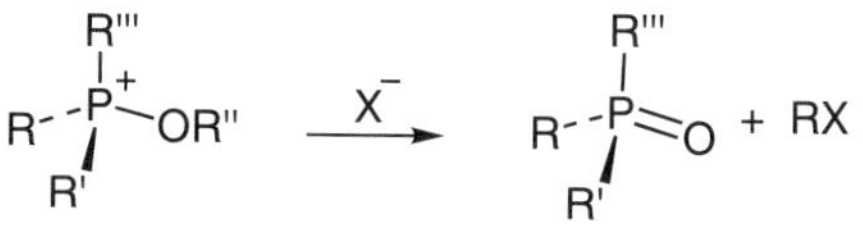

Q. Alkylation of Anions from 4-Coordinate P-H Compounds

Retention with RX

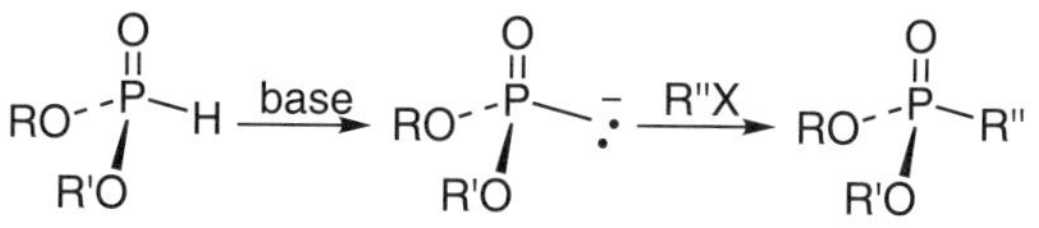

R. Nucleophilic Substitution at 3-Coordinate Phosphorus

Inversion

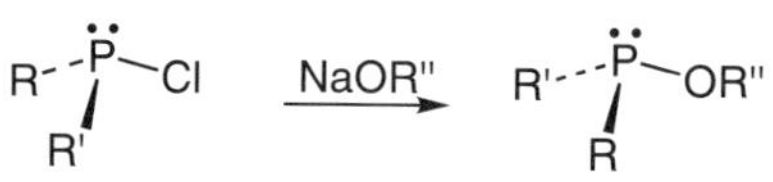

S. Nucleophilic Substitution at Acyclic 4-Coordinate Phosphorus Acid Derivatives

Inversion via 5-Coordinate intermediates or transition state

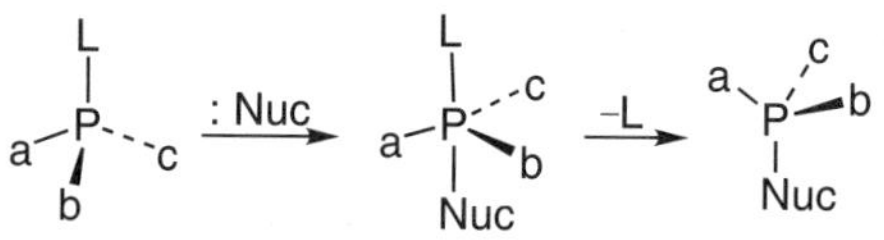

Scheme 9.3 (*continued*)

Retention if the 5-Coordinate species undergoes isomerization

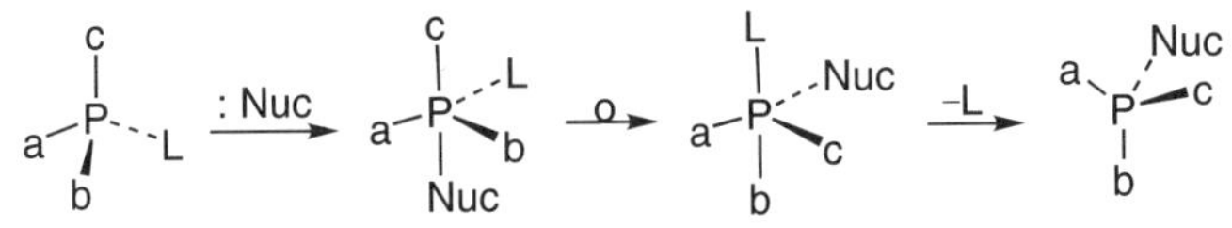

REFERENCES

1. J. Meisenheimer and L. Lichtenstadt, *Ber.* **44**, 356 (1911).
2. W. C. Davies and F. G. Mann, *J. Chem. Soc.*, 276 (1944).
3. K. F. Kumli, W. E. McEwen, and C. A. Vander Werf, *J. Am. Chem. Soc.* **81**, 248 (1959).
4. F. A. Hart and F. G. Mann, *J. Chem. Soc.*, 4107 (1955).
5. L. Horner, H. Winkler, A. Rapp, A. Mentrup, H. Hoffmann, and P. Beck, *Tetrahedron Lett.*, 161 (1961); L. Horner, *Pure Appl. Chem.* **9**, 225 (1964).
6. H. S. Aaron and J. I. Miller, *J. Am. Chem. Soc.* **78**, 3538 (1956).
7. K. M. Pietrusiewicz and M. Zablocka, *Chem. Rev.* **94**, 1375 (1994).
8. D. Valentine Jr., *Asymmetric Synthesis*, Vol. 4, Academic Press, Inc., New York, 1984, Chapter 3.
9. T. Imamoto in R. Engel, ed., *Handbook of Organophosphorus Chemistry*, Marcel Dekker, Inc., New York, 1992, Chapter 3.
10. W. E. McEwen, *Top. Phosphorus Chem.* **2**, 1 (1965).
11. W. E. McEwen and K. D. Berlin, *Organophosphorus Stereochemistry*, John Wiley & Sons, Inc., New York, 1975.
12. H. B. Kagan and M. Sasaki in F. R. Hartley, ed., *The Chemistry of Organophosphorus Compounds*, Vol. 1, John Wiley & Sons, Inc., New York, 1990, Chapter 3.
13. M. J. Gallagher in W. L. F. Armarego, ed., *Stereochemistry of Heterocyclic Compounds*, Part 2, John Wiley & Sons, Inc., New York, 1977, Chapter 5.
14. C. R. Hall and T. D. Inch, *Tetrahedron* **36**, 2095 (1980).
15. Ref. 13, p. 339.
16. L. D. Quin, *The Heterocyclic Chemistry of Phosphorus*, John Wiley & Sons, Inc., New York, 1981.
17. P. Macaudière, M. Caude, R. Rosset, and A. Tambuté, *J. Chromatogr.* **405**, 135 (1987).
18. Y. Okamoto, S. Honda, K. Hatada, I. Okamoto, Y. Toga, and S. Kobayashi, *Bull. Chem. Soc. Jpn.* **57**, 1681 (1984).
19. (a) P. Pescher, M. Caude, R. Rosset, A. Tambuté, and L. Oliveros, *Nouv. J. Chem.* **9**, 621 (1985); (b). P. Pescher, M. Caude, R. Rosset, and A. Tambuté, *J. Chromatogr.* **371**, 159 (1986).
20. A. Tambuté, P. Gareil, M. Caude, and R. Rosset, *J. Chromatogr.* **363**, 81 (1980).
21. R. Luckenbach, *Phosphorus* **1**, 228 (1972).
22. G. Fuchs, L. Doguet, D. Barth, and M. Perrut, *J. Chromatogr.* **623**, 329 (1992).
23. L. Horner and G. Simons, *Phosphorus Sulfur* **19**, 65 (1984).
24. H. Brunner, W. Pieronczyk, B. Schönhammer, K. Streng, I. Bernal, and J. Korp, *Chem. Ber.* **114**, 1137 (1981).
25. H. Takaya, K. Mashima, K. Koyano, M. Yagi, H. Kumobayashi, T. Taketomi, S. Akutagawa, and R. Noyori, *J. Org. Chem.* **51**, 629 (1986).

26. F. Toda, K. Mori, Z. Stein, and I. Goldberg, *J. Org. Chem.* **53**, 308 (1988).
27. R. Bodalski, T. Janecki, Z. Galdecki, and M. Glowka, *Phosphorus Sulfur* **14**, 15 (1982).
28. T. H. Chan, *Chem. Commun.*, 895 (1968).
29. N. K. Roberts and S. B. Wild, *J. Am. Chem. Soc.* **101**, 6254 (1979).
30. (a) A. L. Airey, G. F. Swieger, A. C. Willis, and S. B. Wild, *Inorganic Chem.* **36**, 1588 (1997); (b) V. V. Duninan and E. B. Golovan', *Tetrahedron: Asymmetry* **6**, 2747 (1995).
31. P. Le Maux, H. Bahri, and G. Simonneaux, *J. Chem. Soc., Chem. Commun.*, 1350 (1991).
32. E. Vedejs and Y. Donde, *J. Am. Chem. Soc.* **119**, 9293 (1997).
33. (a) A. Bader, M. Pabel, and S. B. Wild, *J. Chem. Soc., Chem. Commun.*, 1405 (1994); (b) A. Bader, M. Pabel, A. C. Willis, and S. B. Wild, *Inorganic Chemistry* **35**, 3874 (1996).
34. H.-P. Abicht, J. T. Spencer, and J. G. Verkade, *Inorg. Chem.* **24**, 2132 (1985).
35. H. S. Aaron, T. M. Shryne, and J. I. Miller, *J. Am. Chem. Soc.* **80**, 107 (1958).
36. H. S. Aaron, J. Braun, T. M. Shryne, H. F. Frack, G. E. Smith, R. T. Uyeda, and J. I. Miller, *J. Am. Chem. Soc.* **82**, 596 (1960).
37. C. R. Hall and T. D. Inch, *Phosphorus Sulfur* **7**, 171 (1979).
38. H. R. Benschop and G. R. Van den Berg, *Rec. Trav. Chim. Pays-Bas* **87**, 362 (1968).
39. M. J. P. Harger, *J. Chem. Soc., Perkin Trans.* 2, 1505 (1980).
40. P. Kielbasinski and M. Mikolajczyk, *Main Group Chemistry News* **5**, 7 (1997); F. Toda, K. Mori, Z. Stein, and I. Goldberg, *J. Org. Chem.* **53**, 308 (1988).
41. G. Hilgetag and G. Lehmann, *Angew. Chem.* **69**, 506 (1957).
42. C. Donniger and D. H. Hutson, *Tetrahedron Lett.*, 4871 (1968).
43. H. Teichmann and P. Lam, *Z. Chem.* **9**, 310 (1969).
44. J. N. Seiber and H. Tolkmith, *Tetrahedron* **25**, 381 (1969).
45. N. K. Hamer, *J. Chem. Soc.*, 2731 (1965).
46. D. A. A. Akintonwa, *Tetrahedron* **34**, 959 (1978).
47. W. E. McEwen, K. F. Kumli, A. Bladé-Font, M. Zanger, and C. A. Vander Werf, *J. Am. Chem. Soc.* **86**, 2378 (1964).
48. H. L. Boter and D. H. J. M. Platenburg, *Rec. Trav. Chim. Pays-Bas* **86**, 399 (1967).
49. M. Moriyama and W. G. Bentrude, *J. Am. Chem. Soc.* **105**, 4727 (1983).
50. G. Korpiun, R. A. Lewis, J. C. Chickos, and K. Mislow, *J. Am. Chem. Soc.* **90**, 4842 (1968); R. A. Lewis and K. Mislow, *J. Am. Chem. Soc.* **91**, 7009 (1969).
51. Ref. 8, p. 272.
52. W. B. Farnham, R. K. Murray, and K. Mislow, *J. Am. Chem. Soc.* **92**, 5809 (1970).
53. T. Koizumi, H. Amitani, and E. Yoshii, *Synthesis*, 110 (1979).
54. D. B. Cooper, T. D. Inch, and G. J. Lewis, *J. Chem. Soc., Perkin Trans.* 1, 1043 (1974).
55. Ref. 13, p. 60.
56. M. Segi, Y. Nakamura, T. Nakajima, and S. Suga, *Chem. Lett.* 913 (1983).
57. T. Kato, K. Kobayashi, S. Masuda, M. Segi, T. Nakajima, and S. Suga, *Chem. Lett.*, 1915 (1987).
58. S. Jugé and J. P. Genet, *Tetrahedron Lett.* **30**, 2783 (1989).
59. W. J. Richter, *Chem. Ber.* **117**, 2328 (1984).
60. J. M. Brown, J. V. Carey, and M. J. H. Russell, *Tetrahedron* **46**, 4877 (1990).
61. J. V. Carey, M. D. Barker, J. M. Brown, and M. J. H. Russell, *J. Chem. Soc., Perkin Trans.* 1, 831 (1993).
62. T. Imamoto, K. Sato, and C. R. Johnson, *Tetrahedron Lett.* **26**, 783 (1985).
63. J. Michalski and Z. Skrzypczynski, *J. Organomet. Chem.* **97**, C31 (1975); Z. Skrzypczynski and J. Michalski, *J. Org. Chem.* **53**, 4549 (1988).

64. R. K. Haynes, R. N. Freeman, C. R. Mitchell, and S. C. Vonwiller, *J. Org. Chem.* **59**, 2919 (1994).
65. R. Bodalski, E. Rutkowska-Olma, and K. M. Pietrusiewicz, *Tetrahedron* **36**, 2353 (1980).
66. Ref. 7, pp. 1401–1404.
67. K. M. Pietrusiewicz, M. Wieczorek, S. Cicchi, and A. Brandi, *Heteroatom Chem.* **2**, 661 (1991).
68. A. Brandi, P. Cannavo, K. M. Pietrusiewicz, M. Zablocka, and M. Wieczorek, *J. Org. Chem.* **54**, 3073 (1989).
69. A. Brandi, S. Cicchi, and K. M. Pietrusiewicz, *Tetrahedron: Asymmetry* **2**, 1063 (1991).
70. A. Goti, S. Cicchi, A. Brandi, and K. M. Pietrusiewicz, *Tetrahedron: Asymmetry* **2**, 1371 (1991); A. Brandi, S. Cicchi, A. Goti, M. Koprowski, and K. M. Pietrusiewicz, *J. Org. Chem.* **59**, 1315 (1994).
71. K. M. Pietrusiewicz, W. Holody, M. Koprowski, S. Cicchi, A. Goti, and A. Brandi, paper presented at the International Conference on Phosphorus Chemistry, Cincinnati, Ohio, July 12–17, 1998.
72. A. Brèque, J.-M. Alcaraz, L. Ricard, F. Mathey, A. Tambuté, and P. Macaudière, *New J. Chem.* **13**, 369 (1989).
73. L. Horner and M. Jordan, *Phosphorus Sulfur* **8**, 225 (1980).
74. T. Imamoto, T. Oshiki, T. Onozawa, T. Kusumoto, and K. Sato, *J. Am. Chem. Soc.* **112**, 5244 (1990).
75. S. Jugé, M. Stephan, J. A. Laffitte, and J. P. Genet, *Tetrahedron Lett.* **31**, 6357 (1990).
76. A. R. Muci, K. R. Campos, and D. A. Evans, *J. Am. Chem. Soc.* **117**, 9075 (1995).
77. B. Wolfe and T. Livinghouse, *J. Am. Chem. Soc.* **120**, 5116 (1998).
78. E. J. Corey, Z. Chen, and G. J. Tanoury, *J. Am. Chem. Soc.* **115**, 11000 (1993).
79. W. Chodkiewicz, D. Jore, A. Pierrat, and W. Wodzki, *J. Organomet. Chem.* **174**, C21 (1979).
80. J. Neuffer and W. J. Richter, *J. Organomet. Chem.* **301**, 289 (1986).
81. M. J. Burk, J. E. Feaster, W. A. Nugent, and R. L. Harlow, *J. Am. Chem. Soc.* **115**, 10125 (1993).
82. M. J. Burk and M. F. Gross, *Tetrahedron Lett.* **35**, 9363 (1994).
83. R. B. King, J. Bakos, C. D. Hoff, and L. Marko, *J. Org. Chem.* **44**, 3095 (1979).
84. G. Wittig, H. J. Cristau, and H. Braun, *Angew. Chem., Int. Ed. Engl.* **6**, 700 (1967).
85. K. Tani, L. D. Brown, J. Ahmed, J. A. Ibers, M. Yokota, A. Nakamura, and S. Otsuka, *J. Am. Chem. Soc.* **99**, 7876 (1977); S. Otsuka, A. Nakamura, T. Kano, and K. Tani, *J. Am. Chem. Soc.* **93**, 4301 (1971).
86. U. Verfürth and I. Ugi, *Chem. Ber.* **124**, 1627 (1991).
87. J. Omelanczuk, *J. Chem. Soc., Chem. Commun.*, 1718 (1992).
88. Z. Fei, I. Neda, H. Thönnessen, P. G. Jones, and R. Schmutzler, *Phosphorus Sulfur Silicon* **131**, 1 (1997).
89. M. Pabel, A. C. Willis, and S. B. Wild, *Tetrahedron: Asymmetry* **6**, 2369 (1995).
90. J. Omelanczuk and M. Mikolajczyk, *J. Am. Chem. Soc.* **101**, 7292 (1979).
91. M. Mikolajczyk, J. Drabowicz, J. Omelanczuk, and E. Fluck, *J. Chem. Soc., Chem. Commun.*, 382 (1975).
92. Ref. 8, p. 299.
93. C. R. Hall and T. D. Inch, *J. Chem. Soc., Perkin Trans.* 1, 1104 (1979).
94. J. R. Knowles, *Annu. Rev. Biochem.* **49**, 877 (1980).
95. S. J. Abbott, S. R. Jones, S. A. Weinman, and J. R. Knowles, *J. Am. Chem. Soc.* **100**, 2558 (1978).
96. P. M. Cullis and G. Lowe, *J. Chem. Soc., Chem. Commun.*, 512 (1978).

97. G. M. Lowe, B. V. L. Potter, B. S. Sproat, and W. E. Hull, *J. Chem. Soc., Chem. Commun.*, 733 (1979).
98. G. M. Lowe in L. D. Quin and J. G. Verkade, eds., *Phosphorus-31 NMR Spectral Properties in Compound Characterization and Structural Analysis*, VCH Publishers, New York, 1994, Chapter 20.
99. M. R. Webb and D. R. Trentham, *J. Biol. Chem.* **255**, 1775 (1980); M.-D. Tsai, *Biochem.* **19**, 5310 (1980).
100. Ref. 13, p. 411.
101. D. Hellwinkel, *Chem. Ber.* **99**, 3642 (1966).
102. D. Houalla, M. Sanchez, L. Beslier, and R. Wolf, *Org. Magn. Reson.* **3**, 45 (1971).
103. F. Archer, S. Jugé, and M. Wakselman, *Tetrahedron* **43**, 3721 (1987).
104. R. M. Moriarty, J. Hiratake, and K. Liu, *J. Am. Chem. Soc.* **112**, 8575 (1990).
105. N. P. B. Dudman and B. Zerner, *J. Am. Chem. Soc.* **95**, 3019 (1973).
106. P. Kielbasinski, R. Zurawinski, K. M. Pietrusiewicz, M. Zablocka, and M. Mikolajczyk, *Tetrahedron Lett.* **35**, 7081 (1994).
107. A. N. Serreqi and R. J. Kazlauskas, *J. Org. Chem.* **59**, 7609 (1994).

CHAPTER 10

THE LOW- AND HIGH-COORDINATION STATES OF PHOSPHORUS

10.A. GENERAL

Organophosphorus chemistry grew around the basic concept that stable, isolatable structures would have two bonding patterns: that with three covalently bonded atoms and an electron pair on phosphorus (the σ^3, λ^3 state, as in the phosphines), and that with four covalently bonded atoms where one would be a multiply bonded chalcogen (σ^4, λ^5 as in the phosphine oxides) or with four singly bonded atoms with phosphorus bearing a positive charge (σ^4, λ^4, as in phosphonium salts). Structures with double or triple bonds of the p_π-p_π type (thus of low coordination numbers), so common throughout the organic chemistry of carbon and nitrogen, were specifically excluded from any serious consideration, as indeed was true in other types of compounds based on heteroatoms below the first row of eight elements in the Periodic System. Similarly, stable phosphorus compounds with five covalent bonds, thus having a complement of 10 electrons, were not known to exist and were ignored in synthetic studies, although dogma was less compelling in fixing a position that their existence was totally precluded, as with the low-coordination cases. It has only been in the last 3–4 decades that both of these ancient structural restrictions have been demolished, and it has been demonstrated that these structures, as well as species containing six electrons (R—P, R_2P^+), can be generated and in many cases obtained in stable form. Once it was recognized that there could indeed be stable 5-coordinate (as well as 6-coordinate) structures, and that true multiple bonds could be formed to C, N, P and O, a flood of research commenced that continues today. Low- and high-coordination compounds are now well-accepted species that occupy highly important positions in modern organophosphorus chemistry. Indeed, discoveries in these areas are among the most exciting in the whole field. However, there generally are quite specific structural requirements that must be met in the construction of these coordination states, and it must not be construed that synthesis can produce any desired atomic combination. In the multiply bonded species there is a strong tendency for dimerization or oligomerization to occur, which, however, can frequently be blocked by placing very large substituents at the sensitive site. Electronic effects also can be used to achieve stabilization. In the 5-coordinate state, a possible problem is that of dissociation to the ionic form. In the discussions of this chapter, we will see many examples of stabilized structures, as well as of the generation and trapping of the unstable species of transient lifetime, for they too are of prominence in low coordination chemistry.

The discussion to follow will provide information on the major types of structures known for a given coordination state. Some of the more important synthetic methods for obtaining the major species will be outlined here, as will some of their characteristic reactions and properties. Information on important spectral properties for characterization, especially that of ^{31}P NMR, will be given. Phosphenium ions, species now receiving considerable attention as intermediates in synthesis, will receive no further mention in this chapter, because they were encountered in chapters 2.K and 3.A. For a more complete survey of the low-coordination species with multiple and single bonds (the 1-coordinate phosphinidenes and 2-coordinate phosphenium ions), the important monograph by Regitz and Scherer[1] should be consulted. Phosphorus-31 NMR spectra of many of the different species, including high-coordination states (five and six), are covered by Fluck and Heckmann in a 1987 review.[2] A more recent book by Dillon, Mathey and Nixon[3] provides coverage of recent results with carbon-bonded species and is intriguing in emphasizing their relationship to multiply bonded carbon species. The 5-coordinate state is discussed in detail by Holmes in a two-volume treatise.[4] The 6-coordinate state is treated in a recent *Chemical Reviews* article.[5] More specific review articles of smaller scope will be mentioned at appropriate points in the discussion to follow.

10.B. THE 1-COORDINATE STATES (σ^1, λ^3 AND σ^1, λ^1)

10.B.1. The Phosphorus–Carbon Triple Bond (σ^1, λ^3)

10.B.1a. Synthesis. The simplest member of the series, HC≡P, was the first to be detected experimentally and indeed was the first compound known to have any form of multiple bonding of phosphorus to carbon. Phosphaalkyne chemistry is now a well-recognized part of organophosphorus chemistry and has been the subject of several reviews.[6–10] It is even of interest outside the laboratory, because HC≡P has been detected in interstellar space.[11a]

The first experiment that provided a phosphaalkyne molecule consisted of passing PH_3 through an arc between carbon electrodes,[12] whereupon phosphaacetylene (HC≡P) and acetylene were formed in a ratio of 1 : 4. The compound was a colorless gas, stable as a solid below −124°, and gave a black polymer at higher temperatures.

It was characterized by IR spectroscopy (C−H str. 3180 cm^{-1}; C≡P str. 1265 cm^{-1}) and by the formation of CH_3PCl_2 with HCl. Later workers prepared this compound more conveniently by stripping two moles of HCl from CH_3PCl_2 under flash vacuum pyrolysis (FVP) conditions,[13] with removal of HCl to prevent the reverse reaction from occurring. The later experimental conditions of Pellerin, et al.,[14] provide the compound in 30% yield but with significant contamination by CHCl=PCl. The sample was found to have greater stability than first reported. Although at first more of theoretical than of practical importance in chemistry, phosphaacetylene continues to attract attention and has recently found some use as a synthon for heterocyclic systems.[15] A new and greatly improved procedure for its synthesis was described by Guillemin, et al.[16] This consists of the elimination of two moles of HCl from 1,1-dichloromethylphosphine by passing it over solid K_2CO_3 at 620 K, a process described as the Vacuum Gas Solid Reaction (VGSR). The requisite phosphine was synthesized as shown in Eq. 10.1. Phosphaacetylene was obtained in 80% yield from the dichloro precursor.

$$CCl_3\text{-}\overset{O}{\overset{\|}{P}}(OPr\text{-}i)_2 \xrightarrow[\text{2. HX}]{\text{1. }n\text{-BuLi}} HCCl_2\text{-}\overset{O}{\overset{\|}{P}}(OPr\text{-}i)_2 \xrightarrow{AlHCl_2} HCCl_2PH_2 \xrightarrow[620\text{ K}]{K_2CO_3} HC\equiv P \qquad (10.1)$$

Another highly significant discovery in the development of phosphaalkyne chemistry made in 1976[17] was that methylphosphaacetylene ($CH_3C\equiv P$) could be made by stripping HCl from $CH_3CH_2PCl_2$ under FVP conditions (900°, 0.05 mm) in 10% yield as a pure compound, storable at low temperatures (Eq. 10.2).

$$CH_3CH_2PCl_2 \xrightarrow{FVP} CH_3C\equiv P \qquad (10.2)$$

Methylphosphaacetylene was first characterized by photoelectron spectroscopy, but has since been examined more fully and the experimental conditions for its generation improved.[14,18,19] The compound is best preserved at low temperatures, but it has been observed[16] that a 2 to 5% THF solution of the pure compound is stable at room temperature for over a week. Very recently, the VGSR approach has been applied to the synthesis of methylphosphaacetylene as outlined in Eq. 10.3. The yield was 80%.

$$CCl_3\text{-}\overset{O}{\overset{\|}{P}}(OPr\text{-}i)_2 \xrightarrow[\text{2. MeX}]{\text{1. }n\text{-BuLi}} MeCCl_2\text{-}\overset{O}{\overset{\|}{P}}(OPr\text{-}i)_2 \xrightarrow{AlHCl_2} MeCCl_2PH_2 \xrightarrow{VGSR} Me\text{-}C\equiv P \qquad (10.3)$$

Thermal vapor phase eliminations have been used for the synthesis of other phosphaalkynes that are of short life time, including Cl−C≡P (from Cl_3CPCl_2 over zinc,[20] and by VGSR of Cl_3CPH_2)[21] and $CH_2=CH-C\equiv P$,[18] and gas phase intermolecular reactions have been used in the synthesis of HC≡C−C≡P,[22] and N≡C−C≡P.[23] Elimination of HF from CF_3PH_2 with solid KOH at room temperature is a source of F−C≡P.[24] The VGSR method has been used to synthesize R−C≡P where R = Et, n-Bu, Me_3Si, and Cl. All yields from the 1,1-dichlorophosphine precursor were in the 60 to 80% range,[16] and the method is clearly an advance in the synthesis of short-lived phosphaalkynes. Further discussion of the various short-lived species may be found in a recent review.[10]

Although short-lived phosphaalkynes do continue to receive attention,[10] another discovery placed phosphaalkyne chemistry on a very practical basis; in 1981 it was discovered that this triple bond could be kinetically stabilized by placing large substituents on carbon.[25] *tert*-Butylphosphaacetylene (more properly, 2-dimethylpropylidynephosphine) was found to be a stable, distillable liquid (b.p. 61°), easily prepared (Eq. 10.4) by elimination with base of hexamethylsiloxane from compound **1** (itself of interest as a stable phosphaalkene; section 10.C.1), and attention immediately shifted to this new class of stabilized phosphaalkynes. Many have been synthesized, and much is known about their physical and chemical properties. The family should no longer be considered to consist of chemical curiosities, and the $-C \equiv P$ group, firmly established by X-ray analysis[26] (linear, with P—C bond length 1.542 Å), has earned a place among the collection of important phosphorus functional groups.

$$\underset{\underset{\textbf{1}}{t\text{-Bu}\diagup\;\;\;\diagdown\text{SiMe}_3}}{\overset{\text{Me}_3\text{SiO}\diagdown\;\;\;\diagup}{\text{C=P}}} \xrightarrow{\text{NaOH}} t\text{-BuC} \equiv P \; + \; (\text{Me}_3\text{Si})_2\text{O} \qquad (10.4)$$

Most of the stabilized phosphaalkynes have been generated by the hexamethyldisiloxane elimination route, and by 1990, about 15 had been so prepared.[9] The entire synthesis is illustrated in Eq. 10.5, in which it will be seen that acyl halides are the starting materials and can be converted in two steps to the immediate phosphaalkyne precursor.

$$\text{R}-\overset{\text{O}}{\underset{\text{Cl}}{\text{C}}} \; + \; \text{P(SiMe}_3)_3 \longrightarrow \text{R}-\overset{\overset{\text{O}}{\|}}{\text{C}}-\text{P(SiMe}_3)_2 \xrightarrow{1,3\text{-shift}} \underset{\text{R}\diagup\;\;\;}{\overset{\text{Me}_3\text{SiO}\diagdown\;\;\;\diagup\text{SiMe}_3}{\text{C=P}}} \longrightarrow \text{R}-\text{C} \equiv \text{P}$$

(10.5)

In its first form, base was used to promote the siloxane elimination, but an improved procedure consists of simply heating the neat precursor at 120 to 200°, whereupon the phosphaalkyne is distilled from the reaction zone.[27] Most of the stabilized phosphaalkynes have tertiary carbon substituents or di-ortho-substituted aromatics such as mesityl, but a few have suitably branched secondary carbon (e.g., cycloalkyl), and even neopentylphosphaalkyne with a primary carbon has been so generated.

Another elimination route to $C \equiv P$ groups consists of stripping Me_3SiCl from molecules of structure **2** (Eq. 10.6).[28] By this method, phenylphosphaalkyne was first generated, as was $Me_3SiC \equiv P$, but neither are notably stable, and require low-temperature storage.

$$\underset{\underset{\textbf{2}}{\text{Me}_3\text{Si}\diagup\;\;\;\diagdown\text{Cl}}}{\overset{\text{Ph}\diagdown\;\;\;\diagup}{\text{C=P}}} \xrightarrow[\text{high vacuum}]{700°} \text{Ph}-\text{C} \equiv P \; + \; \text{Me}_3\text{SiCl} \qquad (10.6)$$

Amino substituents can act to stabilize the C≡P bond by their electron releasing ability, and this allowed the synthesis of the novel C-amino phosphaalkyne **3**.[29] The method used depends on a rearrangement process, a 1,3 migration of a Me_3Si group from P to N in a presumed intermediate 3, formed as shown in Eq. 10.7. The creation of phosphaalkynes stabilized by electronic effects represented another significant advance in this area, and now many C-amino phosphaalkynes are known (many literature citations may be found in reference 30). Improvements in phosphaalkyne synthesis continue to be made,[3] an example being the Lewis base–induced rearrangement[31] of $R-C\equiv C-PH_2$ to $RCH_2-C\equiv P$.

$$(Me_3Si)_3P + i\text{-}Pr-N=C=O \longrightarrow (Me_3Si)_2P-C\overset{O}{\underset{Me_3Si}{\diagdown}}\diagup^{N-Pr\text{-}i} \rightleftharpoons (Me_3Si)_2P-C\overset{OSiMe_3}{\diagdown}\diagup_{NPr\text{-}i}$$

$$\xrightarrow[-(TMS)_2O]{NaOH} [Me_3Si-P=C=N-Pr\text{-}i] \longrightarrow \underset{\underset{\underset{3}{i\text{-}Pr}}{|}}{Me_3SiN}-C\equiv P \qquad (10.7)$$

10.B.1.b. Properties. Much information is available on the physical properties of phosphaalkynes and is summarized in the reviews cited. Using *t*-butylphosphaacetylene as a model, the C≡P bond is linear with a quite short length, 1.536 Å by microwave spectroscopy[32] and, as noted, 1.542 Å by X-ray analysis.[26] The IR stretching occurs at 1533 cm^{-1}. The ^{31}P NMR signal appears at $\delta -69.2$, and the ^{13}C signal of the C≡P bond at δ 184.8 ($^1J_{PC} = 38.5$ Hz). Other C-alkyl compounds give similar values but heteroatom substitution on C leads to a wide range, e.g., of ^{31}P $\delta -116$ for Cl to $\delta +99$ for Me_3Si, and of ^{13}C from δ 126.4 to 201.9 for the same substituents, respectively. The HC≡P has ^{31}P $\delta -32$ and ^{13}C δ 158.0 ($^1J_{PC} = 56.0$ Hz). Photoelectron spectroscopy has proved useful in characterizing some of these molecules; for the *t*-butyl compound, the first ionization potential (of a π MO) is 9.70 eV, and the second (n MO) is 11.45 eV.

The stable phosphaalkynes are reasonably unreactive as phosphines; they resist atmospheric oxidation, alkylation with alkyl halides, and protonation with aqueous acids. This is consistent with the general picture that the lone pair on multiply bonded phosphorus is far less reactive than in 3-coordinate compounds, and this is indicated by the molecular orbital picture of such species. A discussion of the bonding in low-coordination compounds may be found in reference 11b. Phosphaalkynes do, however, undergo ready addition of hydrogen halides and superacids such as FSO_2OH (with H adding to C) and of halogens. Far more important are the numerous reactions known to form metallic complexes, and others that can be collectively described as cycloadditions of various types. There is already a vast number of papers dealing with both types of reactivities, but fortunately the review literature[11c,33] provides extensive coverage of the research effort. The cycloadditions with phosphaalkynes have had great impact in heterocyclic chemistry; many new ring systems have been created, and new approaches to known systems, especially to phosphinines, have been discovered. Only a brief outline can be presented here.

Using t-butylphosphaacetylene (R−CP) as a model reactant (except as noted), some of the important products, which may be found in the cited reviews, are given in Eqs. 10.8 to 10.19.

Addition of carbenes

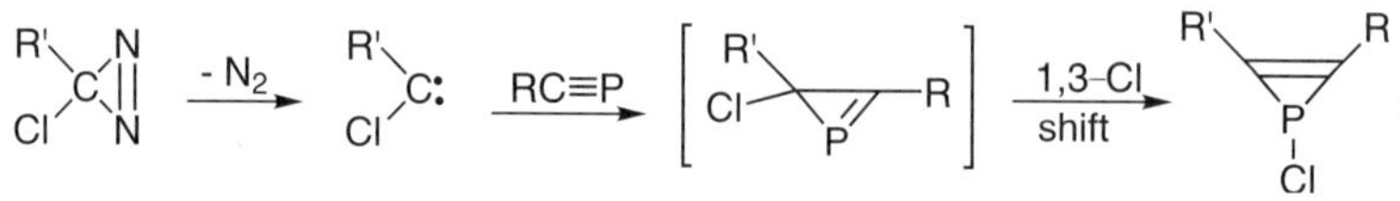

$$(10.8)$$

Addition of silylenes

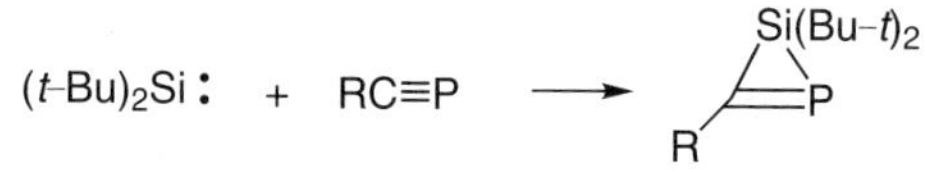

$$(10.9)$$

Addition of phosphinidenes

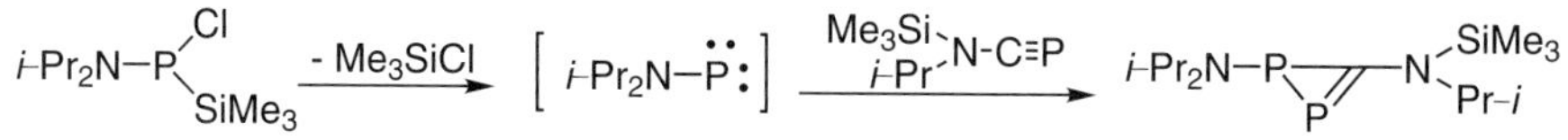

$$(10.10)$$

Dimerization and tetramerization

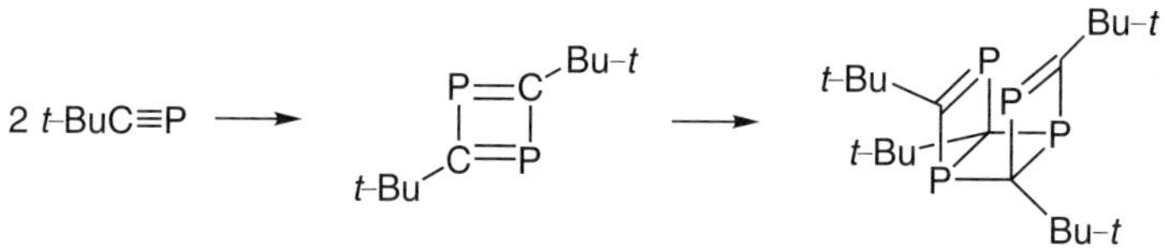

$$(10.11)$$

1,3-Dipolar [3+2] cycloadditions

With diazo compounds

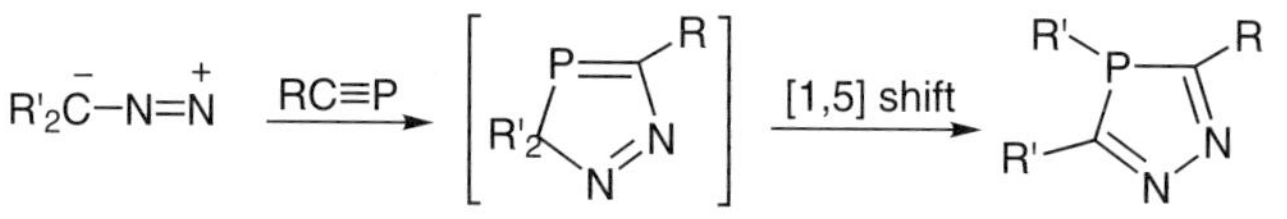

$$(10.12)$$

With azides

$$(10.13)$$

10.B. THE 1-COORDINATE STATES (σ^1, λ^3 AND σ^1, λ^1)   313

With nitrile ylides

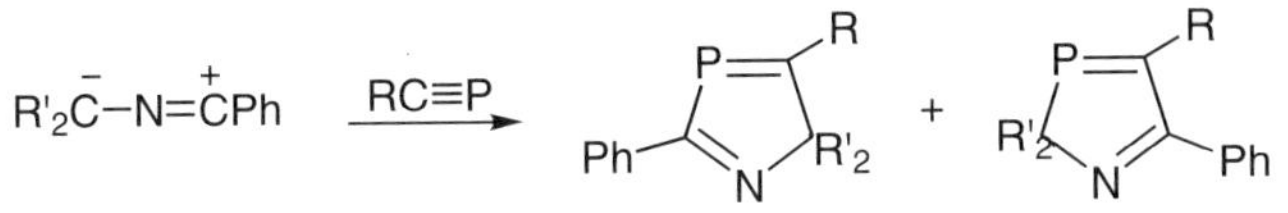

$$R'_2\overset{-}{C}-N=\overset{+}{C}Ph \xrightarrow{RC\equiv P} \text{[structure]} + \text{[structure]} \quad (10.14)$$

With nitrile oxides

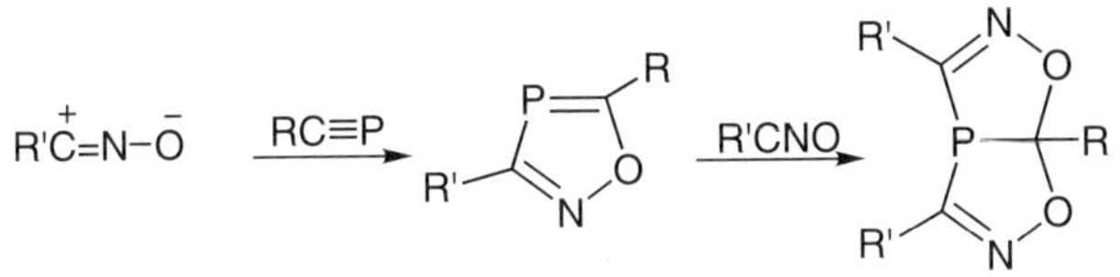

$$R'\overset{+}{C}=N-\overset{-}{O} \xrightarrow{RC\equiv P} \text{[structure]} \xrightarrow{R'CNO} \text{[structure]} \quad (10.15)$$

[4+2] Cycloadditions

With cyclobutadienes (also with azetes and phosphetes)

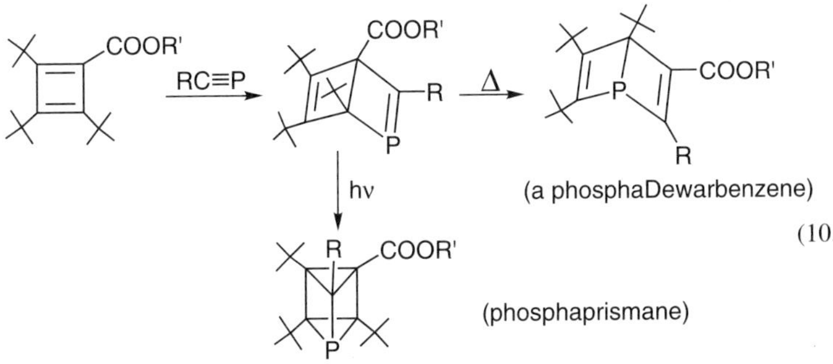

(a phosphaDewarbenzene)

(phosphaprismane)

(10.16)

With phosphole sulfides

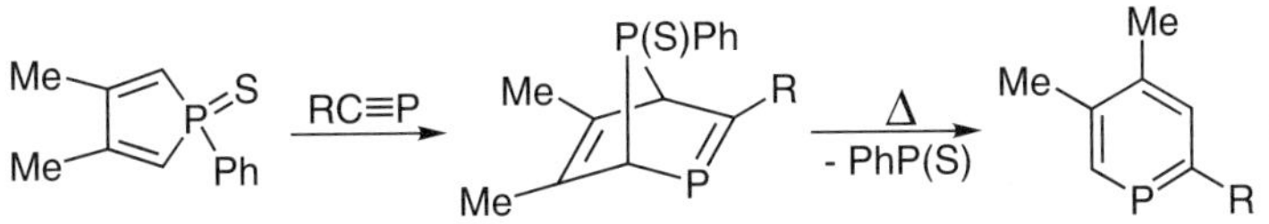

(10.17)

With cyclic dienes

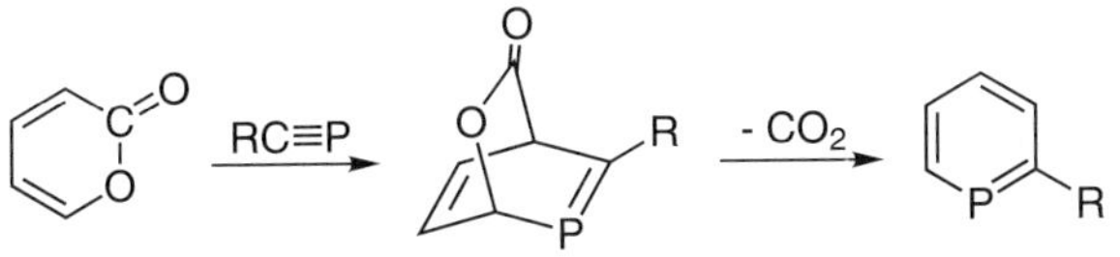

(10.18)

With 1,3- dienes (also heterodienes)

(10.19)

As can be seen, the array of cyclic products is truly impressive, and heterocyclic chemistry has been greatly enriched by the development of phosphaalkyne reactivity.

10.B.2. Phosphinidenes (σ^1, λ^1)

The phosphorus species RP is analogous to a nitrene; it is electron deficient and has the possibility of diradical character (the triplet state) or of possessing two electron pairs (the singlet state). The triplet state is predicted from theory to be much more energetically favored;[34] however, the generation of phosphinidenes has proved to be a major hurdle in organophosphorus chemistry. Many methods have been reported that appear to accomplish this goal, but until very recently (vide infra) the phosphinidene species, if truly formed, has only been detected by trapping reactions. In many cases, it has been proposed as an intermediate in an attempt to account for a reaction product. The products of trapping reactions usually can be explained by other mechanisms that do not involve a totally free phosphinidene, and a careful study is required to eliminate this possibility. Several reviews[11d,35-37] summarize the various reports on the possible intermediacy of phosphinidenes. There is, however, only one report in the literature that does confirm the existence of a free carbon-substituted phosphinidene by direct spectral observation (inorganic phosphinidenes have been detected spectroscopically; H—P is even found in interstellar space). This highly important work of Gaspar,[38] based on a useful synthetic procedure described earlier,[39] was reported in 1994. In this work, mesitylphosphinidene was eliminated from trimesitylphosphirane 4 by irradiation of a frozen methylcyclohexane solution at 254 nm and 77 K (Eq. 10.20). The phosphinidene was directly detected in a liquid helium cryostat (4 K) by electron spin resonance, whereupon a strong broad signal with 11,492 G was observed. This signal is consistent with the predicted triplet form of the phosphindene. The species after the photolysis was also chemically detected by trapping it with diethylacetylene to form a new phosphirane (Eq. 10.20). With this discovery, the field of organic phosphinidene chemistry has been given the firm foundation sought for so long.

One of the early methods proposed to generate phosphinidenes consists of the decomposition of cyclopolyphosphines, which can be considered to be polymers of

$$\text{Mes}-P\overset{\triangleleft}{} \xrightarrow[-CH_2=CH_2]{h\nu, \text{benzene}} [\text{MesP}] \xrightarrow{\text{Et}-C\equiv C-\text{Et}} \underset{\text{Et}}{\overset{\text{Et}}{\diagup}}\hspace{-2pt}\triangleright\hspace{-2pt}P-\text{Mes} \quad (10.20)$$

4   62%

R–P. Indeed, in a mass spectrum of the products formed from $(PhP)_5$ at 400°, a peak for Ph–P is present. The cyclopolyphosphines can be prepared from the reaction of phosphonous dihalides with metals (chapter 3.C), and when heated do serve in some cases as R–P transfer agents. Various substrates have been found to undergo R–P insertion reactions, including the S–S bond (e.g., diethyl disulfide gives $RP(SEt)_2$), the As–As bond, the Ge–H bond, and even the C–C bond,[40] as seen in Eq. 10.21.

$$(PhP)_5 \longrightarrow [\text{PhP}] \xrightarrow{400°} \text{(dibenzofuran-like structure with P–Ph)} \quad (10.21)$$

In some cases, cyclopolyphosphines give products with one or two of the P–P bonds retained, as in the reaction with an acetylene (Eq. 10.22) which gives a mixture of products.

$$(PhP)_5 + PhC\equiv CPh \xrightarrow{240°} \text{[products]}$$

              15%   50%   2–3%

(10.22)

Other methods presumed to give phosphinidene intermediates and their trapping reactions are summarized in reference 11d. Two more recent methods do seem to provide a specially strong case for a phosphinidene intermediate. Fritz and co-workers[41] generated the phosphinophosphinidene **5** (Eq. 10.23), predicted in this case to be the singlet form as a result of the interaction of the attached tertiary phosphine group, and found it to add to alkenes or dienes to give phosphirane derivatives. This is the first instance of this type of reactivity, which is indeed consistent with the postulated singlet reactant.

$$(t\text{-Bu})_2P-P=P\underset{\text{Bu-}t}{\overset{\text{Br}}{\diagup}}\text{Bu-}t \xrightarrow{20°} [t\text{-Bu}_2P-\ddot{P}\colon{}] \longrightarrow \text{cyclohexene adduct: } \text{C}_6\text{H}_{10}\text{>P-P(Bu-}t)_2$$

5

$$\downarrow \text{Me}_2C=CH-CH=CMe_2$$

products: phosphirane and diene adducts with $P-P(Bu-t)_2$

(10.23)

316 THE LOW- AND HIGH-COORDINATION STATES OF PHOSPHORUS

Additions of phosphinidenes, to double and triple bonds have also been observed[39]; in this work the phosphinidenes were generated by photolysis at 254 nm of the cyclic phosphines **4** (Eq. 10.20) and **6** (Eq. 10.24).

$$(10.24)$$

Intramolecular reactions observed for presumed phosphinidenes are also rather convincing that such a species is indeed being generated. Several examples are shown as Eqs. 10.25,[42] 10.26,[43] 10.27,[38,44] 10.28[45] and 10.29.[45]

$$(10.25)$$

$$(10.26)$$

$$(10.27)$$

$$(10.28)$$

$$(10.29)$$

We saw in chapter 8 that the phosphinidene moiety can be bonded in metal coordination compounds. Complexes with terminal phosphinidene units can be conveniently constructed by the fragmentation of complexes of 7-phosphanorbornadienes; the released complex **9** has electrophilic character and can then be reacted with a number of substrates to give valuable products, either by insertion into σ bonds or by cycloaddition with π bonds. It was by the latter process that the phosphirene system was first created (Eq. 10.30).

$$(10.30)$$

The value of the phosphinidene complex in cycloadditions is considerable, and remains under development (e.g., in a demonstration of 1,4-addition of a R—P unit to the benzene ring of [5]metacyclophane,[46] Eq. 10.31).

$$(10.31)$$

An example of an insertion reaction is provided by Eq. 10.32. Other reactions of terminal phosphinidene complexes, which are increasingly of importance in synthetic phosphorus chemistry, are summarized in reviews,[11d,47] and the characteristics of other types of phosphinidene complexes, of less importance in organic chemistry, have also been reviewed.[11e]

$$(10.32)$$

10.C. THE 2-COORDINATE STATE WITH DOUBLE BONDS TO PHOSPHORUS (σ^2, λ^3)

10.C.1. The Phosphorus-Carbon Double Bond

10.C.1.a. Synthesis. The phosphorus-carbon p_π-p_π double bond was first encountered in stable form in the highly delocalized structure **10**,[48] and in phosphabenzenes and their benzo derivatives, as seen in chapter 8.B.

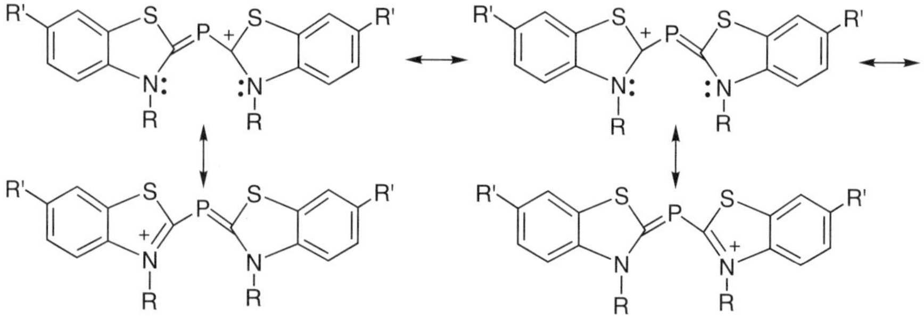

318 THE LOW- AND HIGH-COORDINATION STATES OF PHOSPHORUS

[Structure diagram showing resonance forms of compound 10]

10

It was not until 1976 that examples of isolated P—C double bonds were described. Becker[49] reported that an acylphosphine with a trimethylsilyl group on phosphorus, in cyclopentane solution, underwent a reversible transfer of silyl to carbonyl oxygen below room temperature, leaving phosphorus in doubly bonded form (Eq. 10.33). The phosphaalkenes were stable in the absence of air or water, and were fully characterized by spectral analysis (for R = t-butyl, ^{31}P δ +176.0).

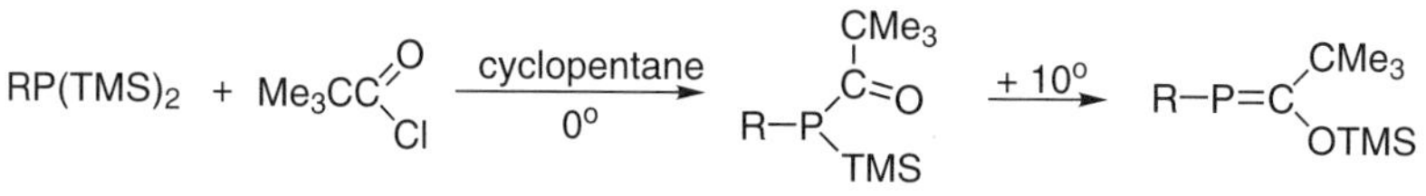

$$RP(TMS)_2 + Me_3CC(O)Cl \xrightarrow[0°]{\text{cyclopentane}} R-P(TMS)-C(CMe_3)=O \xrightarrow{+10°} R-P=C(CMe_3)(OTMS) \quad (10.33)$$

In the same year, it was demonstrated that unstable species such as $H_2C=PCl$, $H_2C=PH$, and $F_2C=PH$ could be generated by gas-phase pyrolytic reactions.[13] But the report that excited the whole phosphorus community and opened the great new research area of thermally stable, isolatable compounds with isolated P—C double bonds was that from F. Bickelhaupt's laboratory in 1978.[50] Bickelhaupt designed a compound with large substituents on both C and P, to retard sterically the polymerization of the double bond. This form of kinetic stabilization remains in wide use, although electronic effects, discovered later, can also be used to provide stabilization. The Bickelhaupt synthesis of an isolated phosphaalkene **11** is surprisingly simple, and consists merely of the dehydrohalogenation with base of a phosphinous chloride (Eq. 10.34), a technique used previously to install a double bond in benzophosphabenzene derivatives. Compound **11** was distilled at 140° (10^{-3} mm) and had m.p. 51–58°, ^{31}P δ +233.

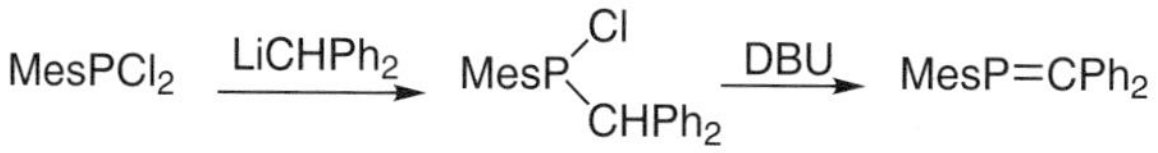

$$MesPCl_2 \xrightarrow{LiCHPh_2} MesP(Cl)(CHPh_2) \xrightarrow{DBU} MesP=CPh_2 \quad (10.34)$$

11

Another important announcement in the same year was that of Issleib et al.,[51] who reported the synthesis of stable phosphaalkenes with C-amino substituents. An example of their synthetic method, which involves a silyl migration from P to C, is shown in Eq. 10.35.

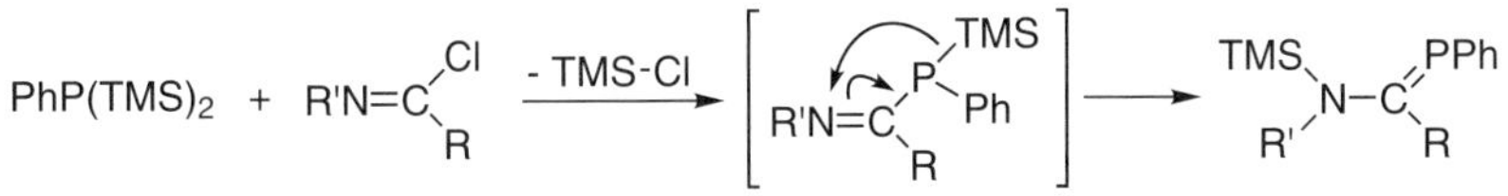

$$PhP(TMS)_2 + R'N=C\begin{smallmatrix}Cl\\R\end{smallmatrix} \xrightarrow{-TMS-Cl} \left[R'N=C\begin{smallmatrix}P(TMS)\\Ph\\R\end{smallmatrix} \right] \longrightarrow \begin{smallmatrix}TMS\\R'\end{smallmatrix}N-C\begin{smallmatrix}PPh\\R\end{smallmatrix} \quad (10.35)$$

Within a few years, many new stable phosphaalkenes were synthesized, and a review[11f] in 1990 provided a list of about 250 compounds that had been characterized. In addition to these, many examples of the P—C double bond conjugated with C=O, C=P, C=C, etc., as well as in cumulene form (e.g., E=C=P—R, where E is C=P, RN, O, etc.) were prepared. The number of known compounds since the 1990 review has swelled greatly, and the whole area of the P—C double bond functional group, as also true for the triple bond, is now a major part of contemporary organophosphorus chemistry. It is surprising how easy it is to generate both of these functional groups, and there should be no reluctance for organic chemists to design work based on these structures.

Although the emphasis here, and in most of the research that has been done, is on stabilized phosphaalkenes, unstabilized phosphaalkenes are also of interest and methods for their generation and detection are known. A list composed in 1994[10] is long and includes, along with other species, the following structures: halo derivatives $F_2C=PH$,[52] $CF_3P=CF_2$,[53] $CH_2=PCl$,[54] $ClP=CCl_2$,[55] and $RClC=PH$[56]; the parent $CH_2=PH$;[57] and P-alkyl and P-phenyl derivatives $RP=CH_2$.[58] Some of these processes are conducted in solution under mild conditions and when trapping agents are present can lead to useful products. Cycloadditions with dienes are common; e.g., $ClP=CCl_2$ is a precursor of 2-chlorophosphinines (Eq. 10.36).

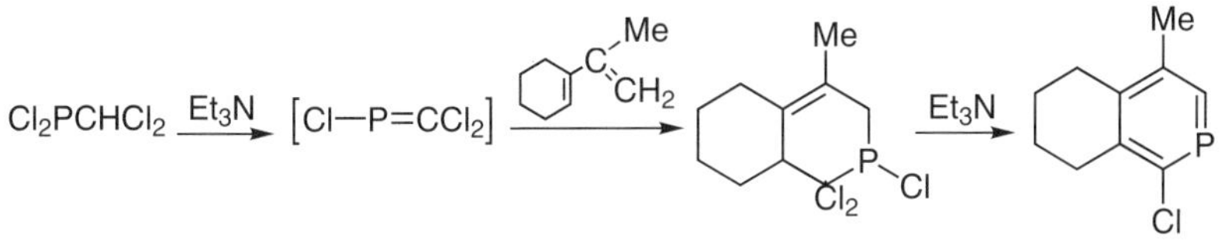

$$Cl_2PCHCl_2 \xrightarrow{Et_3N} [Cl-P=CCl_2] \longrightarrow \text{(cycloadduct)} \xrightarrow{Et_3N} \text{(2-chlorophosphinine)} \quad (10.36)$$

Many methods have emerged over the years for the synthesis of stable phosphaalkenes, and the excellent summary, very recently published,[59] should be consulted for details and references. We will briefly describe here only a few representative methods, but ones that account for the majority of the known compounds. As a general rule, at least one bulky group must be present on C or P to provide kinetic stabilization, although greater stability accompanies additional bulky substitution. Also, many stable compounds are known in which heteroatom substituents on P are present, notably Cl, F, R_2N, R_2P, and Me_3Si. These groups, through either electron attracting or releasing effects, are useful in that they provide significant thermodynamic stabilization of the double bond. Such derivatives are also useful as precursors of other phosphaalkene derivatives through conventional substitution reactions. A frequently encountered compound of this type has the structure Cl—P=C(TMS)(R). In all syntheses of phosphaalkenes, it is possible for E,Z isomers to be formed, and their presence can easily be detected by ^{31}P NMR spectroscopy. Separation into crystalline forms is possible,[60-62] and X-ray structure determination of both isomers has been accomplished, as with the isomers **12** and **13**.[62]

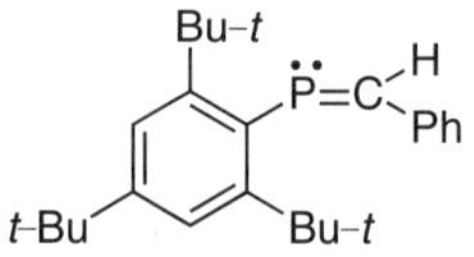

12   13

(Z) ^{31}P δ + 240.5   (E) ^{31}P δ + 256.6

$^1J_{PC}$ = 48.7 Hz $^1J_{PC}$ = 34.5 Hz

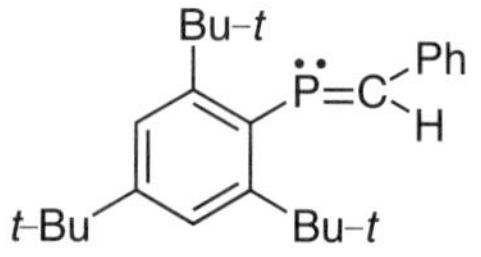

Interconversion of E,Z isomers can be facile in solution, but the structures of many solid compounds have now been established by the X-ray crystallographic method. As is the custom, we will show reaction products without E or Z designation, although in some cases the correct isomer structure is known. The E-isomer usually predominates or may be the exclusive product; it can in some cases be converted to the Z-isomer by UV irradiation.[61]

10.C.1.a.i. The 1,2-Elimination Approach. The original Bickelhaupt method (Eq. 10.34) has been applied many times for the synthesis of new phosphaalkenes and remains standard practice. It is common to provide stabilization by steric protection with large substituents, although electronegative substituents (especially Cl) have also been found to be effective[54a] (e.g., Eq. 10.37). For HX elimination, bases such as DBU, DABCO, Et$_3$N, Li, and Na hexamethyldisilazane have been used.

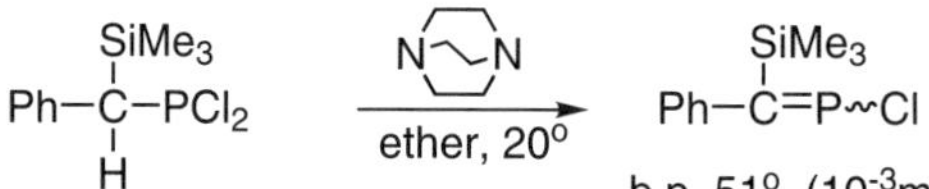

$$\text{Ph-C(SiMe}_3\text{)(H)-PCl}_2 \xrightarrow[\text{ether, 20°}]{\text{N}\frown\text{N}} \text{Ph-C(SiMe}_3\text{)=P}\sim\text{Cl} \quad (10.37)$$

b.p. 51° (10^{-3}mm), ^{31}P δ + 272, + 274 (1:1)

A variation is to eliminate Me$_3$SiX rather than HX, as in Eq. 10.38.[63]

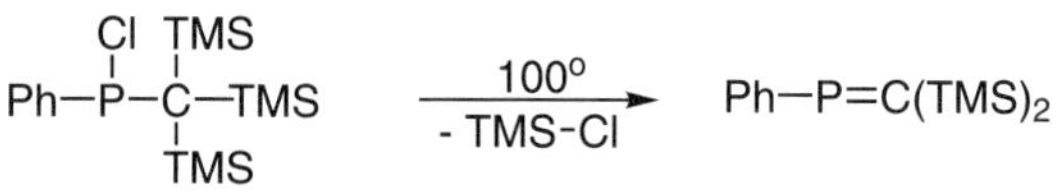

$$\text{Ph-P(Cl)-C(TMS)}_2\text{-TMS} \xrightarrow[\text{- TMS-Cl}]{100°} \text{Ph-P=C(TMS)}_2 \quad (10.38)$$

Haloforms, in the presence of base, are also used in the formation of C-halo phosphaalkenes. These reactions may be visualized as proceeding through the attack of carbenoid intermediates on phosphorus, perhaps followed by elimination processes, but the details are not clear. Two examples are shown in Eqs. 10.39[64,65] and 10.40.[66]

10.C. THE 2-COORDINATE STATE WITH DOUBLE BONDS TO PHOSPHORUS (σ^2, λ^3)

$$\text{Ar-PH}_2 + \text{HCCl}_3 \xrightarrow{\text{KOH}} \text{ArP=C(H)(Cl)} \quad (10.39)$$

(where Ar = 2,4,6-tri-tert-butylphenyl)

$$\text{Ar'-PCl}_2 \xrightarrow{\text{BuLi + HCCl}_3} \text{Ar'-P=CCl}_2 \quad (10.40)$$

(where Ar' = 2,4,6-trimethylphenyl)

10.C.1.a.ii. 1,3-Trimethylsilyl Migration from P to O.

1,3-Trimethylsilyl migration is the original Becker method,[49] which has been developed into a general and widely used procedure (Eq. 10.41). The migration may be referred to as "silatropy"[11f] by its analogy to the better-known 1,3-proton shifts, called prototropy. The migration, which proceeds to the energetically more stable form, is generally brought about thermally. The process is simple in design, and merely requires a preliminary synthesis of a disilyl phosphine for reaction with an acyl halide. The acyl phosphine **14** generally requires only gentle warming for the rearrangement to occur. Phosphaalkenes made in this way are hydrolytically unstable, owing to the necessarily present enol ether type of function.

$$\text{RPLi}_2 + \text{ClSiMe}_3 \longrightarrow \text{RP(TMS)}_2 \xrightarrow{\text{R'COCl}} \text{RP(TMS)(C(O)R')} \longrightarrow \text{R-P=C(R')(OTMS)} \quad (10.41)$$

$$\text{14}$$

10.C.1.a.iii. Condensation Reactions.

This approach may be typified by the reaction of a hindered primary phosphine and a carbonyl compound with elimination of water,[67] in analogy to the conversion of amines to imines. Thus the highly hindered phosphine **15** reacts with benzaldehyde in the presence of a dehydrating agent to form phosphaalkene **16** (Eq. 10.42), quite in contrast to the reaction of unhindered phosphines with carbonyl compounds (chapter 3).

$$\text{Ar-PH}_2 \text{ (15)} + \text{Ph-CHO} \xrightarrow{\text{P}_4\text{O}_{10}} \text{Ar-P=C(H)(Ph)} \text{ (16)} \quad (10.42)$$

A useful variant, which shows one (of several) approaches to C-amino phosphaalkenes, is shown in Eq. 10.43. Here an ortho-esteramide serves as a masked carbonyl compound for the condensation with the phosphine.[68]

$$PhPH_2 + (MeO)_2C{<}^R_{NR'_2} \longrightarrow Ph{\sim}P{=}C{<}^R_{NR'_2} \qquad (10.43)$$

Condensations are also known where silicon species are eliminated,[11f] as in Eq. 10.44.

$$R-P{<}^{TMS}_{TMS} + O{=}C{<}^H_{Ph} \longrightarrow R{\sim}P{=}C{<}^H_{Ph} \qquad (10.44)$$

10.C.i.a.iv. The "Phospha-Wittig" Reaction. This novel process,[69] differs from a conventional Wittig reaction only in that the reactive carbanion site is replaced by a phosphide ion, in complexed form (17). The usual phosphonate group of the Wadsworth–Emmons reaction is seen to be present on this phosphorus and serves as the leaving group (Eq. 10.45). The coordinated group is frequently of the form $M(CO)_5$, where M is W or Mo. The reaction forms the phosphaalkene in coordinated form. A number of examples of this and related reactions are known.[33b]

$$\underset{W(CO)_5}{(EtO)_2\overset{O}{\overset{\|}{P}}{-}\overset{H}{\overset{|}{P}}{-}R} \xrightarrow[-70°]{n\text{-BuLi}} \underset{\underset{\mathbf{17}}{W(CO)_5}}{(EtO)_2\overset{O}{\overset{\|}{P}}{-}\overset{..}{\overset{-}{P}}{-}R} \xrightarrow{R'_2C=O} \underset{W(CO)_5}{R{-}P{=}CR'_2} + (EtO)_2PO_2^- \qquad (10.45)$$

10.C.1.b. Properties. From numerous X-ray structural analyses, it is known[11f] that the average value for the P—C double bond length is about 1.67 Å (cf. to 1.85 Å for P—C and 1.5 Å for P≡C). The molecule is planar about the double bond. It was early noted that the ^{31}P NMR signals were remarkably downfield, and for many molecules with no heteroatoms on C appeared in the range δ +200 to +350. However, substituents can make a very large difference, and to include all shifts that have been reported, the range δ −100 to +450 would have to be considered. Trends and further interpretations of these shifts have been discussed by Fluck and Heckmann,[70] and tables of data are found in the Tebby compilation.[71] As already noted, E,Z isomers can be detected for the phosphaalkenes (as in 18 and 19, whose structures were confirmed by X-ray analysis[62]) and although their shifts can differ significantly, no definitive correlation with the structure has emerged. The coupling constants of ^{31}P to carbon or hydrogen attached to the sp^2 C, however, can be useful; the constant is larger when the lone pair orbital on phosphorus is close to the coupled atom (dihedral angle 0°). This is the same type of correlation discussed in chapter 7 for 3-coordinate compounds, in which conformational rigidity is present. Three-bond couplings also are controlled by the lone pair orientation.

18: $Ar{\sim}P{=}C{<}^{Ph}_H$ ^{31}P δ +256.6; ^{13}C δ 175.87 (J_{PC}=34.49); ^1H δ 8.21 (J_{PH}=25.6)

19: $Ar{\sim}P{=}C{<}^H_{Ph}$ ^{31}P δ +240.5; ^{13}C δ 162.54 (J_{PC}=46.7); ^1H δ 7.87 (J_{PH}=38.1)

10.C. THE 2-COORDINATE STATE WITH DOUBLE BONDS TO PHOSPHORUS (σ^2, λ^3)

Many reactions are known for the phosphaalkene family, and are summarized in review articles.[6–10,11f,33a] We have already seen the very important process of 1,2-elimination of $(Me_3Si)_2O$ to form phosphaalkynes (Eq. 10.4). Here we will consider only the most characteristic of the reactions.

10.C.1.b.i. Reactions at Phosphorus. The lone pair on phosphorus is not readily available for bond formation; however, in certain phosphaalkenes, bonds to O, S, and Se by reaction with ozone, S_8, and Se, respectively, can be achieved. Carbenes also add to phosphorus to form a second P–C double bond. Such additions are common for structure **20**. All products have the general structure **21**, and are of interest as examples of rare stable σ^3, λ^5 bonding (see section 10.D).

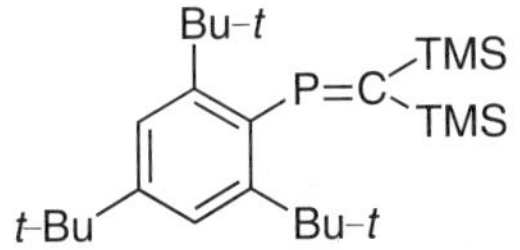

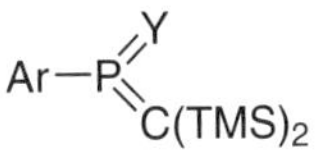

Similarly, the lone pair is not very reactive to alkyl halides, although an example of this reaction is known. Thus phosphaalkene **22** slowly reacts with MeI to form the phosphonium ion **23**, which then adds I⁻ to form the ylide **24** as the main product (Eq. 10.46).

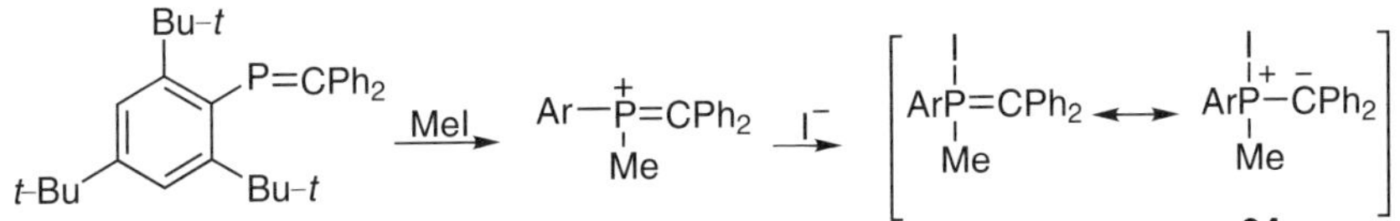

(10.46)

10.C.1.b.ii. 1,2-Additions to the Double Bond. Not only hydrogen halides but also alcohols and N–H compounds can add to the double bond of certain phosphaalkenes. As carbon is the more negative atom in the bond, the proton attaches to it. All products are represented by structure **25** (Eq. 10.47).

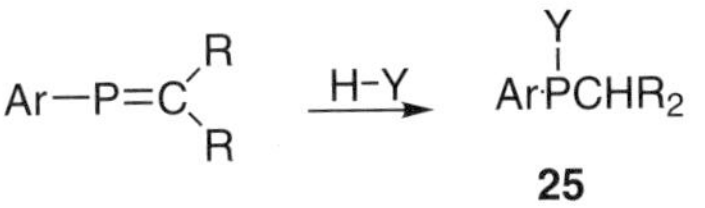

$$Ar-P=C\begin{smallmatrix}R\\R\end{smallmatrix} \xrightarrow{H-Y} Ar\overset{Y}{P}CHR_2 \quad (10.47)$$

25

10.C.1.b.iii. Cycloaddition Reactions. Just as was seen for the C–P triple bond, cycloaddition reactions to form valuable heterocycles constitute a major part of phosphaalkene chemistry. The most important are the [3+2] 1,3-dipolar additions and the [4+2] Diels–Alder type. We will not consider the [2+1] and [2+2] cases, except to mention that the latter reaction is well known from the occurrence of head-to-tail dimerization to 1,3-diphosphetanes; examples are known of both types of reaction and are discussed in the review literature. Equations 10.48,[72] 10.49,[72] 10.50[73] and 10.51[74] illustrate some of the heterocycles that have been formed.

$$\text{ArP=CPh}_2 + \text{Ph}\overset{-}{\text{N}}\text{-N=}\overset{+}{\text{N}} \longrightarrow \underset{\underset{\text{Ar}}{|}}{\overset{\text{Ph}}{\underset{}{\text{N-N}}}}\text{Ph}_2 \tag{10.48}$$

$$\text{ArP=CPh}_2 + \text{Ar'}\overset{+}{\text{C}}\text{=N-}\overset{-}{\text{O}} \longrightarrow \text{structure} \tag{10.49}$$

$$\text{TMS}\sim\text{P=C}\overset{\text{Ad}}{\underset{\text{TMS}}{}} + t\text{-BuC}\overset{-}{\text{H}}\text{-N=}\overset{+}{\text{N}} \longrightarrow \text{structure} \xrightarrow{\text{NaOH}} \text{structure} \tag{10.50}$$

(Ad = 1-adamantyl)

$$\text{ClP=C}\overset{\text{R}}{\underset{\text{TMS}}{}} + \text{Ar}\overset{+}{\text{C}}\text{=N-}\overset{-}{\text{S}} \longrightarrow \text{structure} \xrightarrow{\text{- TMS-Cl}} \text{structure} \tag{10.51}$$

Diels–Alder reactions with acyclic dienes are known but are retarded when a very large stabilizing substituent is present on P; they are better known with the P–Cl (and other P-hetero) derivatives (Eq. 10.52[75]). Cyclopentadiene is especially reactive (Eq. 10.53[76]).

$$\text{Cl-P=C(TMS)}_2 + \text{diene-OTMS} \longrightarrow \text{cyclic product} \tag{10.52}$$

$$\text{Cl}\sim\text{P=C}\overset{\text{Ph}}{\underset{\text{R}}{}} + \text{cyclopentadiene} \longrightarrow \text{bicyclic product (4 diastereomers)} \tag{10.53}$$

10.C.1.b.iv. Coordination Reactions.

Phosphaalkenes have two sites for forming bonds to metals, and many examples are known of both. The lone pair may be used, forming η^1 complexes, or the π electrons may be used, forming η^2 complexes. In some complexes, both sites are involved in the bonding. Many metals can be incorporated in the complexes, and the field has developed very rapidly. Valuable discussions of the synthesis and properties of the complexes are found in the review literature,[6,11f,59] but this branch of organometallic chemistry, somewhat out of the scope of this book, cannot receive further attention here.

10.C.1.c. Phosphapolyenes. It was not long after the discovery of stable phosphaalkenes that it was shown that a second double bond could be conjugated with the P—C double bond. The first example was a 1-phosphaalkadiene, prepared by Appel in 1984[77] (Eq. 10.54).

$$\text{Ar-P(TMS)}_2 + \text{Cl-C(O)CH=CHPh} \longrightarrow \text{ArP=C(OTMS)-C(H)=C(H)Ph} \quad (10.54)$$

(where Ar = 2,4,6-tri-*t*-butylphenyl)

This started another flood of research on the synthesis of various types of phosphadienes and phosphatrienes. Later discoveries were made of "phosphacumulenes," in which the carbon attached to phosphorus is doubly bonded also to another atom, typically C (to give phosphaallenes, —P=C=C) or to O (phosphaketenes, —P=C=O). Some of these structures proved to quite easily prepared and of good stability, and certainly their novelty is very high, making the study of "phosphapolyenes" most fascinating. Appel[11f] pointed out how closely phosphorus resembled carbon in these structures and in their properties, and he pioneered in the development of pericyclic reactions with polyenes in which the analogy to carbon is perhaps at its highest. Thus sigmatropic and electrocyclic mechanisms could be identified, and a "phospha-Cope" process (vide infra) found to prevail; the usual stereochemical demands of these pathways were followed. The analogy of phosphorus to carbon chemistry had hardly been appreciated before the development of phosphapolyene chemistry, it being more common to draw analogies to nitrogen and sulfur. As already noted, the carbon analogy has been recently strongly emphasized by Dillon, Mathey, and Nixon.[3] The many interesting papers on phosphapolyenes were first reviewed by Appel in 1990,[11f] and the Dillon, et al.,[3] brought the review up to date. Given the existence of these excellent reviews, and the fact that much of phosphapolyene chemistry is highly specialized, frequently more of theoretical than practical importance, we will outline here only some of the important structural types. The coordination chemistry of phosphapolyenes is also extensive and the subject of reviews.[3,11f]

1-Phosphabutadienes. In addition to Eq. 10.54, in which strong steric stabilization is required to prevent dimerization, the reaction of Eq. 10.55 may be considered of value in forming a stable product.[78]

$$\text{TMS-CH=CH-CH(Li)-TMS} \cdot \text{TMEDA} + (\text{TMS})_2\text{NPCl}_2 \longrightarrow (\text{TMS})_2\text{N-P(Cl)-C(TMS)=C(H)-C(H)(TMS)} \xrightarrow{\text{DBU}}$$

$$(\text{TMS})_2\text{N-P=C(TMS)-C(H)=C(H)(TMS)} \quad (10.55)$$

2-Phosphabutadienes.[77]

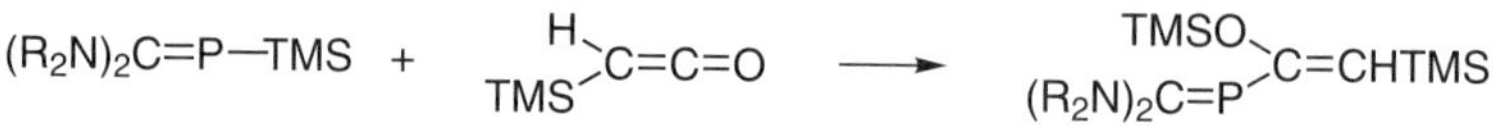

$$(R_2N)_2C=P-TMS + \underset{TMS}{\overset{H}{>}}C=C=O \longrightarrow (R_2N)_2C=P\overset{TMSO}{\underset{}{>}}C=CHTMS \quad (10.56)$$

1,4-Diphosphabutadienes.[79]

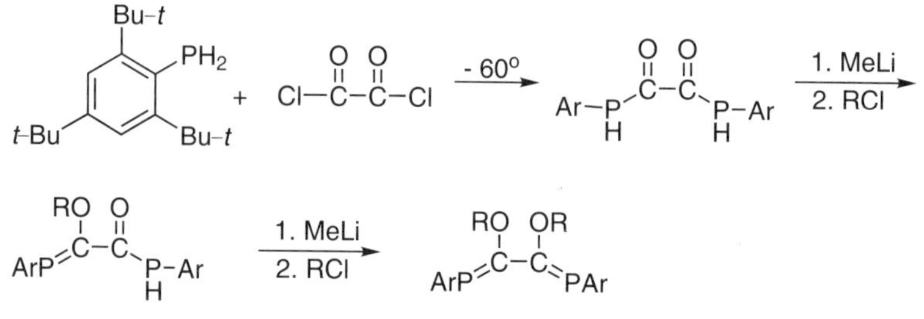

$$(10.57)$$

1,3-Diphosphabutadienes.[80]

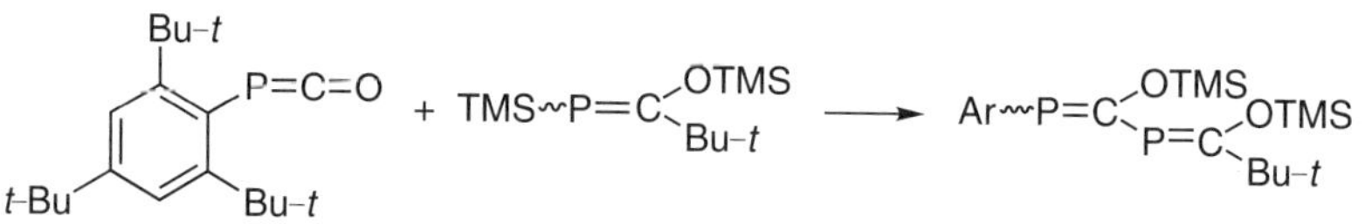

$$(10.58)$$

2,3-Diphosphabutadienes.[81]

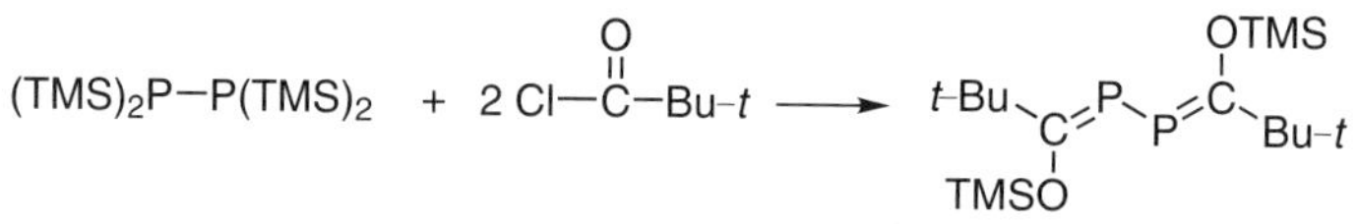

$$(TMS)_2P-P(TMS)_2 + 2\,Cl-\overset{O}{\underset{}{C}}-Bu\text{-}t \longrightarrow t\text{-Bu}\underset{TMSO}{\overset{}{>}}C=P\overset{OTMS}{\underset{}{>}}P=C\overset{OTMS}{\underset{Bu\text{-}t}{>}} \quad (10.59)$$

1-Phosphaallenes.[82]

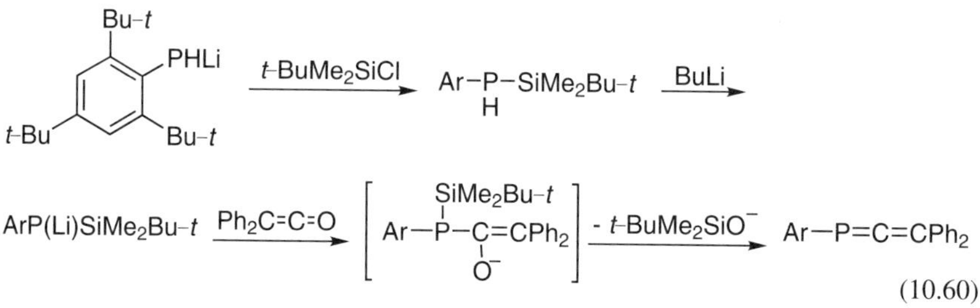

$$(10.60)$$

1,3-Diphosphaallenes.[83]

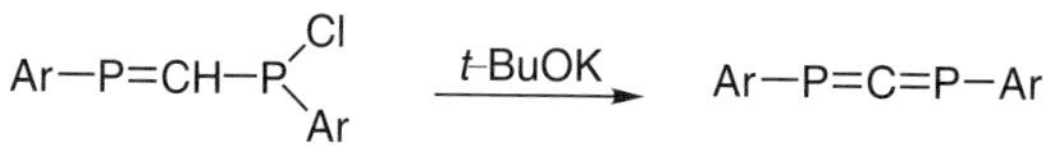

$$Ar-P=CH-P\overset{Cl}{\underset{Ar}{<}} \xrightarrow{t\text{-BuOK}} Ar-P=C=P-Ar \quad (10.61)$$

Phosphaketenes.[84] The oxygen changes the polarization of the P–C bond, making C positive and P negative. This results in a remarkable upfield shifting of the ^{31}P NMR signal (to $\delta -207.4$ in compound **26**); see Eq. 10.62.

10.C. THE 2-COORDINATE STATE WITH DOUBLE BONDS TO PHOSPHORUS (σ^2, λ^3)

$$\text{Ar-P(TMS)}_2 \xrightarrow{\text{COCl}_2} \text{ArP=C=O} \quad (10.62)$$
$$\mathbf{26}$$

(where Ar = 2,4,6-tri-t-butylphenyl)

In many of the reactions of phosphapolyenes, unusual cyclic products are obtained, either through dimerization processes or other cycloadditions. Another fascinating aspect of phosphapolyene chemistry is that a form of the Cope rearrangement is known for certain phospha-hexadiene derivatives. A simple example[85] is that of the rearrangement of tetraphosphahexadiene **27**, which was found to exist in equilibrium with the Cope product **28** (Eq. 10.63). Here the P—P bond is disconnected in a [3,3] sigmatropic rearrangement.

$$\text{PhP(TMS)}_2 \xrightarrow{\text{Cl}_2\text{C=NPh}} \text{Ph-P=C(NPh(TMS))(TMS)-P-Ph} \xrightarrow{\text{Cl}_2\text{C=NPh}}$$

[Structure 27 ⇌ Structure 28] (10.63)

In some other examples (e.g., Eq. 10.64[86]), the C—C bond is disconnected.

[Structure equation 10.64] (10.64)

Other examples of phospha-Cope reactions are discussed in the review by Appel.[11f] The uncovering of this remarkable similarity to carbon chemistry has stimulated much exciting research in the phosphapolyene field, and new discoveries can surely be expected.

10.C.2. The Phosphorus—Nitrogen Double Bond

Iminophosphines were first obtained in stable form in 1973,[87] and therefore stand as the first family of phosphorus compounds recognized to have a stable isolated double bond. The usual requirements for stability are present, and most of the numerous compounds have bulky substituents such as *t*-butyl, mesityl, or 2,4,6-tri-*t*-butylphenyl. Trimethylsilyl substituents are also often encountered. The first iminophosphine had the structure **29**.

$$(TMS)_2N-P=N-TMS \atop TMS$$
29

Iminophosphine chemistry has been reviewed by Niecke,[11g] who cites several earlier reviews, and more recently by Niecke and Gudat.[88]

Numerous X-ray crystallographic analyses have been performed on the iminophosphines, revealing planar geometry about the double bond, with the trans configuration predominating. The ^{31}P NMR shifts are found over a very broad range of δ +100 to +800, although if no heteroatom substituents are present on N and P, the range is much smaller, about δ +300 to +500. Compounds with RO or Cl on P are in the high field end of the range (δ +100 to +200), and N and Si on P have shifts more in the δ +300 to +400 range. The P=N bond has pronounced UV-Vis absorption, quite dependent on the substituents. With only carbon substituents, π to π^* transitions appear around 430 to 440 nm, and n to π^* around 230 nm.

The most important method for the synthesis of iminophosphines remains that of Niecke and Flick[87] involving the 1,2-elimination of a silyl halide from an appropriate aminohalophosphine, as exemplified in Eq. 10.65. A considerable variety exists in the substituents that may be present on P (typically alkyl, aryl, Me$_3$Si, etc).

$$\underset{F}{\overset{R}{>}}P-N\underset{TMS}{\overset{R'}{<}} \xrightarrow[-TMS-F]{\Delta} R-P=N-R' \qquad (10.65)$$

A related process is shown in Eq. 10.66, in which the species being eliminated is a lithium salt.[89] This process requires a bulky substituent in the lithium amide reagent, to avoid nucleophilic substitution at P—X.

Other methods, of less common occurrence in the literature, are summarized in the reviews by Niecke.[11g,88]

$$\underset{Cl}{\overset{H}{ArP-NR}} \xrightarrow{LiNR_2} \underset{Cl}{\overset{Li}{ArP-NR}} \xrightarrow[-LiCl]{\Delta} ArP=NR \qquad (10.66)$$

Iminophosphines undergo a number of reactions. Because the polarity is P$^+$—N$^-$, Lewis bases add to P, and Lewis acids and H$^+$ attack nitrogen; 1,1-additions to P are also possible, as in the reaction with CCl$_4$ (Eq. 10.67[87]). With oxidizing agents such as ozone (Eq. 10.68), sulfur, and selenium, valuable products with σ^3, λ^5 coordination are formed.

$$R_2N-P=N-R' \xrightarrow{CCl_4} R_2N-\underset{CCl_3}{\overset{Cl}{P}}=NR' \qquad (10.67)$$

$$R_2N-P=N-R' \xrightarrow{O_3} R_2N-P\underset{O}{\overset{N-R'}{<}} \qquad (10.68)$$

Cyclic compounds can also be formed by 1,1-additions to the phosphorus atom. Thus dienes undergo [4+1] cycloaddition (Eq. 10.69), rather than Diels–Alder [4+2] cycloaddition to the P=N group.[90]

$$\text{t-Bu-P=N-Ar} + \text{Me}_2\text{C=C=CMe}_2 \longrightarrow \text{product} \qquad (10.69)$$

However, [3+2] dipolar cycloadditions are well known and give rise to novel heterocycles. Thus with an azide[91] the inorganic ring system of **30** is formed (Eq. 10.70), whereas a diazoalkane gives the unusual triazaphospholene **31** (Eq. 10.71).

$$\text{i-Pr}_2\text{N-P=N-Bu-t} \xrightarrow{\text{t-BuN}_3} \textbf{30} \qquad (10.70)$$

$$\text{i-Pr}_2\text{N-P=N-Bu-t} \xrightarrow{\text{t-BuCHN}_2} \textbf{31} \qquad (10.71)$$

[2+2]-Cycloadditions with cumulated double bond systems are also known, and likewise give rise to novel heterocycles. An example[92] is the reaction with carbon dioxide (Eq. 10.72).

$$\xrightarrow{\text{CO}_2} \qquad (10.72)$$

Finally, we must mention the important aspect of coordination chemistry of phosphinimines. Three basic types of complexes may be formed, η^1 at P and N, and η^2 at the π bond. More than one of these bond types may be present in a complex, and this makes the field of great interest and broadness. A review is available for the further consideration of these coordination complexes.[11g]

10.C.3. The Phosphorus-Oxygen Double Bond (and P=S, P=Se)

Despite the importance of the true phosphorus-oxygen double bond with p_π-p_π interaction, no case is known at the present time of a compound with this bonding that is stable at room temperature. Attempts to prepare Y–P=O species (the unfortunate use of the same structure as for the totally different common phosphoryl group was noted in chapter 2.D) lead only to polymeric products, $(-P(Y)-O-)_n$. That the bond does exist, however, is not in doubt. The inorganic compounds

F−P=O, Cl−P=O, Br−P=O, HO−P=O, and H−P=O have been generated and trapped in an argon matrix at 12 to 18 K, and their infrared spectra recorded (reviewed in reference 11h). The stretching of P=O occurs at 1185 to 1292 cm^{-1}, a range rather similar to that of the common phosphoryl group (chapter 7.C). Organic derivatives (R−P=O, oxophosphines) have never been observed directly except as fragments in mass spectra. Attempts to stabilize the species by placing very large substituents on P have not yet proved successful (except in the case of an ArO−P=O derivative, vide infra). Several reactions are known in which there is little doubt this species is being formed in elimination reactions, but the only evidence comes from trapping reactions. These reactions do reveal some of the chemical reactivity of the oxophosphines. In Eq. 10.73, the great electrophilic character of P toward the OH group is seen; indeed, alcohols are the most commonly used trapping reagents. The reaction does suggest a difficulty in working with such species; they are as reactive to water as to alcohols, and traces of water in the system will interfere with the intended trapping reaction.

$$R-P=O \; + \; EtOH \; \longrightarrow \; R-\underset{H}{\overset{\overset{O}{\|}}{P}}-OEt \qquad (10.73)$$

1,1-Cycloadditions occur at P, and these are also useful trapping reactions (Eqs. 10.74 and 10.75).

(10.74)

(10.75)

A simple procedure for generating oxophosphines involves the fragmentation of a heterocyclic phosphine oxide, generally in the phosphirane or 7-phosphanorbornene systems. The first two known examples of each reaction type are shown in Eq. 10.76[93] and 10.77,[94] although several later reports used other related cyclic derivatives[11h]. All of these reactions are conducted in the presence of a trapping reagent. In a more recent experiment on phosphirane fragmentation,[95] the important observation was made that the fragmentation was first-order in the phosphirane even in the presence of a trapping agent (butadiene), in accord with the generation of the low-coordinate species as a separate (slow) step (Eq. 10.78).

$$\text{phosphirane} \xrightarrow{\Delta} t\text{-BuHC=CHBu-}t \; + \; [t\text{-Bu}-\text{P}=\text{O}] \xrightarrow{\text{MeOH}} t\text{-Bu}-\underset{H}{\overset{\overset{O}{\|}}{P}}-\text{OMe}$$

(10.76)

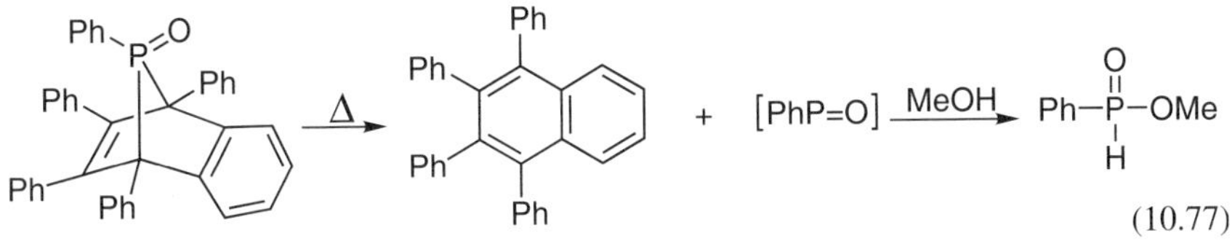

10.C. THE 2-COORDINATE STATE WITH DOUBLE BONDS TO PHOSPHORUS (σ^2, λ^3)

(10.77)

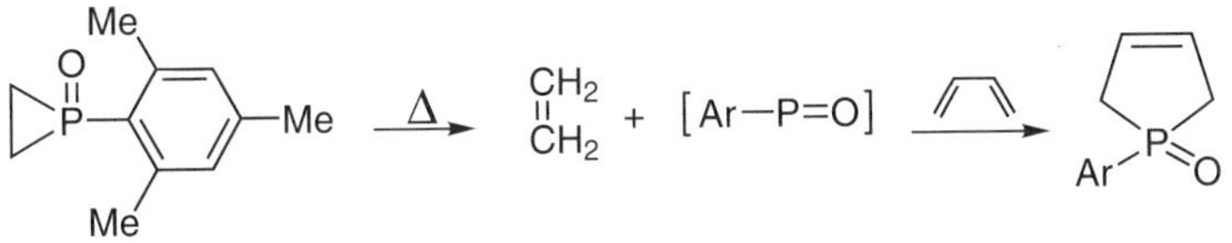

(10.78)

Cowley et al.,[96] describe what appears to be a successful generation of an oxophosphine in the gas phase, as evidenced by the self-trapping of the species according to Eq. 10.79. Here the novel heterocyclic system **32** was synthesized and pyrolyzed at 300° at 10^{-2} mm. The product, formed quantitatively, was the known[97] *sec*-phosphine oxide **33**, formed by insertion into a C—H bond. Insertions into the *t*-butyl group in this family are remarkably easy; compound **33** (^{31}P δ +36.6, $^1J_{PH} = 483.4$ Hz) had been formed on refluxing the same phosphonous dichloride in toluene overnight, followed by hydrolysis.

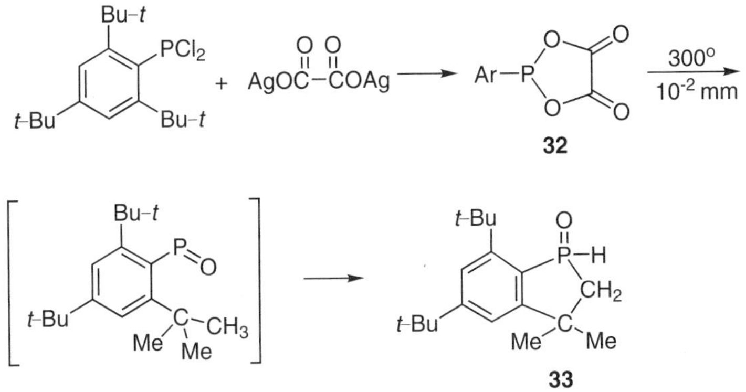

(10.79)

The dehalogenation of phosphonic dihalides with metals has been used to generate oxophosphines; although the expected trapping products are obtained, the process is more complex than the simple heterocyclic fragmentation approach and there is always the possibility that the oxophosphine is not being generated in free form.

The oxophosphine moiety is also found in the form of amide or ester derivatives of phosphenous acid, HO—P=O. The former was first generated in the form of a metal coordination complex by Niecke.[98] The process involved the creation of the species R_2N—P=O by oxidative cleavage of an iminophosphine while coordinated to chromium (Eq. 10.80). The complex formed (**34**) was quite useful in revealing some of the properties of the oxophosphines, albeit in complexed form. Thus the P—O bond length was 1.475 Å, not greatly different from the σ^4, λ^5 phosphoryl group. The P=O stretching occurred at 1200 cm^{-1}, consistent with that of the series X—P=O noted earlier to be in the range 1185 to 1292 cm^{-1}. The ^{31}P NMR shift was quite far downfield, $\delta +319.2$.

$$i\text{-Pr}_2N\text{-P=N-Bu-}t \xrightarrow{Cr(CO)_6} (i\text{-Pr}_2N)\text{P=N-Bu-}t\cdot(CO)_5Cr \xrightarrow{SO_2} (i\text{-Pr}_2N)\text{P=O}\cdot(CO)_5Cr \quad (10.80)$$

$$\mathbf{34}$$

An aryl phosphenite is the first organic derivative with the σ^2, λ^3 P=O group to be directly observed spectroscopically.[99] Here the P=O group was stabilized with a large trialkylphenoxy substituent as in structure **36**; the compound was prepared by pyrolysis in vacuo of the easily synthesized trioxatriphosphorane **35** (a trimer of the phosphenite, first prepared by Chasar, et al.[100]), Eq. 10.81, and collected on a finger at liquid nitrogen temperatures.

$$\text{Ar-O-PCl}_2 \xrightarrow{H_2O} \mathbf{35} \xrightarrow[10^{-5}\,mm]{300\text{-}350°} \mathbf{36} \xrightarrow{MeOH} \begin{cases} \text{ArOH} + [\text{MeO-P=O}] \xrightarrow{MeOH} (\text{MeO})_2\overset{O}{\underset{\|}{P}}\text{-H} \\ \text{Ar-O-}\overset{O}{\underset{\|}{P}}\text{-OMe} \\ \quad\quad\quad\;\; H \quad \mathbf{37} \end{cases} \quad (10.81)$$

Phosphenite **36** was not stable at room temperature, but some spectral properties could be observed at low temperatures. The ^{31}P NMR signal appeared at $\delta +238$, and the stretching of P=O in the infrared at 1235 cm^{-1}. Ultraviolet absorption occurred at 271 and 278 nm. As the sample was allowed to warm up, these signals disappeared due to P=O condensations. Although alcohols reacted with the electrophilic phosphorus in the usual way (forming an H-phosphonate, **37**), the unexpected observation was made that some displacement of the aryloxy group in **36** also occurred, presumably leading to the formation of RO—P=O and then (RO)$_2$PH(O). Similar behavior occurred with water. Compound **36** thus appears to be a useful model for further study of the properties of phosphenites.

The P=S and P=Se bonds have been generated by methods similar to those used for the P=O bond; the 7-phosphanorbornene fragmentation route is especially useful. In neither case has a compound stable at room temperature been obtained; trapping reactions similar to those used for P=O compounds provide the only evidence so far for their existence.[11h]

10.C.4. Double Bonds of Phosphorus to Phosphorus and to Other Elements

Several other elements are now known to form a double bond with phosphorus. This subject lies more within the interests of inorganic chemists and will be treated only briefly here. Surely the most important is the phosphorus–phosphorus double bond (the diphosphene family), first observed by Yoshifuji et al., in 1981.[101] As in other cases, the bond proved to be easily formed, but, of course, there must be present the kinetic stabilizing effect of large substituents on each P. The Yoshifuji process is shown in Eq. 10.82.

$$\text{2,4,6-tri-}t\text{-Bu-C}_6\text{H}_2\text{-PCl}_2 \xrightarrow{\text{Mg}} \text{Ar-P=P-Ar} \quad (10.82)$$

The product was an orange-red crystalline solid, found by X-ray analysis to be planar and with the trans orientation of the substituents. The ^{31}P NMR signal is far downfield, δ +492.4 in C_6D_6. The Yoshifuji method, or variants thereof, has since been used to make a number of symmetrical diphosphenes, all of which had ^{31}P signals at around δ +450 to +600. Methods were later developed for preparing unsymmetrical diphosphenes (whose ^{31}P NMR spectrum consists of two doublets with quite large $^1J_{PP}$ values of 525 to 675 Hz). A few cis-diphosphenes are known and have ^{31}P NMR shifts at lower values (δ +370 to +380). The chemistry of diphosphenes has been well studied and reviewed;[11i,102] the great value of ^{31}P NMR in these studies is also clear from reviews.[103]

Using the kinetic stabilization approach, it has been possible in recent years to synthesize compounds with double bonds of phosphorus with other elements, notably arsenic, silicon, germanium, and tin. Some aspects of this chemistry have been reviewed.[11j]

10.D. THE 3-COORDINATE STATES (σ^3, λ^5)

10.D.1. General

The chemistry of the σ^3, λ^5 state is much less developed than is that of the 1- and 2-coordinate states, and again is an area opened only since the 1970s. Nevertheless, the 3-coordinate structures occupy an important position in low-coordination phosphorus chemistry and deserve brief mention in this survey. It is customary to draw the structure of these species with two double bonds to phosphorus (**38**), but the bonding is probably better expressed as a resonance (**39**) of two dipolar structures (ylides) as shown for the bis(methylene)phosphorane case.

$$\text{R-P}\begin{array}{c}\diagup \text{CH}_2\\ \diagdown \text{CH}_2\end{array} \qquad \text{R-}\overset{+}{\text{P}}\begin{array}{c}\diagup \text{CH}_2\\ \diagdown \text{CH}_2^-\end{array} \longleftrightarrow \text{R-}\overset{+}{\text{P}}\begin{array}{c}\diagup \overset{-}{\text{CH}}_2\\ \diagdown \text{CH}_2\end{array}$$

         38   **39**

The d-orbitals on P do not appear to be involved. X-ray analyses show planarity at phosphorus, with identical bond lengths for the two multiply bonded atoms; for carbon, the length is typically 1.65 to 1.67 Å, nearly the same as found in the phosphaalkenes (1.61 to 1.71 Å).

The 3-coordinate states are never stable at room temperature, unless there are the usual large bulky groups present to prevent intermolecular condensations or intramolecular cyclization. This device has proved to be particularly effective with carbon or nitrogen groups, and a number of stable structures are known. With di-oxo derivatives, the only stabilization can come from substitution on phosphorus and, as will be seen, this does not provide adequate protection. There are no structures of this type presently known to be stable at room temperature.

In this section, we will discuss only the species that have received the most attention, namely R−P(=X)$_2$ where X is C, N, or O. Other related species, for which a stabilized structure has been obtained through the now conventional technique of steric protection with large substituents, include R−P(=C)(=O), R−P(=S)$_2$, R−P(=C)(=N), R−P(=C)(=S), R−P(=C)(=Se), R−P(=N)(=O), R−P(=N)(=S), and R−P(=N)(=Se). Most of these were formed by 1,1-additions to the corresponding 2-coordinate compounds. Information on all of these less-common species can be found in the monograph by Regitz and Scherer.[1]

10.D.2. Bis(methylene)phosphoranes

The first bis(methylene)phosphoranes were synthesized in 1982 by Appel, et al.[104] The method (Eq. 10.83) depends on the addition at low temperature of a carbene fragment to phosphorus of a phosphaalkene. Typically, the carbene precursor is of the form LiC(Cl)R$_2$, where R is usually trimethylsilyl (phenyl is also sometimes used).

$$t\text{-Bu-P=C}\begin{array}{c}\diagup \text{TMS}\\ \diagdown \text{TMS}\end{array} + \text{LiC(Cl)(TMS)}_2 \longrightarrow t\text{-Bu-P}\begin{array}{c}\diagup \text{C(TMS)}_2\\ \diagdown \text{C(TMS)}_2\end{array} \qquad (10.83)$$

The P-chloro and P-alkoxy derivatives can also be made as in Eq. 10.84 and subjected to nucleophilic substitution to obtain other derivatives.

$$\text{Cl-P=C}\begin{array}{c}\diagup \text{TMS}\\ \diagdown \text{TMS}\end{array} \xrightarrow{\text{LiC(Cl)(TMS)}_2} \text{Cl-P}\begin{array}{c}\diagup \text{C(TMS)}_2\\ \diagdown \text{C(TMS)}_2\end{array} \xrightarrow{\text{ROH}} \text{RO-P}\begin{array}{c}\diagup \text{C(TMS)}_2\\ \diagdown \text{C(TMS)}_2\end{array} \qquad (10.84)$$

Another approach, useful in achieving unsubstituted methylene groups on P, consists of the reaction of a phosphaalkene with the dimethylsulfonium methylide[105] (Eq. 10.85).

$$RP=CR_2 + (CH_3)_2S=CH_2 \longrightarrow RP\begin{smallmatrix}CR_2\\CH_2\end{smallmatrix} \qquad (10.85)$$

The ^{31}P NMR shifts are rather downfield as in **40** and **41**, but less so than in the phosphaalkenes that have comparable substituting groups. The ^{13}C NMR shift of the methylene carbon of, e.g., **40** is δ +84.0 ($^{1}J_{PC}$ = 37.2 Hz) and is inbetween that of a true sp^2 C and a –CH$_2^-$ group. This is consistent with the simple representation of the structure by resonance hybrid **39**.

t-Bu–P(C(TMS)$_2$)(C(TMS)$_2$) ^{31}P δ +204.5   MeO–P(C(TMS)$_2$)(C(TMS)$_2$) ^{31}P δ +174
                                  ^{13}C δ 84.0 ^{13}C δ 55.8
40   **41**

Phosphoranes not adequately stabilized have been observed to undergo intramolecular cyclization to form a phosphirane (Eq. 10.86). Other properties are discussed in a review by Appel.^{11k}

$$Ph-P=C\begin{smallmatrix}TMS\\TMS\end{smallmatrix} + LiC(Cl)Ph_2 \longrightarrow Ph-P\begin{smallmatrix}C(TMS)_2\\CPh_2\end{smallmatrix} \longrightarrow Ph-P\begin{smallmatrix}TMS\ TMS\\Ph\ Ph\end{smallmatrix} \qquad (10.86)$$

10.D.3. Bis(imino)phosphoranes

Stable bis(imino)phosphoranes were first reported in 1974 by Niecke.^{106} The method employs a reaction of an azide with an iminophosphine. In one example (Eq. 10.87), [3+2] cycloaddition first occurs to form heterocyclic compound **42**, which then loses N$_2$. Several X-ray analyses have been performed, confirming the planar structure at P. The ^{31}P NMR shift of a typical compound, **43**, is δ +83.4.^{91b}

i-Pr$_2$N–P=N–Bu-t + t-BuN$_3$ ⟶ [t-Bu–N(N=N)–P(i-Pr$_2$N)–N–Bu-t] **42** $\xrightarrow{-N_2}$ i-Pr$_2$N–P(=NBu-t)(=NBu-t)

$$(TMS)_2CH-P\begin{smallmatrix}NTMS\\NTMS\end{smallmatrix} \qquad (10.87)$$
43

10.D.4. Dioxophosphoranes; Metaphosphoric Acid Derivatives

Only one dioxophosphorane (Cl–PO$_2$) has ever been directly detected, and no stable compounds with this structural feature are known. Nevertheless, the subject is of great interest. It first attracted attention because of a possible biological role for the anion of metaphosphoric acid (HO–PO$_2$) in phosphorylations (for recent reviews, see reference 107 and 108, the latter speaking against the possibility for the existence of free PO_3^- in water). More recently, they received attention because practical applications can be seen for metaphosphoric acid derivatives as laboratory phosphorylaing agents. Several reviews cover the subject of dioxophosphoranes in detail.[11-1,109,110]

The dioxophosphorane group is generally expressed by structure **44**, but a better representation might be the resonance hybrid **45**, which includes charges on the P and O atoms.

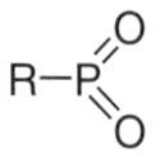

44 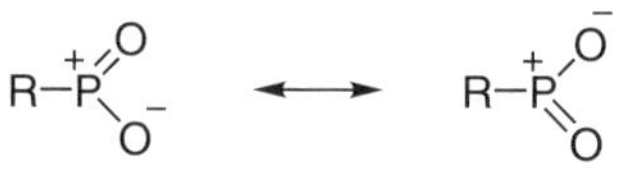   **45**

Theoretical treatment of the bonding is included in a very recent paper by Chesnut[111] that is concerned also with the bonding in the 4-coordinate phosphoryl group (chapter 2.D). To review, he (and others) believes the bond to be properly represented by the structure $R_3P^+-O^-$, with the back-bonding due to ionic interaction. From this view, an expression for the dioxophosphorane group might be **46**.

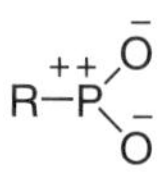

46

The only observable, although unstable, compound with the dioxophosphorane structure is the chloro derivative, $Cl-PO_2$, which was generated in the gas phase and preserved in an argon matrix.[112] Its IR spectrum had a peak for P—O stretching at 1443 cm^{-1}. The photoelectron spectrum was later recorded at 450 K,[113] and there is no doubt that the dioxophosphorane group can be created. In all other attempts made so far to generate an organic dioxophosphorane, only intermolecular condensation products of the monomer have been observed. Cases are known, however, in which the powerfully electrophilic dioxophosphorane (and oxothiophosphorane) group is trapped on generation in solution (usually at low temperatures) by formation of a Lewis salt with aprotic amines or ethers. These Lewis salts function as donors of the dioxophosphorane in other reactions. Some postulated Lewis salts are shown as structures **47**,[114] **48**,[115] **49**[116] and **50**.[117]

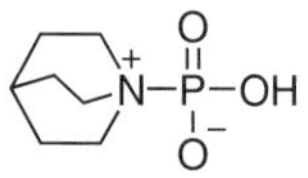

47
^{31}P δ + 10.2

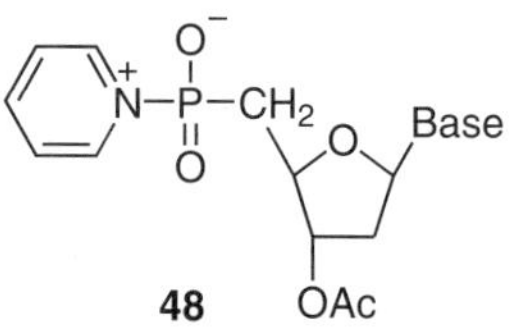

48
^{31}P δ - 5.1

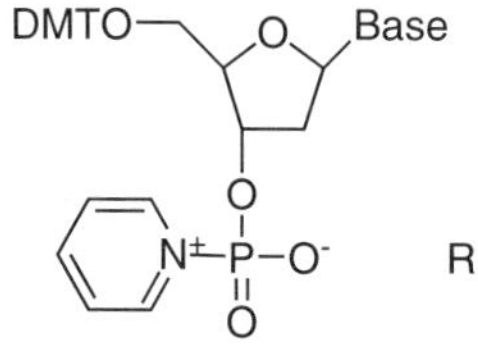

49
^{31}P δ + 12

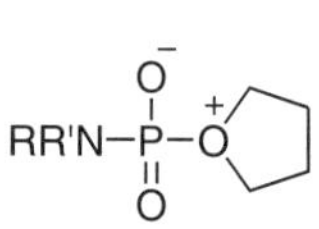

50
a, MesHN, ^{31}P δ +8.5
b, Et$_2$N, ^{31}P δ +12

10.D. THE 3-COORDINATE STATES (σ^3, λ^5)

Of greatest interest in organophosphorus chemistry are the esters **51** and amides **52** of metaphosphoric acid, the thio-ester **53**, and the aryldioxophosphoranes **54**, all of transient existence.

RO–P(=O)(O)   R₂N–P(=O)(O)   RO–P(=S)(O)   R–P(=O)(O)

51   **52**   **53**   **54**

Methods proposed for the generation of these dioxophosphoranes are numerous and are well covered in the reviews cited. Four methods are selected for display here, the first[118] (Eq. 10.88) of historical interest as the earliest demonstration of the creation of an alkyl metaphosphate in an organic solvent, the others because they are relatively simple and employ starting materials that are reasonably accessible (Eq. 10.89[119] and 10.90[120]) or because of the versatility of the approach and extensive documentation (Eq. 10.91[121]). These reactions are all conveniently carried out in solution; other methods do exist that are conducted at high temperatures in the gas phase (reviewed in reference 111).

$$\underset{\substack{\text{Br Ph O}\\ \text{| | ||}\\ \text{CH}_3\text{CH-C-P-OH}\\ \text{| |}\\ \text{Br OMe}}}{} \xrightarrow{\text{base}} \underset{\substack{\text{↑Br Ph O}\\ \text{| | ||}\\ \text{CH}_3\text{CH-C-P-O}^-\\ \text{| |}\\ \text{Br OMe}}}{} \longrightarrow \text{CH}_3\text{CH=CPhBr} + \left[\text{MeO-P}{=}\text{O}_2\right]$$

(10.88)

$$\text{Ph-C(=O)Cl} + (\text{MeO})_3\text{P} \longrightarrow \text{Ph-C(=O)-P(=O)(OMe)}_2 \xrightarrow{\text{H}_2\text{NOH}} \text{Ph-C(=NOH)-P(=O)(OMe)}_2 \xrightarrow[\text{acetone}]{\text{NaI}}$$

$$\text{Ph-C(=NOH)-P(=O)(OMe)(O-H)} \xrightarrow{-\text{HCl}} \text{PhC}{\equiv}\text{N} + \left[\text{MeO-P}{=}\text{O}_2\right]$$

(10.89)

$$\underset{\substack{\text{X R'}\\ \text{|| /}\\ \text{Y-P-N}\\ \text{| \textbackslash}\\ \text{O R}\\ \text{|}\\ \text{H}}}{} \rightleftharpoons \underset{\substack{\text{X H R'}\\ \text{|| | /}\\ \text{Y-P-N}\\ \text{| \textbackslash}\\ \text{O}^- ^+\text{R}}}{} \xrightarrow{\Delta} \left[\text{Y-P}{=}\text{X}{=}\text{O}\right] + \text{R'RNH}$$

R = 1- adamantyl, R' = H
R = R' = Et
Y = RO or Ph
X = O or S

(10.90)

[bicyclic structure with Y–P(=X), H, H, N-Ph, O] $\xrightarrow{\text{ArCO}_3\text{H}}$ [bicyclic structure with Y–P(=X)(=O), H, H, N-Ph, O] $\xrightarrow{\Delta \text{ or } h\nu}$ [phthalimide-like structure with N-Ph] + $\left[\text{Y-P}{=}\text{X}{=}\text{O}\right]$

(10.91)

The known chemistry of the dioxophosphorane species is completely dominated by the electrophilic phosphorus center; no cycloaddition or metal complexation reactions involving the oxygen atoms are yet known. The −OH group is the most common nucleophilic species used in combination with the dioxophosphorane group, and the many examples known (Scheme 10.1) range from the reaction with simple alcohols to form phosphates from RO−PO$_2$ or phosphonates from R−PO$_2$, to more complex alcohols such as cholesterol[110] or carbohydrate derivatives[100] and even to the free OH groups on the surface of solids[122] such as silica gel (to form structural feature **55**), zeolites, and cellulose powder. The OH group of phosphorus acids is especially easily phosphorylated to form mixed anhydrides such as **56**.[123]

The NH group can also be phosphorylated, and even the aromatic ring, when suitably activated as in the dialkyl anilines[117,124] (Eq. 10.92) or in pyrroles[125] can be C-phosphorylated to form phosphonic acids. The dioxophosphoranes can react with oxygen functions such as carbonyl groups to form enol phosphates[126] (Eq. 10.93) and with epoxides to form 1,3,2-dioxaphospholane derivatives[127] (Eq. 10.94).

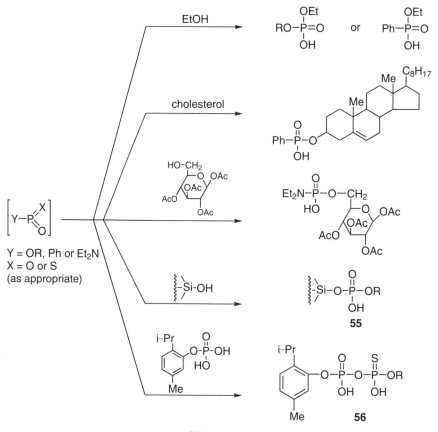

Scheme 10.1

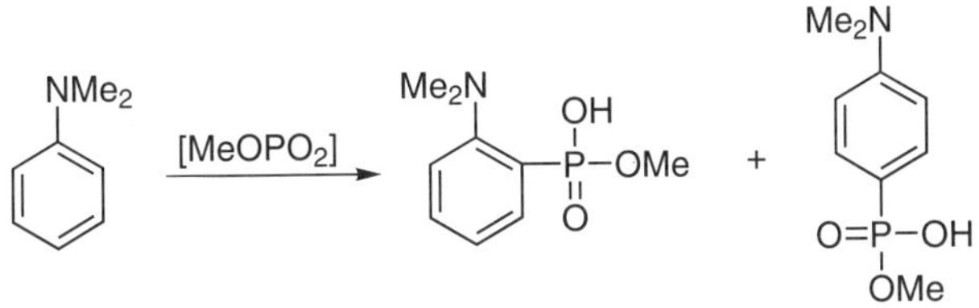

10.E. COMPOUNDS WITH 5 AND 6-COORDINATE PHOSPHORUS

(10.92)

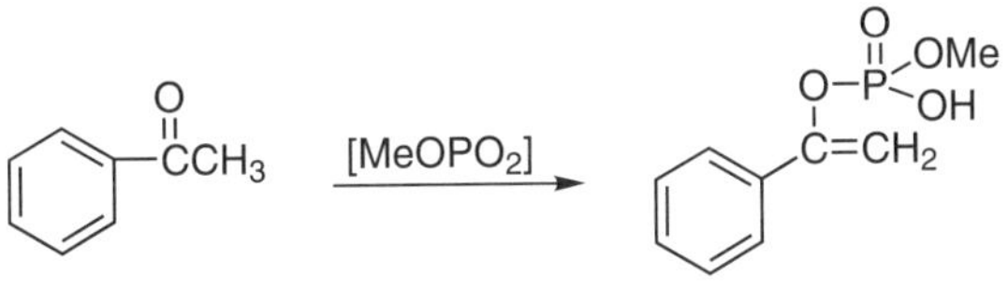

(10.93)

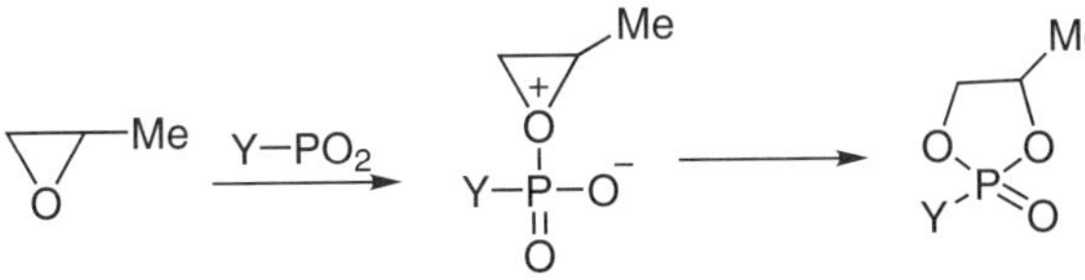

(10.94)

Thus, even though the dioxophosphorane species have so far defied isolation or even observation attempts, a useful chemistry is developing for them.

10.E. COMPOUNDS WITH 5 AND 6-COORDINATE PHOSPHORUS

10.E.1. General

In chapter 2.F, the bonding and some general characteristics of compounds with the 5-coordinate state (the phosphorane family) were presented, and in later chapters several examples were seen of reaction products and reaction intermediates (or transition states) with this structural feature. In this section, we will add only a brief survey of some important types of 5-coordinate compounds of interest to organophosphorus chemists. Useful reviews from this standpoint are those of Hellwinkel,[128] Smith[129] and Burgada and Setton.[130]

10.E.2. Compounds with Five Bonds to Carbon or Hydrogen

Historically, compounds with bonds only to carbon and hydrogen were the first of the 5-coordinate structures to be synthesized. Standing as a landmark in organophosphorus chemistry, Wittig[131] in 1948 showed that pentaphenylphosphorane could be formed by the reaction of phenyllithium with tetraphenylphosphonium iodide, and this process has since been used to prepare several pentaarylphosphoranes (Eq. 10.95). Pentaphenylphosphorane is a crystalline solid, melting with decomposition at 124°. Decomposition in solution is rapid at 130°, and leads to a product mixture (benzene, biphenyl, triphenylphosphine, etc.) indicative of a free radical process.

$$Ar_4P^+ + ArLi \longrightarrow Ar_5P \qquad (10.95)$$

The compound has the trigonal bipyramidal structure[132] with apical bond lengths of 1.987 Å and equatorial lengths of 1.850 Å. The ^{31}P NMR signal in ether or THF appears at δ −89, a highly shielded position that is quite characteristic of the 5-coordinate state.[133]

Another widely used method to prepare pentaarylphosphoranes consists of the reaction of phosphineimines (Eq. 10.96) or methiodides of phosphineimines (Eq. 10.97) with aryllithiums.

$$Ar_3P=N-Tos \xrightarrow{Ar'Li} Ar_3Ar'P\overset{Li}{\underset{|}{-}}NTos \xrightarrow{Ar'Li} Ar_3Ar'_2P \qquad (10.96)$$

$$Ar_3\overset{+}{P}=NMePh \xrightarrow{Ar'Li} Ar_3Ar'P-NMePh \xrightarrow{Ar'Li} Ar_3Ar'_2P \qquad (10.97)$$

Attempts to synthesize simple alkyl or alkylarylphosphoranes by these methods lead only to unstable products, consistent with the general requirement that electronegative groups are required for stabilization. An exception has been found in the case of derivatives of the phosphacubane system; structure **57** is a stable compound,[134] with ^{31}P NMR δ −90.

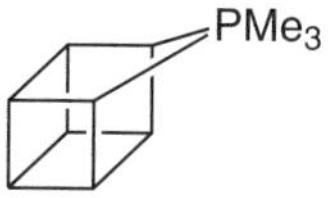

57

Spirocyclic structures (e.g., **59**) are also known with a P-alkyl substituent, and are easily formed from the attack of an alkyllithium on the spirocyclic phosphonium salt **58**[135] (Eq. 10.98).

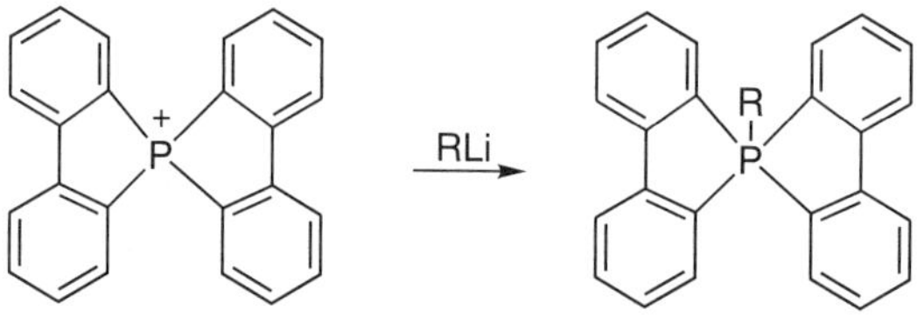

58   **59**   (10.98)

H-Phosphoranes are rare, the best example being that formed by addition of hydride (from $LiAlH_4$ or $NaBH_4$) to a spirocyclic phosphonium salt[136] (Eq. 10.99). The H-phosphoranes have some unusual properties; the hydrogen can either be removed with strong base (*t*-butyllithium) to form a phosphoranide ion, or with a protonic acid to release H_2 and regenerate the phosphonium ion. It is even possible

58 $\xrightarrow{\text{LiAlH}_4}$ [tetraphenyl H-phosphorane structure] (10.99)

to observe loss of H· with formation of the violet phosphoranyl radical, a process that occurs simply on placing the H-phosphorane in inert solvents.

10.E.3. Pentaoxyphosphoranes

A quite stable noncyclic structure of the pentaoxyphosphorane type is the pentaphenoxy compound, m.p.103–104°, ^{31}P NMR δ −85.7, made by the reaction of phenol with PCl$_5$.[137] Derivatives with simple alkyl groups are known but are of much lower stability. The pentamethoxy derivative has ^{31}P NMR δ −66.[133] Penta(neopentoxy)phosphorane is a more stable white solid, m.p. 79–80°, ^{31}P δ −72[138a] These structures are conveniently made by the addition of peroxides to tertiary phosphites[138b] (Eq. 10.100); alkyl phenylsulfenates are also useful reagents.[138a]

$$(\text{EtO})_3\text{P} \xrightarrow[\text{EtO-S(O)Ph}]{\text{EtO-OEt}} (\text{EtO})_5\text{P} \qquad (10.100)$$

By far the best known type of pentaalkoxyphosphorane is that in which two oxygen atoms are connected by an ethylenic bridge, thus to possess a 1,3,2-dioxaphospholene moiety in the phosphorane. These structures are easily formed by the reaction of tertiary phosphites (both alkyl and aryl) with α-dicarbonyl compounds, first reported by Birum and Dever;[139] ortho-quinones also perform well in this process.[140] The discovery of this reaction (Eq. 10.101) was a major factor in the rapid development of phosphorane chemistry, and much information was gathered on the characteristics of the 5-coordinate state from studies with these molecules.

$$(\text{RO})_3\text{P} + \begin{array}{c}\text{O=C-R'}\\\text{O=C-R'}\end{array} \longrightarrow \begin{array}{c}\text{1,3,2-dioxaphospholene}\end{array} \qquad (10.101)$$

Yet another simple process to attain the pentaalkoxyphosphorane state is the reaction of a tertiary phosphite with two moles of a simple carbonyl compound (aldehyde or ketone). Here the carbonyl groups are reductively coupled at carbon, and the carbonyl oxygens add to the phosphite to give the dioxaphospholane moiety in a phosphorane (Eq. 10.102). This process also has been used to make a great number of phosphoranes.

$$(RO)_3P \;+\; \begin{matrix} O=CHR' \\ O=CHR' \end{matrix} \;\longrightarrow\; (RO)_3P\begin{matrix} O-CHR' \\ | \\ O-CHR' \end{matrix} \quad (10.102)$$

We should note that the geometry of many of the phosphoranes with dioxy rings has been determined, and it is generally found that four- and five-membered rings of all types prefer to occupy an equatorial and an apical position of the trigonal bipyramid, whose 90° bond angle is more compatible with that inside a small ring. Six-membered cyclic pentaoxyphosphoranes can be either apical-equatorial or diequatorial;[130] they are usually apical-equatorial, with the ring in the boat or twist-boat shape.[141] Even in seven- and eight-membered rings, an apical-equatorial preference is expressed,[142] although some exceptions are known (e.g., reference 142b).

10.E.4. Mixed-Atom Phosphoranes

The presence of strongly electronegative groups on phosphorus makes possible the construction of a very large number of compounds with alkyl substituents, which we noted to be rare in phosphoranes with all-carbon substitution. Thus phosphonites and phosphinites may take the place of tertiary phosphites in the reactions with α-dicarbonyl compounds or with simple carbonyl compounds, leading to mixed carbon-oxygen substitution patterns (e.g., as in Eq. 10.103). In such mixed-atom systems, a general rule is that the most stable structure has the most electronegative atoms in the apical positions (as seen in compound **60**), but in many cases the molecules are fluxional due to the low energy barriers for isomerization by the pseudorotation or turnstile rotation processes.

$$PhP(OMe)_2 \;+\; \begin{matrix} O=C-Me \\ | \\ O=C-Me \end{matrix} \;\longrightarrow\; \text{60} \quad (10.103)$$

60

Amino groups also act to stabilize phosphoranes, as do halogens. The fluorine atom is the most apicophilic of all substituents; it is particularly important as a stabilizing group and leads to quite stable structures even with alkyl substituents on phosphorus. Its apicophilicity is so strong that it can force a five-membered ring to relinquish any apical-equatorial preference and adopt a di-equatorial position, as in structure **61**.

10.E. COMPOUNDS WITH 5 AND 6-COORDINATE PHOSPHORUS

[Structure 61: cyclic phosphorane with three F substituents]

61

Fluorine is a common substituent in phosphorane chemistry. Fluorophosphoranes have played a significant role in studies of the fluxional character of the 5-coordinate state, because ^{19}F NMR is an additional tool of convenience in determining apical or equatorial dispositions. The fluxional character of phosphoranes has been extensively studied and is discussed in chapter 2.F, and in more detail in the review articles cited.

The mixed-atom phosphorane structure can also be derived from intramolecular interaction of an electron pair on a donor atom with phosphorus in the 4-coordination state. This can be illustrated with the atrane type of structure[143] as in Eq. 10.104, although other examples are also known.

$$\text{[atrane structure]} \xrightarrow{H^+} \text{[protonated atrane structure]} \qquad (10.104)$$

10.E.5. 6-Coordinate Phosphorus Compounds

Many stable compounds have been made in recent years in which phosphorus has six attached groups. The earlier work was reviewed in 1972;[128] more recent reviews are those by Wong, et al.,[5] (in which several other reviews are cited) and by Holmes.[144] In general, the octahedral structure, with two apical and four equatorial bonds, is adopted. Phosphorus is known in neutral, anionic (the most common), or cationic forms in this coordination state, and the creation of new structures continues to be an active part of "hypervalent" phosphorus chemistry. Many of the known compounds can be viewed as Lewis salts arising from a donor group (neutral or ionic) interacting with 5-coordinate phosphorus. Some of the concepts of the 5-coordinate state are useful here; highly electronegative groups prefer the apical positions and are important in stabilizing the structure. Fluxional character can be present,[145] and ^{31}P NMR shifts are found at very high field. There are relatively few compounds with carbon substituents on phosphorus; some structures are shown as **62** to **67**. The ^{31}P data cited for **62** to **66** were taken from Lamondé, et al.,[146] and for **67** from Shevchenko, et al.[147]

$\text{MePCl}_5^- \text{N}(n\text{-C}_5\text{H}_{11})_4^+$ $\qquad$ $\text{PhPCl}_5^- \text{N}(n\text{-C}_5\text{H}_{11})_4^+$

62, ^{31}P δ −208 $\qquad$ **63**, ^{31}P δ −223

$\text{PhPCl}_4\text{CN}^- \;\text{NEt}_4^+$ $\qquad$ $\text{MeP(CN)}_5^- \;\text{NEt}_4^+$

64, ^{31}P δ −224.4 $\qquad$ **65**, ^{31}P δ −333.7

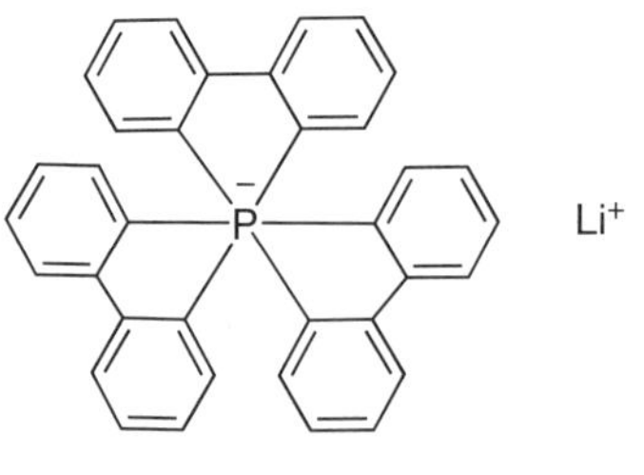

66, ^{31}P δ −181

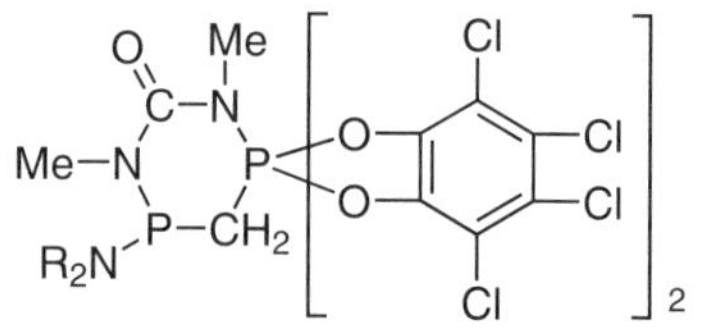

67, ^{31}P δ −115

Six-coordinate compounds are receiving attention at present for two reasons that are of interest in organophosphorus chemistry. The first of these is that they are being recognized as transient intermediates in certain reactions of 5-coordinate structures, adding a new dimension to considerations of reaction mechanisms. An example is that of Evans and co-workers,[148] in which a 5-coordinate starting material (**68**) is used in the Wittig olefination reaction. An anion is readily formed at the methylene group attached to P, and condensation with an aromatic aldehyde leads to the expected Wittig olefin (Eq. 10.105). Examination of the reacting mixture by ^{31}P NMR, however, revealed the intermediacy of a group of compounds with high-field shifts (δ −106.1 to −116.8) suggestive of 6-coordinate intermediates (possibly isomers of **69**).

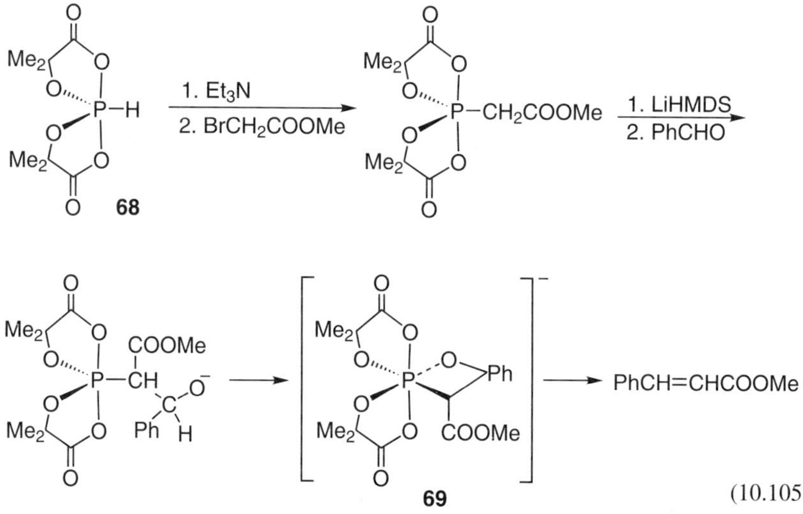

(10.105)

The second development, which in fact depends on the concepts of the first, is control of olefin stereochemistry offered by the 6-coordinate oxaphosphetane intermediate derived from a 5-coordinate starting material in the Wittig reaction.[149] The anion from phosphorane **70** was found to give excellent selectivity for the formation of Z-olefins, whereas conventional Wittig reactions generally give mostly the E-isomer (Eq. 10.106). The anion when condensed with an assortment of aromatic aldehydes gave high yields of olefins, with Z : E values exceeding 96 : 4. It is presumed that the 6-coordinate structure **71** is an intermediate, and its collapse leads to the Z-olefin.

(10.106)

REFERENCES

1. M. Regitz and O. J. Scherer, eds., *Multiple Bonds and Low Coordination in Phosphorus Chemistry*, Georg Thieme Verlag, Stuttgart, Germany, 1990.
2. E. Fluck and G. Heckmann in J. G. Verkade and L. D. Quin, eds., *Phosphorus-31 NMR Spectroscopy in Stereochemical Analysis*, VCH Publishers, Deerfield Beach, FL, 1987, Chapter 2.
3. K. B. Dillon, F. Mathey, and J. F. Nixon, *Phosphorus: The Carbon Copy*, John Wiley & Sons, Inc., New York, 1998.
4. R. R. Holmes, *Pentacoordinated Phosphorus*, 2 vols., American Chemical Society, Washington, DC, 1980.
5. C. Y. Wong, D. K. Kennepohl, and R. G. Cavell, *Chem. Rev.* **96**, 1917 (1996).
6. J. F. Nixon, *Chem. Rev.* **88**, **1327** (1988).
7. J. F. Nixon, *Chem. Soc. Rev.* **24**, 319 (1995).
8. L. N. Markovski and V. D. Romanenko, *Tetrahedron* **45**, 6019 (1989).
9. M. Regitz, *Chem. Rev.* **90**, 191 (1990); *Angew. Chem., Int. Ed. Engl.* **27**, 1484 (1988).
10. A. C. Gaumont and J. M. Denis, *Chem. Rev.* **94**, 1413 (1994).
11. Ref. 1, (a) M. Regitz, p. 58, (b) W. W. Schoeller, pp. 5–32, (c) M. Regitz, pp. 89–105, (d) F. Mathey, pp. 33–47, (e) G. Huttner and H. Lang, pp. 48–57, (f) R. Appel, pp. 157–219, (g) E. Niecke, pp. 293–320, (h) L. D. Quin and J. Szewczyk, pp. 352–366, (i) M. Yoshifuji, pp. 321–337, (j) F. Bickelhaupt, pp. 287–292, (k) R. Appel, pp. 367–374, (l) M. Meisel, pp. 415–442.
12. T. E. Gier, *J. Am. Chem. Soc.* **83**, 1769 (1961).
13. M. J. Hopkinson, H. W. Kroto, J. F. Nixon, and N. P. C. Simmons, *J. Chem. Soc., Chem. Commun.*, 513 (1976).

14. B. Pellerin, J. M. Denis, J. Perrocheau, and R. Carrié, *Tetrahedron Lett.* **27**, 5723 (1988).
15. E. P. O. Fuchs, M. Hermesdorf, and M. Regitz, *J. Organomet. Chem.* **338**, 329 (1989).
16. J. C. Guillemin, T. Janati, P. Guenot, P. Savignac, and J. M. Denis, *Angew. Chem., Int. Ed. Engl.* **30**, 196 (1991).
17. M. J. Hopkinson, H. W. Kroto, J. F. Nixon, and O. Ohashi, *Chem. Phys. Lett.* **42**, 460 (1976).
18. N. P. C. Westwood, H. W. Kroto, J. F. Nixon, and N. P. C. Simmons, *J. Chem. Soc., Dalton Trans.*, 1405 (1979).
19. H. W. Kroto, J. F. Nixon, and N. P. C. Simmons, *J. Mol. Spectrosc.* **77**, 270 (1979).
20. T. J. Dennis, S. Firth, H. W. Kroto, G. Y. Matti, C.-Y. Mok, R. J. Suffolk, and D. R. M. Walton, *J. Chem. Soc., Faraday Trans.* **87**, 917 (1991).
21. S. Lacombe, G. Pfister-Guillouzo, J. C. Guillemin, and J. M. Denis, *J. Chem. Soc., Chem. Commun.*, 403 (1991).
22. H. W. Kroto, J. F. Nixon, and K. Ohno, *J. Mol. Spectrosc.* **90**, 512 (1981).
23. T. A. Cooper, H. W. Kroto, J. F. Nixon, and O. Ohashi, *J. Chem. Soc., Chem. Commun.*, 333 (1980).
24. H. W. Kroto, J. F. Nixon, N. P. C. Simmons, and N. P. C. Westwood, *J. Am. Chem. Soc.* **100**, 446 (1978).
25. G. Becker, G. Gresser, and W. Uhl, *Z. Naturforsch. B* **36**, 16 (1981).
26. A. N. Chernega, M. Y. Atipin, Y. T. Struchkov, M. F. Meidine, and J. F. Nixon, *Heteroatom Chem.* **2**, 665 (1991).
27. W. Rösch, U. Vogelbacher, T. Allspach, and M. Regitz, *J. Organomet. Chem.* **306**, 39 (1986).
28. (a) R. Appel, G. Maier, H. P. Reisenauer and A. Westerhaus, *Angew. Chem., Int. Ed. Engl.* **20**, 197 (1981); (b) B. Solouki, H. Bock, R. Appel, A. Westerhaus, G. Becker, and G. Uhl, *Chem. Ber.* **115**, 3747 (1982).
29. R. Appel and M. Poppe, *Angew. Chem., Int. Ed. Engl.* **28**, 53 (1989).
30. J. Grobe, D. Le Van, M. Hegemann, B. Krebs, and M. Läge, *Heteroatom Chem.* **5**, 337 (1994).
31. J. C. Guillemin, T. Janati, and J. M. Denis, *J. Chem. Soc., Chem. Commun.*, 415 (1992).
32. H. Oberhammer, G. Becker and G. Gresser, *J. Mol. Spectrosc.* **75**, 283 (1981).
33. Ref. 3, (a) Chapter 4, (b) pp. 111–113.
34. M. T. Nguyen, M. A. McGinn, and A. F. Hegarty, *Inorg. Chem.* **25**, 2185 (1986); S.-J. Kim, T. P. Hamilton, and H. F. Schaefer III, *J. Phys. Chem.* **97**, 1872 (1993).
35. U. Schmidt, *Angew. Chem., Int. Ed. Engl.* **14**, 523 (1975).
36. B. Weber and M. Regitz in M. Regitz, ed., *Methoden der Organischen Chemie (Houben-Weyl), Band E(2) (Organischen Phosphorverbindungen)*, Georg Thieme Verlag, Stuttgart, Germany, 1982, pp.15–19.
37. Ref. 3, Chapter 2.
38. X. Li, S. I. Weissman, T.-S. Lin, P. P. Gaspar, A. H. Cowley, and A. I. Smirnov, *J. Am. Chem. Soc.* **116**, 7899 (1994).
39. X. Li, D. Lei, M. Y. Chiang, and P. P. Gaspar, *J. Am. Chem. Soc.* **114**, 8526 (1992).
40. A. Ecker and U. Schmidt, *Monatsh. Chem.* **102**, 1851 (1971).
41. G. Fritz, T. Vaahs, H. Fleischer, and E. Matern, *Angew. Chem., Int. Ed. Engl.* **28**, 315 (1989); *Z. Anorg. Allg. Chem.* **570**, 54 (1989).
42. M. Yoshifuji, T. Sato, and N. Inamoto, *Chem. Lett.*, 1735 (1988).
43. A. H. Cowley, F. Gabbaï, R. Schluter, and D. Atwood, *J. Am. Chem. Soc.* **114**, 3142 (1992).
44. X. Li, D. Lei, M. Y. Chiang, and P. P. Gaspar, *Phosphorus Sulfur Silicon* **76**, 71 (1993).

45. R. A. Aitken, A. H. Cowley, F. P. Gabbaï and W. Masamba, *Phosphorus Sulfur Silicon* **111**, 182 (1996); R. A. Aitken, W. Masamba, and N. J. Wilson, *Tetrahedron Lett.* **38**, 8417 (1997).
46. M. J. van Eis, C. M. D. Komen, F. J. T. de Kanter, W. H. de Wolf, K. Lammertsma, F. Bickelhaupt, M. Lutz, and A. L. Spek, *Angew. Chem. Int. Ed. Engl.* **37**, 1547 (1998).
47. Ref. 3, Chapter 3.
48. K. Dimroth and P. Hoffmann, *Angew. Chem., Int. Ed. Engl.* **3**, 384 (1964); *Chem. Ber.* **99**, 1325 (1966).
49. G. Becker, *Z. Anorg. Allg. Chem.* **423**, 242 (1976).
50. T. C. Klebach, R. Lourens, and F. Bickelhaupt, *J. Am. Chem. Soc.* **100**, 4886 (1978).
51. K. Issleib, H. Schmidt, and H. Meyer, *J. Organomet. Chem.* **160**, 47 (1978).
52. H. Eshtiagh-Hosseini, H. W. Kroto, J. F. Nixon, S. Brownstein, J. R. Morton, and K. F. Preston, *J. Chem. Soc., Chem. Commun.*, 653 (1979).
53. H. E. Hosseini, H. W. Kroto, J. F. Nixon, and O. Ohashi, *J. Organomet. Chem.* **296**, 351 (1985).
54. (a) R. Appel and A. Westerhaus, *Angew. Chem., Int. Ed. Engl.* **19**, 556 (1980); (b) S. Lacombe, B. Pellerin, J. C. Guillemin, J. M. Denis, and G. Pfister-Guillouzo, *J. Org. Chem.* **54**, 5958 (1989).
55. P. Le Floch and F. Mathey, *Tetrahedron Lett.* **30**, 817 (1989); H. T. Teunissen, J. Hollebeek, P. J. Nieuwenhuizen, B. L. M. van Baar, F. J. J. de Kanter, and F. Bickelhaupt, *J. Org. Chem.* **60**, 7439 (1995).
56. C. Grandin, E. About-Jaudet, N. Collignon, J. M. Denis, and P. Savignac, *Heteroatom Chem.* **3**, 337 (1992); J. C. Guillemin, M. Le Guennec, and J. M. Denis, *J. Chem. Soc., Chem. Commun.*, 988 (1989).
57. B. Pellerin, P. Guenot, and J. M. Denis, *Tetrahedron Lett.* **28**, 5811 (1987).
58. L. D. Quin, A. N. Hughes and B. Pete, *Tetrahedron Lett.* **28**, 5783 (1987); A. C. Gaumont, B. Pellerin, J. L. Cabioch, X. Morise, M. Lesvier, P. Savignac, P. Guenot, and J. M. Denis, *Inorg. Chem.* **35**, 6667 (1996).
59. Ref. 3, Chapter 5.
60. M. Yoshifuji, K. Toyota, K. Shibayama, and N. Inamoto, *Chem. Lett.*, 1653 (1983).
61. M. Yoshifuji, K. Toyota, and N. Inamoto, *Tetrahedron Lett.* **26**, 1727 (1985).
62. R. Appel, J. Menzel, K. Knoch, and P. Volz, *Z. Anorg. Allg. Chem.* **534**, 100 (1986).
63. E. Niecke, W. W. Schoeller, and D.-A. Wildbredt, *Angew. Chem., Int. Ed. Engl.* **20**, 131 (1981).
64. R. Appel, C. Casser, M. Immenkeppel, and F. Knoch, *Angew. Chem., Int. Ed. Engl.* **23**, 895 (1984).
65. R. Appel and M. Immenkeppel, *Z. Anorg. Allg. Chem.* **553**, 7 (1987).
66. S. J. Goede and F. Bickelhaupt, *Chem. Ber.* **124**, 2677 (1991).
67. V. D. Romanenko, A. V. Ruban, M. I. Povolotskii, L. K. Polyachenko, and L. N. Markovskii, *Zhur. Obshch. Khim.* **56**, 1186 (1986); *Chem. Abstr.* **106**, 50305 (1987).
68. A. S. Ionkin and B. A. Arbusov, *Izv. Akad. Nauk, SSSR Ser. Khim.* 1641 (1990); *Chem. Abstr.* **113**, 212148 (1990).
69. A. Marinetti and F. Mathey, *Angew. Chem., Int. Ed. Engl.* **27**, 1382 (1988).
70. Ref. 2, pp. 76–84.
71. A. Schmidpeter and K. Karaghiosoff in J. C. Tebby, ed., *Handbook of Phosphorus-31 Nuclear Magnetic Resonance Data*, CRC Press, Boca Raton, FL, 1991, Chapter 2.
72. T. A. Van der Knaap, T. C. Klebach, F. Visser, R. Lourens, and F. Bickelhaupt, *Tetrahedron* **40**, 991 (1984).
73. T. Allspach, M. Regitz, G. Becker, and W. Becker, *Synthesis*, 31 (1986).

74. G. Märkl and W. Hölzl, *Tetrahedron Lett.* **29**, 4535 (1988).
75. M. Abbari, P. Cosquer, F. Tonnard, Y. Y. C. Yeung Lam Ko, and R. Carrié, *Tetrahedron* **47**, 71 (1991).
76. R. Appel, J. Menzel, and F. Knoch, *Chem. Ber.* **118**, 4068 (1985).
77. R. Appel, F. Knoch, and H. Kunze, *Chem. Ber.* **117**, 3151 (1984).
78. R. H. Neilson, B. A. Boyd, D. A. Dubois, R. Hani, G. M. Scheide, J. T. Shore, and U. G. Wettermark, *Phosphorus Sulfur* **30**, 463 (1987).
79. R. Appel, J. Hünerbein, and N. Siabalis, *Angew. Chem., Int. Ed. Engl.* **26**, 779 (1987); R. Appel, B. Niemann, W. Schuhn, and N. Siabalis, *J. Organomet. Chem.* **347**, 299 (1988).
80. R. Appel, P. Fölling, W. Schuhn, and F. Knoch, *Tetrahedron Lett.* **27**, 1661 (1986).
81. R. Appel, V. Barth, and F. Knoch, *Chem. Ber.* **116**, 938 (1983).
82. M. Yoshifuji, K. Toyota, K. Shibayama, and N. Inamoto, *Tetrahedron Lett.* **25**, 1809 (1984); M. Yoshifuji, K. Toyota, N. Inamoto, K. Kirotsu, T. Higuchi, and S. Nagase, *Phosphorus Sulfur* **25**, 237 (1985).
83. H. H. Karsch, H.-U. Reisacher, and G. Müller, *Angew. Chem., Int. Ed. Engl.* **23**, 618 (1984).
84. R. Appel and W. Paulen, *Angew. Chem., Int. Ed. Engl.* **22**, 785 (1983).
85. R. Appel and V. Barth, *Angew. Chem., Int. Ed. Engl.* **18**, 469 (1979).
86. R.Appel, V. Barth, and M. Halstenberg, *Chem. Ber.* **115**, 1617 (1982).
87. E. Niecke and W. Flick, *Angew. Chem., Int. Ed. Engl.* **12**, 585 (1973).
88. E. Niecke and D. Gudat, *Angew. Chem., Int. Ed. Engl.* **30**, 217 (1991).
89. E. Niecke, R. Rüger, and W. W. Schoeller, *Angew. Chem., Int. Ed. Engl.* **20**, 1034 (1981).
90. V. D. Romanenko, A. P. Drapailo, A. V. Ruban, and L. N. Markovski, *Zhur. Obshch. Khim.* **57**, 1402 (1987), *Chem. Abstr.* **108**, 186860 (1988); E. Niecke and M. Lysek, *Tetrahedron Lett.* **29**, 605 (1988).
91. S. Pohl, E. Niecke and H.-G. Schäfer, *Angew. Chem., Int. Ed. Engl.* **17**, 136 (1978); E. Niecke and H.-G. Schäfer, *Chem. Ber.* **115**, 185 (1982); E. Niecke, A. Seyer, and D.-A. Wildbredt, *Angew. Chem., Int. Ed. Engl.* **20**, 675 (1981).
92. U. Dressler, E. Niecke, S. Pohl, W. Saak, W. W. Schoeller, and H.-G. Schäfer, *J. Chem. Soc., Chem. Commun.*, 1086 (1986).
93. H. Quast and M. Heuschmann, *Angew. Chem., Int. Ed. Engl.* **17**, 867 (1978).
94. J. K. Stille, J. L. Eichelberger, J. Higgins, and M. E. Freeburger, *J. Am. Chem. Soc.* **94**, 4761 (1972).
95. P. P. Gaspar, A. M. Beaty, X. Li, H. Qian, and J. C. Watt, paper presented at the International Conference on Phosphorus Chemistry, Cincinnati, Ohio, July 12–17, 1998.
96. A. H. Cowley, F. P. Gabbaï, S. Corbelin, and A. Decken, *Inorg. Chem.* **34**, 5931 (1995).
97. M. Yoshifuji, I. Shima, K. Ando, and N. Inamoto, *Tetrahedron Lett.* **24**, 933 (1983).
98. E. Niecke, M. Engelmann, H. Zorn, B. Krebs, and G. Henkel, *Angew. Chem., Int. Ed. Engl.* **19**, 710 (1980).
99. L. D. Quin, S. Jankowski, A. G. Sommese, P. M. Lahti, and D. B. Chesnut, *J. Am. Chem. Soc.* **114**, 11009 (1992); L. D. Quin and A. S. Ionkin, *J. Org. Chem.* **60**, 5186 (1995).
100. D. W. Chasar, J. P. Fackler, A. M. Mazany, R. A. Komoroski, and W. J. Kroenke, *J. Am. Chem. Soc.* **108**, 5956 (1986).
101. M. Yoshifuji, I. Shima, N. Inamoto, K. Hirotsu, and T. Higuchi, *J. Am. Chem. Soc.* **103**, 4587 (1981).
102. Ref. 3, Chapter 7.
103. M. Yoshifuji in L. D. Quin and J. G. Verkade, eds., *Phosphorus-31 NMR Spectral Properties in Compound Characterization and Structural Analysis*, VCH Publishers,

Inc., New York, 1994, Chapter 14; H.-P. Schrödel and A. Schmidpeter, *Phosphorus Sulfur Silicon* **129**, 69 (1997).
104. R. Appel, J. Peters, and A. Westerhaus, *Angew. Chem., Int. Ed. Engl.* **21**, 80 (1982).
105. T. Baumgartner, B. Schinkels, D. Gudat, M. Nieger, and E. Niecke, *J. Am. Chem. Soc.* **119**, 12410 (1997).
106. E. Niecke and W. Flick, *Angew. Chem., Int. Ed. Engl.*, **13**, 134 (1974).
107. G. R. Thatcher and R. Kluger, *Adv. Phys. Org. Chem.* **25**, 99 (1989).
108. W. P. Jencks in E. N. Walsh, E. J. Griffith, R. W. Parry, and L. D. Quin, eds., *Phosphorus Chemistry*, American Chemical Society, Washington DC, 1991, Chapter 8.
109. F. H. Westheimer, *Chem. Rev.* **81**, 313 (1981).
110. L. D. Quin, *Coordination Chem. Rev.* **137**, 525 (1994).
111. D. B. Chesnut, *J. Am. Chem. Soc.* **120**, 10504 (1998).
112. R. Ahlrichs, C. Ehrhardt, M. Lakenbrink, S. Schunk, and H. Schnökel, *J. Am. Chem. Soc.* **108**, 3596 (1986).
113. M. Meisel, H. Bock, B. Solouki, and M. Kremer, *Angew. Chem., Int. Ed. Engl.* **28**, 1373 (1989).
114. F. Ramirez and J. F. Marecek, *Tetrahedron* **35**, 1581 (1979).
115. D. G. Knorre, A. V. Lebedev, A. S. Levina, A. I. Rezvukhin, and V. F. Zarytova, *Tetrahedron* **30**, 3073 (1974).
116. A. Ballmark and J. Stawinski, *Tetrahedron Lett.* **37**, 5739 (1996).
117. L. D. Quin, C. Bourdieu, and G. S. Quin, *Tetrahedron Lett.* **31**, 6473 (1990).
118. A. C. Satterthwait and F. H. Westheimer, *J. Am. Chem. Soc.* **100**, 3197 (1978).
119. E. Breuer, R. Karaman, H. Leader, and A. Goldblum, *J. Chem. Soc., Chem. Commun.*, 671 (1987).
120. L. D. Quin and S. Jankowski, *J. Org. Chem.* **59**, 4402 (1994).
121. L. D. Quin, S. Jankowski, G. S. Quin, A. Sommese, J. S. Tang, and X.-P. Wu in E. N. Walsh. E. J. Griffith, R. W. Parry, and L. D. Quin, *Phosphorus Chemistry*, American Chemical Society, Washington, DC, 1992, Chapter 9.
122. L. D. Quin in L. D. Quin and J. G. Verkade, eds., *Phosphorus-31 NMR Spectral Properties in Compound Characterization and Structural Analysis*, VCH Publishers, New York, 1994, Chapter 32.
123. L. D. Quin, P. Hermann, and S. Jankowski, *J. Org. Chem.* **61**, 3944 (1996).
124. C. H. Clapp and F. H. Westheimer, *J. Am. Chem. Soc.* **96**, 6710 (1974).
125. L. D. Quin and B. G. Marsi, *J. Am. Chem. Soc.* **107**, 3389 (1985).
126. A. C. Satterthwait and F. H. Westheimer, *J. Am. Chem. Soc.* **102**, 4464 (1980).
127. R. Bodalski and L. D. Quin, *J. Org. Chem.* **56**, 2666 (1991).
128. D. Hellwinkel in G. M. Kosolapoff and L. Maier, eds., *Organic Phosphorus Compounds*, Vol. 3, John Wiley & Sons, Inc., New York, 1972, Chapter 5B.
129. D. J. H. Smith in D. H. R. Barton and W. D. Ollis, eds., *Comprehensive Organic Chemistry*, Vol. 2, Pergamon Press, Oxford, UK, 1979, Chapter 10.4.
130. R. Burgada and R. Setton in F. R. Hartley, ed., *The Chemistry of Organophosphorus Compounds*, Vol. 3, John Wiley & Sons, Inc., New York, 1996, Chapter 3.
131. G. Wittig and M. Rieber, *Naturwissenschaften* **35**, 345 (1948).
132. P. J. Wheatley and G. Wittig, *Proc. Chem. Soc.*, 251 (1962).
133. J. F. Brazier, L. Lamandé and R. Wolf in J. C. Tebby, ed., *Handbook of Phosphorus-31 Nuclear Magnetic Resonance Data*, CRC Press, Boca Raton, FL, 1991, Chapter 18.
134. E. W. Turnblom and T. J. Katz, *J. Am. Chem. Soc.* **95**, 4292 (1973).
135. D. Hellwinkel, *Chem. Ber.* **98**, 576 (1965).
136. D. Hellwinkel, *Angew. Chem., Int. Ed. Engl.* **5**, 968 (1966).

137. F. Ramirez, A. J. Bigler, and C. P. Smith, *J. Am. Chem. Soc.* **90**, 3507 (1968).
138. (a) L. L. Chang, D. B. Denney, D. Z. Denney, and R. J. Kazior, *J. Am. Chem. Soc.* **99**, 2293 (1977); (b) D. B. Denney and D. H. Jones, *J. Am. Chem. Soc.* **91**, 5821 (1969); D. B. Denney, D. Z. Denney, B. C. Chang, and K. L. Marsi, *J. Am. Chem. Soc.* **91**, 5243 (1969).
139. G. H. Birum and J. L. Dever (to Monsanto Chemical Co.), U.S. Pat. 2,961,455 *C.A.* **55**, 8292 (1961), Nov. 22, (1960).
140. F. Ramirez, *Pure Appl. Chem.* **9**, 337 (1964).
141. K. C. K. Swamy, S. D. Burton, J. M. Holmes, R. O. Day, and R. R. Holmes, *Phosphorus Sulfur Silicon* **53**, 437 (1990).
142. (a) S. D. Burton, K. C. K. Swamy, J. M. Holmes, R. O. Day, and R.R. Holmes, *J. Am. Chem. Soc.* **112**, 6104 (1990); (b) T. K. Prakasha, R. O. Day, and R. R. Holmes, *Inorg. Chem.* **31**, 725 (1992).
143. J. G. Verkade, *Acc. Chem. Res.* **26**, 483 (1993).
144. R. R. Holmes, *Acc. Chem. Res.* **31**, 535 (1998); *Chem. Rev.* **96**, 927 (1996).
145. R. G. Cavell and L. Vande Griend, *Inorg. Chem.* **22**, 2066 (1983).
146. L. Lamondé, M. Koenig, and K. Dillon in J. C. Tebby, ed., *Handbook of Phosphorus-31 Nuclear Magnetic Resonance Data*, CRC Press, Boca Raton, FL, 1991, Chapter 19.
147. I. V. Shevchenko, P. G. Jones, A. Fisher, and R. Schmutzler, *Heteroatom Chem.* **3**, 177 (1992).
148. M. J. Bojin, S. Barkallah, and S. A. Evans Jr., *J. Am. Chem. Soc.* **118**, 1549 (1996).
149. S. Kojima, R. Takagi, and K.-Y. Akiba, *J. Am. Chem. Soc.* **119**, 5970 (1997).

CHAPTER 11

ORGANOPHOSPHORUS CHEMISTRY IN BIOLOGY, AGRICULTURE, AND TECHNOLOGY

11.A. BIOLOGICAL PHOSPHATES

11.A.1. General

Phosphoric acid dervatives are so fundamentally important in the chemistry of living systems, in so many ways, that one may well ask, as F. H. Westheimer did in 1987[1] in a fascinating essay, "Why Nature Chose Phosphates." Indeed, if Nature had not chosen phosphates as essential chemicals in many critically important biological processes and materials, it is difficult to imagine any other chemical types that would be able to meet the manifold demands of living systems as we know them. There are many ways in which organophosphorus compounds (mostly esters of phosphoric acid, but some amides and anhydrides) are involved in biological systems; a presentation of some of the structural types of biophosphates is necessary to complete the beginning study of organophosphorus chemistry as offered in this book. Any text on biochemistry will provide much more information on details of synthesis, metabolic pathways, energy transfer, nucleic acid structure and function, etc.

The biological alcohols that are bound in biophosphates are of extraordinarily diverse types, as we show in the representative structures to follow. However complex, the synthetic principles and chemistry of biophosphates resemble those of simpler phosphates, being dominated by nucleophilic substitution processes. Of course, highly specialized techniques are required in the handling of these sometimes delicate materials, and in the protection of other reactive sites in complex alcohols. At physiological pH around 7, phosphomonoesters exist in monoanion or dianion form, and phosphodiesters as anions, but the molecular forms will be shown in this presentation. Cell membranes are impervious to ionic species, and thus important phosphodiesters such as the nucleic acids remain within the cell. The ionic form is important for another reason: it is much more stable to simple hydrolysis than is the molecular form, although of course enzymes (phosphoesterases) can catalyze the hydrolysis.

11.A.2. Nucleotides and Nucleic Acids

In nucleotides and nucleic acids, the alcoholic group is found in the form of a nucleoside, which consists of a D-ribose or D-2-deoxyribose unit to which is bound, at the 1-position, certain nitrogen heterocycles of the pyrimidine or purine type (the "base.", designated B in the structures shown). The phosphate group is attached to the hydroxy group at the 5-position to form the corresponding nucleotides as in structures **1** (from D-ribose) and **2** (from D-2-deoxyribose), which are seen to be monoalkyl phosphates (phosphomonoesters).

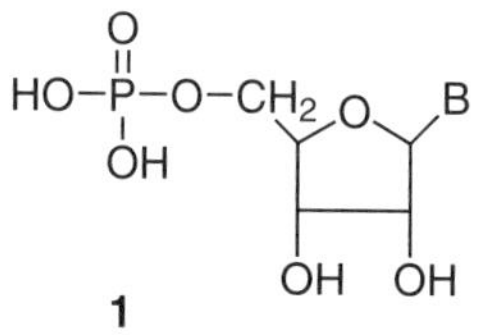

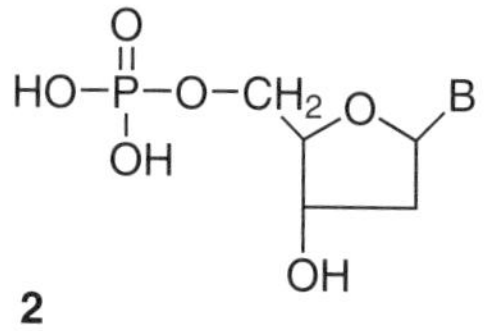

The hydroxy group at the 3-position of the nucleotide is available also for bonding to phosphorus, and it does so by using the phosphate group of another nucleotide, forming an internucleotide bond as in structure **3**. This basic structural feature is repeated many times to form polynucleotides, better known as nucleic acids (ribonucleic acid, RNA, and deoxyribonucleic acid, DNA), substances so well known as to require no further comment.

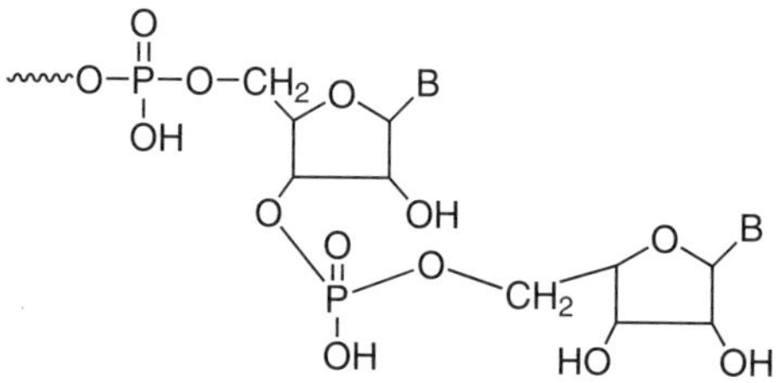

To the organophosphorus chemist, these substances are recognizable as dialkyl phosphates, or phosphodiesters, and they may be synthesized by specialized application of some of the fundamental reactions of phosphoric acid chemistry. Some examples of the prominent procedures for nucleotide and oligonucleotide synthesis (the phosphoramidite and the H-phosphonate methods) were described earlier (chapters 3.B and 5.A, respectively).

Nucleotides are also found in another biologically important phosphodiester form, in which the monophosphate group bonds intramolecularly to the hydroxy group at its pentose 3-position. This gives rise to a cyclic structure shown as **4** (cyclic adenosine monophosphate, AMP), in which phosphorus is seen to be present in the familiar 1,3,2-dioxaphosphorinane ring system.

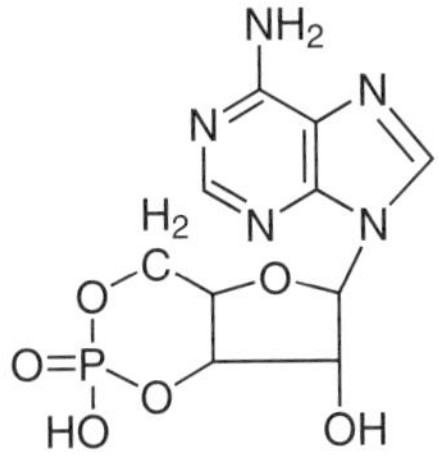

4

11.A.3. Membrane Phospholipids

In membrane phospholipids, the alcohol group bonding to phosphorus is offered by glycerol, forming the phosphoglycerides, or by the aminoalcohol sphingosine, forming the sphingomyelins. Some less common structures of different types are also known. Phosphoglycerides are 1-glyceryl phosphates, in which the 2- and 3-hydroxy groups are esterified by a variety of long-chain fatty acids. A modification is found in the plasmalogens, in which the 3-ester group is replaced by an ether group of form $-OCH=CHR$. The parent phosphate of the phosphoglycerides has structure **5** and is known as phosphatidic acid (R in structures **5** to **11** is a long chain). Although it is a minor component of the membrane phospholipids, it is a key intermediate in the biosynthesis of more complicated phosphodiesters formed with other alcohols. Prominent alcohols found in these phosphodiesters are choline (giving phosphatidyl choline, **6**), inositol (as in **7**), the aminoacid serine as in **8**, ethanolamine as in **9**, and glycerol as diphosphatidyl glycerol (**10**).

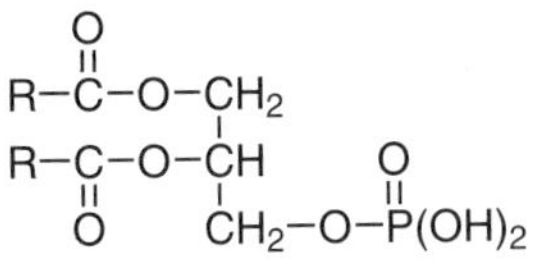

5

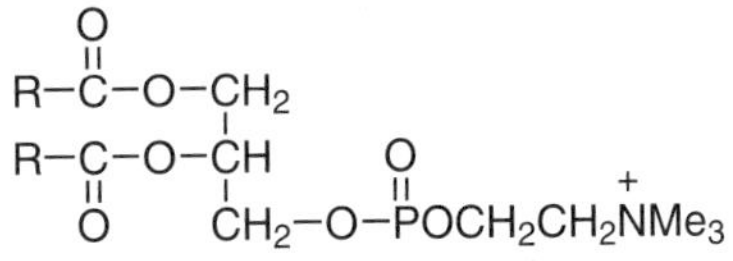

6

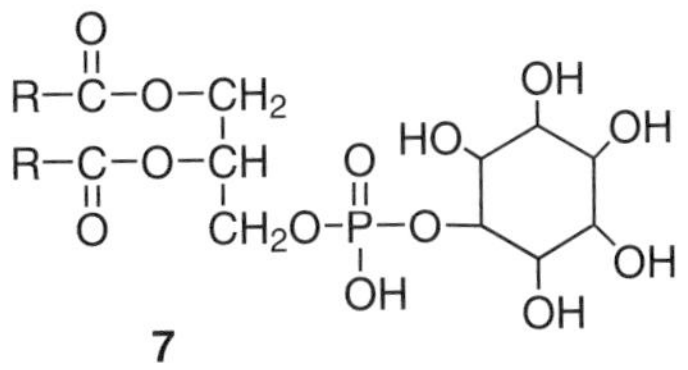

7

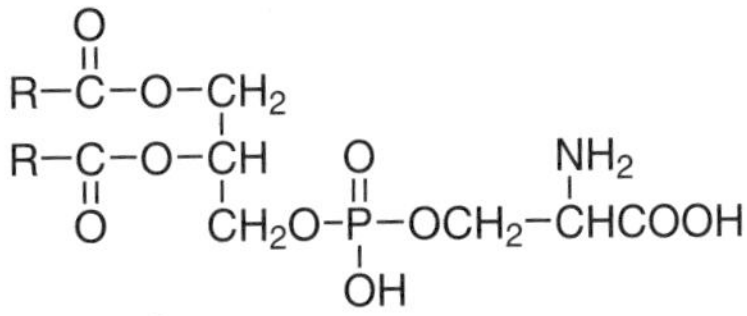

8

$$
\begin{array}{c}
\text{O} \\
\text{R}\overset{\|}{\text{C}}\text{-O-CH}_2 \\
\text{RC-O-CH} \quad \text{O} \\
\overset{\|}{\text{O}} \quad \text{CH}_2\text{O-}\overset{\|}{\text{P}}\text{-OCH}_2\text{CH}_2\text{NH}_2 \\
\text{OH} \\
\mathbf{9}
\end{array}
$$

structure **10**: diphosphatidylglycerol (cardiolipin) with two RC(=O)-O-CH$_2$-CH(O-C(=O)R)-CH$_2$-O-P(=O)(OH)- groups linked through a central glycerol.

In a sphingomyelin (**11**), the phosphocholine unit that is present in phosphatidyl choline is found attached to a hydroxy group of the sphingosine molecule.

structure **11**: CH$_3$(CH$_2$)$_{12}$CH=CH-CH(OH)-CH(NHC(=O)R)-CH$_2$-O-P(=O)(OH)-OCH$_2$CH$_2$NMe$^+$

Phospholipid chemistry is a highly developed field, and numerous techniques for phospholipid synthesis have been devised and presented in the biochemical literature.

11.A.4. Protein Phosphates

The polypeptide chain of proteins can carry a variety of attached groups, known as prosthetic groups, and among these is phosphate. An example is the important substance casein, obtained from milk, which is known to be a mixture of phosphoproteins. The phosphate group could in principle be attached to hydroxy or free amino groups found on aminoacids in the polypeptide chain, but it is the former that has been determined experimentally to carry phosphate. Thus we can describe phosphoproteins as having phosphomonoester character, albeit as only a small part of the molecular structure. The polypeptide chain of enzymes can also carry phosphate groups, as in glycogen synthase b, in which the attachment is known to be at the hydroxy group of a serine unit.

11.A.5. Sugar Phosphates

We can illustrate the importance of phosphoric acid derivatives in sugar chemistry by examining the intermediates in the highly important processes of glucose metabolism (glycolysis) to 3-carbon units, and of photosynthesis that incorporates carbon dioxide into carbohydrate molecules. The presentation here, as for phospholipids, leaves many sugar phosphates and metabolic intermediates unmentioned.

In the initiation of glycolysis, the hydroxy group at the 6-position of glucose is converted to a monophosphate (**12**) by the action of ATP. This compound undergoes isomerization to fructose-6-phosphate (**13**), which is then converted to fructose 1,6-

diphosphate (**14**). It is this substance that undergoes cleavage to 3-carbon phosphates, dihydroxyacetone phosphate (**15**) and glyceraldehyde 3-phosphate (**16**). The latter is the precursor of other 3-carbon phosphates, such as 2,3- and 1,3-diphosphoglyceric acid (**17**), 3-phosphoglyceric acid (**18**), 2-phosphoglyceric acid (**19**), and finally phosphoenol pyruvic acid (PEP; **20**). To complete glycolysis, the latter is hydrolyzed to pyruvate with release of phosphate. As an enol phosphate, PEP is hydrolyzed readily with release of energy, and this is an important feature of glycolysis.

In photosynthesis, the first detectable intermediate is 3-phosphogylceric acid, which is formed by reaction of carbon dioxide with ribulose 1,5-diphosphate (Eq. 11.1). The unstable intermediate is cleaved to produce two moles of 3-phosphoglycerate (**21**).

In subsequent processes, 1,3-diphosphoglyceric acid (**17**), glyceraldehyde 3-phosphate (**16**), and finally fructose 6-phosphate (**13**) and glucose 6-phosphate (**12**) make appearances.

11.A.6. Phosphagens: Energy Rich Phosphates

Critical to many biochemical pathways is the energy balance, and certain phosphoric acid derivatives play a major role in driving some processes by energy release that accompanies the cleavage of a phosphate group and transfer to a nucleophilic substrate. Probably the best known of the energy-rich phosphates is adenosine triphosphate (ATP; **22**), which can transfer the terminal phosphate group to a substrate with the release of significant energy (the free energy of hydrolysis is -7.3 kcal/mol). This is hardly a surprise to organophosphorus chemists, who understand that a pyrophosphate group is highly reactive to nucleophiles in an exothermic process.

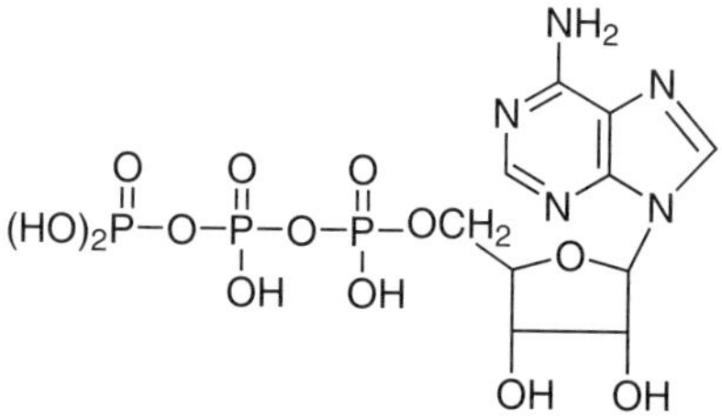

Similarly, an enol phosphate like PEP (**20**) would be highly reactive to nucleophiles with energy release, and this too is an important biological phosphagen. Indeed, it has the highest value for the free energy of hydrolysis (-14.8 kcal/mol) of any of the phosphagens. Other important phosphagens include the mixed anhydride acetyl phosphate (**23**) and the phosphoramidates (**24** and **25**) derived from the guanidine-containing compounds creatine and arginine, respectively.

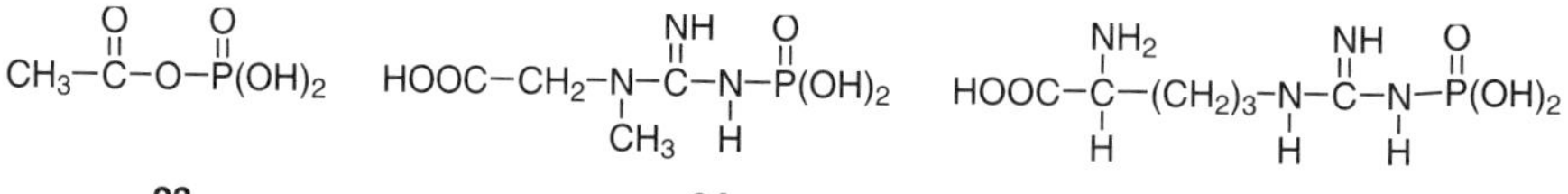

11.A.7. Other Types of Biophosphates

Life processes depend on several other types of phosphates of diverse structures. A few of the more important ones are shown here: structure **26** (nicotinamide adenine dinucleotide phosphate (NADP$^+$), one of several phosphomonoesters functioning as co-enzymes and vitamins); **27** (myoinositol 1,2,4-triphosphate, the intracellular "second-messenger" in control of the liberation of calcium ion from an intracellular store; **28** (phytic acid, myoinositol hexaphosphate, a major form of phosphorus in plants; 29 (isopentenyl pyrophosphate, critical in the biosynthesis of squalene and then cholesterol). Compound **30**, anatoxin-a(s), has quite different activity; it is a potent natural poison, isolated from a blue-green alga, which acts as cholinesterase inhibitor.

11.B. THE NATURAL PRODUCTS CHEMISTRY OF THE C–P BOND

The extensive involvement of phosphoric acid derivatives in living systems has been known for many years, but the possibility that phosphorus could be present in Nature in other molecular forms was completely overlooked until 1959, when Horiguchi and Kandatsu[2] reported the exciting discovery of 2-aminoethylphosphonic acid (AEPA; **31**) in hydrolysates of lipids of mixed protozoa growing in sheep rumen. The amount was far from trivial; from 203 g of protozoa was isolated 63 mg of crystalline AEPA.

$$NH_2CH_2CH_2-\overset{O}{\underset{\|}{P}}(OH)_2$$

31

Later work confirmed that the compound was present in individual protozoa such as the common *Tetrahymena pyriformis* and *Paramecium tetraurelia*, but an observation showing a much broader occurrence in Nature was its discovery, sometimes in significant amounts, in numerous marine invertebrate animals.[3,4] Sea anemones, members of the Phylum *Coelenterata* (*Cnidaria*), are particularly rich in AEPA; in *Metridium dianthus*, for example, AEPA accounts for a remarkable 0.99% of the dry weight.[4] Molluscs, including some found in fresh water, are another prominent source of AEPA. It has also been detected in some bacteria and phytoplankton. Chordates, however, have very little or none of the compound, and although there are reports on its presence in mammalian (including human) organs, the amounts are quite small, and it has to be considered that their presence may well be the result of ingestion from food sources. The literature on naturally occurring C–P compounds has grown enormously, much of it in the biochemistry area. There are now numerous reviews to guide the reader into this fascinating aspect of phosphorus chemistry. The early work was covered by Quin;[5a] more extensive treatment has been provided in books by *The Role of Phosphonates in Living Systems* by Hilderbrand and Henderson[5b] and *Biochemistry of Natural C–P Compounds* by Hori, Horiguchi and Hayashi.[5c] Another review is that by Thayer.[5d]

AEPA is seldom found in free form in animals; it is primarily present in lipids, where it can be found as a glyceride (as in **32**, the phosphonate equivalent of phosphatidyl ethanolamine, **9**) or as a sphingolipid bound to ceramide, as in **33**. Another lipid form is shown as **34**, in which the phosphonate group is bonded at C-6 of galactose, which also bears a ceramide group at C-1.

Other complex structures containing AEPA connected to sugars are also known.[6] Natural lipids that are derivatives of phosphonic acids rather than phosphoric acid are collectively known as phosphonolipids, and they have been extensively studied in the years since their first isolation.

The phosphonolipids were the first bound form of AEPA to be characterized, but it was recognized early[4] that there was a protein involvement for the compound as well. When the lipids are extracted from animals by conventional techniques, a largely proteinaceous residue is left, which after acid hydrolysis is found to contain a relatively large amount of AEPA among the various aminoacids. The bonding in isolated proteins, however, was later discovered to involve an association with galactose, as well as a polypeptide, and these substances are now known as phosphonoglycoproteins. Another bound form of AEPA (along with its 1-hydroxy derivative) is known as lipophosphonoglycan, a highly complex bound mixture of various sugars, fatty acids, inositol, sphingosines, and glycerine. Other complex structures containing AEPA connected to sugars are also known.

Several other phosphonic acids that are derivatives of AEPA have been isolated from hydrolysates of living organisms. These include the three possible N-methylated derivatives (**35** to **37**), and the 1-hydroxy (**38**), and 2-carboxy (**39**) derivatives. It is a characteristic of all of these compounds, as is true for any phosphonic acid, that the C—P bond is strongly resistant to acid or base hydrolysis; the natural phosphonic acids are typically released from bound forms by hydrolysis in 6N HCl at 100° for 24 h without undergoing cleavage of the C—P bond. Phosphonoacetaldehyde (**40**), a possible biosynthetic precursor of AEPA, has also been isolated from the microorganism *Bacillus cereus* and, except for some products of bacterial fermentation (vide infra), appears to be the only non-amino phosphonic acid so far detected in a living system. There are indications in the literature that other C—P compounds are present, but have not yet been isolated.

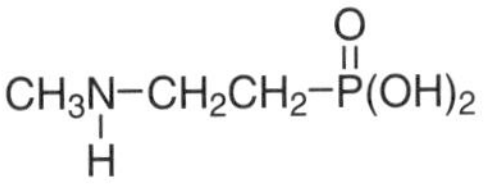

$$CH_3\underset{H}{N}-CH_2CH_2-\overset{O}{\underset{}{P}}(OH)_2 \qquad (CH_3)_2NCH_2CH_2-\overset{O}{\underset{}{P}}(OH)_2 \qquad (CH_3)_3\overset{+}{N}-CH_2CH_2-\overset{O}{\underset{}{P}}(OH)_2$$

35 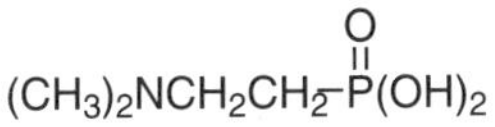   **36** 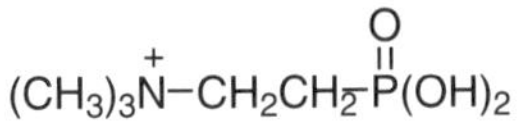   **37**

$$NH_2CH_2\underset{OH}{C}H\overset{O}{\underset{}{P}}(OH)_2 \qquad NH_2\underset{COOH}{C}HCH_2\overset{O}{\underset{}{P}}(OH)_2 \qquad H\overset{O}{\underset{}{C}}-CH_2\overset{O}{\underset{}{P}}(OH)_2$$

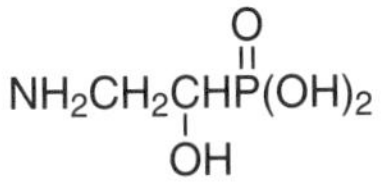

38 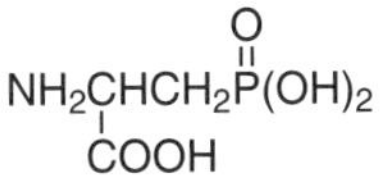   **39** 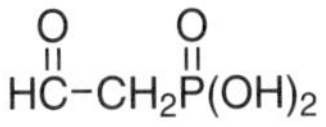   **40**

An entirely different source of C—P compounds of natural origin was first recognized in 1969.[7] From the products in a fermentation broth of the bacterium *Streptomyces fradiae* was isolated a new phosphonic acid that had the properties of an antibacterial antibiotic. The compound was named Phosphonomycin, or Fosfomycin, and was determined to be (−)-(1R,2S)-1,2-epoxypropylphosphonic acid, **41**.

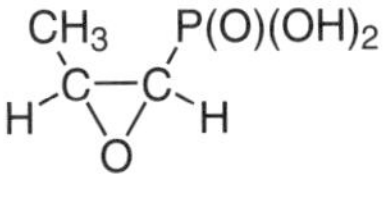

$$\underset{H}{\overset{CH_3}{\diagdown}}\overset{}{\underset{}{C}}\overset{P(O)(OH)_2}{\diagup}\underset{H}{\overset{}{\diagup}}\underset{O}{\overset{}{\diagdown}}C$$

41

The discovery of Fosfomycin was an extremely important event in phosphorus chemistry; with an occasional exception, such as the discovery of antibacterial activity in phosphorus counterparts of the sulfa drugs (*para*-aminobenzenephosphonic acid,[8a] and later in its amide derivatives[8b]), phosphorus compounds had been largely ignored by medicinal chemists seeking new agents against infectious diseases, and Fosfomycin brought fresh attention to the field. Fosfomycin is active against both Gram-positive and Gram-negative bacteria, and its effectiveness is comparable to that of the well-known antibiotics Tetracycline and Chloramphenicol.[7] It is nontoxic, and has been marketed and used clinically.

The possibility that phosphorus could appear bound to carbon among products of fermentation had been overlooked for the first two to three decades of antibiotics research, but the discovery of Fosfomycin triggered new activity in this area. Several novel phosphonic acids, as well as phosphinic and even H-phosphinic acids, were soon discovered in fermentation broths of other *Streptomyces* forms, as well as of other bacteria, and more seem likely to follow. Most of the characterized compounds are aminophosphonic or aminophosphinic acids (structures **42** to **48**) with substantial antibiotic activity, discussed in reviews.[5b,5c]

$$CH_3-\overset{O}{\underset{OH}{P}}-CH_2CH_2-\underset{NH_2}{CH}\underset{CH_3}{CHNH}\underset{CH_3}{CH}\underset{}{CNHCHCOOH}$$

42 (Bialaphos)

$$H_2N-\underset{CH_3}{CH}CONH-\underset{CH_2COOH}{CHCONH}\underset{COOH}{CHCH}=CHCH_2-\overset{O}{\underset{OH}{P}}-OH$$

43 (Plumbemycin A)

$$CH_3CON-CH_2CH_2CH_2-\overset{O}{P}(OH)_2$$
$$\underset{OH}{}$$

44

$$O=\overset{H}{C}-\underset{OH}{N}-CH_2CH_2CH_2-\overset{O}{P}(OH)_2$$

45 (Fosmidomycin)

$$O=\overset{H}{C}-NCH_2 \underset{OH}{} \overset{H}{\underset{H}{C=C}} \overset{P(OH)_2}{\underset{}{}}$$
$$\overset{}{\underset{O}{}}$$

46

$$CH_3CO-NCH_2CHCH_2-\overset{O}{P}(OH)_2$$
$$\underset{OH}{} \underset{OH}{}$$

47

$$H_2N-\underset{NH}{\overset{}{C}}NH(CH_2)_3-CHCONH-\underset{NHR}{N}-\overset{O}{\underset{H_3C}{P}}\overset{OH}{\underset{OH}{CH}}-\overset{O}{C}-OCH_3$$

48

The tripeptide **42** releases the free phosphinic acid phosphinothricin (**49**) on hydrolysis, and this compound, although devoid of antibiotic activity, has found commercial use as the herbicide Glufosinate.

$$HOOC\underset{NH_2}{C}-HCH_2CH_2-\overset{O}{\underset{OH}{P}}-CH_3$$

49

It was an important discovery when H-phosphinates (**50** to **53**) were found in fermentation broths, because they are unique in being the only naturally occurring organophosphorus compounds with phosphorus in an oxidation state lower than that of an alkylphosphonic acid. Thus **53** may be considered as a reduced form of AEPA. The ease of oxidation of synthetic H-phosphinates would have suggested that their survival in biological material would be unlikely.

11.B. THE NATURAL PRODUCTS CHEMISTRY OF THE C–P BOND

$$\text{HOOCCHCH}_2\text{CH}_2-\overset{\overset{O}{\|}}{\underset{\underset{OH}{|}}{P}}-H \quad \text{HOOCCHCH}_2-\overset{\overset{O}{\|}}{\underset{\underset{OH}{|}}{P}}-H \quad \text{HOOCCHCH}_2-\overset{\overset{O}{\|}}{\underset{\underset{OH}{|}}{P}}-H$$
$$\underset{NH_2}{} \qquad \underset{OH}{} \qquad \underset{NH_2}{}$$

50   **51**   **52**

$$\text{H}_2\text{NCH}_2\text{CH}_2\overset{\overset{O}{\|}}{\underset{\underset{OH}{|}}{P}}-H$$

53

Fosfomycin (**41**) as well as H-phosphinate **51** can be considered as members of another subclass of C–P antibiotics, which are phosphonic acids without amino substituents. Other structures of this type are Fosfonochlorin (**54**), an inhibitor of bacterial cell wall growth, and the more recently discovered[9] Phosphonothrixin (**55**), with herbicidal activity.

$$\text{ClCH}_2\overset{\overset{O}{\|}}{\text{C}}-\overset{\overset{O}{\|}}{\text{P}}(\text{OH})_2 \qquad\qquad \text{CH}_3\overset{\overset{O}{\|}}{\text{C}}-\underset{\underset{CH_2OH}{|}}{\overset{\overset{OH}{|}}{\text{C}}}-\text{CH}_2-\overset{\overset{O}{\|}}{\text{P}}(\text{OH})_2$$

54   **55**

It is now obvious that a surprising structural diversity exists among naturally occurring C–P compounds, and new opportunities are there for discoveries of additional ones. Modifying the natural structures to enhance biological activity would also be an attractive research activity, and there are already reports in this area. A recent example is the synthesis[10a] of derivatives of Fosmidomycin (**45**) and the related **46**; the former has high antibacterial activity, associated with inhibition of an enzyme in the nonmevalonate pathway for terpenoid biosynthesis.[10b]

Considerable successful work has been done on elucidating the biosynthetic origin of aminophosphonic acids and on the isolation of enzymes capable of forming and cleaving the C–P bond. The pathway to AEPA in *T. pyriformis* involves the highly unusual and high-energy process of rearranging phosphoenolpyruvate to phosphonopyruvate (Eq. 11.2). This equilibrium lies far to the side of PEP. An enzyme, PEP phosphomutase, has been isolated that catalyzes this interconversion;[11,12] the carbonyl is aminated and reduced to give the amino group, and decarboxylation also occurs.

$$(\text{HO})_2\overset{\overset{O}{\|}}{\text{P}}-\text{O}-\underset{\underset{COOH}{|}}{\text{C}}=\text{CH}_2 \;\rightleftharpoons\; \text{O}=\underset{\underset{COOH}{|}}{\text{C}}-\text{CH}_2-\overset{\overset{O}{\|}}{\text{P}}(\text{OH})_2 \;\xrightarrow{-\text{CO}_2}\; \text{O}=\underset{\underset{H}{|}}{\text{C}}-\text{CH}_2-\overset{\overset{O}{\|}}{\text{P}}(\text{OH})_2 \quad (11.2)$$

Other studies[13] on the biosynthesis of Bialaphos (**42**) resulted at about the same time in the isolation of another enzyme along with its substrate, with the new structure carboxyphosphonoenol pyruvate (CPEP; Eq. 11.3). The enzyme CPEP phosphonomutase converts CPEP to a H-phosphinic acid, which then goes on to form the phosphinic acid (**49**) moiety bound in Bialaphos. Later work led to other biosynthetic discoveries, and in a recent discussion[11b] of the status of the field, it was pointed out that three completely different enzymatic strategies are now known to be involved in C−P cleavages.

$$CH_2=C-O-\overset{O}{\underset{OH}{P}}-COOH \xrightarrow{CPEP \ phosphonomutase} CH_2-\overset{O}{P}H(OH) \rightarrow \rightarrow CH_2-\overset{O}{P}\overset{Me}{\underset{OH}{}}$$
$$\underset{COOH}{} \quad \underset{COOH}{C=O} \quad \underset{COOH}{C=O}$$
$$CPEP$$

(11.3)

We should remember from chapter 2.K that the conversion of phosphate to phosphonate is very rare but is not unknown in the synthetic laboratory; for example, it has been accomplished by Wiemer and co-workers[14] by the reactions shown in Eq. 2.36.

It is not the purpose of this discussion to develop these matters of biochemistry, fascinating though they may be. The biochemical literature offers speculation on the function of C−P compounds in living systems, but this aspect of natural C−P compounds awaits further clarification. Several reviews[5] are available that go into these and other aspects of the chemstry of the natural aminophosphonic acids.

The natural products chemistry of the C−P bond goes beyond the bounds of the global terrestrial and marine environments and into the Universe. In exciting new discoveries, the molecule phosphaacetylene, HCP, as noted in chapter 10.B, has been detected in interstellar space, and alkylphosphines, said to be formed regioselectively by photochemical addition of phosphine to alkenes, have been observed in the atmospheres of Jupiter and Saturn. Methylphosphonic acid, as well as some higher alkylphosphonic acids, have been detected by gas chromatography–mass spectroscopy in water extracts of the numerous organic compounds brought to Earth on the Murchison meteorite in 1969.[15] These discoveries have fueled speculation on the possibility of delivery of C−P compounds to Earth from the heavy meteoritic bombardment of eons ago, and the possible importance of carbon-bound phosphorus in the creation of early life forms. It has been suggested that early living systems may have had genetic molecules based on phosphonates rather than the DNA and RNA of present systems,[16] and experiments are even being designed to test some of the possible ways that C−P bonds could have been formed in the prebiotic stages of chemical evolution on Earth.[17] Other thoughts are presented in a fascinating essay by Yagi.[18] More developments in this intriguing new aspect of organophosphorus chemistry seem sure to come, and along with the numerous discoveries of C−P compounds in present-day living systems and their metabolic products, phosphorus chemistry can make a strong claim for inclusion among the natural products chemistry of organic molecules, so long dominated by nitrogen, oxygen, and sulfur compounds.

11.C. ORGANOPHOSPHORUS COMPOUNDS IN MEDICINE

After being largely ignored for many years as a source of medicinal compounds, organophosphorus chemistry has at last achieved an important and well-recognized place in the search for new drugs. Instrumental in bringing increased attention to the field were the discoveries mentioned earlier of the antibiotic C−P acid derivatives (compounds **41** to **55**) in bacterial fermentation broths. The acids were of relatively simple, easily synthesized structures, quite unlike the highly complex structures of most antibiotics, and derivatives are easily made for testing. Medicinal organophosphorus chemistry is now an active and exciting area, with many opportunities for fresh research.

Probably the first organophosphorus compound to receive acclaim as a valuable chemotherapeutic agent is the anticancer drug cyclophosphamide (**56**). This compound is chiral at phosphorus, and the stereoisomers have been obtained by resolution techniques[19]. It has been found that the (−)-stereoisomer is a better antitumor agent than is the (+) isomer. Cyclophosphamide can be viewed as a member of the family of "nitrogen mustard" alkylating agents, and as is true for other members of this family, it exhibits strong activity against a broad range of human cancers. This activity was first discovered in 1958,[20] and the compound remains in wide clinical use to this day. It was later discovered that cyclophosphamide undergoes metabolism to lose the 1,3,2-oxazaphosphorinane ring, through a process involving the hydroxy derivative **57** in a key step that leads to conversion to the true active agent, compound **58** (Eq. 11.4). Cyclophosphamide, therefore, should properly be described as a prodrug, one that releases the active material through enzymatic oxidation in the living system.

$$\underset{\mathbf{56}}{\text{cyclophosphamide}} \longrightarrow \underset{\mathbf{57}}{\text{hydroxy derivative}} \longrightarrow \underset{}{\text{acrolein}} + \underset{\mathbf{58}}{\text{phosphoramide mustard}} \tag{11.4}$$

High-level anticancer activity has been found in a number of phosphorus compounds of quite different structural types, and there is much current research in this area. Among others, examples of active compounds that show some of the structural diversity are derivatives of choline, including ether derivatives (**59**) of lysophosphatidyl choline, which are cytotoxic against tumor cells; Miltefosine (**60**), a choline phosphate; a phosphonyl derivative of choline, **61**; amino acid derivatives, such as N-phosphonoacetyl-L-aspartic acid (PALA) (**62**); and modified nucleotide derivatives, such as the unsymmetrical pyrophosphate **63**.

In the design of these anticancer drugs and other bioactive phosphonic acid derivatives, several rationales are kept in mind. An obvious one is that an exact phosphonate replica of a known biologically active phosphate could inhibit the process in which the phosphate is involved, which is an application of antimetabolite theory. The CH_2 group attached to P is isosteric with an O atom of a phosphate, meaning it has very similar size and bond angles, and a phosphonate and corresponding phosphate molecule are virtually superimposable. The high stability of the C—P bond would block any important natural processes involving hydrolysis of a phosphate ester group. The acidity difference between a phosphonic acid (slightly weaker) and a phosphomonoester is of no consequence in regard to the first ionization, because both acids will be completely ionized at physiological pH. The weaker second ionization of the phosphonic acid, however, could present a significant difference from a monophosphate. An adjustment in the phosphonic acid acidity, without creating a steric problem, can be made by placing the small fluorine atom on the α carbon. This increases the acidity by the needed amount of 0.5 to 1.0 pK units. A phosphonic acid modified to increase its similarity to a phosphoric acid in this way is said to be isopolar with the natural substance.

A second rationale is that a phosphonic acid designed to resemble a naturally occurring carboxylic acid might inhibit the biochemical work of this acid. To this end, phosphonic acid analogues of all of the natural aminoacids have been synthesized. Yet another concept driving some of the current research with phosphonic acids is that the tetrahedral structure of the phosphonate group resembles that of the transition state (or intermediate) in the attack of nucleophiles on the acyl group, which is of course of major importance in many enzymatic-mediated hydrolysis processes. Phosphonic and phosphinic acids and phosphonamidates designed as transition state analogs or mimics might act as inhibitors of hydrolytic enzymes such as esterases and amidases, and thus have useful chemotherapeutic properties. PALA (**62**) is an example of a rationally designed mimic of the transition state of aspartate transcarbamylase, involved in pyrimidine biosynthesis and, as noted, is a potent anticancer drug. Fosinopril (**64**), with valuable antihypertensive activity, can also be viewed as a transition state analogue. Many other examples are now known for which these three rationales for the design of new drugs have been effective. An excellent review of these rationales and the results of their application in the creation of new phosphorus-based chemotherapeutic agents has been presented by Engel.[21]

Applications of some of these rationales can be observed in the very important current work on the structural modification of nucleotides and their derivatives. Two objectives can be present in such work: (1) The construction of molecules in which a carbon substituent takes the place of O in phosphates, pyrophosphates, and triphosphates. Compound **65**, which is isosteric and isopolar with ATP, was synthesized for this purpose, and other phosphonic acid analogues are useful in mechanistic studies of biochemical processes involving nucleotide derivatives, (2) Developing molecules that are antiviral agents, for example as has been achieved by modifying the antiviral agent AZT 5′-phosphate (**66**) through formation of the isosteric phosphonate **67**.[22] Many examples of antiviral nucleotides with phosphorus groups are now known.

Antiviral activity in phosphorus compounds is not restricted to the modified nucleotide family. Indeed, the first active compound to be discovered had the very simple structure of trisodium phosphonoformate (**68**, first synthesized in 1924), and the discovery[23] of its potent activity in 1978 was also a factor in bringing attention to phosphorus compounds as medicinal agents. This compound is in clinical use under the name Foscarnet, and is known to inhibit viral DNA polymerase; it is a useful agent in the treatment of *Herpes simplex*, and is also active against HIV and the Epstein–Barr virus.

Application of the principle of replacing O by CH_2 to create new bioactive phosphorus agents has resulted in a major breakthrough in the treatment of bone diseases such as osteoporosis and Paget's disease.[24] It has been found that inorganic pyrophosphate inhibits the formation and dissolution of hydroxyapatite in bone. The dissolution process (resorption) leads to major medical problems. The idea arose that bisphosphonic acids (**69**) are isosteric with pyrophosphoric acid (**70**) but are hydrolytically stable and, if attracted to bone, might block the resorption process.

$$\text{HO-}\overset{\overset{O}{\|}}{\underset{\underset{OH}{|}}{P}}\text{-[CH}_2\text{]-}\overset{\overset{O}{\|}}{\underset{\underset{OH}{|}}{P}}\text{-OH} \qquad \text{HO-}\overset{\overset{O}{\|}}{\underset{\underset{OH}{|}}{P}}\text{-[O]-}\overset{\overset{O}{\|}}{\underset{\underset{OH}{|}}{P}}\text{-OH}$$

69   **70**

Many bisphosphonic acid derivatives, with substituents on the methylene carbon, have been found to exhibit the desired effect on bone, and a few are in clinical use for the treatment of bone diseases. The drugs Etidronate (**71**) and Clodronate (**72**) were among the earliest to find clinical use, but recently attention has shifted to derivatives with aminoalkyl side chains, as found in Pamidronate (**73**), Alendronate (**74**), and Risedronate[25] (**75**). The field is currently very active, and exciting new discoveries in bisphosphonate bone-disease therapy can be anticipated.

71: $(HO)_2P(O)-C(OH)(CH_3)-P(O)(OH)_2$

72: $(HO)_2P(O)-C(Cl)_2-P(O)(OH)_2$

73: $(HO)_2P(O)-C(OH)(CH_2CH_2NH_2)-P(O)(OH)_2$

74: $(HO)_2P(O)-CH(CH_2\text{-pyridyl})-P(O)(OH)_2$

75: $(HO)_2P(O)-C(OH)((CH_2)_3NH_2)-P(O)(OH)_2$

Activity against bone disease is not the only useful medicinal property of bisphosphonates; as ester derivatives, some have been found be active as antidepressants (e.g., **76**) and antihypercholesterolemic agents (**77**).

76: HO-C$_6$H$_4$-O-C(CH$_3$)(P(OCH$_2$OCCMe$_3$)$_2$(O))$_2$

77: farnesyl-type chain with terminal $-P(OH)_2(=O)$ groups (geminal bisphosphonic acid)

Organophosphorus compounds exhibit a number of other useful forms of bioactivity. Some that are of current importance are the clinically useful antihypertensive agent Fosinopril (**64**), the naphthylmethylphosphonate **78**, which is an inhibitor of insulin receptor tyrosine kinase,[26] and the phosphinopeptide analogue

78: naphthyl-CH(P(O)(OCH$_2$OCCH$_3$)$_2$)(OCH$_2$OCCH$_3$)

(**79**) of glutathionylspermidine (GSP) with potent trypanocidal activity through inhibition of GSP synthetase.[27]

$$\text{H}_3\overset{+}{\text{N}}-\underset{\text{H}}{\overset{\text{COO}^-}{\text{C}}}-\text{CH}_2\text{CH}_2-\overset{\text{O}}{\text{C}}-\underset{\text{H}}{\text{N}}-\text{CH}_2-\overset{\text{O}}{\text{C}}-\underset{\text{H}}{\text{N}}-\text{CH}_2-\overset{\text{O}}{\underset{|}{\text{P}}}-\text{O}^-$$

$$(\text{CH}_2)_4\overset{+}{\text{NH}}_2(\text{CH}_2)_4\overset{+}{\text{NH}}_3$$

79

Finally, we note another application of organophosphorus chemistry in the development of medicinals, their use as prodrugs to improve drug delivery to particular biological targets. We observed this behavior in the case of cyclophosphamide, but it is a widely studied technique to improve drug activity. Solubility properties have also been modified by attaching phosphorus groups; steroidal drugs, for example, are poorly soluble in water, but phosphorylation of a free OH group and formation of a sodium salt greatly improves the water solubility. Enzymatic hydrolysis then regenerates the steroidal drug. Organophosphorus drugs themselves can be converted to prodrugs with improved bioavailability. This is of particular importance in the antiviral field, where many active compounds contain phosphorus acid groups. These can be converted to ester form to improve accessibility to the target, following which hydrolysis releases the active drug. Fosinopril (**64**) is another example of a prodrug. A recent review cites many examples of the use of this approach with phosphorus drugs.[26]

11.D. AGRICULTURAL APPLICATIONS OF ORGANOPHOSPHORUS COMPOUNDS

11.D.1. General

Before World War II, there were almost no practical applications for synthetic organophosphorus compounds. During the 1930s and into the course of the war, research groups in England and Germany independently discovered that certain phosphoryl derivatives possessed exceedingly high levels of mammalian toxicity, a property not previously known, and one that was developed on both sides for potential use in warfare. Although this horrible prospect of wholesale human destruction never came to pass, the discoveries nevertheless have had enormous impact on post-war life, an impact that continues to this very day. It was discovered that the mammalian toxicity level could be greatly reduced by proper structural modification, but that toxicity to insects was very high and of potentially great practical value in the protection of crops. The chemical industry rapidly responded to these observations and commercialized organophosphorus compounds, primarily phosphoric acid derivatives, as insecticidal agents. Over the years, many organophosphorus compounds have been made and used in very large quantities in agriculture, not only as insecticides but also later as herbicides and in other applications.

Although other post-war applications for organophosphorus compounds emerged, it can be argued that the insecticide industry brought the greatest attention to this field and inspired new research into industrial applications of these compounds.

Even though the wartime discoveries of toxic phosphorus compounds were put to good use for humanity, we are nevertheless plagued to the present by the threat of the use of these same chemicals (and some later discovered ones) against humans. The compounds are not of complex structure and some are relatively easily made. Thus the threat has developed of their manufacture by terrorist groups and rogue nations, and indeed examples of the realization of this threat are well known. It is not the purpose of this chapter to develop the field of phosphorus chemical warfare agents; an excellent summary of the chemistry of these agents and of the attempts to ban their use by international treaties and disarmament has been presented by Black and Harrison.[28] However, all organophosphorus chemists should be aware of the structures that have become associated with high toxicity and, of course, avoid their synthesis unless stringent safety control of the experimental work and avoidance of environmental poisoning are practiced. To this end, we show here the structures of the most commonly mentioned agents. Close derivatives of these structures are also likely to have considerable toxicity and should be avoided. The structural requirements for high toxicity are quite severe, however, and in fact very few phosphorus compounds out of the many thousands that are known are in this category. The popular impression sometimes encountered that phosphorus compounds in general are highly dangerous and poisonous is without any basis, although as mentioned in chapter 2.L occasionally unexpected forms of toxicity among phosphorus compounds might be encountered.

All of the toxic phosphoryl compounds referred to above act as phosphorylating agents; they react with the OH group of the serine moiety of the enzyme acetylcholinesterase, which functions in neural transmission by hydrolyzing acetylcholine. This compound transmits the neural signal across the myoneural gap separating the nerve ending from the muscle; cholinesterase destroys acetylcholine, and if it fails to do so, the signal to the muscle is not terminated and muscular paralysis occurs. Thus the toxic phosphorylating agents have become known as anticholinesterase agents; by chemically changing the serine moiety at the esteratic site of the enzyme, the activity of the enzyme is lost. The English wartime discovery was primarily of anticholinesterase compounds of structures typified by **80**, di-isopropyl fluorophosphate (which in fact was later found to have clinical value in the treatment of glaucoma, e.g., as Floropryl offered by Merck, Sharpe, and Dohme).

$$((CH_3)_2CHO)_2-\overset{\overset{O}{\|}}{P}-F$$

80

In Germany, several agents of even higher toxicity were discovered, the principal ones (and the ones constituting the major present-day threat) being the phosphoric acid derivative Tabun (**81**), and the methylphosphonates Sarin (**82**, also known as GB) and Soman (**83**).

11.D. AGRICULTURAL APPLICATIONS OF ORGANOPHOSPHORUS COMPOUNDS

$$(CH_3)_2N-\overset{\overset{O}{\|}}{\underset{CN}{P}}-OC_2H_5 \qquad CH_3-\overset{\overset{O}{\|}}{\underset{F}{P}}-OCH(CH_3)_2 \qquad CH_3-\overset{\overset{O}{\|}}{\underset{F}{P}}-O-\overset{\overset{CH_3}{|}}{CH}C(CH_3)_3$$

81   **82**   **83**

After the war, an even more toxic methylphosphonate with structure **84** (known as VX) was discovered. Its toxicity to humans, expressed by the LD_{50} value (the concentration needed to kill 50% of a species), is 0.009 mg/kg body weight and is said to be the most toxic synthetic poison known.

$$CH_3-\overset{\overset{O}{\|}}{\underset{OCH_2CH_3}{P}}-SCH_2CH_2N(i\text{-}Pr)_2$$

84

An excellent description of the discovery and mode of biological action of the phosphorus cholinesterase inhibitors has been given by Toy and Walsh;[29] synthetic procedures are well described in the review by Black and Harrison.[28]

11.D.2. Organophosphorus Insecticides

The enzyme acetylcholinesterase is also involved in the biochemistry of insects, and its inhibition can cause their death. Phosphorylating agents in huge numbers have been synthesized with the objective of minimizing the mammalian toxicity problem while maintaining high toxicity to insects. Phosphorus compounds have distinct advantages in the pesticide market; they are relatively easy to make, and they biodegrade readily by hydrolysis, so that the problems of residual activity, so serious with the chlorinated hydrocarbon pesticides, are avoided. Parathion (**85**) was one of the first commercially produced insecticides; its mammmalian toxicity (LD_{50}) is 55 mg/kg, which is rather low and requires careful handling and application in the field. The O-methyl derivative proved to be less toxic (300 to 400 mg/kg). Both have been used extensively, with a production peak in 1966 of about 55,000,000 lb of the two combined. Interest in these compounds has greatly declined with the introduction of safer agents. Indeed, many compounds are now known that are relatively harmless to humans yet with excellent toxicity to insects. One of the best of these is the well-known garden insecticide Malathion (**86**) with LD_{50} of 4100 mg/kg. Phosmet (**87**) has LD_{50} greater than 4640 mg/kg.

$$O_2N-\underset{}{\underset{}{\bigcirc}}-O-\overset{\overset{S}{\|}}{P}(OEt)_2 \qquad (MeO)_2\overset{\overset{S}{\|}}{P}-S-\underset{\underset{\overset{\|}{O}}{CH_2-COEt}}{\overset{\overset{O}{\|}}{CH-COEt}} \qquad (MeO)_2\overset{\overset{S}{\|}}{P}SCH_2N\underset{}{\underset{}{\bigcirc}}$$

85   **86**   **87**

A summary by Toy and Walsh[29b] of the major phosphorus insecticides produced commercially is of great value in appreciating the structural requirements for useful activity. With only three exceptions out of 40, all of the phosphorus insecticides are derivatives of phosphoric acid, the majority of which have thiophosphoryl groups. There are many other highly active insecticidal compounds that have not been commercialized, and it is indeed remarkable that such a wide range of toxicity and insecticidal potency can result from manipulations of the four substituents attached to phosphorus. The three nonphosphate compounds are phosphonates **88** to **90**, which as a class are generally considered to be more expensive to produce than phosphates. Note, however, that **90** is simply the adduct of dimethyl H-phosphonate with chloral.

$C_6H_5-\overset{S}{\underset{OEt}{P}}-O-\underset{}{\bigcirc}-NO_2$ $C_2H_5-\overset{S}{\underset{OEt}{P}}-S-\underset{}{\bigcirc}$ $\overset{MeO}{\underset{MeO}{}}\overset{O}{P}-\overset{OH}{\underset{}{C}}HCCl_3$

88 (EPN)   **89** (Fonophos)   **90** (Dipterex)

Without exception, all active compounds are methyl or ethyl esters; were free —OH or —SH groups present, they would be ionized at physiological pH, upsetting the necessary balance between water and lipid solubility. Phosphorus may bear either a phosphoryl or thiophosphoryl group. A good leaving group must be present, and this is usually aryloxy (as in Parathion, **85**) or heteroaryloxy (as in Diazinon, **91**), or vinyloxy (as in DDVP, **92**). Indeed, such compounds appear to be the only *O,O,O*-trialkyl phosphates found among the commercialized insecticides. Other good leaving groups include thioxy (as in Amiton, **93**) or phosphate (thus a pyrophosphate moiety, as in Schradan, **94**). The leaving group must be carefully selected so that the molecule will survive long enough in the aqueous environment so as to reach the enzyme and perform phosphorylation on the serine unit. A particularly common configuration is that of an *O,O,S*-trialkyl phosphorodithioate (as in Malathion, **86**, and Phosmet, **87**).

$(EtO)_2-\overset{S}{P}-O-\underset{}{\text{pyrimidinyl}}$ $(MeO)_2\overset{O}{P}-OCH=CCl_2$ $(EtO)_2\overset{S}{P}-SCH_2CH_2NEt_2$

91 (Diazinon)   **92** (DDVP)   **93** (Amiton)

$\overset{Me_2N}{\underset{Me_2N}{}}\overset{O}{P}-O-\overset{O}{P}\overset{NMe_2}{\underset{NMe_2}{}}$

94 (Schradan)

Much of the ingenuity in the development of new insecticides is found in the selection of the substituents to be placed on O, S, or N functions. Relatively few reaction types are used in the manufacture of organophosphorus insecticides. A useful summary of these, with references to particular compounds, may be found in the review by Eto.[30] The basic reactions of phosphate chemistry, many of which have been seen elsewhere in this book, are employed. Displacement of halogen from P with various nucleophiles is a useful process. The phosphorodithioate structure $(RO)_2P(S)(SH)$ is particularly easily approached by reaction of phosphorus pentasulfide with alcohols. The acidic SH group may then add to activated double bonds (as to diethyl maleate, to make Malathion, Eq. 11.5), or it may be alkylated, e.g., with chloromethyl derivatives (**95**, Eq. 11.6). Reaction of the acidic SH group with formaldehyde and a nucleophile is also practiced.

$$\text{MeOH} + P_4S_{10} \longrightarrow (\text{MeO})_2\overset{\overset{S}{\|}}{P}-SH \xrightarrow{EtO_2CC=CCO_2Et} \underset{\underset{86}{}}{\overset{EtO_2CCH-CH_2CO_2Et}{\underset{\overset{\|}{S}}{S-P(OMe)_2}}} \quad (11.5)$$

$$(\text{MeO})_2\overset{\overset{S}{\|}}{P}-SH + \underset{\mathbf{95}}{\text{ClCH}_2N\text{-ring}} \longrightarrow (\text{MeO})_2\overset{\overset{S}{\|}}{P}-S-CH_2N\text{-ring} \quad (11.6)$$

Thiophosphoryl derivatives are also prepared from the dithio derivatives, by chlorination to replace —SH by —Cl, followed by displacement of Cl with nucleophiles (Eq. 11.7).

$$(EtO)_2\overset{\overset{S}{\|}}{P}-SH \xrightarrow{Cl_2} (EtO)_2\overset{\overset{S}{\|}}{P}-Cl \xrightarrow[\text{base}]{HO-C_6H_4-NO_2} \underset{85}{(EtO)_2\overset{\overset{S}{\|}}{P}-O-C_6H_4-NO_2} \quad (11.7)$$

Another useful process is the Perkow reaction (chapter 3), which involves the reaction of a trialkyl phosphite with an α-chloro carbonyl compound. Especially important is the use of chloral, which leads to the insecticide DDVP (**92**, Eq. 11.8).

$$(MeO)_3P + Cl_3CC(=O)H \longrightarrow \underset{92}{(MeO)_2\overset{\overset{O}{\|}}{P}-OCH=CCl_2} \quad (11.8)$$

11.D.3. Organophosphorus Compounds as Herbicides

Phosphorus compounds were late entries in the field of organic herbicides, and to this date only a few compounds have attained major commercial importance. The first of these to be discovered, however, is now dominant in the field of herbicides; it is the well-known Glyphosate (sold under the tradename Round-up by the discovering company, Monsanto). Glyphosate (applied to the plant as a salt) has the structure **96** and is named *N*-(phosphonomethyl)glycine. It is a broad-spectrum, nonselective herbicide that is active against many notorious weeds. It has very low mammalian toxicity, and undergoes biodegradation in the environment. An account of the discovery of Glyphosate may be found in the recent review by Sikorski and Logusch;[31] a thorough study of this important herbicide is provided in a later book by Franz, et al.[32] The compound was first synthesized by a version of the Mannich reaction shown in Eq. 11.9, in which phosphorous acid condenses with formaldehyde and glycine in the presence of HCl. However, other methods have been developed that are superior to the Mannich route, and are outlined by Sikorski and Logusch.[31] It is stated there that the Monsanto commercial process makes use of iminodiacetic acid rather than glycine in the Mannich reaction. This is a relatively cheap reactant and serves to block nitrogen from secondary reactions. The blocking $-CH_2COOH$ group is easily removed by a number of techniques.

$$(HO)_2\overset{O}{\overset{\|}{P}}-H + CH_2O + NH_2CH_2CO_2H \xrightarrow{HCl} (HO)_2\overset{O}{\overset{\|}{P}}-CH_2NHCH_2CO_2H \quad (11.9)$$
$$\mathbf{96}$$

Glyphosate is known to act by the inhibition of the plant enzyme 5-enolpyruvoyl-shikimate-3-phosphate synthetase (EPSPS), which is involved in the biosynthesis of aromatic aminoacids and other aromatic compounds in plants (but not in mammals and other members of the animal kingdom). It is a competitive inhibitor of phosphoenolpyruvate (PEP), and it has been proposed that it may function as a transition state mimic that occupies the site on the enzyme where the protonated form of PEP normally would fit. The proposed similarity between Glyphosate and protonated PEP is illustrated by structures **97** and **98**.

97

98 (ionic form)

Much research has gone into the modification of the Glyphosate structure to produce other herbicidal compounds, but so far no superior structure seems to have been commercialized.

The discovery of the antibiotic phosphinothricylalanylalanine (**42**) was noted in section 11.B. This compound was later shown to be a very effective herbicide and is on the market as Bialaphos. The tripeptide is first hydrolyzed in the plant to release phosphinothricin (**49**) as the active material, functioning as an inhibitor of glutamine synthetase. Pure phosphinothricin is also commercially available, under the name Glufosinate. Glutamine synthetase catalyzes the conversion of glutamate to glutamine, which is the source of ammonia in amino acid synthesis, and is also involved in nucleic acid synthesis. When the enzyme is inhibited, ammonia builds up in the plant and causes its death. The photorespiratory pathway is also blocked. Although glutamine synthetase is involved in mammalian biochemistry, neither Bialaphos nor phosphinothricin show any significant mammalian toxicity.

Phosphinothricin may be synthesized in several ways. The discoverers of the antibiotic[33] employed the Michaelis–Arbusov reaction shown in Eq. 11.10; however, other methods have been devised for commercial synthesis and are discussed by Sikorski and Logusch.[31]

$$CH_3P(OEt)_2 + \begin{array}{c} BrCH_2CH_2-CHCO_2Me \\ | \\ NHCOCF_3 \end{array} \longrightarrow \begin{array}{c} O \\ \| \\ CH_3PCH_2CH_2-CHCO_2Me \\ | \quad\quad\quad\quad | \\ OEt \quad\quad\quad NHCOCF_3 \end{array} \xrightarrow{H_2O}$$

$$\begin{array}{c} O \\ \| \\ CH_3PCH_2CH_2-CHCO_2H \\ | \quad\quad\quad\quad | \\ OH \quad\quad\quad NH_2 \end{array} \qquad\qquad (11.10)$$

Many other phosphorus compounds show valuable herbicidal activity, and much current research effort is going on in this area. In addition to the phosphorus-containing amino acid derivatives, other structural types are of interest, such as are seen in compounds **99** (Betasan), **100**, and **101**. The latter is in use as a cotton defoliant.

$$\begin{array}{c} S \\ \| \\ (i\text{-}PrO)_2P-SCH_2CH_2NHSO_2Ph \end{array} \qquad \begin{array}{c} O \quad Cl \\ \| \quad | \\ PhP-CHCOOPr\text{-}i \\ | \\ OPr\text{-}i \end{array} \qquad (BuS)_3P=O$$

          99   **100**   **101**

11.D.4. Other Uses of Phosphorus Compounds in Agriculture

Crops are molested by other types of organisms than insects, and phosphorus compounds play a role in their control as well. Thus the various phosphorus insecticides are used to control nematodes (small worms) and members of the separately classified *Acarina* that include mites and ticks. Such active compounds are known as nematocides and acaracides. Fungi also do major damage to crops, and phosphorus-based fungicides are sold commercially. An example is the simple compound Phosetyl, the monoethyl ester of phosphorous acid, $EtO-P(O)H(OH)$. It is used as the sodium or aluminum salt.

Phosphorus compounds can also exhibit valuable plant growth regulatory (PGR) activity, and some are available commercially for this purpose. 2-Chloroethylphosphonic acid (Ethephon) was one of the first shown to have useful PGR activity. It is used as a ripening agent, especially for tomatoes and pineapples, and has other useful effects on plants. As a ripening agent, it acts through degradation to ethylene (Eq. 11.11), a well-known agent for fruit ripening. Note the proposed intermediacy of the low-coordinate metaphosphate ion in this degradation.

$$Cl\text{-}CH_2\text{-}CH_2\text{-}\overset{\overset{O}{\|}}{P}\overset{O^-}{\underset{O^-}{\diagup}} \longrightarrow Cl^- + H_2C=CH_2 + \left[\overset{O}{\underset{O}{\diagup}}P\text{-}O^-\right] \xrightarrow{H_2O} (HO)_2PO_2^-$$

(11.11)

Another valuable PGR agent has structure **102**, which increases the yield of sucrose from sugarcane. It is known as glyphosine (marketed as Polaris by Monsanto).

$$HOOCCH_2N\underset{CH_2PO(OH)_2}{\overset{CH_2PO(OH)_2}{\diagup}}$$

102

11.E. THE IMPORTANCE OF ORGANOPHOSPHORUS COMPOUNDS IN TECHNOLOGY

11.E.1. General

The last few decades have seen the development of many practical uses for organophosphorus compounds in addition to those in medicine and agriculture. In this section, a catalog will be supplied of some of the most important phosphorus chemicals manufactured commercially for nonbiological purposes, with a note on these uses. The list should not be considered to be complete, but it is representative of the great variety of structures finding commercial significance and of the many technical uses where phosphorus chemicals have value. A broader description of these applications is provided by Toy and Walsh,[29] and in encyclopedias of technology (such as the *Kirk-Othmer Encyclopedia*[34]). When these uses are combined with those for organophosphorus compounds in medicine and agriculture, it becomes obvious just how important these compounds have become in serving the needs of the chemical industry and of humanity.

Two of these numerous technical uses require special comment. Phosphorus chemicals are among the most important of commercial flame retardants, and many phosphorus compounds are on the market. This is an active area of current research and is a very important subfield of industrial organophosphorus chemistry. Recent reviews[29c,35] are available to provide information on the mechanism of flame retardancy and on the structure and synthesis of commercial products. Some of the most important structural types exhibiting flame retardancy will be noted in the catalog of compounds to follow. The second technical use requiring preliminary comment is that of the construction of phosphine-containing metal coordination compounds as catalysts for various commercial processes. This application lies more in the realm of metal coordination chemistry, but it is of great current research interest and also stands as a subfield of organophosphorus chemistry with which students of organophosphorus chemistry should be aware. Much current activity goes into the design of new optically active tertiary phosphines as ligands in complexes for use in homogenous asymmetric syntheses. Some of the useful phosphines of this type were mentioned in chapter 9, and some examples of the practical applications of these and other phosphine complexes are mentioned in section 11.E.2.a. A 1990 review[36] is available that provides broad coverage of the structures and syntheses of the numerous phosphines studied as ligands for this purpose. Catalysts for other commercial processes, such as the important hydroformylation of alkenes (the Oxo process), also contain phosphines but of less intricate structures; indeed, triphenylphosphine is a common ligand in such catalysts.

11.E.2. Nonbiological Commercial Phosphorus Chemicals

11.E.2.a. Tertiary Phosphines. Many metal complexes of triphenylphosphine and other tertiary phosphines are known that are useful as catalysts for various processes. One of the earliest complexes to find commercial use is the Reppe compound $Ni(CO)_2(PPh_3)_2$, which effects the polymerization of alkenes and acetylenes. Triphenylphosphine is also a ligand in the Wilkinson catalyst $RhCl(PPh_3)_3$ for the homogeneous hydrogenation of alkenes. Similar catalysts using this phosphine are employed in the hydrogenation of acetylenes and carbonyl compounds; in the Oxo hydroformylation of alkenes with hydrogen and CO, a useful catalyst has the structure $Rh(CO)(Ph_3P)_{2 \text{ or } 3}$. This type of catalyst can lead to very high selectivity (about 10:1) for the attachment of CO to the terminal carbon in this process. Another major discovery in hydrogenation catalysis was that asymmetry could be transferred to carbon when the catalyst contained an optically active phosphine. The first example of such a phosphine ligand, discovered by Knowles,[37] was compound **103**, for which several synthetic approaches were seen in chapter 9.

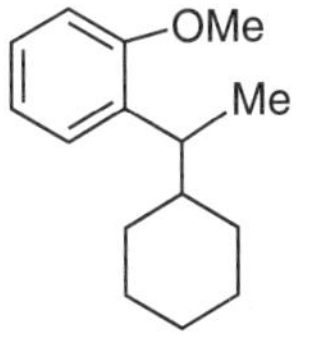

103

Knowles used the catalyst to prepare optically active L-dopa (used to treat Parkinson disease), according to Eq. 11.12, a process that has been commercialized.

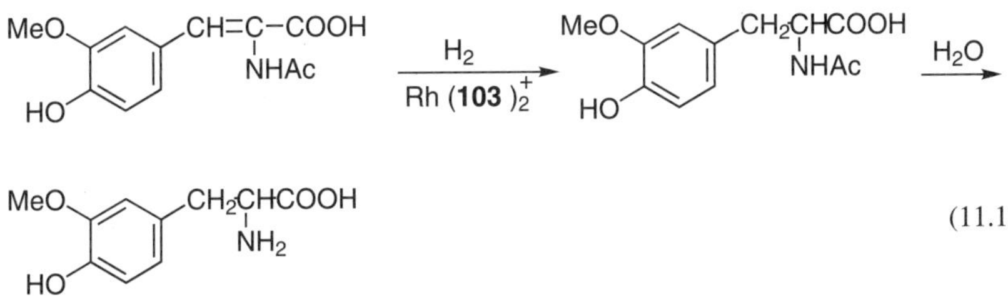

(11.12)

A later discovery was that the diphosphine **104** (DIPAMP, chapter 9) as the R,R-stereoisomer gave even more effective catalysts, and much research then followed on the synthesis of optically active diphosphines for use in asymmetric syntheses.

Kagan and Sasaki[36] provide the structures for about 140 monophosphines and diphosphines that are optically active due either to an asymmetric phosphorus or a carbon atom and that have been used to effect asymmetric hydrogenations. It is typical to receive an enantiomeric excess well over 90% with some of these catalysts; those described recently by Burk, et al., (e.g., DuPHOS, **105**, described in chapter 9) provide 99% ee.

These observations of the great effectiveness of chiral phosphine complexes in asymmetric hydrogenations have led to the exploration of their use in related chiral syntheses, which are of major importance in the pharmaceutical industry, (the subject of a review by Burk, et al.[38]). More recent accomplishments with the DuPHOS-rhodium system include the catalytic hydrogenations of certain enamines to give optically active amines with 99% ee,[39] of alkylidene succinates to give optically active succinates with 94% ee[40], and of enol esters to give optically active saturated esters.[41] Reactions other than hydrogenation can be performed to give chiral products; a recent example is the hydrosilylation of alkenes with $HSiCl_3$ catalyzed by a palladium complex of a L-menthylphosphetane derivative.[42] In addition to chiral syntheses, new applications with commercial value continue to emerge for phosphine-containing catalyst systems. More recent examples are the addition of alcohols to alkynes catalyzed with a methylgold(I) complex of tertiary phosphines,[43] and the ring-opening metathesis polymerization (the ROMP reaction) of strained cyclic alkenes in water solution with a ruthenium complex of dicyclohexyl(2-trimethylammonioethyl)phosphine.[44]

Phosphonites with optically active carbon substituents can also be used in asymmetric synthesis. A recent example is the construction of bis-phosphonites from the reaction of phosphinous chlorides or esters with carbohydrate derivatives that have two open OH groups.[45] It is claimed that the products, used in Rh-catalyzed hydrogenations, give the highest enantioselectivity of any ligand derived from a natural product as the chiral center.

11.E.2.b. Tetrakis(hydroxymethyl)phosphonium Chloride.
The reaction of phosphine with formaldehyde and HCl gives rise to $(HOCH_2)_4P^+Cl^-$, known as THPC. It and the hydroxide are widely used as flame retardants for cotton fabrics.

11.E.2.c. Phenylphosphinic Acid, PhPH(O)OH.
Made commercially by the hydrolysis of $PhPCl_2$, phenylphosphinic acid is used as a stabilizer and color-preventing agent in Nylon.

11.E. THE IMPORTANCE OF ORGANOPHOSPHORUS COMPOUNDS IN TECHNOLOGY

11.E.2.d. Trialkyl and Triaryl Phosphites. Triphenyl phosphite is used as a stabilizer in alkyd resins and vinyl plastics, and tri(nonylphenyl) phosphite is used to stabilize rubber-styrene polymeric materials. Tri(iso-octyl) phosphite is manufacured as a stabilizer for poly(vinyl chloride).

11.E.2.e. Dibutyl H-Phosphonate (Dibutyl Phosphite). Added to petroleum-derived oils that are used in hydraulics systems and for gear lubrication, dibutyl H-phosphonate prevents wear on interacting metal surfaces.

11.E.2.f. Trialkylphosphine Oxides. Long-chain derivatives, such as trioctylphosphine oxide (made by the radical addition of PH_3 to 1-octene (chapter 3), followed by peroxide oxidation) are quite useful in the selective extraction of uranium salts from water solution into kerosene solution.

11.E.2.g. Nitrilotris(methylenephosphonic acid). This compound was introduced some years ago as an alternative to metallic phosphates in detergent formulations. It is a highly effective sequestering agent, forming chelate complexes with iron, magnesium, and calcium. It is prepared as seen in Eq. 11.13.

$$(HO)_2\overset{O}{\overset{\|}{P}}-H + NH_4Cl + CH_2O \xrightarrow{HCl} N[CH_2P(O)(OH)_2]_3 \qquad (11.13)$$

11.E.2.h. Salts of Alkylphosphonic Acids. A number of soluble salts of phosphonic acids are used as corrosion inhibitors. A new entry in this old field is the water-soluble telomeric product obtained by radical addition of $H-P(O)(ONa)_2$ to disodium maleate with H_2O_2 as initiator.[46] The product is a mixtue of the 1:1 adduct (phosphonosuccinic acid) and a 1:2 adduct (1-phosphonobutane-1,2,3,4-tetracarboxylic acid).

11.E.2.i. Alkylphosphonic Acid Esters. Dimethyl methylphosphonate finds use as a flame retardant and viscosity reducer. Some other important commercial flame retardants that are phosphonic acid derivatives are shown as structures **106** to **109**.

106 (Fyrol 6): $(EtO)_2\overset{O}{\overset{\|}{P}}CH_2N(CH_2CH_2OH)_2$

107 (Antiblaze 19)

108 (Antiblaze 78): $ClCH_2CH_2-\overset{O}{\overset{\|}{P}}(OCH_2CH_2Cl)_2$

109 (Fyrol Bis-Beta): $CH_2=CH-\overset{O}{\overset{\|}{P}}(OCH_2CH_2Cl)_2$

11.E.2.j. Trialkyl Phosphates. Tributyl phosphate is used in uranium extraction and in hydraulics fluids. It can also be employed as a defoamer. Tris(2-ethylhexyl) phosphate finds use as a plasticizer for vinyl polymer sheets, to which it also imparts flame resistance. Phosphates with haloalkyl substituents, such as **110** to **112**, are widely used flame retardants.

O=P[OCH$_2$CH(Br)CH$_2$(Br)]$_3$
110
(Firemaster T23P, Fyrol HB-32)

O=P(OCH$_2$CH$_2$Cl)$_3$
111
(Fyrol CEF, Union Carbide 3CF)

$$O=P\left[OCH\begin{matrix}CH_2Cl\\CH_2Cl\end{matrix}\right]_3$$
112
(Fyrol FR-2)

11.E.2.k. Triaryl Phosphates. Triaryl phosphates are very stable thermally and find extensive use in technology. They are employed as plasticizers, gasoline additives, lubricants, hydraulics fluids, etc. Triphenyl and tricresyl phosphate are especially well known in technology. The latter at one time was an important flame retardant, but now aryl groups with longer alkyl substituents are more commonly used for this purpose. 2-Ethylhexyl diphenyl phosphate is manufactured for use as a nontoxic plasticizer in food-wrapping plastics. It is also valuable as an aircraft hydraulics fluid, in part because it has a very high spontaneous ignition temperature (>1000°).

11.E.2.l. Sodium Bis(2-ethylhexyl) Phosphate. This salt is a useful wetting agent and extractant for uranium salts.

11.E.2.m. Mixed Monoalkyl and Dialkyl Phosphates. These compound types are prepared industrially as a mixture by the reaction of phosphorus pentoxide with alcohols. The mixture finds extensive use as an acid catalyst (e.g., in the condensation of urea and formaldehyde). In the form of salts with ammonia or amines, the mixture is used as a textile lubricant, flame retardant, corrosion inhibitor, gasoline additive, and emulsifier. By adjusting the reactant ratio in their manufacture, it is possible to retain some of the P—O—P links of P$_2$O$_5$, and this type of product finds use as a wetting agent, plastics stabilizer, heavy metal sequesterant, and extractant for uranium salts.

11.E.2.n. Thiophosphoric Acids and Salts. *O,O*-Dialkyl phosphorodithioates, (RO)$_2$P(S)SH, are formed when alcohols react with phosphorus pentasulfide. These acids are used as lube oil additives and as agents for ore flotation. Zinc salts are also effective in these applications, and in addition can be used as corrosion inhibitors.

11.F. SOME MAJOR USES OF ORGANOPHOSPHORUS COMPOUNDS AS REAGENTS IN ORGANIC SYNTHESIS

11.F.1. General

The emphasis in this chapter has so far been given to phosphorus-containing compounds that are of biological or practical value in various capacities. Many phosphorus-based reactions, however, lead to valuable phosphorus-free organic products and can be developed into useful synthetic methods in organic chemistry. The example that all students are familiar with is the Wittig synthesis of alkenes from ylides and carbonyl compounds, of such importance and scope that Georg Wittig received the Nobel Prize in Chemistry in 1979 for its development. There are, however, numerous other applications of phosphorus compounds in organic synthesis, and some of the more important of these are discussed in this section. For a fuller treatment, the 1979 monograph by Cadogan, *Organophosphorus Reagents for Organic Synthesis*,[47] should be consulted.

11.F.2. Alkene Synthesis Using Ylides and Phosphonate Carbanions

Probably the first, and still the best known, example of the use of organophosphorus compounds as reactants in the preparation of organic compounds is that of the synthesis of alkenes by the Wittig reaction (Eq. 11.14). The related Wadsworth–Emmons (W–E) condensation—the name advocated by Johnson[48]—with α-carbanions of phosphonates (Eq. 11.15), frequently those stabilized by electron withdrawing groups, is also an excellent method for alkene synthesis. It was presented in chapter 5.F.

$$Ph_3\overset{+}{P}C-\bar{C}HR + O=CR'_2 \longrightarrow Ph_3\overset{+}{P}\underset{-O}{\overset{CHR}{\diagdown}}CR'_2 \longrightarrow Ph_3P\underset{O}{\overset{CHR}{\diagdown}}CR'_2 \longrightarrow \begin{array}{c} Ph_3P=O \\ + \\ RCH=CR'_2 \end{array} \quad (11.14)$$

$$(RO)_2\overset{O}{\overset{\|}{P}}-\bar{C}HCOOR + O=CR'_2 \longrightarrow (RO)_2\overset{O}{\overset{\|}{P}}-CHCOOR \longrightarrow (RO)_2\overset{O}{\overset{\|}{P}} + R'_2C=CHCOOR \quad (11.15)$$

Both the Wittig and the W–E reactions received extensive coverage in the Cadogan monograph[47] and in the more recent comprehensive book by Johnson,[48] but many other reviews have been prepared and some are cited in the well-known text by March, *Advanced Organic Chemistry*.[49] Suffice it to say at this point that these reactions are of fundamental importance in organic chemistry, as is made clear by March in his concise treatment of the subject, and are widely used not only to construct particular alkenes without problems of double bond location but also as a means of coupling carbon chains. Innumerable syntheses of complex natural products and pharmaceutical compounds have been effected with these reagents.

The most common phosphine precursor of the Wittig reagent is simply triphenylphosphine, but other phosphines can also be used.[50] They have been reacted with halides in great variety. Ylides are formed readily with a number of bases, varying in strength from butyllithium, through sodium hydride, sodium amide, and potassium hexamethyldisilazide/18-crown-6, to sodium hydroxide and alkoxides. The ylides are rarely isolated in alkene synthesis (but this can be done if desired) and are used directly in the condensation with carbonyl compounds. The solutions are said to be "salt-free" if lithium bases are not used in the synthesis. The aldehyde or ketone may have various structures in the aliphatic, alicyclic, and aromatic families, and many substituents can be tolerated (double and triple bonds even in conjugation, OH, OR, NR_2, nitro or haloaryl, ester, etc.). Even cycloalkenes can be formed by the Wittig method, and the reaction also succeeds with the carbonyl of ketenes and isocyanates, as well as with imines and even nitroso groups to form C=N bonds. Many of these considerations also apply to the use of phosphonate carbanions in the W–E synthesis, although some restrictions do apply.

The mechanism of the Wittig olefination reaction has been widely studied, and considerable evidence points to the initial formation of a four-membered cyclic adduct, having the oxaphosphetane structure (Scheme 11.1). These intermediates can be detected by ^{31}P NMR spectroscopy; that from the reaction of $Ph_3P=CHMe$ with benzaldehyde, for example, gives two signals at δ −61.2 and δ −61.6 due to cis, trans isomerism from the two C-substituents. ^{13}C and ^1H NMR have also been employed in the examination of these intermediates; the spectral work is discussed in detail in a review by Vedejs and Marth.[51] Coupling constants have been especially helpful in deducing the geometry of the oxaphosphetane ring, which generally is puckered with the 5-coordinate phosphorus in a nearly trigonal bipyramidal conformation. X-ray data have been collected that support these conclusions.

The formation of isomeric oxaphosphetanes leads to a serious problem with the Wittig reaction (shared with the W–E reaction): if E,Z isomers can be formed, generally a mixture of the two will result. Strong biasing in one direction can be experienced, however, and this can be influenced by the structure of the reactants, the solvent, and the presence of salts in the reaction mixture. Wittig reagents from

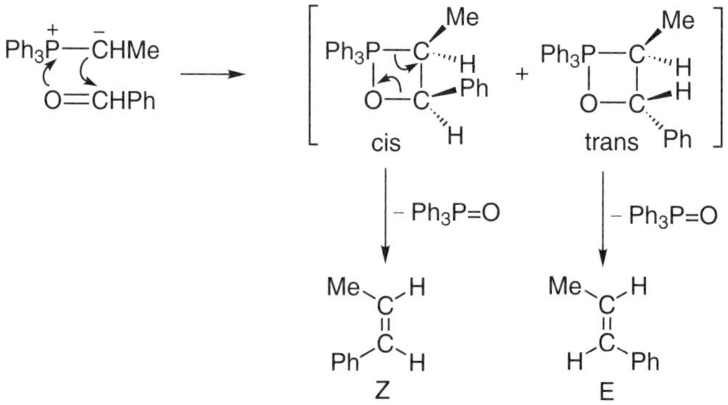

Scheme 11.1

triarylphosphines with electron-withdrawing groups on the ylidic carbon generally give a preponderance of the E alkene, whereas with nonstabilized ylides the Z alkene is favored. Moderate stabilization, as with the phenyl group, generally leads to little isomer preference. These effects have been associated with nonbonded interactions in the oxaphosphetane and to the degree of puckering of the ring in the transition state. These ideas are developed by Vedejs and Marth.[52]

In the W–E reaction, the E isomer generally predominates, but again the regiochemistry can be controlled by the choice of phosphonate derivative; we have already noted this for phosphonamides, and for phosphonate groups attached to substituents on 5-coordinate phosphorus (chapter 10.E), where the Z-isomer is the major product. The Z-isomer also is highly favored when the phosphonate has 2,2,2-trifluoroethyl substituents, as in the synthesis of unsaturated esters[53] from carboethoxymethylphosphonates (**113**) and of vinylphosphonates[54] from the reaction of the diphosphonate **114** with aldehydes (Eq. 11.16).

$$\text{YCH}_2\text{-P(O)(OCH}_2\text{CF}_3)_2 \xrightarrow{\text{KHMDS / 18-Crown-6}} \text{Y}\overline{\text{CH}}\text{-P(O)(OCH}_2\text{CF}_3)_2 \xrightarrow{\text{RCHO}} \begin{array}{c} R \\ H \end{array}\!\!\!C\!=\!C\!\!\!\begin{array}{c} Y \\ H \end{array}$$

113, Y = COOCH$_2$CH$_3$
114, Y = P(O)(OCH$_2$CF$_3$)$_2$

(11.16)

The regiochemical effects in alkene formation by these reactions are discussed in reviews.[47,48,55]

11.F.3. Replacements of Oxygen by Sulfur with Thiophosphorus Compounds

Thiophosphorus compounds find important use in the synthesis of organic compounds with the C=S group. Another application is the replacement of epoxy oxygen by sulfur. A review covers this aspect of thiophosphorus chemistry in detail.[56] The replacement of carbonyl oxygen by sulfur is generally accomplished with the commercially available Lawesson reagent (**115**), easily prepared as shown in Eq. 11.17.

$$\text{PhOMe} + \text{P}_4\text{S}_{10} \xrightarrow{\Delta} \text{Lawesson's reagent } \mathbf{115}$$

(11.17)

The Lawesson reagent can be viewed as a dimer of the σ^3, λ^5 monomer **116**, with which it appears to be in equilibrium and which may be the active form (Eq. 11.18).

$$\text{Ar-P(S)(S)-S-P(S)(S)-Ar} \rightleftharpoons 2\,\text{ArP(S)}_2 \xrightarrow{R_2C=O} R_2\overset{+}{C}-O-\overset{S}{\underset{S}{P}}-\text{Ar} \longrightarrow$$

$$R_2C\overset{O}{\underset{S}{\diamond}}\overset{S}{\underset{Ar}{P}} \longrightarrow R_2C=S + \left[\text{ArP}\overset{O}{\underset{S}{=}}\right] \longrightarrow \text{trimer} \qquad (11.18)$$

The most straightforward uses of the reagent are to synthesize thioketones from ketones, thioamides from amides and lactams (including β-lactams), thiohydrazides from hydrazides, and thioesters from esters and lactones, all with considerable structural variety. The reactions are typically conducted in refluxing toluene or benzene. Many examples are given in reference 46. With certain difunctional derivatives of the amide structures, secondary reactions can occur that give rise to heterocycles containing sulfur in the ring; again many examples are known.[47] As noted in Eq. 11.19, the ability of the thioamide to enolize is generally implicated in ring formation: the resulting −SH group acts as a nucleophile toward an acyl group to close the ring.

$$(11.19)$$

In Eq. 11.20, the thioenol and another −SH group undergo oxidation to form a disulfide link; this is followed by migration of the alkyl group on N to S (an example of the Dimroth rearrangement).

$$(11.20)$$

The replacement of carbonyl oxygen by sulfur can also be effected by reaction with O,O-dialkyl phosphorodithioates, $(RO)_2P(S)SH$, which we have seen to be readily available from the reaction of alcohols with phosphorus pentasulfide. The O−S interchange may be visualized as in Eq. 11.21, in which a 5-coordinate intermediate with a four-membered ring is involved.

$$R_2C=O + (R'O)_2\overset{S}{\underset{}{P}}-SH \longrightarrow R_2\overset{OH}{\underset{}{C}}-S-\overset{S}{\underset{}{P}}(OR')_2 \longrightarrow \underset{HS}{\overset{R'O}{\cdots}}\overset{S-CR_2}{\underset{OR'}{P-O}} \longrightarrow$$

$$(R'O)_2\overset{S}{\underset{}{P}}-OH + R_2C=S \qquad (11.21)$$

The thionation seems to be best known with ketones and amides. Another notable reaction is that with nitriles, forming primary thioamides directly (Eq. 11.22).

$$RC\equiv N + (R'O)_2\overset{S}{\overset{\|}{P}}-SH \longrightarrow R-\overset{S}{\overset{\|}{C}}NH_2 + (R'O)_2\overset{S}{\overset{\|}{P}}-S-\overset{S}{\overset{\|}{P}}(OR')_2 \quad (11.22)$$

The dithiophosphates are also useful in converting oxiranes to thiiranes, presumably by the mechanism shown in Eq. 11.23.

$$(11.23)$$

11.F.4. Ester Formation with the Mitsunobu Reaction.

We observed in chapter 3.C that triphenylphosphine forms a complex with dialkyl azodicarboxylates that can serve to couple the carboxylic acid and alcohol components of an ester. This process has become known as the Mitsunobu reaction, and since it occurs under mild conditions, it has proved useful in the synthesis of esters of complex and sensitive structure. The synthesis is illustrated in Eq. 11.24.

$$Ph_3P + EtO_2C-N=N-CO_2Et \longrightarrow Ph_3\overset{+}{P}-\overset{CO_2Et}{\overset{|}{N}}-\overset{-}{N}-CO_2Et \xrightarrow{R\overset{O}{\overset{\|}{C}}-OH}$$

$$Ph_3\overset{+}{P}-\overset{CO_2Et}{\overset{|}{N}}-NHCO_2Et \xrightarrow{R'OH} Ph_3\overset{+}{P}-OR' \xrightarrow{RCOO^-} RCOOR' + Ph_3P=O \quad (11.24)$$

11.F.5. Alkyl Halides from Tertiary Phosphine-Halogen Adducts.

These halophosphonium halide adducts are useful reagents for the conversion of alcohols (primary and secondary) to alkyl halides (Eq. 11.25). The reactions are conducted under mild conditions, and because carbocation formation is not involved, skeletal rearrangements are not a complication. Cadogan[47] gives broad coverage of alkyl halide formation by this technique.

$$Ph_3P \xrightarrow{Br_2} Ph_3\overset{+}{P}Br \ Br^- \xrightarrow{ROH} \overset{Br}{\overset{|}{Ph_3P}}-O-R \xrightarrow{-Br^-} Ph_3\overset{+}{P}-O-R \ (Br^- \longrightarrow Ph_3P=O + RBr$$

$$(11.25)$$

11.F.6. Deoxygenations with Tertiary Phosphines and Phosphites.

These compounds are excellent reducing agents and are used to remove oxygen from amine oxides, sulfoxides, ozonides, nitrones (to form Schiff bases), epoxides (forming alkenes), and others. A discussion of these and other uses can be found in the Fieser and Fieser *Reagents for Organic Synthesis*,[57] especially in Volume 1, but in later editions also.

REFERENCES

1. F. H. Westheimer, *Science* **235**, 1173 (1987).
2. M. Horiguchi and M. Kandatsu, *Nature* **184**, 901 (1959).
3. J. S. Kittredge, E. Roberts, and D. G. Simonsen, *Biochemistry* **1**, 624 (1962).
4. L. D. Quin, *Science* **144**, 1133 (1964).
5. (a) L. D. Quin, *Top. Phosphorus Chem.* **4**, 23 (1967); (b) R. L. Hilderbrand, ed., *The Role of Phosphonates in Living Systems*, CRC Press, Boca Raton, FL, 1983; (c) T. Hori, M. Horiguchi, and A. Hayashi, eds., *Biochemistry of Natural C—P Compounds*, Japanese Association for Research on the Biosynthesis of C—P Comounds, Maruzen, Kyoto, Japan, 1984; (d) J. S. Thayer, *Appl. Organomet. Chem.* **3**, 203 (1989).
6. S. Araki, S. Abe, S. Odani, S. Ando, N. Fujii, and M. Satake, *J. Biol. Chem.* **262**, 14141 (1987); S. Araki, S. Abe, S. Ando, K. Kon, N. Fujiwara, and M. Satake, *J. Biol. Chem.* **264**, 19922 (1989).
7. D. Hendlin, et al., *Science* **166**, 122 (1969).
8. J. D. Thayer, H. J. Magnuson, and M. S. Gravett, *Antibiot. Chemother.* **3**, 256 (1953); G. O. Doak and L. D. Freedman, *J. Am. Chem. Soc.* **76**, 1621 (1954).
9. T. Kimura, K. Nakamura, and E. Takahashi, *J. Antibiot.* **48**, 1130 (1995).
10. (a) E. Öhler and S. Kanzler, *Phosphorus Sulfur Silicon* **112**, 71 (1996); (b) T. Kuzuyama, T. Shimizu, S. Takahashi, and H. Seto, *Tetrahedron Lett.* **39**, 7913 (1998).
11. (a) E. Bowman, M. McQueeney, R. J. Barry, and D. Dunaway-Mariano, *J. Am. Chem. Soc.* **110**, 5575 (1988); (b) D. Dunaway-Mariano, paper presented at the International Conference on Phosphorus Chemistry, Cincinnati, Ohio, July 12–17, 1998.
12. H. M. Seidel, S. Freeman, H. Seto, and J. R. Knowles, *Nature* **335**, 457 (1988); H. M. Seidel, S. Freeman, C. H. Schwalbe, and J. R. Knowles, *J. Am. Chem. Soc.* **112**, 8149 (1990).
13. T. Hidaka and H. Seto, *J. Am. Chem. Soc.* **111**, 8012 (1989).
14. T. Calogeropoulou, G. B. Hammond, and D. F. Wiemer, *J. Org. Chem.* **52**, 4185 (1987).
15. (a) J. C. Guillemin, S. LeSerre, and L. Lassalle, *Adv. Space Res.* **19**, 1093 (1997); (b) G. W. Cooper, W. M. Onwo, and J. R. Cronin, *Geochim. Cosmochim. Acta* **56**, 4109 (1992); *Chem. Abstr.* **118**, 9598 (1993).
16. M. Balter, *Science* **273**, 870 (1996).
17. R. M. de Graaf, J. Visscher, and A. N. Schwartz, *Nature* **378** (1995).
18. T. Yagi in ref. 5c, p. 116.
19. R. Kinas, K. Pankiewicz, and W. J. Stec, *Bull. Acad. Pol. Sci.* **23**, 981 (1975); T. Kawashima, R. D. Kroshefsky, R. A. Kok, and J. G. Verkade, *J. Org. Chem.* **43**, 1111 (1978).
20. H. Arnold and F. Bourseaux, *Angew. Chem.* **70**, 539 (1958).
21. R. Engel, ed., *Handbook of Organophosphorus Chemistry*, Marcel Dekker, Inc., New York, 1992, Chapter 11.

22. H. Tanaka, M. Fukui, K. Haraguchi, M. Masaki, and T. Miyasaka, *Tetrahedron Lett.* **30**, 2567 (1989); D. H. R. Barton, S. D. Géro, B. Quiclet-Sire, and M. Samadi, *Tetrahedron Lett.* **30**, 4969 (1989).
23. F. Helgstand et al., *Science* **201**, 819 (1978).
24. O. L. M. Bijvoet, H. A. Fleisch, R. E. Canfield, and R. G. G. Russell, eds., *Bisphosphonates on Bones*, Elsevier, Amsterdam, 1995; H. Fleisch, *Drugs* **42**, 919 (1991).
25. F. H. Ebetino, paper presented at the International Conference on Phosphorus Chemistry, Cincinnati, Ohio, July 12–17, 1998.
26. J. P. Krise and V. J. Stella, *Adv. Drug Delivery Rev.* **19**, 287 (1996).
27. S. Chen, C.-H. Lin, C. T. Walsh, and J. K. Coward, *Bio-org. Med. Chem. Lett.* **7**, 505 (1997).
28. R. M. Black and J. M. Harrison in F. R. Hartley, ed., *The Chemistry of Organophosphorus Compounds*, Vol. 4, John Wiley and Sons, Inc., New York, 1996, Chapter 10.
29. A. D. F. Toy and E. N. Walsh, *Phosphorus Chemistry in Everyday Living*, 2nd. ed., American Chemical Society, Washington, DC, 1987, (a) Chapter 18, (b) Chapter 19, (c) Chapter 10.
30. M. Eto in R. Engel, ed., *Handbook of Organophosphorus Chemistry*, Marcel Dekker, Inc., New York, 1992, Chapter 16.
31. J. A. Sikorski and E. W. Logusch in R. Engel, ed., *Handbook of Organophosphorus Chemistry*, Marcel Dekker, Inc., New York, 1992, Chapter 15.
32. J. E. Franz, M. K. Mao, and J. A. Sikorski, *Glyphosate: A Unique, Global Herbicide*, American Chemical Society, Washington, DC, 1997.
33. E. Bayer, K. H. Gugel, K. Hägele, H. Hagenmaier, S. Jessipow, W. A. König, and H. Zähner, *Helv. Chim. Acta* **55**, 224 (1972).
34. *Kirk-Othmer Encyclopedia of Chemical Technology*, 4th. ed., John Wiley & Sons, Inc., 1996, pp. 792–795.
35. E. D. Weil in R. Engel, ed., *Handbook of Organophosphorus Chemistry*, Marcel Dekker, Inc., New York, 1992, Chapter 14.
36. H. B. Kagan and M. Sasaki in F. R. Hartley, ed., *The Chemistry of Organophosphorus Compounds*, Vol. 1, John Wiley & Sons, Inc., New York, 1990, Chapter 3.
37. W. S. Knowles and M. J. Sabacky, *J. Chem. Soc., Chem. Commun.*, 1445 (1968); W. S. Knowles, M. J. Sabacky, and B. D. Vineyard, *J. Chem. Soc., Chem. Commun.*, 10 (1972).
38. M. J. Burk, M. F. Grose, G. P. Harper, C. S. Kalberg, J. R. Lee, and J. P. Martinez, *Pure Appl. Chem.* **68**, 37 (1996).
39. M. J. Burk, G. Casy, and N. B. Johnson, *J. Org. Chem.* **63**, 6084.
40. M. J. Burk, C. S. Kalberg, and A. Pizzano, *J. Am. Chem. Soc.* **120**, 4345 (1998).
41. M. J. Burk, C. S. Calberg, and A. Pizzano, *J. Am. Chem. Soc.* **120**, 4345 (1998).
42. A. Marinetti and L. Ricard, *Organometallics* **13**, 3956 (1994).
43. J. H. Teles, S. Brode, and M. Chabana, *Angew. Chem., Int. Ed. Engl.* **37**, 1415 (1998).
44. D. M. Lynn, B. Mohr, and R. H. Grubbs, *J. Am. Chem. Soc.* **120**, 1627 (1998).
45. T. V. RajanBabu, T. A. Ayers, G. A. Halliday, K. K. You, and J. C. Calabrese, *J. Org. Chem.* **62**, 6012 (1997).
46. K. P. Davis, P. A. T. Hoye, M. J. Williams, and G. Woodward, *Phosphorus Sulfur Silicon* **109**, 197 (1996).
47. J. I. G. Cadogan, ed., *Organophosphorus Reagents in Organic Synthesis*, Academic Press, London, 1979.
48. A. W. Johnson, *Ylides and Imines of Phosphorus*, John Wiley & Sons, Inc., New York, 1993.

49. J. March, *Advanced Organic Chemistry*, 4th ed., John Wiley & Sons, Inc., New York, 1992, pp. 956–963.
50. A. W. Johnson and R. B. Lacount, *Tetrahedron* **9**, 130 (1960); H. J. Bestmann and O. Kratzer, *Chem. Ber.* **95**, 1894 (1962).
51. E. Vedejs and C. F. Marth in L. D. Quin and J. G. Verkade, eds., *Phosphorus-31 NMR Spectral Properties in Compound Characterization and Structural Analysis*, VCH Publishers, Inc., New York, 1994, Chapter 23.
52. E. Vedejs and C. F. Marth, *J. Am. Chem. Soc.* **110**, 3948 (1988).
53. W. C. Still and C. Gennari, *Tetrahedron Lett.* **24**, 4405 (1983).
54. A. A. Davis, J. J. Rosén, and J. J. Kiddle, *Tetrahedron Lett.* **39**, 6263 (1998).
55. B. E. Maryanoff and A. B. Reitz, *Chem. Rev.* **89**, 863 (1989).
56. R. A. Cherkasov, G. A. Kutyrev, and A. N. Pudovik, *Tetrahedron* **41**, 2567 (1985).
57. L. F. Fieser and M. Fieser, *Reagents for Organic Synthesis*, Vol. 1, John Wiley & Sons, Inc., New York, 1967.

INDEX

Abramov reaction, 66, 146
Acetylcholinesterase inhibitors, 34, 368–371
Adenosine triphosphate, ^{31}P NMR spectrum, 201
2-Aminoethylphosphonic acid, 357–362
Aminophosphines
  ^{31}P NMR spectra, 189–190
  from phosphinous halides, 58, 60
  from phosphonous dihalides, 53, 62–64
  from phosphorus trichloride, 64–68
  as precursors of iminophosphines, 328
  as reactants in nucleotide synthesis, 68
Amiton, 370
Antibiotics
  phosphinic acids, 360–361
  phosphonic acids, 359–362
Anticancer phosphorus compounds, 363–365
Antiviral phosphorus compounds, 365
Arbusov rearrangement, 61, 145, see also Michaelis–Arbusov reaction
AZT 5′-phosphate, 365

Betasan, 373
Bialaphos, 360, 372–373
BINAP ligand, 75
Biphilic addition to 3-coordinate phosphorus
  of dialkylperoxides, 32, 67
  of dienes, 56
  of α, β-unsaturated carbonyl compounds, 56
Biosynthesis
  of 2-aminoethylphosphonic acid, 361
  of bialaphos, 361–362
Biophosphates, 351–357
Biphosphine disulfides, 88, 115
Biphosphines, 86–88
Bismethylenephosphonic acids and esters
  in bone disease, 151, 365–366
  synthesis, 151
Bis(methylene)phosphoranes, 334–335
Bonding
  in dioxophosphoranes, 336
  to phosphorus, modes, 8–10, 21–23
  in the phosphoryl group, 12–15
  in ylides, 15–17
  in trigonal bipyramidal phosphorus, 17–20
Bone disease, bisphosphonates in treatment of, 365–366

Carbon–phosphorus bond, thermal stability, 11–12
Clayton–Jensen reaction, 149
Clodronate, 366
Conjugate addition reactions, see Michael reaction

387

6-Coordinate phosphorus
  ^{31}P NMR, 343
  in Wittig reactions, 344–345
  syntheses, 343–345
Cycloaddition, see McCormack reaction; 1,3-dipolar cycloadditions
  of α-dicarbonyl compounds and 3-coordinate phosphorus, 32
  of iminophosphines and dienes, 329
  of oxophosphines with dienes dicarbonyl compounds, 330
  with phosphaalkynes, 313
  with phosphalkenes, 323
Cyclophosphamide, 363

DDVP insecticide, 370
Dialkyl phosphites, see H-phosphonates
Diazinon, 370
Diels–Alder reactions
  of vinylphosphine oxides, 107–108
  with phosphaalkenes, 323–324
  with phosphaalkynes, 314
1,3,2-Dioxaphosphoranes
Dioxophosphoranes, 335–339
Diphosphenes, 333
1,3-Dipolar cycloadditions
  of iminophosphines, 329, 335
  of phospholene oxides, 286–287
  of phosphaalkenes, 324
  of phosphaalkynes, 313
Dipterex, 370
DiPAMP ligand, 283, 291, 375–376
Doak–Freedman reaction, 150–151
DuPHOS ligands, 292, 376

Electrolysis, of phosphonium ions, 122
Electron spin resonance spectroscopy, 226–227
EPN insecticide, 370
Etidronate, 366

Flame retardants, 374, 376, 377
Floropryl, 368
Fonophos, 370
Foscarnet, 365
Fosfomycin, 359–360
Fosfonochlorin
Fosinopril, 366
Fosmidomycin, 360

Glufosinate, 360, 372–373
Glyphosate, 372
Glyphosine, 374

Halophosphines, see phosphinous halides; phosphonous dihalides
Halophosphonium halides
  reactions, 55, 60, 100, 125–126
  syntheses, 50, 55, 100, 124–126
Heterophosphonium ions, 84, 103, 121, see also halophosphonium ions
  ^{31}P NMR spectra, 193–196
  alkoxy derivatives, 126–127
  amino derivatives, 128–129
  aryloxy derivatives, 128
  equilibrium with phosphorane structure, 123–124
  detection by ^{31}P NMR as intermediates, 194–196
  optically active, 288, 294
Horner reaction, 110
Hypophosphorous acid esters, 138, 147

Iminophosphines, 327–329, 332
Iminophosphoranes
  reactions, 76, 340
  syntheses, 83, 104
Infrared spectroscopy, 217–219
Insecticides, see phosphorus insecticides

Literature of phosphorus chemistry, 35–37

Lawesson reagent, 102, 115–116, 156, 381–382

Malathion, 369, 371
Mass spectra
  of phosphines, 220–221
  of phosphine oxides, 221
  of phosphorus acids and esters, 221–223
McCormack reaction, 31, 56, 120, 251–252
Medicinal phosphorus compounds, 362–367
Metal complexes
  of phosphaalkenes, 324
  reactions in the coordination sphere of, 238, 245–248, 254, 317
  in racemate resolutions, 277–278
  of phosphines as catalysts, 375–376
Metaphosphoric acid derivatives, 336–339
Methyldichlorophosphine, see methylphosphonous dichloride
Methylphosphonous dichloride, synthesis, 49
Michael reactions
  of hypophosphorous acid esters, 138
  of phosphines, 71, 73–74, 80
  of phosphinothioites, 116
  of phosphinoyl carbanions, 110
  of H-phosphonates, 146
  of secondary phosphines, 293
  of secondary phosphine oxides, 99, 107, 112
  of thiophosphoric acids, 371
  of vinylphosphine oxides, 107, 110
Michaelis–Arbusov reaction
  detection of intermediates by ^{31}P NMR, 194
  of phosphinites, 61, 100, 286
  of phosphites, 143–144
  of phosphonites, 145
  to form bismethylene phosphonates, 151–152

Michaelis–Becker reaction, 146
Miltefosine, 363
Mitsunobu reaction, 84, 154, 383
  detection of intermediates by ^{31}P NMR, 195

Naturally occurring C–P compounds, 357–362
Nomenclature of phosphorus compounds
  general rules, 41–43
  of hetero-substituted phosphines, 64–65
  of phosphonous acid derivatives, 62
Nucleophilic substitution on phosphoryl compounds
  addition–elimination mechanism, 29–30, 140
  elimination–addition mechanism, 30–31
Nuclear magnetic resonance
  of ^{13}C
    of aliphatic P compounds, 213–214
    of aromatic P compounds, 216
    of bis(imino)phosphoranes, 335
    of bis(methylene)phosphoranes, 335
    description, 212
    of phosphaalkenes, 322
    of phosphaalkynes, 311
    stereochemical control of coupling to ^{31}P, 215–216
    of unsaturated P compounds, 217
  of ^{1}H
    of P-alkyl groups, 206
    of P-aryl groups, 207
    description, 204–205
    diastereotopic protons, 208
    of H–P groups, 209–210
    of iminophosphines, 328
    long-range effects, 209
    of P-methyl groups, 206
    of phosphaalkenes

Nuclear magnetic resonance (*Continued*)
  of phosphaalkynes, 311
  stereochemical control of coupling to ^{31}P, 210–212
  of unsaturated groups, 207
  of ^{17}O
    phosphoryl groups, 223–226
  of ^{31}P
    of acids with P—C bonds, 196–198
    of bis(methylene)phosphoranes, 335
    of 6-coordinate phosphorus, 343
    description and theory, 169–172
    in detection of reaction intermediates, 194–196
    of halophosphines, 188–189
    of heterophosphonium ions, 193–196
    of hetero-substituted phosphines, 189–191
    of iminophosphines, 328
    of phosphaalkenes, 322
    of phosphaalkynes, 311
    of phosphaketenes, 326
    of phosphenite, 332
    of phosphoniun salts, 192–193
    of phosphoranes, 340, 341
    of phosphoric acid derivatives, 198–201
    of primary phosphines, 180–183
    of secondary phosphines, 173–180, 184–185
    structural influences on, 173–180
    of tertiary phosphines, 185–188
    of tertiary phosphine oxides, 191–192
    of tertiary phosphine sulfides, 191
Nucleophilic substitution on 3-coordinate phosphorus, 31, 52, 59
Nucleotides, 68, 147, 352–353, 365

1,2-Oxaphospholane derivatives, 56, 66, 97, 106

Oxidation
  of phosphines, 82–83, 101
  of phosphine sulfides, 116–117
Oxophosphines, 330

PALA, 363
Pamidronate, 366
PAMP ligand, 279, 283, 289, 291
Parathion, 369
Perkow reaction
  for synthesis of DDVP, 371
  of phosphinites, 61
  of phosphites, 144–145
Phenyldichlorophosphine, *see* phenylphosphonous dichloride
Phenylphosphinic acid, 137, 142, 376
Phenylphosphonous dichloride
  synthesis, 45–46
Phosetyl, 373
Phosmet, 369
Phospha-acylium ions, 28
Phosphaalkenes
  ^{31}P NMR spectra, 322
  reactions, 310, 322–327
  syntheses, 57, 85, 317–322, 334–335
Phosphaalkynes
  ^{31}P NMR spectra, 311
  reactions, 311–314
  syntheses, 308–311
Phosphaallenes, 326
Phospha–Cope reactions, 327
Phosphagens, 356
Phosphaketenes, 326–327
7-Phosphanorbornadienes, metal complexes, 317
7-Phosphanorbornenes, 96–97, 116, 331, 337
Phosphapolyenes, 325–327
  ^{16}O, ^{17}O, ^{18}O-Phosphate, 296–297
Phospha–Wittig reaction, 322
Phosphazenes, 154–155
Phosphenites, 332
Phosphetes, 247–265
Phosphenium ions, 27–28

Phosphepins, 264–265
Phosphetanes, 47, 97, 106, 100, 241–247
Phosphetenes, 247–265
Phosphide ions, 71–73, 78, 84, 85
Phosphine imines, see iminophosphoranes
Phosphines
  basicity, 78–79
  conjugation in, 78
  optically active, 277–280, 288–294, 299
  properties and structure, 77–78
  primary
    ^{31}P NMR spectra, 180–183
    reactions, 51, 321, 322
  secondary
    optically active, 279–280
    ^{31}P NMR spectra, 184–185
    reactions
      syntheses, 69, 71, 72, 75
    synthesis, 70–73, 76
  tertiary
    ^{31}P NMR spectra, 185–186
    optically active, 277–279, 288–299
    reactions, 79–86
    syntheses, 71–76
Phosphine–boranes, optically active, 289–291
Phosphine ligands in metal complexes, 55, 375–376
Phosphine oxides
  basicity, 103
  conjugation in, 106–107, 108–109
  inductive effect of, 108–109
  inductive effect of, 108–109
  nucleophilicity, 103–104
  optically active, 274–276, 282–288, 299
  primary, syntheses, 101
  secondary
    acidity, 112
    chiral, 285
    optically active, 286
    reactions, 99, 112, 113

    syntheses, 51, 53, 58, 100–101
  tertiary
    ^{31}P NMR spectra, 191–192
    optically active, 274–276, 282–288
    reactions, 110
    syntheses, 61, 81, 96–102, 111–113
Phosphine selenides
  reactions, 76
  structure and bonding, 114
  syntheses, 115
Phosphine sulfides
  conjugative effect, 117
  as nucleophiles, 117
  optically active, 276–277
  ^{31}P NMR spectra, 191
  reactions, 117
  structure and bonding, 114
  syntheses, 83, 102, 114–116
Phosphine tellurides, 114
Phosphinic acids and esters
  acidity, 137
  derivatives, 60, 138–140
  naturally occurring, 360, 362
  optically active, 281, 282, 283, 284, 289, 291
  syntheses, 147
H-Phosphinic acids, 137
  esters, 138, 142, 148, 147, 148
  natrually occurring, 360–361
  optically active esters, 284
Phosphinidenes, 314–317
Phosphinines, 103, 260–264, 319
  heterosubstituted, 264
Phosphinites
  optically active, 294
  reactions, 57–61
  syntheses, 57–61
Phosphinous acid derivatives, see phosphinites, aminophosphines, phosphinous halides
Phosphinous Halides
  optically active, 293, 294
  reactions, 53–55, 57, 58–59, 143
  syntheses, 50–51, 60

Phosphinoyl carbanions
  Michael reactions of, 110
  reactions of, 109–110, 112
  resonance in, 109
Phosphinothricin, see Glufosinate
Phosphiranes, 72, 235–238, 315, 331
Phosphirenes, 238–241, 312
Phosphites
  dialkyl (see H-Phosphonates)
  trialkyl
    reactions, 65–68, 145–146, 149, 337, 341–342
    syntheses, 64–65
Phosphoenolpyruvate phosphomutase, 361
Phospholane derivatives, 112, 120, 141, 249–250, 316
Phospholene derivatives, 50, 55, 56, 70, 75, 99, 100, 104, 108, 116, 120, 251–252, 316
  optically active, 287
Phospholes
  1-H, 74, 252–255
  2-H, 255
  hetero-substituted, 255–257, 312, 313
Phospholipids, 353–354
Phosphonate carbanions
  generation, 149, 158–162
  reactants in nucleophilic substitution, 149, 160–161
  reaction with carbonyl compounds, 158–159
H-Phosphonates
  in nucleotide synthesis, 136
  in phosphonate synthesis, 145
  reactions
    with Grignard reagents, 51, 101
    with trimethylsilyl chloride, 66
  synthesis, 51, 135–137
  tautomeric equilibria, 135–136
Phosphonic acids and esters
  acidity, 133–134

  carbanions from, see phosphonate carbanions
  commercial applications, 377
  derivatives, 138–141
  as flame retardants, 377
  as medicinal agents, 362–367
  naturally occuring, 134–135, 357–362
  optically active, 284
  pesticidal activity, 368–374
  reactions, 76, 337
  syntheses, 54, 143–153
Phosphonic dichlorides
  reactions, 54, 76, 98
  syntheses, 76, 149
Phosphonins, 265
Phosphonites
  optically active, 290
  reactions, 309
Phosphonitrilic chloride, 154
Phosphonium ions
  ^{31}P NMR spectra, 192–193
  commercial applications, 376
  optically active, 277, 288
  reactions, 98–99, 120–123, 340
  syntheses, 79, 81, 118–120
Phosphonolipids, 358–359
Phosphonomycin, see Fosfomycin
Phosphonothrixin, 361
Phosphonous dihalides
  ^{31}P NMR spectra, 188–189
  reactions, 51–57, 62–64, 143, 309, 316, 318, 319, 320, 321, 331, 333
  synthesis, 45–49
Phosphonous acid derivatives, see phosphonous dichlorides, aminophosphines, phosphonites
Phosphoramidites, see aminophosphines
Phosphoranes
  fluxional character, 18–19, 343
  optically active, 297–298
  ^{31}P NMR spectra, 340–341

stereochemistry, 17–19
syntheses, 66, 67, 81, 105, 122, 339–343
Phosphoric acid esters
acidity, 138
commercial applications, 377, 378
derivatives, 138–140
as flame retardants, 378
optically active, 294–297, 298
^{31}P NMR spectra, 198–201
syntheses, 153–155
Phosphorinanes, 48, 54, 70, 73, 74, 138, 148, 257–260
Phosphorus-31 NMR, see Nuclear magnetic resonance
Phosphorus–carbon double bond, see Phosphaalkenes
Phosphorus fungicides, 373
Phosphorus herbicides, 371–373
Phosphorus heterocycles, multicyclic, 265–267
Phosphorus insecticides, 367–371
Phosphorus–phosphorus bond
in biphosphines and biphosphine sulfides, 86–88
cleavage of, 50
in cyclopolyphosphines, 86
Phosphorus plant growth regulators, 373–374
Phosphorus radicals, 28–29, 85, 102, 118–119
Pischimuka reaction, 157–158
Plumbemycin A, 360
Prodrugs with phosphorus, 367
Protein phosphates, 354
Pseudorotation, of trigonal bipyramidal phosphorus, 18–19
Pudovik reaction, 146

Racemate resolution
with enzymes, 298–299
of phosphine oxides, 274–276
of phosphine sulfides, 276–277

of phosphines, 277–280
of phosphonium salts, 277
of phosphorus acids, 280–282
Radicals, see Phosphorus radicals
Rearrangements
Arbusov, 61
Beckmann-type, 32
of 3-coordinate P(OH) to P(O)H, 20–21
of enol phosphates to phosphonates, 33
Lossen-type, 32
Phospha–Cope, 327
1,2-phosphinoyl shifts, 33, 111–112
[3,3] sigmatropic shifts, 327
[1,5] sigmatropic shifts
Reduction,
of halophosphonium ions, 50
of phosphine oxides, 74–76
of phosphinic chlorides, 50
of phosphine sulfides and selenides, 50, 116–117
of phosphonium ions, 122
with silanes, 74–76
with tertiary phosphines, 384
with tertiary phosphites, 384
Reppe complex, 375
Risedronate, 366
Roundup, see Glyphosate

Safety, 34–35
Schradan, 370
Selenophosphonic acids and derivatives, 54
optically active, 294
syntheses, 59
Staudinger reaction, 60, 83
Stereoselective reactions, 299–303
Sugar phosphates, 354

Tautomeric shift of 3-coordinate P-OH derivatives, 20–21, 60

Thiophosphonic acids and derivatives
  optically active, 280–281, 282, 294
  syntheses, 54
Thiophosphoric acids and derivatives
  acidity, 155
  commercial applications, 378
  optically active, 295, 297
  ^{16}O, ^{17}O, ^{18}O-thiophosphate, 297
  reactions, 156–157, 382–383
  syntheses, 156–158
  tautomeric equilibria, 155–156
Toxicity of phosphorus compounds, 34–35, 367–369
Trigonal bipyramidal phosphorus, 17–21
2,4,6-Tri-*tert*-butylphenylphosphonous dichloride, synthesis, 48

Turnstile rotation, of trigonal bipyramidal phosphorus, 19

Ultraviolet–visible spectroscopy, 227–229

Wadsworth–Emmons reaction, 158–159, 379, 381
Wilkinson catalyst, 375
Wittig reaction, 379–381

X-ray diffraction analysis, 226